HOW TO
DESIGN AND
BUILD
ELECTRONIC
INSTRUMENTATION
2ND EDITION

D1427803

C.

Other TAB Books by the Author

HOW TO
DESIGN AND
BUILD
ELECTRONIC
INSTRUMENTATION
2ND EDITION

JOSEPH J. CARR

TAB TAB BOOKS Inc.
Blue Ridge Summit, PA 17214

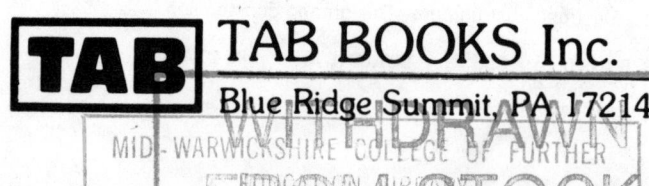

SECOND EDITION

FIRST PRINTING

Library of Congress Cataloging in Publication Data

Carr, Joseph J.
    How to design and build electronic instrumentation.

    Bibliography: p.
    Includes index.
    1. Electronic instruments—Design and construction.
I. Title.
TK7878.4.C37   1986   621.3815'4   85-27735
ISBN 0-8306-9560-5
ISBN 0-8306-0460-X (pbk.)

Contents

Introduction

T HE FIRST EDITION OF THIS BOOK DEALT MOSTLY WITH ANALOG-IN-
strumentation methods and circuits. That need is still with us, but the
advent of the microprocessor in the meantime caused a revolution in instrumentation design. In this edition I have maintained the material on
analog methods because they are very much with us still. After all, most
transducers and other signal sources are still analog in nature. In many
cases, you will want to convert the signal voltage or current as fast as possible before inputting it to the computer for further processing. In other cases,
memory or speed limitations in the computer (or certain other reasons)
make it advisable to use analog methods other than amplification and scaling external to the computer. Examples of such processing include filtering, integration, and differentiation.

The coverage of microcomputers and microprocessors is expanded
because those machines are now ubiquitous. Computer prices have
dropped from thousands of dollars for a Z80 computer and a few thousand
bytes of memory to almost ridiculous levels. One powerful microcomputer,
the famous Timex 1000, was treated as a toy in the marketplace (and in
fact, I bought mine at *Toys 'R Us*), and when the company dropped out
of the computer market those machines sold for as little as $9.95. Another
computer company offered a $50 trade-in value for any competitor's ma-
chine when its machine was bought. Many people went out and bought
Timex machines for $10 and turned them in on the other machine for a
$50 discount—unopened. Interestingly enough, that "toy" computer was
more powerful than O.E.M. machines used in several instrumentation
designs. Today, IBM PC, XT, and AT machines are more capable than
machines which used to cost the same amounts and that were popular

when the first edition was written. The ubiquitous Apple II series is also now improved and lower in cost than ever before.

You can now take a ni-cad-battery-powered computer into the field for data-acquisition purposes. In those cases, previous methods required some form of data collection such as an audio cassette recorder. The analog data was converted into audio tones, stored on audio cassette, and then carried back to the laboratory where it was reconverted into audio tones. These tones were converted back to analog data signals, and either input directly to an analog-to-digital-converter-equipped computer or "digitized" from a paper recording of the data. Of course, each data-conversion point introduces error and distortion! Today, you take the computer and a low-power A/D converter into the field and record the data directly onto disks. It's truly an exciting development.

Chapter 1

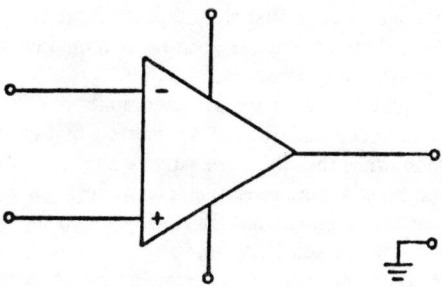

Operational Amplifiers

A N OPERATIONAL AMPLIFIER IS AN INTEGRATED-CIRCUIT GAIN BLOCK that has the following properties:

- Infinite open-loop gain
- Zero output impedance
- Infinite input impedance
- Infinite bandwidth

Of course, we do not seriously expect real operational-amplifier devices to meet these specifications, but the approximations that are available on the market are good enough to make the operational amplifier one of the easiest-to-apply and most exciting linear ICs in the designer's arsenal of tricks. In due course I will explain some of the more useful implications of these properties.

PRACTICAL CHARACTERISTICS

An operational amplifier will have an open-loop voltage gain (A_{vol})—that is, gain *without feedback*—considerably greater than any gain normally encountered in the closed-loop configuration. The output-voltage transfer function of the typical operational amplifier is given by

$$E_{out} = \pm A_v E_{in} \qquad (1.1)$$

where

E_{out} is the output voltage

A_v is the closed-loop voltage gain
E_{in} is the input voltage.

The $\pm$ symbol implies that the output voltage may assume either polarity, and this requires that the operational-amplifier power supply be able to offer dual-polarity voltages.

Figure 1-1 shows a typical operational-amplifier power-supply circuit. Two batteries are used; one produces a voltage that is positive with respect to common, while the other is negative with respect to common. Voltage levels for most common, operational amplifiers run from about 1.5 to 3.5 volts on the minimum end and 15 to 40 volts on the upper end. Very common is a $\pm$ 18-volt specification.

In most of my examples the batteries will actually be electronic dc power supplies operated from the ac power mains. These must be regulated and well filtered or problems will result. In cases where stable potentials are required it will be necessary to use some sort of regulator circuit, some examples of which are given in a later chapter.

A typical operational-amplifier-circuit symbol is shown in Fig. 1-2. In actual practice you may find that the VCC (positive dc power supply) and VEE (negative dc power supply) terminals are not shown on the schematic. This is to make the diagram less-crowded, therefore easier to read. You should not forget, though, that the power-supply terminals are to be assumed present regardless of whether or not they are shown in the diagram.

Most operational amplifiers have a single-ended-output terminal, but the inputs are differential with respect to ground. The implication of this is that the two input terminals will have an *equal but opposite* effect on the output voltage. If the same grounded, referenced voltages are applied to the input terminals their respective effects will cancel, producing a zero-output potential. This situation is shown in Fig. 1-3.

One input is labeled (–) and is called the inverting input. It will produce an output voltage that has the polarity opposite that of the input voltage. That is to say it is 180° out of phase with the input.

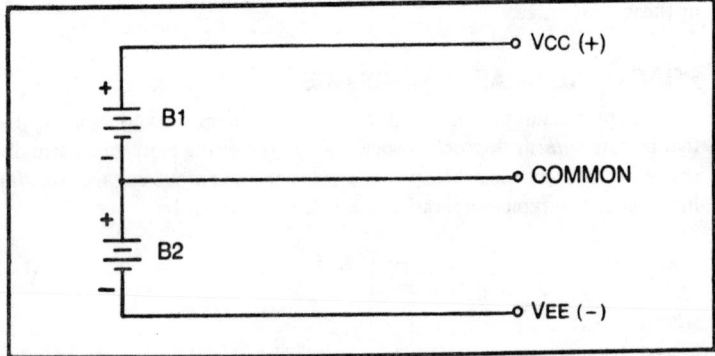

Fig. 1-1. Power supply configuration for operational amplifiers.

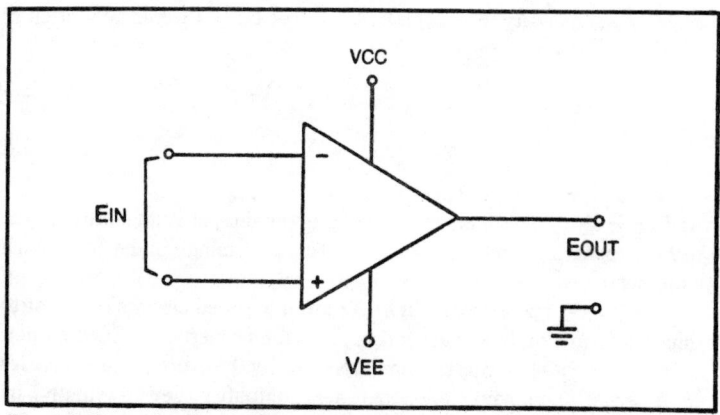

Fig. 1-2. Operational amplifier circuit symbol.

The other input is labeled (+) and is called the noninverting input. You might guess from the foregoing that the (+) input produces an output voltage that is in phase with the input voltage; the two polarities are the same.

The opposite natures of the two input terminals are the reason why *common-mode* voltages (i.e., E_3 in Fig. 1-3) produce zero net effect on the output voltage. One way of viewing this situation is that the two outputs algebraically add and their sum is zero. In due course you will see a more substantial explanation.

Differential amplifiers will not produce an output signal unless they are presented with an input condition in which the quantity $(E_1 - E_2)$ is nonzero. Equation 1.1 may be rewritten as

$$E_{out} = -A_v E_1 \tag{1.2}$$
$$= +A_v E_2 \tag{1.3}$$
$$= A_v (E_2 - E_1) \tag{1.4}$$

Equation 1.4 more nearly defines the transfer properties of the operational amplifier. In the remainder of this book I will adopt a convention

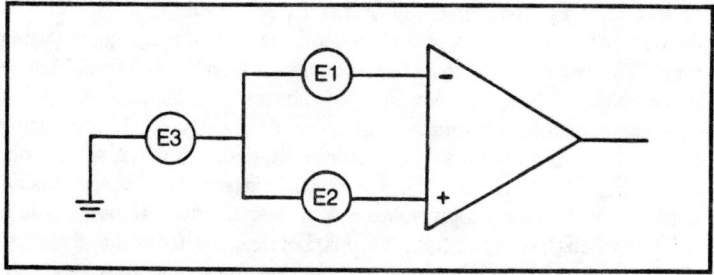

Fig. 1-3. Voltage $E_2 + E_1$ is seen as a differential voltage, while E_3 is common mode.

3

whereby the quantity (E_2-E_1) is called E_{in}, so Eq. 1.4 would be written as

$$\text{if} \qquad E_{in} = E_2 - E_1 \qquad\qquad (1.5)$$

$$\text{then} \qquad E_{out} = A_v E_{in} \qquad\qquad (1.6)$$

And E_{in} is always assumed to be the differential-input voltage; that is, the voltage appearing *across* the $(+)$ and $(-)$ input terminals of the operational amplifier.

Since most operational amplifiers are integrated circuits it is best to think of them simply as gain blocks with the properties defined earlier. These properties allow us to completely specify the overall characteristics of a stage employing the operational-amplifier device simply by manipulating the feedback loop.

Although you really should not be too worried over the internal aspects of the IC operational amplifier, a certain minimal cognizance is in order to help make you a more clever designer. To that end I will do a quick survey of the inner workings of a typical, low-cost, operational-amplifier IC.

Figure 1-4 shows the partial schematic of an IC operational amplifier. Transistors Q1 and Q2 form a differential pair. The base of Q1 is the noninverting $(+)$ input, while the base of Q2 forms the inverting $(-)$ input. The emitters of these two transistors are connected together and then connected to a *constant current source* (CCS). Although graphics simplification has caused me to show this CCS symbolically, it is actually two or more transistors biased in such a way as to keep the collector current of one of them constant over wide excursions of load impedance. The output from the differential pair is taken from the collector of Q2—point A. In all cases the relationship between the three currents shown is such that

$$I_{CCS} = I1 + I2 \qquad\qquad (1.7)$$

When input voltages E_1 and E_2 are equal, current I_{CCS} will split into two equal parts, making $I1 = I2$. In this circumstance the voltage at point A will take on its quiescent, or resting, value. This is approximately one half of V_{CC}. Let us assume, though, that E_2 is greater than E_1. This is equivalent to saying that a positive potential is applied to the noninverting input. This will cause Q1 to be biased on harder than Q2; so I1 will become greater than I2. Since the current I_{CC} is constant, an increase in I1 must be accompanied by a decrease in I2 in order to satisfy Eq. 1.7. Reducing current I2 means there is less collector-emitter current in Q2; so the voltage drop across the Q2-collector resistor (R2) is less. The voltage appearing at point A is the supply voltage, V_{CC}, less the drop across R2; so it will rise when the drop across R2 drops. Hence, a positive voltage applied to the noninverting input creates a positive-going voltage at point A. Similarly, had a negative-going potential been applied to the noninverting

4

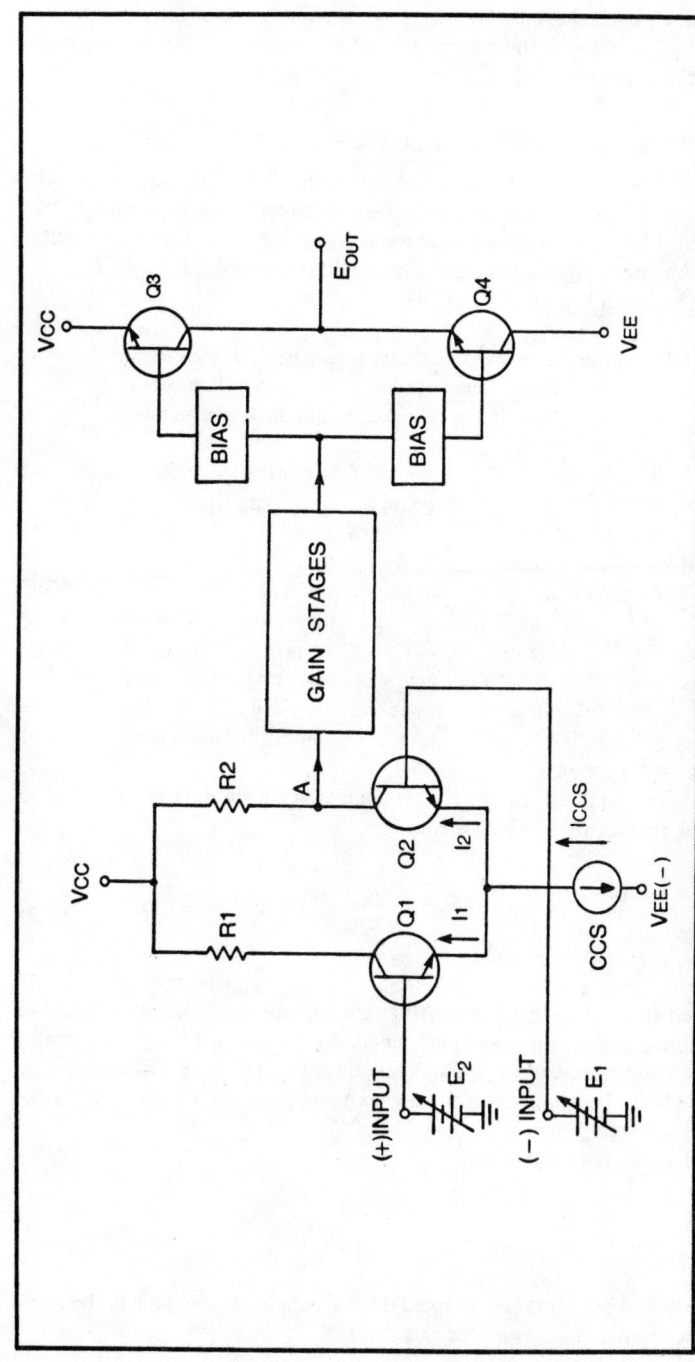

Fig. 1-4. Partial schematic of a typical operational amplifier.

input, we would have found the voltage at point A also negative-going. Applying potentials to the inverting input will have similar effects, but opposite those described for the noninverting input.

SIMPLE INVERTING FOLLOWERS

Figure 1-5 shows the circuit for a simple inverting follower, a very common—perhaps the most common—operational-amplifier circuit. In this circuit I have grounded the noninverting input for the sake of simplicity.

Another property of operational amplifiers is applicable only if the device has both (+) and (−) inputs:

> The respective inputs to a differential amplifier will "stick together."
> That is to say that placing a potential on one input allows us to treat
> the other input as if that potential had also been applied there.

In the case of my simple inverting follower the (+) input is connected to ground. Because of the foregoing property I can treat the inverting input as if it *also* were grounded. Since the input is really a high-impedance input we call this condition a *virtual ground*.

The inputs to an operational amplifier will neither *sink* nor *source* current, and we treat the input junction as if it were grounded when writing equations. There are two ways to use these properties to write the actual transfer equations. For the present I will consider what I call the Kirchhoff's-law method. For purposes of my discussion I will consider only the properties given earlier, plus the fact that the output terminal may be viewed as a voltage source.

When an input voltage (E_{in}) is applied, a current (I_{in}) will flow in the input resistor (R_{in}). This current is

$$I_{in} = \frac{E_{in}}{R_{in}} \tag{1.8}$$

which is simply Ohm's law. Although we can treat the input junction mathematically as if it were grounded, we must realize that it is actually a high impedance, and that the input will not accept or generate a current. To satisfy Kirchhoff's current law we must generate a current (I_f) to exactly cancel I_{in}.

By Kirchhoff's current law

$$I_{in} + I_f = 0 \tag{1.9A}$$

$$I_{in} = -I_f \tag{1.9B}$$

The operational-amplifier output terminal supplies the voltage to the feedback resistor that generates this current

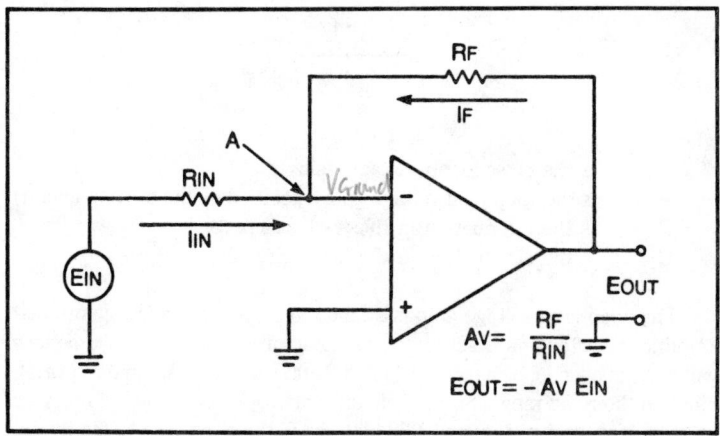

Fig. 1-5. Inverting follower circuit.

$$I_f = \frac{E_{out}}{R_f} \tag{1.10}$$

Because the equality of Eq. 1.9B exists, I may substitute both Eq. 1.8 and Eq. 1.10 into Eq. 1.9B, which gives me

$$\frac{E_{in}}{R_{in}} = \frac{-E_{out}}{R_f} \tag{1.11}$$

A transfer equation expresses the output voltage in terms of the input voltage and a term that tells us what the intervening circuitry does. In the case of the operational amplifier it amplifies; so I can obtain the transfer equation by solving Eq. 1.11 for E_{out}.

$$E_{out} = - \frac{R_f E_{in}}{R_{in}} \tag{1.12}$$

The term (R_f/R_{in}) in Eq. 1.12 gives me the voltage gain of the inverting follower and is equivalent to the term A_v in Eq. 1.6. A very important point to remember is that the voltage gain of an inverting follower is given by

$$A_v = \frac{-R_f}{R_{in}} \tag{1.13}$$

I may also view the operational amplifier from the vantage point of elementary feedback theory. The basic equation for gain in a feedback amplifier is given by

7

$$A_v = \frac{A_{vol}C}{1 + A_{vol}B} \qquad (1.14)$$

where

A_v is the closed-loop voltage gain
A_{vol} is the open-loop voltage gain (specified by the op-amp maker)
B is the attenuation in the feedback path
C is the attenuation in the input circuit.

The open-loop voltage gain is a function of the particular operational amplifier selected and is set by the manufacturer. Some low-cost devices will have an open-loop gain of only about 20,000, but most are in the several-hundred-thousand range. Some premium-grade operational amplifiers have an open-loop gain well over 1,000,000, while the average 741-family device boasts about 50,000.

The two attenuation terms are due to the feedback and input resistances, R_f and R_{in}. Consider Fig. 1-6(A). Point A in this circuit is the same as point A in Fig. 1-5. We can view the operational amplifier as a voltage source, and this is designated E_{out} in Fig. 1-6(A). We can find the voltage component of E_{out} that appears from point A to ground by using the standard voltage-divider equation

$$E_A = (E_{out}) \; \frac{R_{in}}{R_{in} + R_f} \qquad (1.15)$$

The factor $R_{in}/(R_{in} + R_f)$ is known as the attenuation factor and is the

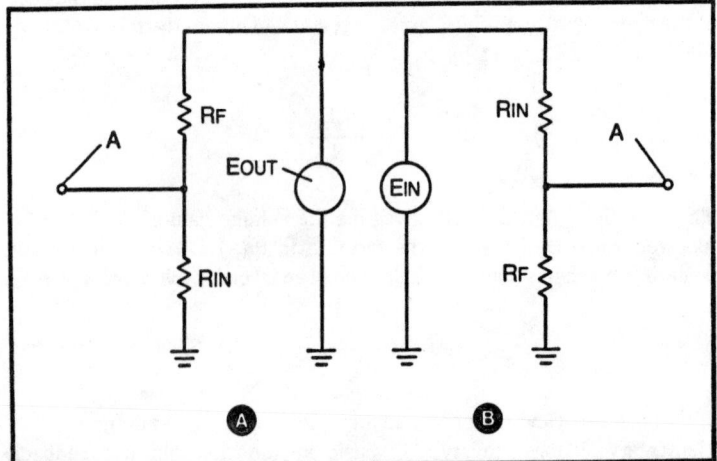

Fig. 1-6. Equivalent circuits for the potentials seen by point A in Fig. 1-5.

value of B in Eq. 1.14. Similarly, from the point of view of the input voltage, the voltage from point A to ground is merely the voltage drop across R_f caused by the input voltage. Again, by the voltage-divider equation

$$E_A = (E_{in}) \frac{R_f}{R_{in} + R_f} \qquad (1.16)$$

The factor $R_{in}/(R_{in} + R_f)$ is the attenuation factor for the input and completely specifies C in Eq. 1.14. By plugging Eq. 1.15 and Eq. 1.16 into Eq. 1.14 I arrive at the feedback expression for the inverting follower

$$A_v = \frac{A_{vol} \left[\dfrac{R_f}{R_{in} + R_f} \right]}{1 + A_{vol} \left[\dfrac{R_{in}}{R_{in} + R_f} \right]} \qquad (1.17)$$

I have already established that the numerical value of A_{vol} is typically very high, even in low-cost, bargain-basement operational amplifiers. I can, therefore, write the quantity

$$\frac{A_{vol}}{1 + A_{vol}} \approx \frac{A_{vol}}{A_{vol}} \approx 1 \qquad (1.18)$$

While this may appear algebraically unpalatable, it is reasonable for all reasonable values of the B and C terms. Even the poorest operational amplifier of the lower end of the 741-class will have an A_{vol} on the order of 20,000; so Eq. 1.18 becomes 20,000/20,001, or 0.99995, close enough to unity for all practical purposes. This makes Eq. 1.17 somewhat simpler

$$A_v = \frac{\left[\dfrac{R_f}{R_{in} + R_f} \right]}{\left[\dfrac{R_{in}}{R_{in} + R_f} \right]} \qquad (1.19)$$

Eq. 1.19 reduces to little more than a division-of-fractions problem; so I may invert the denominator and multiply

$$A_v = \left[\frac{R_{in} + R_f}{R_{in}} \right] \cdot \left[\frac{R_f}{R_{in} + R_f} \right] \qquad (1.20)$$

9

Since the $(R_{in} + R_f)$ terms cancel out, I may rewrite Eq. 1.20 as

$$A_v = \frac{R_f}{R_{in}} \qquad (1.21)$$

You can see that the closed-loop voltage gain for an inverting follower is given by the ratio of the feedback resistance to the input resistance regardless of the technique that is used to make the analysis.

A similar situation obtains for the noninverting input, an example of which is shown in Fig. 1-7. Note that, in this case, the input voltage is applied directly to the noninverting inputs; so no attenuation of the input voltage occurs. This eliminates the C term in the feedback equation (Eq 1.14). For the noninverting follower, then, I may write

$$A_v = \frac{A_{vol}}{1 + A_{vol}B} \qquad (1.22)$$

which becomes, when the B term is inserted

$$A_v = \frac{A_{vol}}{1 + A_{vol} \left[\dfrac{R_{in}}{R_f + R_{in}} \right]} \qquad (1.23)$$

In this case, also, the $A_{vol}/(1 + A_{vol})$ term is nearly unity, so Eq. 1.23 becomes

$$A_v = \frac{1}{\left[\dfrac{R_{in}}{R_{in} + R_f} \right]} \qquad (1.24)$$

Again, I invert and multiply because Eq. 1.24 is another division-of-fractions problem.

$$A_v = \left[\frac{R_{in} + R_f}{R_{in}} \right] (1) \qquad (1.25)$$

$$A_v = \left[1 + \frac{R_f}{R_{in}} \right] \qquad (1.26)$$

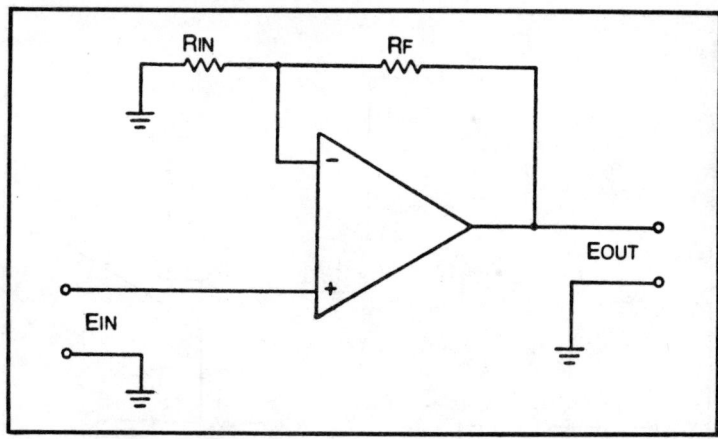

Fig. 1-7. Noninverting follower.

Equation 1.26 is the voltage-gain factor for a noninverting follower. I know from Eq. 1.6 that

$$E_{out} = A_v E_{in} \qquad (1.27A)$$

so

$$E_{out} = \left[1 + \frac{R_f}{R_{in}}\right] (E_{in}) \qquad (1.27B)$$

SINGLE-POWER-SUPPLY OPERATION

There are many applications for operational amplifiers which require operation from a power supply with only one polarity, usually positive with respect to ground. This could come about where the power supply is a constraining specification, such as automotive or portable equipment, or where an operational amplifier or two is to be sneaked into a design in which it is but a minor feature. In that case the use of a second power supply would prove prohibitively expensive. Single-supply operation could also come about because someone else cannot think in terms of more than one supply without blowing a fuse—it happens, believe it or not!

So, now we will consider how this thing is done despite the fact that it is not the best procedure if at all avoidable.

The simplest approach, and that used by most designers, is shown in Fig. 1-8(A). This circuit is basically the inverting follower which we considered earlier, but the VEE-power-supply terminal is connected to ground instead of a negative power supply—bless economy. In the original inverting-follower circuit this would unbalance things and cause the op-amp output

11

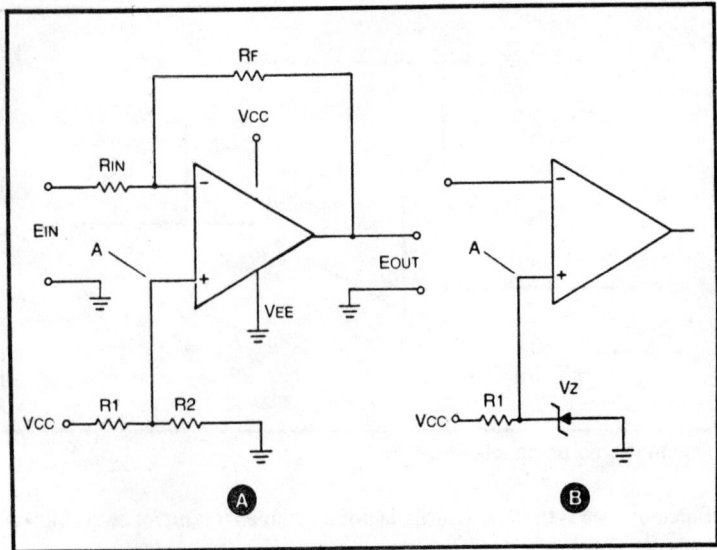

Fig. 1-8. Single-supply operation of an op amp.

to saturate against one power-supply rail limit. A resistor-voltage divider, though, will raise the noninverting input to some positive potential, and that changes the operational amplifier's operating point. Of course, and this is the reason why I dislike the single-supply method so intensely; it also raises the resting output voltage to the same level. Instead of varying around zero, the output signal will go through its excursions about some positive-voltage level.

An alternative method for achieving the same thing is shown in Fig. 1-8(B). In this circuit a zener diode sets the potential at the noninverting input.

There is a certain minimum voltage, different for each class of operational amplifier, that must be applied to the noninverting input. This can be determined from the manufacturer's data sheet for the device or by examination of its internal schematic. The value of this minimum is determined by the number of PN junctions between an input terminal and the V_{EE}-power-supply terminal. These junctions *must* be biased or the device will not operate at all; so a voltage drop is created. Generally speaking, this information can be computed from data-sheet specifications. It is the difference between the maximum allowable value for V_{EE} and the maximum allowable negative-input voltage. In other words, the voltage is given by

$$V_A = V_{EE} - E_{in\,(max-)} \tag{1.28}$$

If this information is not given in the spec sheet, you may be able to

12

examine the internal-circuit diagram of the device and count the number of PN junctions present. Be sure to count both base-emitter and base-collector junctions. Multiply the total number of PN junctions by 0.9 volts. In most common operational amplifiers the resultant voltage will be somewhere between 2.7 and 3.5 volts. You will, though, find a few micropower types that offer values as low as 1.5 volts.

It has become common practice to make the point-A voltage at least one-third of Vcc and often as high as one-half Vcc. In the latter case make R1 = R2; and in the former, R1 = 2 R2. Choose your values such that they are between 1000 and 20,000 ohms.

Another method, shown in Fig. 1-9, uses a bridge circuit consisting of resistors R1, R2, R3, and R4. This is actually the same technique as in the previous example, but it also preserves the differential nature of the operational amplifier's two input terminals. The same rules apply as were required for Fig. 1-8(A), but additionally we require that R1 = R2 and R3 = R4. These resistors should be matched manually or with an ohmmeter or be purchased as precision types. The parallel combination of R1 and R3 must be included as part of R_{in} when making closed-loop voltage-gain calculations.

The last method that I will consider is shown in Fig. 1-10. Here I am using voltage regulators to derive potentials for Vcc and Vee from a single-polarity main source. In Fig. 1-10(A) you see the use of zener-diode regulators. Two are required, one each for Vcc and Vee. Capacitors are not used for filtering, as we are already operating from a dc source, but

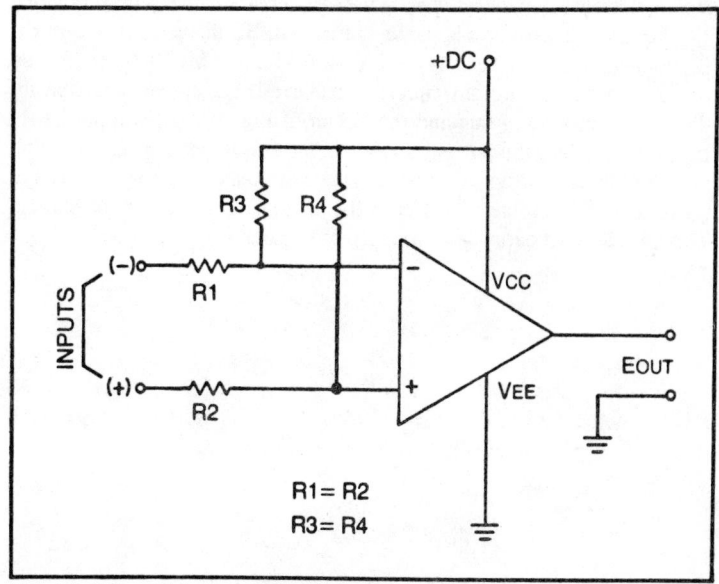

Fig. 1-9. Differential single-supply operation of an op amp.

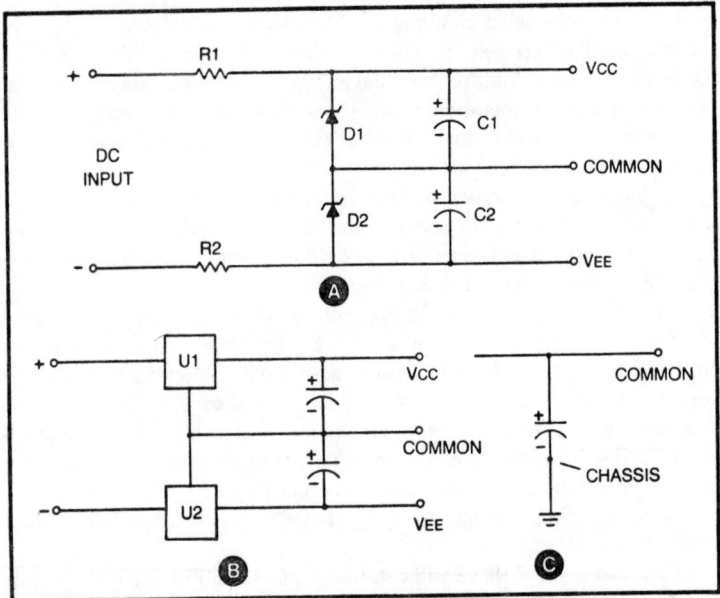

Fig. 1-10. At (A), a zener diode supply; at (B), a three-terminal IC voltage regulator supply; and at (C), ac bypassing between floating and chassis grounds.

instead for *decoupling* and the preservation of stability. These capacitors, incidentally, are not always strictly necessary but are *always* good practice.

The alternate method is shown in Fig. 1-10(B) and uses three-terminal, IC, voltage regulators such as the LM340/LM320 or MC7800/MC7900 series. Again, decoupling capacitors are preferred. The op-amp common and the chassis ground in this and the circuit of Fig. 1-10(A) will not be the same point unless the power supply feeding this is floating. It is possible to make them the same point for ac signals by connecting a bypass capacitor of several microfarads between the op-amp common and the chassis. This is called a floating, or counterpoise, ground.

Chapter 2

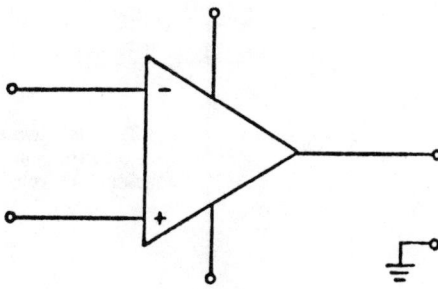

Digital Electronics

D IGITAL-ELECTRONIC-LOGIC INTEGRATED CIRCUITS OFFER THE DE-
signer powerful tools at very low cost. If you are already familiar
with analog instrumentation circuits using operational amplifiers and analog-
function modules, then knowledge of digital electronic techniques will make
you able to expand your horizons by at least an order of magnitude.

WHAT IS DIGITAL?

Digital signals are voltage levels that are allowed to assume either of
two values. Figure 2-1 shows such a situation where 0 volts is one permissible level and +5 volts is the other. None of the voltages between or
beyond these limits are allowed. The levels of Fig. 2-1 have become industry standard because they are the voltages recognized by the ubiquitous
TTL family of IC logic devices.

Digital logic recognizes only two different states, and these are called
true and false in classical logic or 1 and 0 in electronic applications. *Positive
logic* assigns logical-level 0 to the 0-volts condition, and logical-level 1 to
the +5-volts condition. *Negative logic*, on the other hand, uses exactly the
opposite convention; logical-level 0 is +5 volts, and logical-level 1 is 0 volts.
Unless specified otherwise in some particular discussion, always assume
that positive logic is being used (a convention I will follow in this book).
The TTL and CMOS device names assume that positive logic is being used.
In all of my discussions:

Logic level	Voltage
0	0 volts
1	+5 volts

15

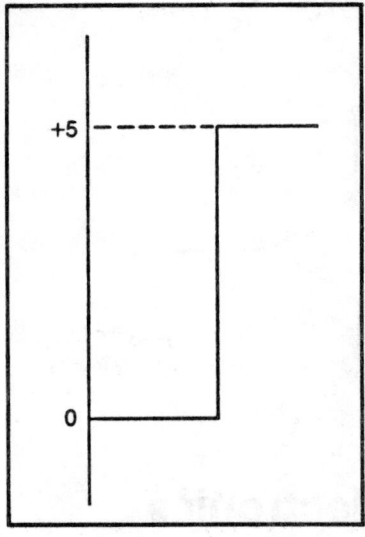

Fig. 2-1. Only two signal levels are allowed in digital circuits; here, 0 volts and +5 volts are used.

Before going any further into digital electronics, I should bow to custom and discuss a little binary arithmetic, the mathematics of base-2. The only digits recognized by the binary system are 0 and 1. (It should pique your interest that these digits are also assigned to digital-logic elements, and for good reason.)

In the common decimal system (base-10) there are 10 different digits: 0, 1, 2, 3, 4, 5, 6, 7, 8, and 9. If we want to express a number greater than 9, we must use a position-weighted notation in which the position of a digit gives it additional value. Take as an example, the number 569. In actuality, this is a position-weighted notation of a specific decimally-related sum of quantities, or

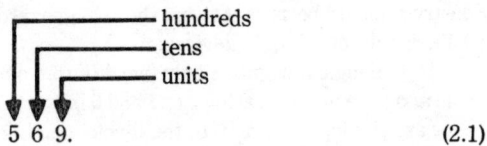

$$5 \ 6 \ 9. \qquad\qquad (2.1)$$

In the position immediately to the left of the decimal point, the value of the digit is the same as its natural value; it is unchanged. In this case, the value is (9 × 1). In the next position, each digit represents 10 times its natural value. In this case the 6 actually represents 6 × 10 or 60. In the next position, each digit represents a quantity 100 times its natural value; so in this case the 5 represents 5 × 100 or 500. The representation 569, then, is actually a shorthand way of writing

$$(5 \times 100) + (6 \times 10) + (9 \times 1) = 569 \qquad (2.2)$$

or

$$(5 \times 10^2) + (6 \times 10^1) + (9 \times 10^0) = 569 \qquad (2.3)$$

We also use the same concept, that of position adding weight, in binary-numbers notation. In that case, though, each position may be occupied by only two different digits, 0 and 1. The general expression for the weighting system is

$$2^n + 2^{(n-1)} + \ldots + 2^3 + 2^2 + 2^1 + 2^0 \qquad (2.4)$$

Both number systems can be used to represent any given quantity. For example, find the decimal equivalent of the binary number 1110010.

$$\begin{array}{ccccccc} 1 & 1 & 1 & 0 & 0 & 1 & 0 \end{array} \qquad (2.5)$$
$$(2^6) + (2^5) + (2^4) + 0 + 0 + (2^1) + 0 \qquad (2.6)$$
$$= 64 + 32 + 16 + 0 + 0 + 2 + 0 \qquad (2.7)$$
$$= 114_{10}$$

(The subscript 10 in 2.7 indicates that 114 is a base-10 number.)

Table 2-1 shows the binary representation of decimal numbers up

Table 2-1. Binary Equivalents of Decimal Numbers to 33.

Decimal	Binary	Decimal	Binary
0	000000	17	010001
1	000001	18	010010
2	000010	19	010011
3	000011	20	010100
4	000100	21	010101
5	000101	22	010110
6	000110	23	010111
7	000111	24	011000
8	001000	25	011001
9	001001	26	011010
10	001010	27	011011
11	001011	28	011100
12	001100	29	011101
13	001101	30	011110
14	001110	31	011111
15	001111	32	100000
16	010000	33	100001

(The leading zeros are not strictly necessary but were included here to balance the appearance of the chart).

through 33. The rules of binary arithmetic are very simple and are as follows

$$\text{For addition;} \begin{cases} 0 + 0 = 0 \\ 1 + 0 = 1 \\ 0 + 1 = 1 \\ 1 + 1 = 0 \quad \text{plus carry 1 to} \\ \qquad\qquad\qquad \text{next place to left} \end{cases} \tag{2.8}$$

(This chart represents all meaningful combinations.)

Example

Add the two binary numbers 101 and 001

$$\begin{array}{r} 1\,0\,1 \\ +\ 0\,0\,1 \\ \hline 1\,1\,0 \end{array} \tag{2.9}$$

Example

Add the two binary numbers 010110 and 001011

$$\begin{array}{r} \text{carries} \quad 1\ 1\ 1\ 1 \\ 0\ 1\ 0\ 1\ 1\ 0 \\ +\ 0\ 0\ 1\ 0\ 1\ 1 \\ \hline 1\ 0\ 0\ 0\ 0\ 1 \end{array} \tag{2.10}$$

Subtraction in digital computers and logic circuits must be done surreptitiously. These circuits, in even the biggest, number-crunching computers, can only perform one arithmetic operation, addition. To subtract, we have to fool the thing into thinking it is adding. This is done by *two's complement* arithmetic.

The complement of a binary number is its inverse. That is to say, the complement of 0 is 1, and the complement of 1 is 0. When we complement a binary number, of however long a length, we merely change all of the ones to zeros and all of the zeros to ones.

Digit	Complement	
0	1	(2.11)
1	0	

The two's complement of a binary number is formed by complementing the binary number and then adding 1 to the least significant digit (2^0). For example, find the two's complement of 1011.

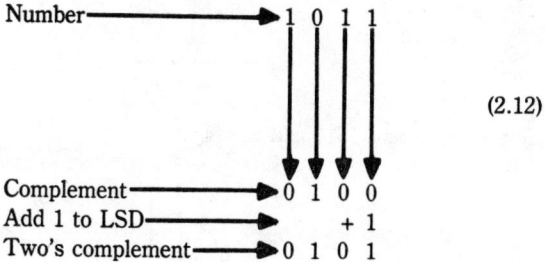

$$(2.12)$$

Number $\longrightarrow$ 1 0 1 1

Complement $\longrightarrow$ 0 1 0 0
Add 1 to LSD $\longrightarrow$ + 1
Two's complement $\longrightarrow$ 0 1 0 1

Example

Subtract 0100 from 1011.

$$\begin{array}{r} 1\ 0\ 1\ 1 \\ -\ 0\ 1\ 0\ 0 \\ \hline ? \end{array} \qquad (2.13)$$

First, find the two's complement of 0100, and then add it to 1001.

$$\begin{array}{c} 0\quad1\quad0\quad0 \\ \downarrow\quad\downarrow\quad\downarrow\quad\downarrow \\ 1\quad0\quad1\quad1 \\ +\ 1 \end{array} \qquad (2.14)$$

two's complement 1 1 0 0

Add this result to 1011

$$\begin{array}{r} 1\ 0\ 1\ 1 \\ \text{carry} \quad +\ 1\ 1\ 0\ 0 \\ \hline 1 \longleftarrow \quad 0\ 1\ 1\ 1 \longleftarrow \text{answer} \end{array} \qquad (2.15)$$

Let's check this answer by also performing the same addition in decimal notation. First, convert the binary numbers 0100 and 1011 to their respective decimal equivalents and then add them in the ordinary manner.

$$\begin{array}{rcl} 1\ 0\ 1\ 1 & & 11_{10} \\ -\ 0\ 1\ 0\ 0 & \longrightarrow & -4_{10} \\ \hline 1 \quad\quad 0\ 1\ 1\ 1 & & +7_{10} \end{array} \qquad (2.16)$$

The binary answer was 0111 plus a "carry 1." $0111_2 = 7_{10}$; so the answer is correct. Note that a "carry 1" tells us that the result is positive. Let us try an example in which the answer would be negative, like subtracting 1011 from 0100.

$$\begin{array}{r} 0\ 1\ 0\ 0 \\ -\ 1\ 0\ 1\ 1 \\ \hline ? \end{array} \qquad (2.17)$$

First, complement 1011 and then add 1

$$\begin{array}{c} 1\ 0\ 1\ 1 \\ \downarrow\ \downarrow\ \downarrow\ \downarrow \\ 0\ 1\ 0\ 0 \\ +\ 1 \\ \hline 0\ 1\ 0\ 1 \end{array} \leftarrow \begin{array}{l} \text{two's complement} \\ \text{of 1011} \end{array} \qquad (2.18)$$

add the two's complement to 0100

$$\begin{array}{r} 0\ 1\ 0\ 0 \\ +\ 0\ 1\ 0\ 1 \\ \hline 1\ 0\ 0\ 1_2\ =\ 9_{10} \end{array} \qquad (2.19)$$

But the decimal version is

$$\begin{array}{r} 4_{10} \\ -\ 11_{10} \\ \hline -\ 7_{10} \end{array} \qquad (2.20)$$

Note that in Eq. 2.19 the binary answer had *no carry bit*; this tells us that the resultant is *negative*; so the correct representation is given by complementing the result and adding 1

$$\begin{array}{ll} \text{(result from Eq. 2.19)} & 1\ 0\ 0\ 1_2 \\ & \qquad\downarrow \\ \text{Complement} & 0\ 1\ 1\ 0_2 \\ \text{Add 1} & \qquad +\ 1 \\ & \overline{1\ 1\ 1\ 1_2} \\ & 0\ 1\ 1\ 1_2\ =\ -7 \end{array} \qquad (2.21)$$

LOGIC GATES

The basic building blocks of almost all digital-electronic-logic circuits, even unto the most complex computers, are a small collection of similar, but strikingly different, *gates*.

Figure 2-2 shows not an actual gate but the most elementary logic element. This circuit is a complementer, or, as it is called more frequently, an inverter. It takes the level at the input and complements it. A 1 becomes a 0, and a 0 becomes a 1. The truth table for the inverter is also shown

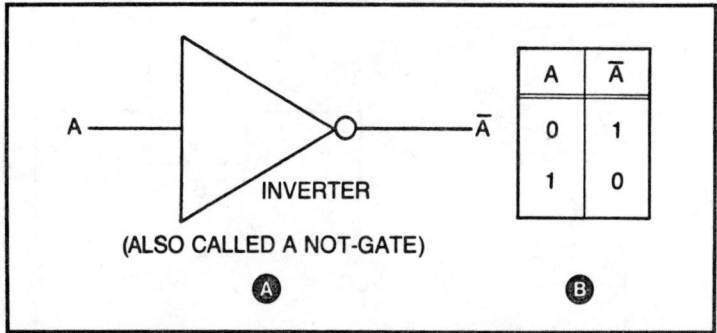

A	Ā
0	1
1	0

INVERTER

(ALSO CALLED A NOT-GATE)

Ⓐ   Ⓑ

Fig. 2-2. At (A), the symbol for an inverter, or NOT gate, logic element; and (B), the truth table for the inverter.

in Fig. 2-2. If A is the input, then the complement at the output is called A, which is read not-A, or bar-A. The former is the more common expression. Any time you see a circle at the output or input of a logic-element symbol, it is taken to mean an inverted condition. A triangle symbol without the circle at the output is called a buffer. It would have a truth table such as:

$$
\begin{array}{cc}
A & \overline{A} \\
0 & 0 \\
1 & 1
\end{array}
$$

in which case, though, the bar over the output A is incorrect. Therefore, whenever there is no inversion, delete the bar. This is nothing more than a logical method for saying A is A.

The elementary OR gate is shown in Fig. 2-3. An OR gate is one which will give an output if either input A *or* input B is HIGH (logical 1). The truth table is shown in Fig. 2-3(C), while the schematic symbol is shown in Fig. 2-3(A).

An equivalent circuit is shown in Fig. 2-3(B). We can represent digital bits (0 or 1) with a switch because a switch also has but two states; it is a *binary device*. Here we call "off" logical 0, and "on" logical 1. The output condition in Fig. 2-3(B) is the voltage appearing across the resistor at C. This point will be high (1) if either switch A *or* switch B is closed or if they both are closed.

A related but more complex type is the exclusive-OR gate, which gives you a 1 output if either A is 1 or B is 1 but not if both are 1. The symbol and truth table appear in Fig. 2-4.

Figure 2-5(A) shows the NOR gate, which is an OR gate with an inverted output. Again note the use of a circle at the output end of the circuit symbol. NOR stands for not-OR, and the gate could be represented by an OR gate followed by an inverter (the so-called NOT gate).

21

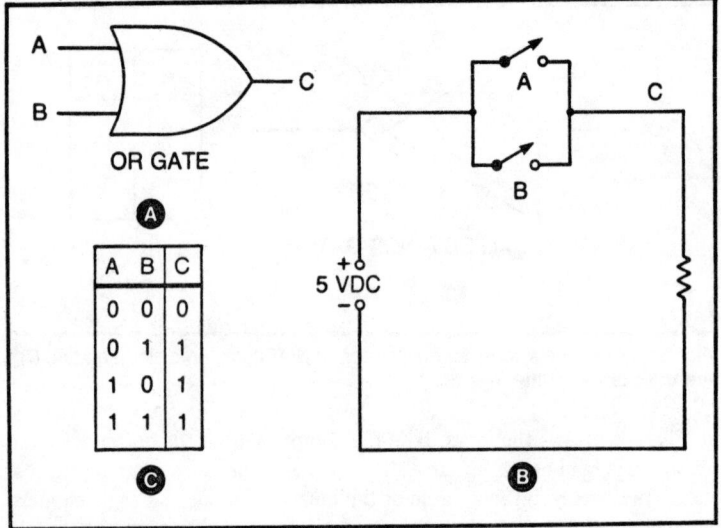

Fig. 2-3. At (A), the symbol for an OR gate; (B), switch circuit equivalent of the OR gate; and (C), the truth table for the OR gate.

The equivalent circuit is shown in Fig. 2-5(B). The output across C will be logical zero if either switch A or switch B is closed. The truth table for this action is shown in Fig 2-5(C). It shows that we will get a 1 output *only* if both inputs are 0.

Figure 2-6(A) shows the AND gate. It produces a 1 output *only* if input A *and* input B are also 1. The truth table is shown in Fig. 2-6(C).

A circuit equivalent is shown in Fig. 2-6(B). The AND gate is represented by a pair of switches (for inputs A and B) in series. There will be a voltage across C only if both switch A *and* switch B are closed.

The inverted version is the NAND gate of Fig. 2-7(A). The circuit symbol is the same as for the AND gate except that the inverting circle is present at the output. This type of gate can be represented by a pair of series-connected, SPST switches placed across the load. Point C will be zero only

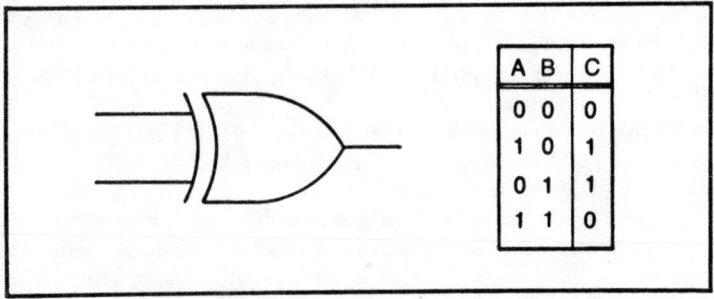

Fig. 2-4. The symbol and truth table for the exclusive OR gate.

22

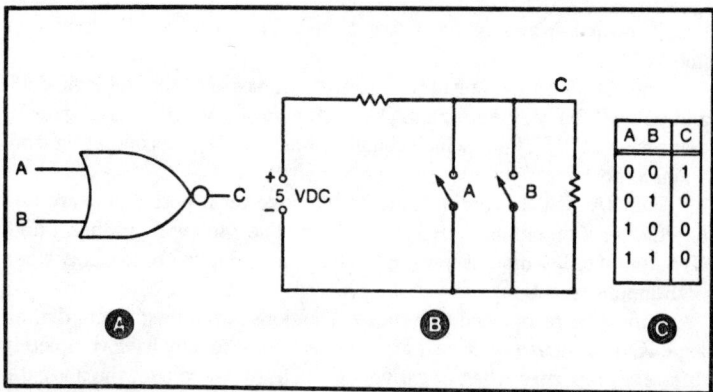

Fig. 2-5. At (A), the symbol for a NOR gate; at (B), the switch circuit equivalent of the NOR gate; and at (C), the truth table for the NOR gate.

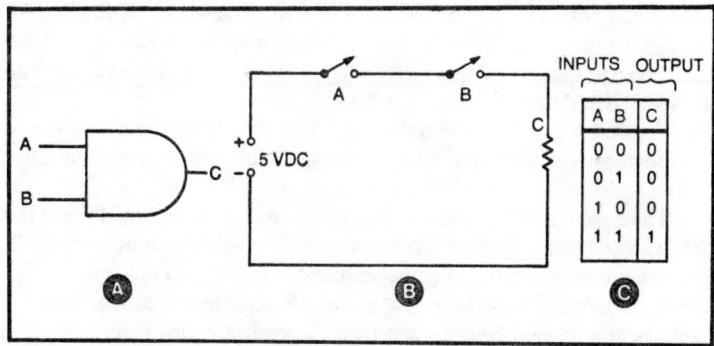

Fig. 2-6. The symbol for an AND gate appears at (A); the switch circuit equivalent at (B); and the truth table at (C).

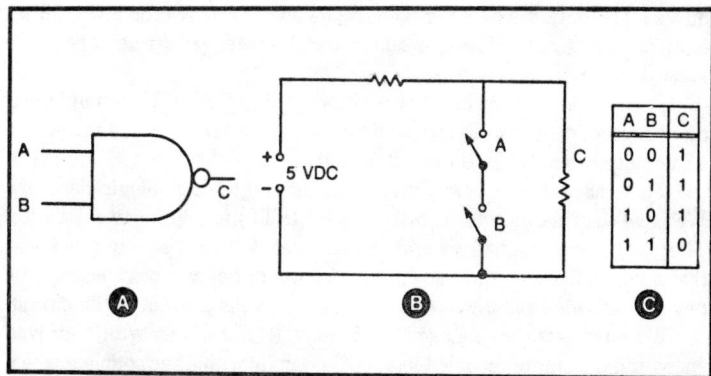

Fig. 2-7. The symbol for a NAND gate is at (A), with the switch circuit equivalent at (B), and the truth table at (C).

23

if both switch A and switch B are closed. The word NAND stands for not-AND.

The NAND gate is one of the real work-horses of the digital-logic field, and you will be seeing it again. One of the most popular NAND gate IC devices is the TTL 7400 which contains four NAND gates operating from a single power supply.

The NAND gate can be made into an inverter if both inputs are tied together or if one input is tied to +5 volts. The 7400 will produce a high (1) output if either input is low (logical 0). The output will be low only when both inputs are high.

It must be recognized that the designations normally given to digital-logic IC devices use positive logic. The same device may have completely different properties when negative logic is used. Of course, the not-gate, or inverter, will be the same no matter which logic convention is used, but the other gates will take on completely different properties. In general, it is true to state that a positive-logic, AND gate is a negative-logic, OR gate and a positive-logic, NAND gate is a negative-logic, NOR gate. Although it tends to confuse the issue, many digital IC catalogues will list gates such as the TTL 7400 as NAND/NOR gates. In positive logic, though, they are NAND gates.

Figure 2-8 is a *timing diagram* that shows the output response of our different types of gates for various combinations of repetitive-input-pulse trains.

The example given in Fig. 2-8(A) shows the response properties of the AND gate. Output C will have a positive pulse only when both inputs A and B are also positive. In distinct contrast is the NAND response of Fig. 2-8(B). Remember that the output of the NAND gate will be high if either input is low. It is low only when both A *and* B are also high.

The responses of the OR and NOR gates are shown in Figs. 2-8(C) and 2-8(D), respectively. The output of the OR gate will be high whenever there is a high condition at either input. The NOR gate, on the other hand, produces a low output whenever both inputs are low. It will be low if either input is high. Again we see the dual nature between positive and negative logic.

The response of the inverter is shown in Fig. 2-8(E). This simple, but often crucial, device will produce a high output when the input is low and a low output whenever the input is high.

There have been several different families of IC digital-logic elements. We have seen the resistor-transistor-logic (RTL) family for two decades. They were slow and had several bad features, but their saving grace was that no combination of input or output shorts or opens would destroy the device—provided there was no more than 3.6 volts present in the circuit.

We later saw the diode-transistor-logic (DTL) family with improved speed and the emitter-coupled-logic (ECL) family with its ac coupling which was, and still is, one of the fastest families. The really big boom in digital logic came following the introduction of the transistor-transistor-logic (TTL,

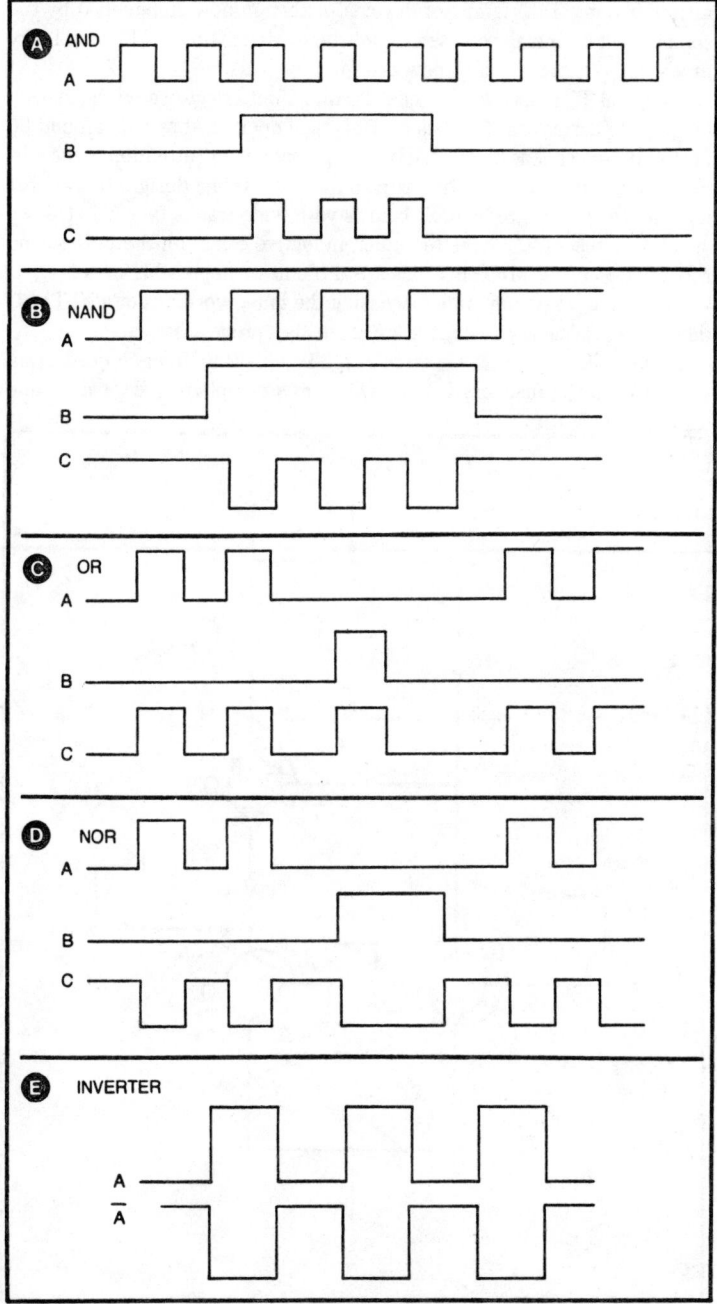

Fig. 2-8. Timing diagrams for some common gates.

or in the slang T²L) family of devices. Although now superseded by the complementary-metal-oxide-semiconductor (CMOS) family, TTL is still very much in evidence, even in new designs.

Typical TTL speeds are about 20 MHz, but selected devices, as well as those in certain subfamilies of TTL, will operate at speeds around 80 MHz. At high speeds (over 1 MHz) though, layout and attention to the wiring used can play an important part in the success of a design. It is almost mandatory that printed-circuit boards with wide tracks be used at those speeds. Wires simply have too much inductive effect on the pulses, and this creates many problems that are difficult or impossible to solve.

I will not waste any time discussing the inner workings of all TTL IC devices, but a look at a typical inverter may prove instructive.

The TTL inverter is shown in Fig. 2-9. This is a direct-coupled, transistor circuit. Transistors Q2 and Q3 form a complementary pair, while

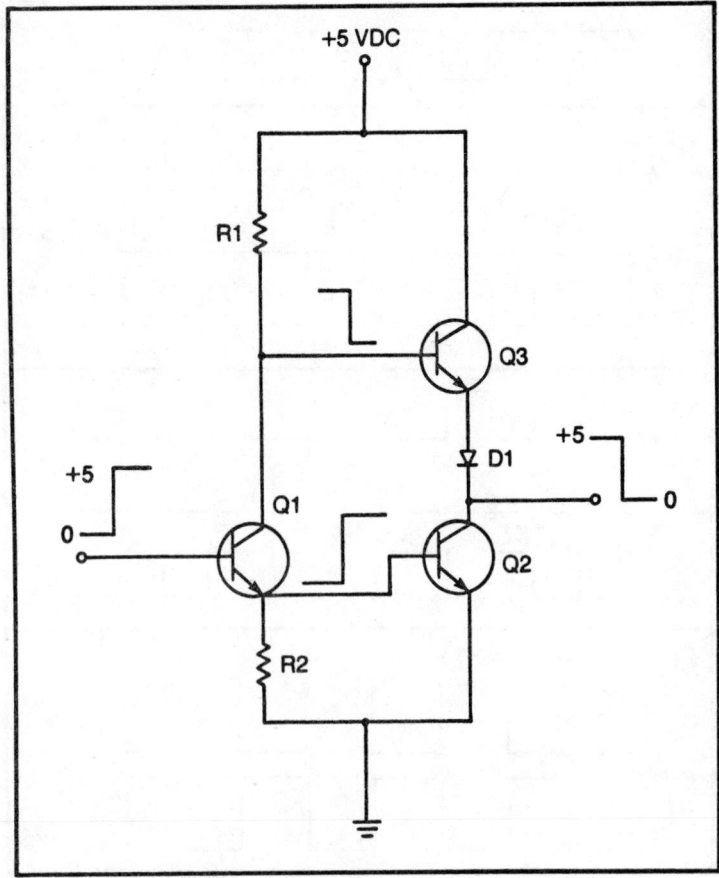

Fig. 2-9. Typical TTL inverter internal circuit.

26

Q1 is an inverting amplifier at the collector and a noninverting amplifier at the emitter. A high (+ 5 volt) signal applied to the base of Q1 will saturate this transistor. The potential at the collector of Q1 then drops, while that at the emitter rises. This condition removes bias from transistor Q3; so it is cut off. Transistor Q2, on the other hand, is biased hard "on". Under this condition, Q2 is also saturated; so its collector will be at a potential of 0 volts or nearly so. The collector of Q2, though, is also the device-output terminal. This means that we have a condition where a high input causes the output terminal to be low.

When the input signal applied to the base of Q1 is low, Q1 will be cut off. This makes the emitter voltage at Q1 zero and the collector voltage high. Under this condition Q2 is cut off, and Q3 is turned on hard. This passes Vcc (+) through Q3 and D1 to produce a + 5-volt high condition at the output. Table 2-2 lists some of the more popular TTL devices. Most of these can be purchased for relatively little money, especially from those companies that specialize in industrial surplus. Be wary, though, if some companies seem to offer parts at a tremendous discount compared with other companies advertising in the same market. Some of them go to the semiconductor manufacturers and offer to buy all the rejects shoveled out the back door at pennies a pound. Some of these are reputable enough to test them and sell only those that come close to meeting the specs, but others use just plain dastardly practices in that they sell junk as first grade stuff. Unfortunately, bargains do crop up from time to time; so it is possible for some companies to offer bargain-basement stuff at low, low prices. The best bet is to try a few companies, get burned a little, and then stick with the companies that seem to be good. The best source of company addresses are the hobby electronics, ham radio (not CB generally), and personal computer (hobby) magazines. The ads in these are often worth the price of the magazine, especially if you are the type who can flip out over an exciting new IC device.

CMOS GATES, ETC.

Complementary-metal-oxide-semiconductor (CMOS) digital-logic integrated circuits are the latest family of devices, although one is cautioned about prematurely writing the TTL's obituary. CMOS devices are based on the MOS field effect transistor instead of the bipolar transistors used in TTL and earlier families.

For now let us content ourselves with a naive description of the MOSFET-based-logic element. Two different types of MOSFET are used in these. The p-channel turns on when the gate (i.e., input) is zero with respect to the source, while the n-channel turns on when the gate is positive with respect to the source.

An example of a simple CMOS inverter is shown in Fig. 2-10. The circuit is designed with an n-channel and a p-channel series connected in a complementary-symmetry circuit. Output is taken from their mutual junc-

Table 2-2. Typical TTL device numbers.

7400	Quad, two-input NAND gate
7401	Quad, two-input NAND gate with open-collector output
7402	Quad, two-input NOR gate
7403	Quad, two-input NAND gate with open-collector
7404	Hex inverter
7405	Hex inverter, with open-collector
7406	Hex inverter, with open-collector, good to 30 volts
7407	Non-inverting driver, open-collector, good to 30 volts
7408	Quad, two-input AND gate
7409	Quad, two-input AND gate (with open collector output)
7410	Triple, three-input NAND gate
7414	Hex Schmitt trigger
7416	Inverting hex driver, good to 15 volts
7417	Noninverting hex driver, good to 15 volts
7420	Dual, four-input NAND gate
7430	Single, eight-input NAND gate
7432	Quad, two-input OR gate
7437	Quad, two-input NAND buffer
7440	Dual, four-input NAND buffer
7442	BCD-to-1-of-10 decoder, TTL compatible outputs
7445	BCD-to-1-of-10 decoder, high current (80 mA), 30-volt outputs
7446	BCD-to-1-of-10 decoder, Nixie outputs
7447	BCD-to-seven segment decoder, 40-mA, 30-volt outputs
7473	Dual JK flip-flop
7474	Dual type-D flip-flop
7475	Quad-latch (4×1 bit memory)
7476	Dual JK flip-flop
7483	Four-bit full adder
7485	Two 2 × four bit magnitude comparator
7486	Quad exclusive-OR gate
7489	sixteen × four-bit memory
7490	Decodable decade (divide-by-10) counter
7492	Divide-by-12 counter
7493	Divide-by-16 counter
7495	Four-bit PIPO shift register
7496	Five-bit PIPO shift register

tion at the source of one and the drain of the other. Input signal is applied simultaneously to *both* gates.

When the input signal is low (0 volts), we find transistor Q1 turned off and Q2 turned on. This allows the Vcc potential to appear at the output terminal. When the input signal snaps high we find that the situation reverses, Q1 is on and Q2 is off. This places the output terminal at ground potential (0 volts).

If we view the output stage of the CMOS inverter as a series circuit consisting only of the channel resistances of transistors Q1 and Q2, the situation as seen by the output terminal will be that shown in Fig. 2-11.

In Fig. 2-11(A) you see the situation existing when the input is low, in which case Q1 is off and Q2 is on. This makes R_{Q1} high and R_{Q2} low. By normal and ordinary voltage-divider action this places the output high. The situation reverses, however, when the input is high. This is shown in Fig. 2-1(B). In this case Q1 is on and Q2 is off. This also reverses the relative resistances of the transistor's channels, making R_{Q1} low and R_{Q2} high. Once again, by voltage-divider action, you find the output of the device low.

Figure 2-11 demonstrates one of the really unique aspects of CMOS devices. The typical CMOS output will have a low-resistance path to Vcc (+) in the high condition and a similar path to ground in the low state. The turned-off transistor in each output state will present an extremely high resistance; so the power supply always sees a high impedance to ground whenever the output is in a stable state.

The only time this resistance drops appreciably is when the output stage is in transition from one state to the other. At those times the respective on/off roles of the transistors are changing to the reverse condition. When the output is stable in either high or low states, the power supply sees an

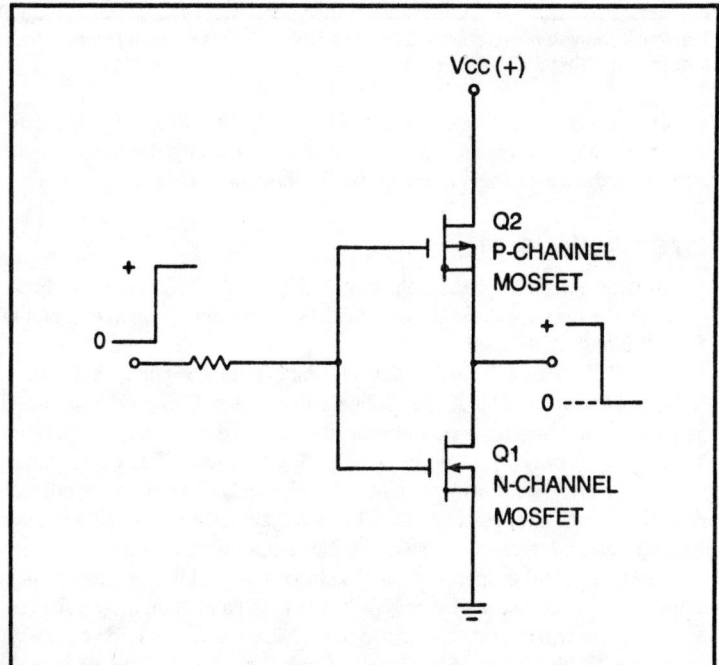

Fig. 2-10. CMOS inverter circuit.

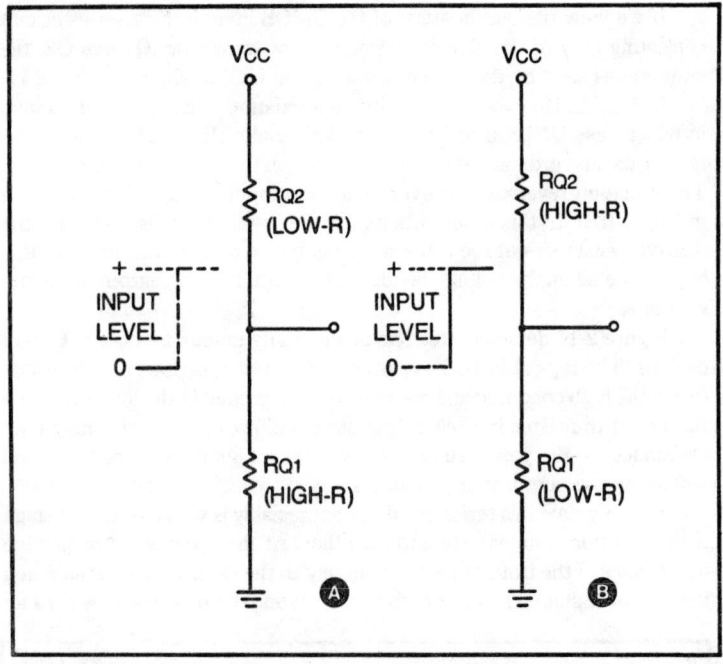

Fig. 2-11. Equivalent resistor circuits for a CMOS inverter—with (A) input low, and (B) input high.

extremely high resistance to ground. In the stable condition, then, the device will draw only microamperes of current. The only time appreciable current is drawn in this circuit is *during transition times*.

CMOS VERSUS TTL

Now let us contrast the behavior of TTL and CMOS devices. First, compare the circuits of the TTL and CMOS inverters by again examining Figs. 2-9 and 2-10, respectively.

The TTL stage will always draw current in not-inconsiderable amounts. A certain common TTL IC NAND gate draws 8 mA when the output is high and 22 mA when the output is low. A CMOS equivalent, on the other hand, asks for only 15 μA on high and 170 μA on low. This conservation of power is one reason why the CMOS devices have become very popular recently. A computer that uses all TTL, for example, might require at least 10 amperes and perhaps as much as 150 amperes at 5 volts dc.

Another useful property of CMOS chips is their ability to operate normally over a wide range of supply voltages (VCC) and logic-level voltages. Most of them will operate over a range of at least + 4.5 to + 15 volts, with some able to handle up to + 18 volts dc at VCC.

The TTL IC, on the other hand, will not operate well, or at all in many cases, unless the supply voltage is very nearly +5 volts dc.

Also, the noise immunity of the CMOS device is superior because the device will not change state unless the input potential reaches a threshold of +Vcc/2. If the supply is +12 volts, the device will remain stable unless a logic level of Vcc /2, or 6 volts, is offered. The TTL IC, though, must see an input of something over 4.5 volts, or it is unsure of the proper state. Noise on the signal can easily add to the normal logic level to produce this ambiguous state. It takes a much larger noise signal to confuse CMOS.

Because of this sensitivity to both supply and logic-level voltages, the heavy current demands of the TTL chip will often cause sufficient IR drop in the printed-circuit tracks to create problems. When a TTL output drops low, there is a sudden and very sharp rise in current demand, and this will drop the terminal voltage at the device's Vcc (+) input pin.

Besides the possibility of starving the chip, this creates "glitches," or spurious pulses, that raise havoc with things like gates, counters, and flip-flops.

The PC board containing even modest numbers of TTL devices must have a generous sprinkling of 0.001-μF-to-1-μF bypass capacitors just to alleviate this type of problem. The usual ratio is at least one capacitor for every two chips unless one particular chip is relatively remote (several inches constitutes "remote") from the others.

One must not be penny wise, etc., in the matter of bypass capacitors. They are dirt cheap and are easily obtained everywhere. Therefore, one is advised to use a very conservative protocol in the matter of bypassing TTL integrated circuits. Use a 0.001 μF (10 volt or higher) at *each* Vcc (+) terminal, and solder it directly to the Vcc pin—*not near it, but to it.* Also, for every few devices along the same power-line conductor install a 0.1 μF capacitor. At the power-supply entrance for +5 volt (like on the edge connector or terminal, if wire board is used) solder a 1-μF, tantalum, electrolytic capacitor. Of course, it should go without saying, these capacitors should have their other ends soldered to ground. They are used to locally store a charge to be dumped into the circuit whenever a nearby TTL device undergoes a transition. These capacitors will also bypass any glitches carried down the PC power line.

CMOS AND STATIC DAMAGE

The main problem when working with CMOS devices is that static electricity built up on your body, tools, bench, and other devices, or transient charges generated as you connect test equipment can easily destroy CMOS chips. Although they are relatively safe when connected into their circuit, they can be extremely vulnerable when out of the circuit.

Occasionally you will hear that the CMOS static problem is overrated, and in some ways it is true that too much has been made of this problem. Although some more-recent devices have diode-protected terminals, one must be ready to pay the cost, if sloppy handling ruins a CMOS device.

I have destroyed some; so view dimly those claims that the problem is overrated.

The cost of sloppiness can be quite high. Although common CMOS devices are relatively cheap (many of those in the 4000 series), some special-purpose, CMOS devices are extremely expensive. One digital clock chip costs around $15, for example, while a certain microcomputer chip sells for almost $100! A hidden cost, incidentally, is the cost in time, money, and aggravation spent debugging a project where a single CMOS function has been destroyed by static. So if you want to ignore the problem, feel free, but let's hear no sobbing later when your $150 A/D converter goes up in smoke.

Why So Sensitive?

A representation of the typical CMOS-chip innards, a MOSFET transistor, is shown in Fig. 2-12(A). The gate electrode is insulated from the

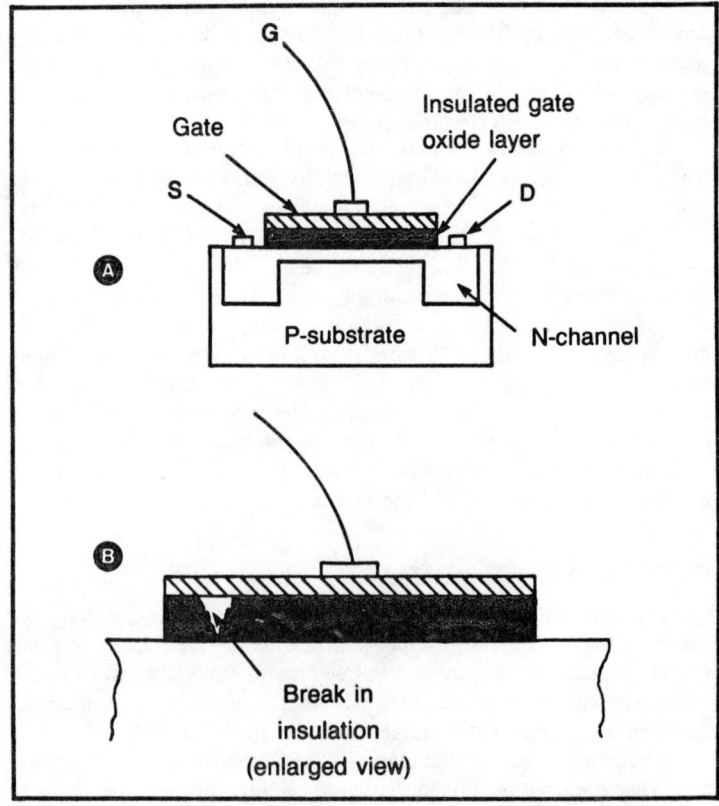

Fig. 2-12. At (A), a typical MOSFET structure of CMOS input transistors. At (B), a possible failure mode due to static in careless handling.

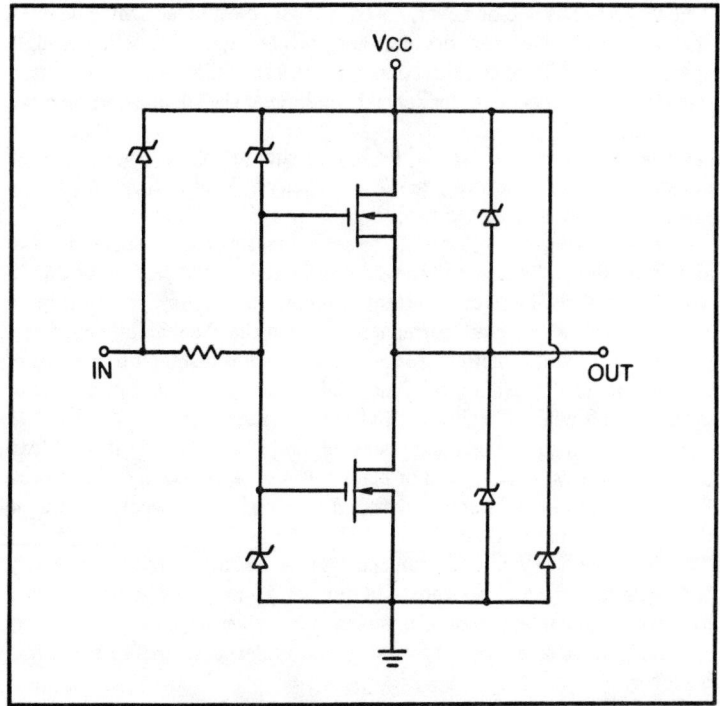

Fig. 2-13. Diode protection to diminish the effects of static on CMOS ICs.

n-channel by a metal-oxide layer that may be as little as 1/10,000 inch thick. The breakdown voltage of so thin a layer is typically 80 volts. With the possibility of a buildup of several hundred volts of static electricity on your body, tools, and work surfaces, it is easy to see how, without proper precautions, you could destroy the CMOS device.

The failure mechanism of a CMOS-MOSFET is shown by the structure diagram in Fig. 2-12(B). When the insulated layer of metal oxide is breeched metal ions are drawn in through the gate, and the channel and gate become shorted together.

A more insidious fact, though, is that the action may be delayed because a sufficient number of metal ions may not enter the breech immediately. Eventually the accumulation of ions will build up to a point where the gate is shorted, and this destroys the device. It may explain a "spontaneous" device failure several weeks or even months after construction.

The Cure: Ground Everything

Even though some CMOS devices feature built-in zener diodes to protect against static damage, some safeguards are still recommended. An example of the circuit for a protected CMOS inverter is shown in Fig. 2-13.

This protection is not, however, absolute. Also, most older CMOS devices, and many of those current devices that provide a special-function or high-speed operation, have no protection. I suggest the following manufacturer-recommended procedures for special handling of CMOS and certain other IC devices.

A *grounded environment* is the key to eliminating damage to CMOS devices. This will allow any static electricity to drain off harmlessly to ground *before* it gets to the device.

For example, a large cookie sheet or piece of do-it-yourself sheet aluminum should be fastened to your workbench. If the amount of CMOS work done is small and intermittent, then opt for the former, but if more frequent work is expected, permanently mount the sheet aluminum. Some people use a "static blotter" made of sheet metal fastened to a piece of convenient-sized Masonite or some similar material. This can be placed on the bench when working with CMOS devices.

You must also ground all of your tools and yourself to the sheet. Flexible ground wires, made of red or black test-lead wire, seem to work best. For tools, I prefer one screwdriver and a pair of metal tweezers because they can be made to take solder easily.

The ground to your body can consist of a connection to a metal watch band with an alligator-clip lead, or better yet, a wire permanently soldered to a low-cost, metal-banded, IC bracelet. Remember to scrape off the cheap chrome plating to expose the base metal underneath before soldering, however.

The metal surface on the work bench should be earth-grounded, *but not directly!* Accidentally contacting the 115-volt power mains while grounded directly will most likely be fatal. Instead, use a 1-megohm, 1/2-watt resistor between earth ground and the counterpoise ground consisting of the metal plate on the bench and your body.

The earth end of the resistor should be connected to a metal, cold-water pipe (*not* the plastic type!), a ground rod driven into the earth, or the ground of a nearby ac-power terminal. *Do not* use the *neutral* of the ac line, only ground. The power-line ground is the third wire and is the round hole in the wall outlet (in the USA), not one of the slots. The metal box containing the outlet assembly is also grounded, as is the metal screw holding on the cover plate.

If you have an empty outlet box already in place, you may make a permanent ground connection by drilling a hole in a plain, metal, outlet-box cover to accept a 1/4-inch phone jack (see Fig. 2-14). A 1-megohm resistor was wired inside the box from the main terminal of the jack to the ground terminal. The flexible ground wire from the bench ground sheet is then connected to the tip terminal of a standard phone plug.

This is a good idea *only* if you first *remove* any ac wiring that is in the box. If this prospect scares you, hire an electrician to do the job because the consequences of leaving them can be even worse—your demise from electrocution! If you leave the ac wires in the box, taping them back or

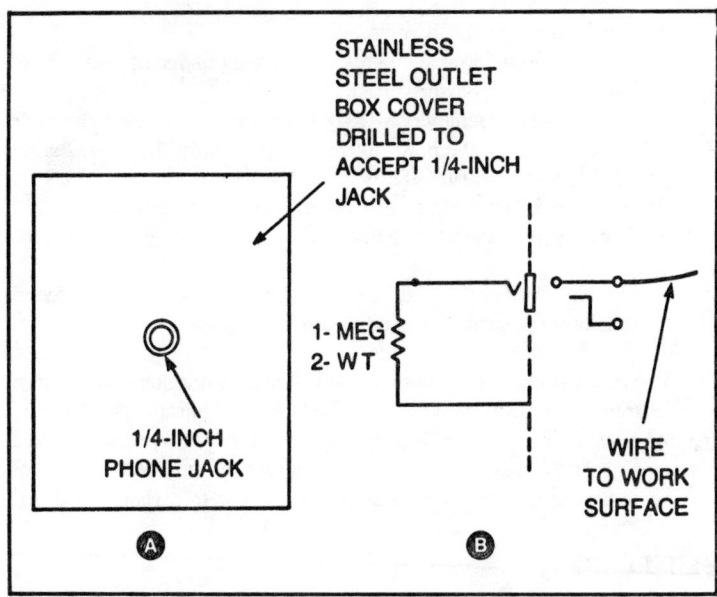

Fig. 2-14. Wallplate grounding system for handling CMOS ICs.

covering them with wire nuts, they may accidentally come loose and fry you. How would you like to be a 4,000-watt fuse? Such an arrangement would be a real widow(er)-maker; so do it right or leave it alone.

CMOS RULES OF THUMB

Rules of thumb are suggestions you give everybody because experience has taught that they are reliable, but they are only occasionally followed by those who think they know it all. Anyway here are a few:

1. Always handle CMOS devices, or equipment containing CMOS devices, in a grounded environment such as we have described.
2. CMOS devices are shipped from the factory either wrapped in metal foil or with their leads embedded in a black, conductive, foam-rubber-like mess. The PC boards containing CMOS devices are often shipped from the manufacturer in black, conductive bags. Some others are shipped in conductive, clear-plastic containers. *Leave them* in whatever container or foam base they are received, and remove them only when ready to use, but not before creating the static-free, grounded environment.
3. When removing CMOS from the shipping medium, ground yourself first. Before actually touching the device or PC board, touch the container or foam base first. This equalizes pin-voltage differences that could arise from residual static.

35

4. Avoid touching the IC pins.
5. Use only soldering implements that have grounded tips (i.e., those with three-wire power cords).
6. Do not insert or remove CMOS devices from the circuit with power turned on. Turn the power off, insert or remove the device as required, and then turn the power back on.
7. Avoid wearing nylon or any other synthetic clothing.
8. Use only grounded test equipment (i.e., the three-wire power cord again).
9. Handle CMOS-equipped PC boards by the edges only, and avoid touching the pins of any edge connectors used.

A navy chief once told me that every safety regulation ever written was written in blood. In the world of CMOS we can paraphrase that saying and claim that every handling rule ever written was written in melted silicon; so it behooves you to follow our silly little regulations, unless of course, you want to build an expensive little silicon-to-carbon converter.

FLIP-FLOPS

Flip-flops are used as switches, pulse creators, and as circuits to perform certain other digital-logic functions. Most flip-flops can be built from NAND- and NOR-gate combinations but are usually bought as separate ICs.

Some textbooks list the monostable and astable multivibrators under the heading flip-flop, but in this case I chose to place them in the chapter on waveform generators, a sneaky way to boost the length of that chapter.

Examples of bistable RS (reset-set) flip-flops are shown in Figs. 2-15 and 2-16. These are made from cross-coupled NOR and NAND gates, respectively. Although both are usually labeled similarly in circuit diagrams, they have quite different properties. The two truth tables are also shown in the respective figures.

Also note that these flip-flops have two output terminals, labeled Q

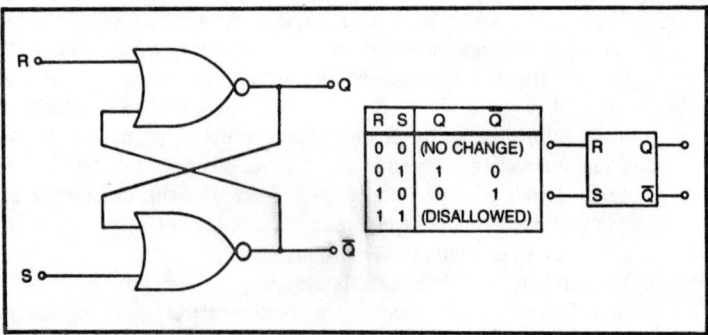

Fig. 2-15. An RS flip-flop circuit using NOR gates, with truth table and circuit symbol.

36

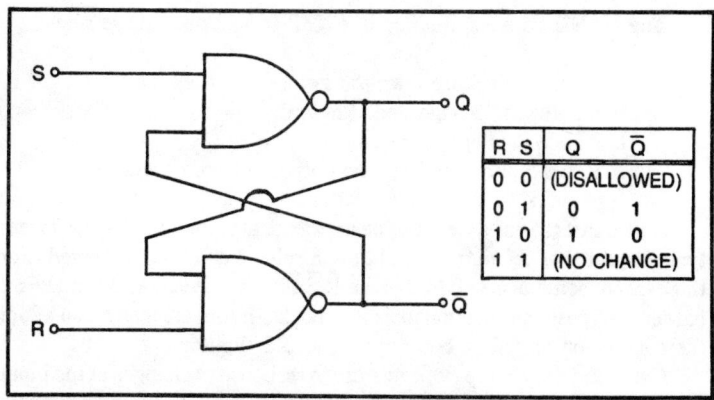

Fig. 2-16. An RS flip-flop consisting of NAND gates.

and $\overline{Q}$ ($\overline{Q}$ reads not-Q). These are complementary outputs; so one will be high when the other is low, and vice versa.

The NOR gate RS flip-flop of Fig. 2-15 will follow these rules:

1. If both inputs are zero, Q and $\overline{Q}$ will remain in their previous states.
2. If the set input is made high, the Q output will be high and the $\overline{Q}$ will be low. The flip-flop will stay in this condition regardless of any subsequent changes in the set input.
3. If the reset input is made high, the Q output will be low, and the $\overline{Q}$ will be high. The flip-flop will remain in this condition regardless of any subsequent changes in the reset input.
4. If both inputs are simultaneously made high, the flip-flop will not know what to do. Since this results in an unpredictable, ambiguous output, it must henceforth and forever more be considered a *disallowed state*.

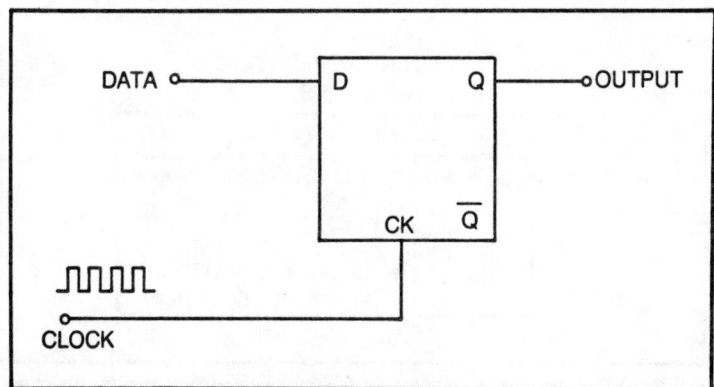

Fig. 2-17. Circuit symbol for a type-D flip-flop.

37

The NAND gate RS flip-flop of Fig. 2-16 will obey these rules:

1. If both R and S are low, you get a disallowed state.
2. If S is low, Q is high and $\overline{Q}$ is low.
3. If R is low, $\overline{Q}$ is high and Q is low.
4. If R and S are both high, no output change occurs.

Notice that the rules for this type are identical to the rules for the former type, except that the output condition in rules 2 and 3 are reversed. For this reason, some people call this the $\overline{R}\text{-}\overline{S}$ flip-flop. Because this makes it harder to typeset service manuals and books, however, many also adopt the convention of calling both circuits an RS flip-flop.

Gates and the RS flip-flops operate in response to changes at the input and will do so at the time the changes occur. This is called *asynchronous* behavior. A class of flip-flops called clocked flip-flops operate only in re-

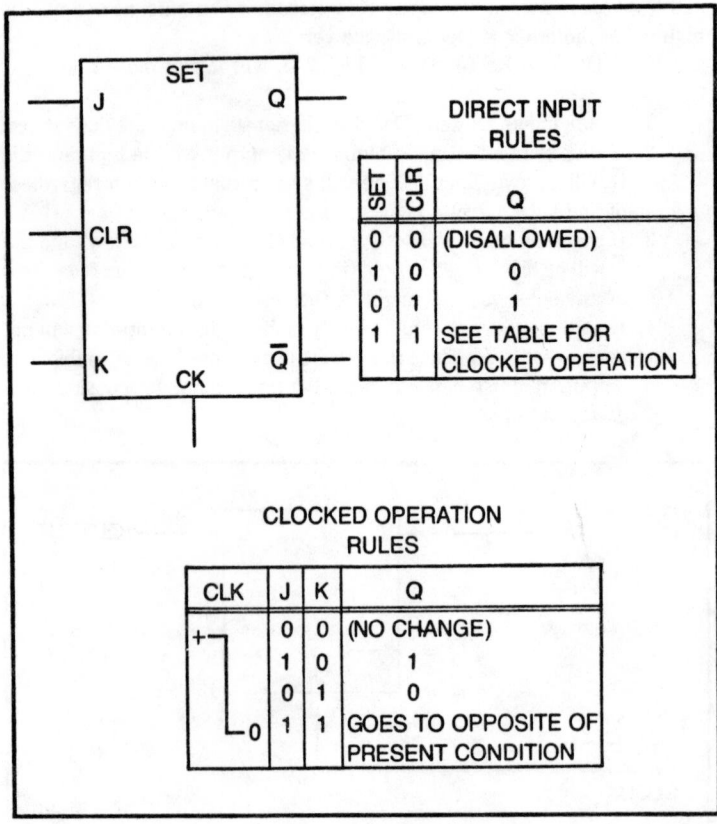

Fig. 2-18. The circuit symbol for a JK flip-flop, with truth tables for direct-set rules and for clocked operation rules.

Table 2-3. Typical CMOS Digital ICs.

4000	Dual, three-input NOR gate
4001	Quad, two-input NOR gate
4002	Dual, four-input NOR gate
4006	18-bit SISO shift register
4007	Inverter
4008	Four-bit full-adder
4009	Hex inverter
4010	Hex buffer
4011	Quad, two-input NAND gate
4012	Dual, four-input NAND gate
4013	Dual, type-D, flip-flop
4014	8-bit, PISO shift register
4015	Dual, four-bit, SIPO, shift register
4016	Quad analog switch
4017	Decade counter, 1-of-10 decoded outputs
4018	Programmable counter
4019	4PDT data selector/router
4020	fourteen-bit binary counter
4021	eight-bit PISO shift register
4022	Octal (base-8) counter
4023	Triple, three-input NAND gates
4024	Seven-bit binary counter
4025	Triple, three-input NOR gate
4026	Decade counter, w/seven-segment decoder
4027	Dual JK flip-flop
4028	BCD-to-1-of-10 decoder
4029	Synchronous up/down counter, base-10 or base-16.
4030	Quad exclusive-OR gate
4031	Sixty-four-bit SISO shift register
4032	Triple, full-adder

sponse to input changes at times dictated by a system clock (pulse train). Clocked logic circuits are said to operate *synchronously*.

An example of a type-D flip-flop is shown in Fig. 2-17. The data applied to the D input will be transferred to the Q output only when the clock terminal (CK) is high. This occurs only on positive transitions of the clock pulse. This clock pulse must have sharp, well-defined edges, and must be noise-free if errorless operation is to be realized. The Q output will remain at the condition dictated by the data present at the last clock pulse until a new clock pulse is received. Data update can occur only at a clock-pulse time.

The JK flip-flop is a little more complicated and is shown in Fig. 2-18. It has two types of input, direct and clocked. The direct inputs force the output to a predetermined condition, while the clocked inputs follow a protocol that depends upon the conditions at the J and K inputs.

If the set input is made high and the clear is low, the output will be low. If the set is low and the clear is high, the output is high.

Clocked operation results if both set and clear are tied high. Under this condition, the flip-flop responds to changes at the J and K inputs but will only permit a change at the negative transition of the clock pulse, the falling edge. The truth tables are also given. The JK flip-flop is by its nature a divide-by-2 binary counter.

Table 2-3 lists some of the common 4000 series CMOS digital ICs.

Chapter 3

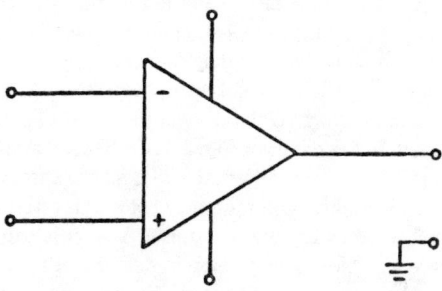

Special IC Devices

I N THIS CHAPTER, I WILL EXAMINE SOME SPECIAL-FUNCTION INTEGRAT-
ed circuits which have found particularly important applications in electronic instrumentation circuits. Of course, the galaxy of special offerings
from the manufacturers is much larger than this or any other text can offer; so please don't have a heart attack if I seem to have left out your particular favorite. This could happen by several processes. It could, for
example, be too similar to a device I did include. It might be less useful
(in my judgement) than you thought, or I may simply be ignorant of it or
its usefulness.

Although may of these devices contain internal operational amplifiers
or perform tasks often delegated to operational amplifiers, I am avoiding
them in this chapter—no matter how clever or useful. Special or particularly
nice operational amplifiers are covered elsewhere in this book.

TIMERS AND COUNTERS

Figure 3-1 shows the internal block diagram of the type-555 integrated-
circuit timer. This low cost IC is usually supplied in an 8-pin mini-DIP
package and has deservedly become one of the most popular of the miscellaneous IC devices.

One reason for the popularity of the 555 is its extreme flexibility. It
is neither TTL nor CMOS but is of a bipolar design. One principal difference
between this bipolar chip and TTLs (which also are bipolar) is that the
555 can be operated over a wide range of supply voltages. It is happy with
dc supply voltages anywhere between 5 and 15 volts, but somewhere between 9 and 12 volts seems optimum. When the output of the 555 is in

the low condition it can *sink* up to 200 milliamperes of current. Alternatively, when the output terminal is in the high state the 555 will *source* the same amount of current. The range of applications for the 555 is wide, limited only by cleverness and imagination. In the astable mode the 555 will act as a signal source, while in the monostable mode it operates as a one-shot. It will also function as a timer or delay stage.

The 555 timer is one IC that will prove even more useful once you are familiar with its inner workings and logic functions. To accomplish this end I will use modified versions of Fig. 3-1 in a block-diagram analysis explaining the operation of the different 555 configurations.

The monostable multivibrator, or one-shot as it is often called, remains off until it is triggered by an external pulse. After it is triggered the one-shot will produce a single, constant-amplitude pulse of fixed time duration. Once the allotted time has run its course the one-shot reverts to its dormant state until another trigger pulse is received. The one-shot is called a *monostable* circuit because it has but one stable state.

Figure 3-2 shows a monostable multivibrator using the 555 timer IC. Figure 3-2(B) is the block diagram, while Fig. 3-2(A) shows how this circuit appears in a schematic diagram. The heart of the 555 timer is an RS flip-flop that is controlled by a pair of voltage comparators. An RS flip-flop, you should recall from the last chapter, is a bistable circuit; that is, it has two stable states. In its initial state the Q output is low and the not-Q output is high. If a pulse is applied to the FF-SET S input the situation reverses itself, and the not-Q output becomes low, while the Q is high. A comparator (of which, more later) has two input terminals. Its output will go high when both input terminals see the same voltage.

Under initial conditions at time t0, the not-Q terminal of the RS flip-flop is high, and this biases transistor Q1 hard "on", placing IC pin 7 effectively at ground potential. This keeps capacitor C1 discharged. Also, amplifier U1 is an inverting type; so the output terminal (pin 3) is initially in the low state.

Resistors R_a, R_b, and R_c are inside the IC and are of equal value, nominally 5000 ohms. These form a voltage divider that is used to control the voltage comparators. The inverting (–) input of comparator 1 is biased to a potential of

$$E_1 = (Vcc) \frac{R_b + R_c}{R_a + R_b + R_c} = \frac{2}{3} (Vcc) \qquad (3.1)$$

This means that the output of comparator 1 will go high when the control voltage applied to IC pin 5 is equal to 2/3 Vcc. Similarly, the same voltage divider is used to bias comparator 2. The voltage applied to the noninverting input of the second comparator is given by

$$E_2 = (Vcc) \frac{R_c}{R_a + R_b + R_c} = \frac{1}{3} (Vcc) \qquad (3.2)$$

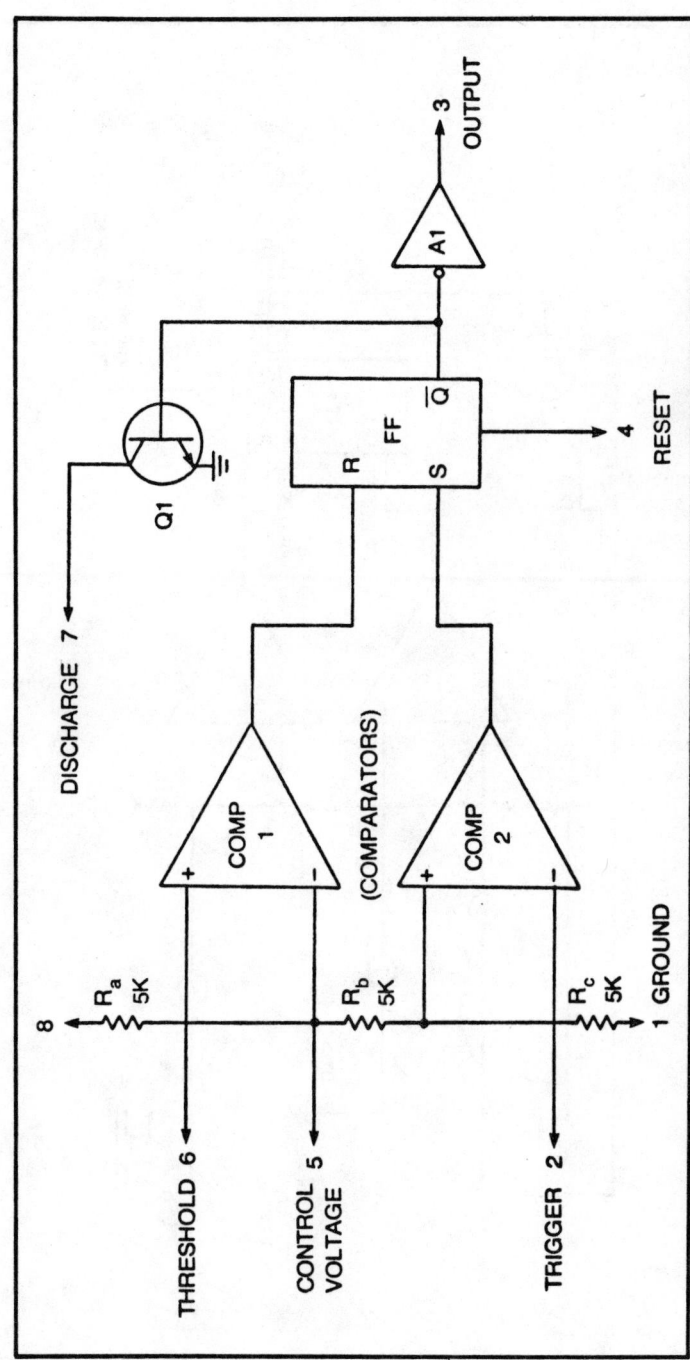

Fig. 3-1. Block diagram of the 555 internal circuitry.

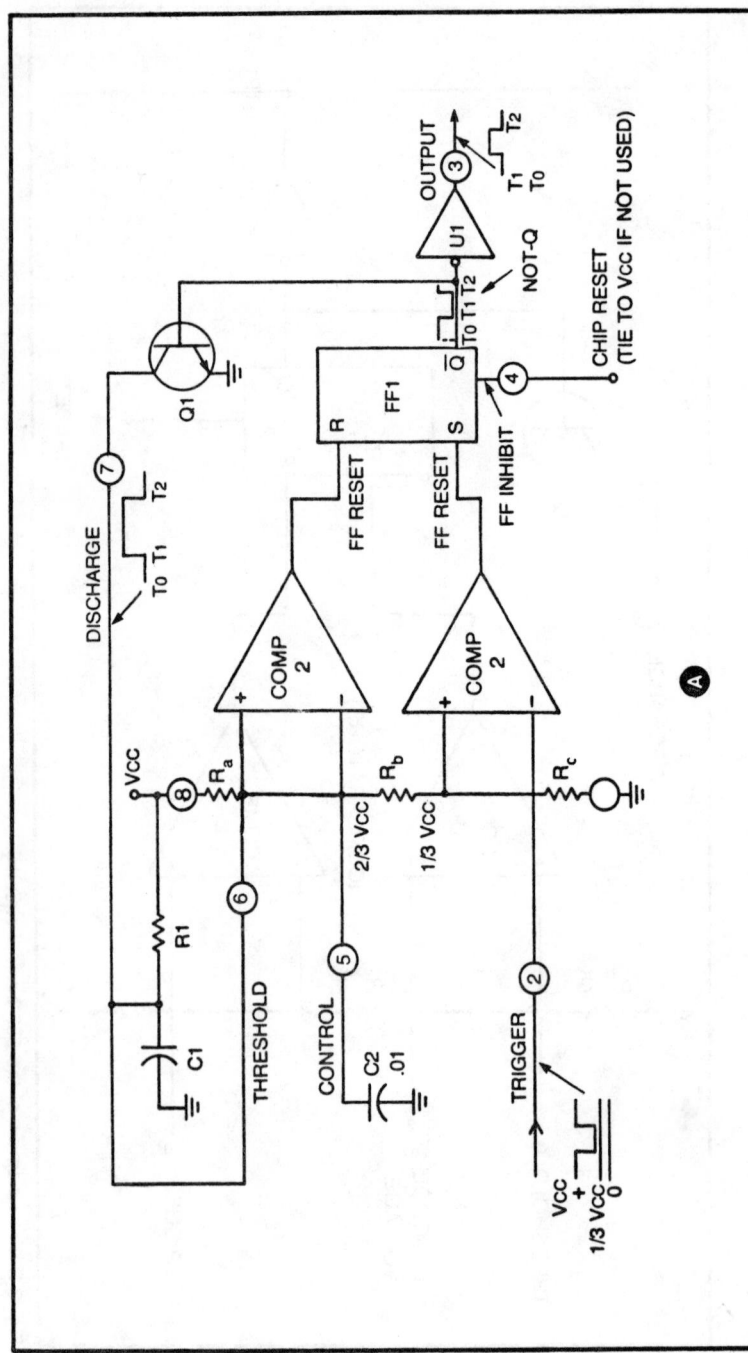

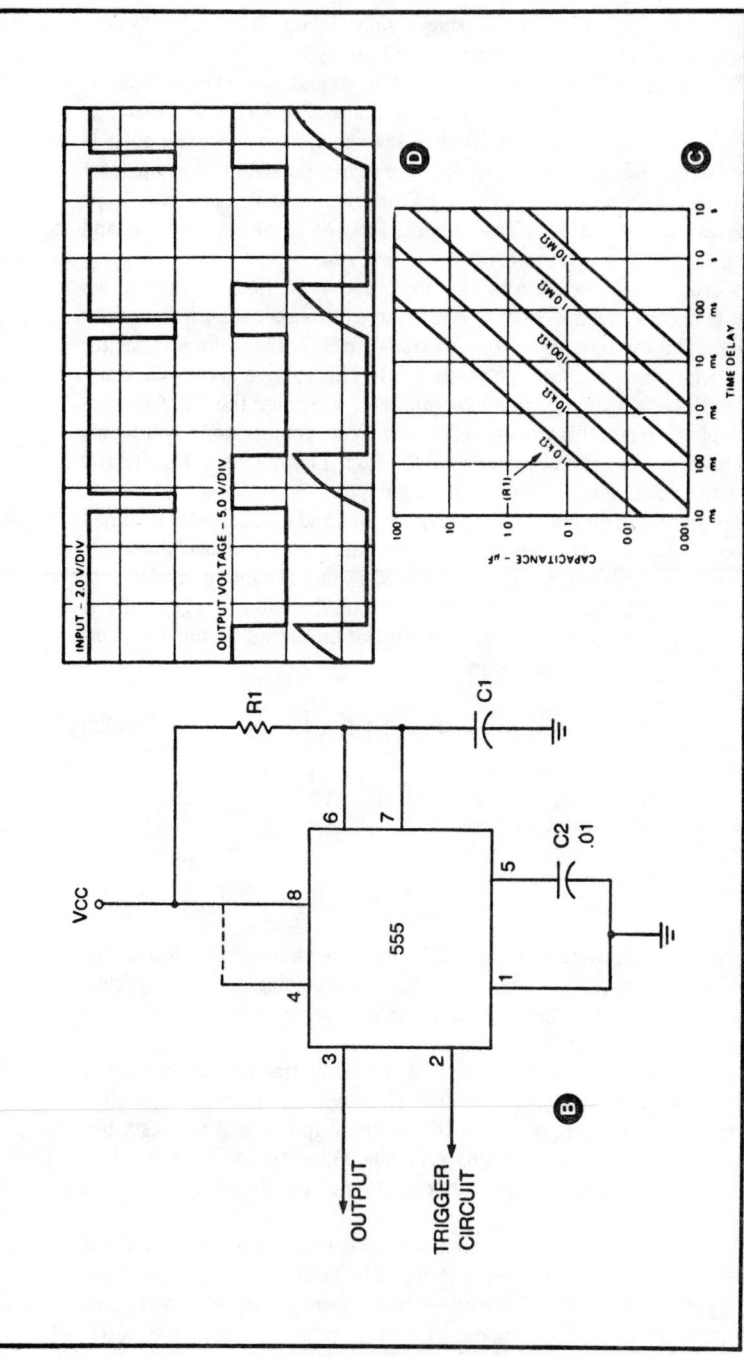

Fig. 3-2. At (A) the circuit relationship of internal and external circuitry for 555 monostable operation; (B) shows the schematic symbol; (C) is a graph showing time delay as a function of R and C; and (D) shows timing circuit waveforms.

45

When the voltage applied to the trigger input (IC pin 2) drops to 1/3 Vcc, the output of the second comparator will go high.

The control voltage is, in this case, the voltage across capacitor C2. This capacitor charges through resistor R_a and will reach 2/3 Vcc less than one millisecond after power is applied. If a short, negative-going pulse is applied to the trigger input at time t1 in our figure, the output of comparator 2 will snap to its high state as soon as the trigger pulse amplitude drops to a level equal to 1/3 Vcc. This puts the flip-flop in the set condition and causes the not-Q output to drop to the low state.

A drop to the low state by the not-Q output (at time t1) causes two things to occur simultaneously. One is to force the output of buffer amplifier U1 high, and the other is to turn off transistor Q1. This allows capacitor C1 to begin charging through resistor R1. The voltage across C1 is applied to the noninverting input of comparator 1 through the threshold terminal, pin 6. When this voltage reaches 2/3 Vcc, comparator 1 will toggle to its high state and will reset the RS flip-flop. This occurs at time t2 and forces the not-Q output again to its high state.

The output of amplifier U1 again goes low, and transistor Q1 is turned back on. Whenever Q1 is on, we find that capacitor C1 is discharged. At this point the cycle is complete, and the 555 timer is again in its dormant state. The output terminal will remain low until another trigger pulse is received. The approximate length of time that the output terminal remains in the high condition is given by

$$t = (t2 - t1) = 1.1 \ R1C1 \tag{3.3}$$

where
   t   is the time duration in seconds
   R1 is in ohms
   C1 is in farads.

This function is graphed in Fig. 3-2(C) for times between 0.01 and 10.0 seconds with values of R1 and C1 that are easily obtainable. The time relationship between the trigger pulse, output pulse, and the voltage across capacitor C1 is shown in Fig. 3-2(D).

If the reset terminal is not used, it should be tied to Vcc to prevent noise pulses from jamming the flip-flop. If, however, negative-going pulses are applied simultaneously to the trigger input (pin 2) and the reset terminal (pin 4), the output pulse will terminate. When this occurs the output terminal drops back to the low state even though the time duration (t) has not yet expired.

An astable multivibrator is similar in many respects to the monostable variety except that it is self-retriggering. The astable multivibrator, then, has *no* stable states, so will swing back and forth between high and low output conditions. This action produces a wave train of square wave pulses.

An example of a 555 astable multivibrator configuration is shown in Fig. 3-3. Again we have a version of Fig. 3-1 for block-diagram analysis in Fig. 3-3(B) and the schematic diagram in Fig. 3-3(A). As in the previous case, the inverting input of comparator 1 is biased to 2/3 Vcc and the noninverting input of comparator 2 is held to a level of 1/3 Vcc through the action of the resistor voltage divider R_a, R_b, and R_c. The remaining two comparator inputs are strapped together and are held at a voltage determined by the time constant C1 (R1 + R2). Under initial conditions the not-Q output of the RS flip-flop is high. This turns on transistor Q1, keeping the junction of resistors R1 and R2 at ground potential. Capacitor C1 has been charged, but when Q1 is turned on it will discharge through resistor R2. When the voltage across capacitor C1 drops to a level of 2/3 Vcc, the output of comparator 1 goes high, and that resets the flip-flop. This action again turns on Q1 and allows C1 to discharge to 1/3 Vcc. Capacitor C1, then, alternately charges to 2/3 Vcc and discharges to 1/3 Vcc. Figure 3-3(C) shows the relationship between high and low times. The high time, t1, is given by

$$t1 = 0.693 (R1 + R2) C1 \qquad (3.4)$$

and the low time by

$$t2 = 0.693 (R2) C1 \qquad (3.5)$$

The total period of the waveform, t, is the sum of t1 and t2, and is given by

$$t_{sec} = t1 + t2 \qquad (3.6)$$

$$= 0.693 (R1 + 2R2) C1 \qquad (3.7)$$

In any electrical circuit, or physical system for that matter, the frequency of an oscillation is the reciprocal of the period. In this case the frequency of oscillation is

$$f_{Hz} = 1/t_{sec} \qquad (3.8)$$

$$= \frac{1}{0.693 (R1 + 2R2) C1} \qquad (3.9)$$

$$= \frac{1.44}{(R1 + 2R2) C1} \qquad (3.10)$$

Equation 3.10 is shown graphically in Fig. 3-3(D) for frequencies between 0.1 and 100,000 hertz, using easily obtainable component values. The relationship between the C1 voltage and the output state is shown in Fig. 3-3(E). The duty cycle (also called duty factor) is the percentage of

47

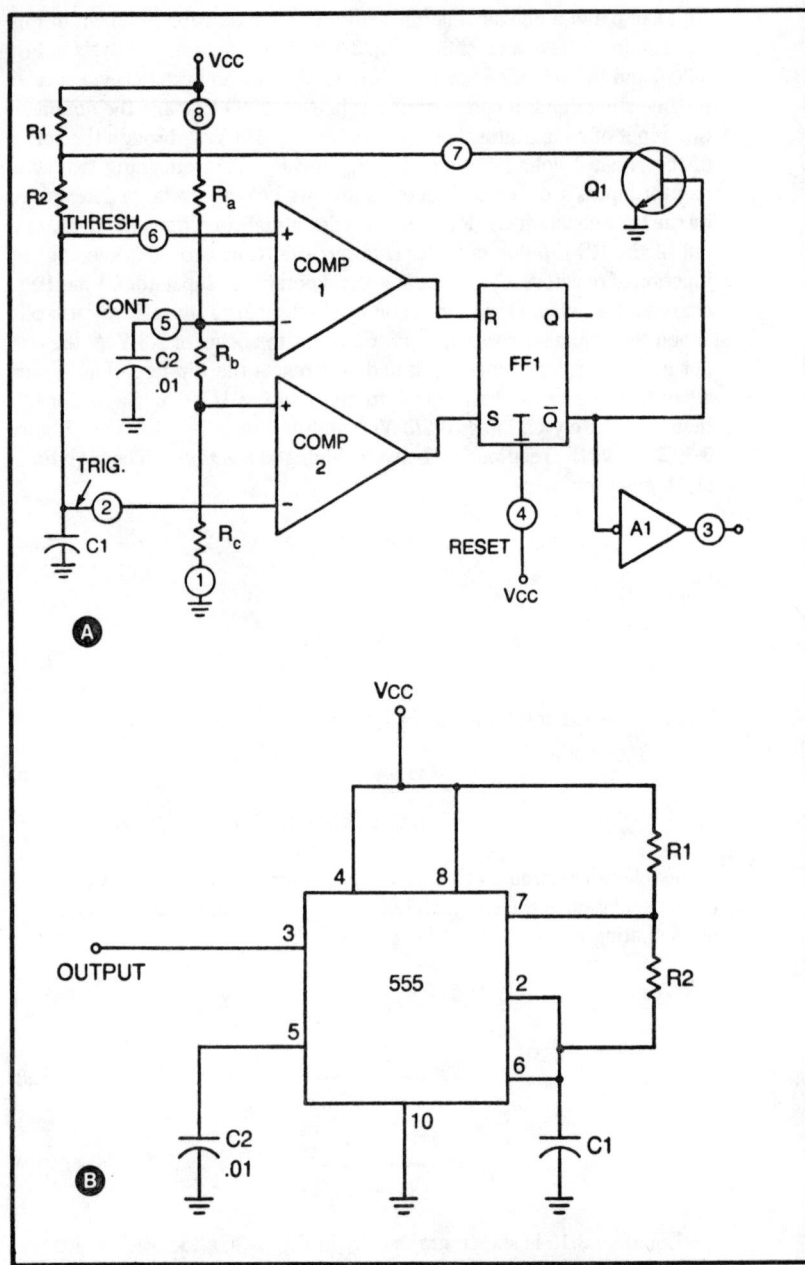

Fig. 3-3. At (A), the circuit for astable operation; (B) as it appears in circuit diagrams; (C) duty cycle of the output wave; (D) graph showing frequency of operation as a function of R and C; and (E) is the astable timing waveforms.

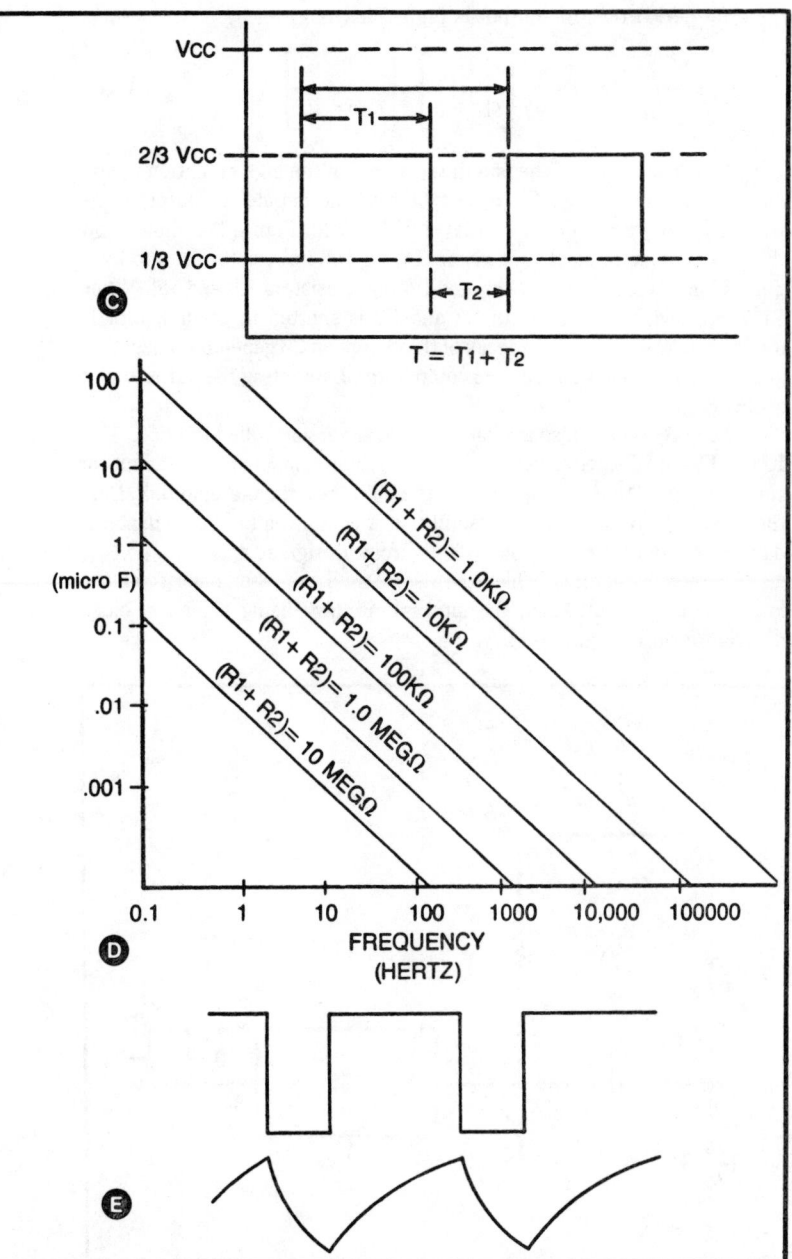

49

the total period that the output is high. That is given by

$$DF = \left[\frac{t1}{t1 + t2}\right] = \left[\frac{R2}{R1 + R2}\right] \tag{3.11}$$

The trigger input of the 555 timer is held in the high condition in normal, dormant operation. To trigger the chip and initiate the output pulse we must bring pin 2 down to a level of 1/3 Vcc. Figure 3-4 shows how this can be done manually. Resistor R4 is a pull-up resistor used to keep pin 2 high. Capacitor C3 is charged through resistors R3 and R4. When S1 is pressed, the junction of R3 and C3 is shorted to ground, rapidly discharging C3. The sudden decay of the charge on C3 generates a negative-going pulse at pin 2 that triggers comparator 2, initiating the output-pulse sequence.

The 555 timer is also available in a dual version called the 556. This IC is a 14-pin DIP package containing two timers, independent except for supply voltage, that are identical to the 555. It is, then, a dual 555. Both the 555 and 556 timers are so useful that it is an even bet you will someday make use of them if you do any circuit design at all.

Another useful timer, which can also be used as an 8-bit binary counter, is the Exar type XR-2240, the internal circuitry being shown in block diagram form in Fig. 3-5(A).

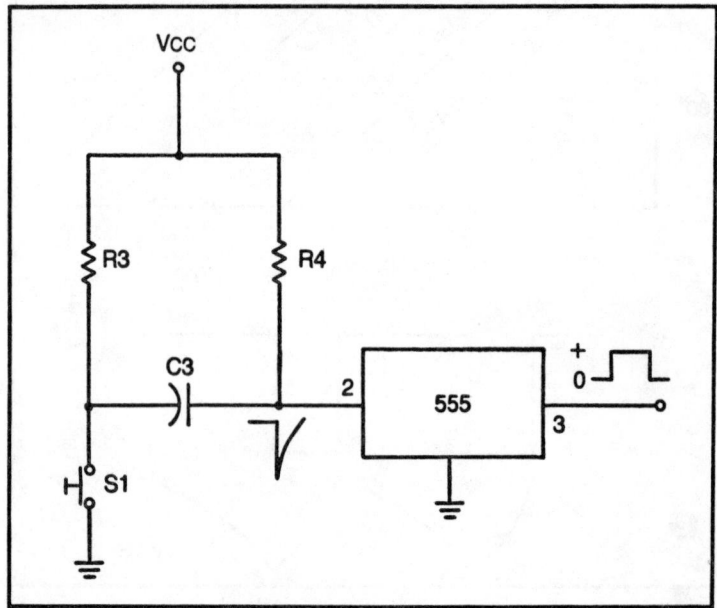

Fig. 3-4. Triggering requires a negative-going pulse that can be generated by a differentiator and pull-up resistor R4.

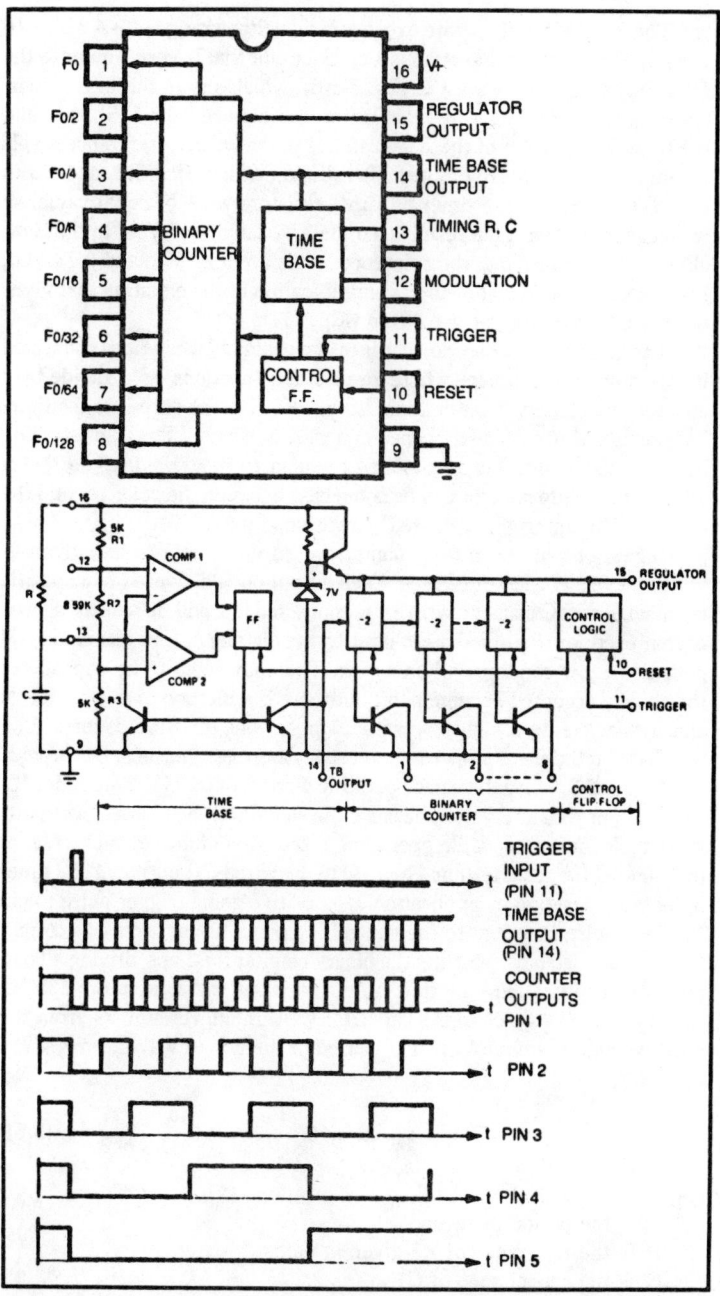

Fig. 3-5. At (A), a block diagram of the Exar XR-2240 (Courtesy Exar Corporation); at (B), a partial schematic of internal circuitry; and at (C), timing waveforms.

51

The XR-2240 will operate over a supply-voltage range of +4.5 to +18 volts dc. The time-base section is a clock circuit that is very similar to the 555. This fact may be seen in Fig. 3-5(B), which shows the internal circuitry of the XR-2240. One main difference between the 555 timer and the time-base portion of the XR-2240 lies in the relative reference levels created by the respective, internal, voltage dividers (R1, R2, and R3 in Fig. 3-5(B)). In the 555 timer all three resistors were of equal value; so we had comparator input terminals of 0.33 Vcc and 0.66 Vcc. In the Exar chip, on the other hand, the reference levels are 0.27 Vcc and 0.73 Vcc, respectively. One result of this is simplification of the equation that gives the period of the output waveform (Eq. 3.11).

The binary-counter section consists of a chain of JK flip-flops connected in the standard manner—where each stage functions as a divide-by-2 counter. The binary-counter chain is connected to the output of the time-base section through an NPN open-collector transistor. The transistor collector is also connected to IC pin 14 (called *time base output*) so that a 20-kilohm pull-up resistor can be connected between the collector and the output of the internal, regulated, power supply (pin 15).

Digital outputs from this counter are, in the usual fashion, given as voltage levels at a set of IC pins. Each output bit is delivered to a specific terminal of the IC package where it is connected to a pull-up resistor similar to that used for the time-base-output terminal. The output terminals will generate a low condition when active. This may seem to be opposite to the usually accepted arrangement; but there is a method to this madness, and it does create a highly versatile, stable, long-duration counter. You will find that these are properties not usually associated in single-IC designs.

Figure 3-6 shows the basic operating circuit for the XR-2240 timer IC. This chip proves interesting because the sole difference between circuits for astable and monostable operation is the 51-kilohm feedback resistor (R3) linking the reset terminal (pin 10) to the wired-OR outputs. The timer is set into operation by application of a positive-going trigger pulse to pin 11. This pulse is routed to the control logic and has several jobs to perform simultaneously: resetting the binary-counter flip-flops, driving all outputs low, and enabling the time-base circuit. As was true in the 555 IC, this timer works by charging capacitor C1 through resistor R1 from the positive voltage source Vcc. The period of the output waveform is given by

$$t = R \times C \tag{3.11}$$

where
    t is the period in seconds
    R is the resistance of R1 given in ohms
    C is the capacitance of C1 in farads.

The pulses generated in the time-base section are counted by the binary-counter section, and the output stages change states to reflect the current count. This process will continue until a positive-going pulse is applied to the reset terminal.

Figure 3-5(C) shows the relationship between the trigger pulse, time-base pulses, and the various output states. The reason for the open-collector, output circuit is to allow the user to wire a permanent-OR output so that the output duration can be programmed. Each binary output is wired in the usual power-of-2 sequence: 1, 2, 4, 8, 16, 32, 64, 128 . . .. If these are wired together the output will remain low as long as any *one* output is low. This allows you to program the output duration from 1 t to 255 t (where t is as defined in Eq. 3.11) by connecting together those outputs which sum to the desired time period. For example, let us design a timer with a 57-second time delay. In binary notation decimal 57 is equal to

$$32 + 16 + 8 + 1 = 57 \qquad (3.12)$$

If you set $t = R1C1 = 1$ second and wire together the pins on the XR-2240 corresponding to these weights, a 57-second time delay will be realized. The base diagram to this IC shows you that these pins are numbers 1, 4, 5, and 6. If you short together those four pins, the counter output will remain low for 57 seconds following each trigger pulse.

You could change the time base or the wired-OR terminals to change output periods. Of course, if the time-base frequency were doubled, the

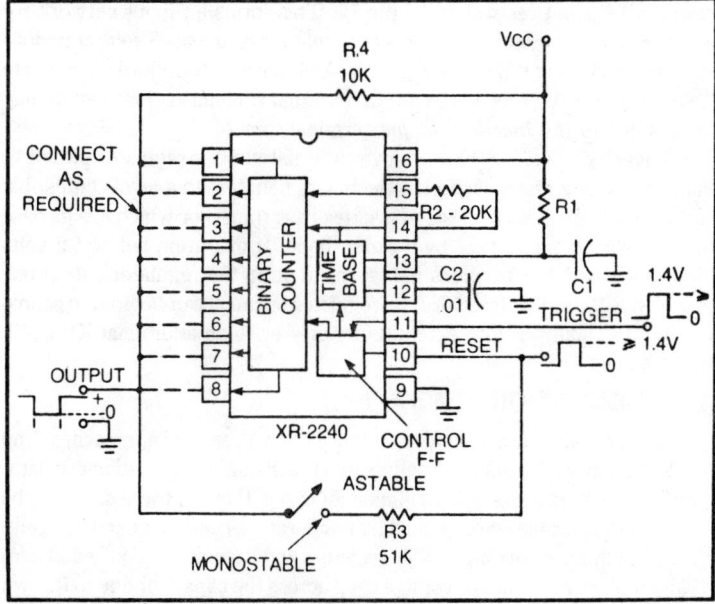

Fig. 3-6. Basic circuit for operation of the XR-2240.

counter would reach the desired state in half the time. This feature allows programming of the XR-2240 to time durations that might prove difficult to achieve using conventional circuitry.

Each output must be wired to Vcc through a pull-up resistor of 10 kilohms, unless, of course, the wired-OR output configuration is used. In that case a single 10-kilohm resistor is used.

Current through the output terminals must be kept at a level of 5 milliamperes or less—this then serves as a general guide to the selection of pull-up resistors.

The amplitude of reset and trigger pulses must be at least two PN-junction-voltage drops ($2 \times 0.7 = 1.4$ volts). In most practical applications it might be wise to use pulses greater than 4 volts in amplitude, or standard TTL levels, in order to guard against the possibility that any particular chip may be a little difficult to trigger near minimum values or that outside factors conspire to actually reduce pulse amplitude at a critical moment.

Synchronization to an external time base or modulation of the pulse width is possible by manipulating pin 12. In normal operation this pin, which is the noninverting input of comparator 1, is bypassed to ground through a 0.01-μF capacitor so that noise signals will not interfere with operation. A voltage applied to pin 12 will vary the pulse width of the signal generated by the time base. This voltage should be approximately $+2$ to $+5$ volts for a time-base-change multiplier of approximately 0.4 to 2.25, respectively.

If it is desired to synchronize the internal time base to an external reference, connect a series RC network consisting of a 0.1-μF capacitor and a 5.1-kilohm resistor to IC pin 12. This forms an input network for sync pulses, and these should have an amplitude of at least 3 volts at periods between 0.3 t and 0.8 t (see Fig. 3-7). Another way to link the count rate to an external reference is to use an external time base. This signal may be applied to the *time base output* terminal, pin 14.

Each Exar XR-2240 has its own internal voltage-regulator circuit to hold the dc potentials applied to the binary counters to a level compatible with TTL logic. This consists of a series-pass transistor which has its base held to a constant voltage by a zener diode. If operation below 4.5 volts dc is anticipated, it becomes necessary to strap the regulator-output terminal (pin 15) to Vcc (pin 16). The regulator terminal can be used to source up to 10 milliamperes to external circuitry or to an additional XR-2240.

LONG-DURATION TIMERS

Long-duration timers can be built using any of several approaches. You could, for example, connect a unijunction transistor in a relaxation-oscillator configuration or use a 555. In almost all cases, though, there seems to be an almost inevitable error created by temperature-coefficient and inherent-tolerance limits of the high-value resistors and capacitors required. There is also a certain amount of voltage drop across the capacitor due to its own internal-leakage resistances and the impedance of the circuit in which it

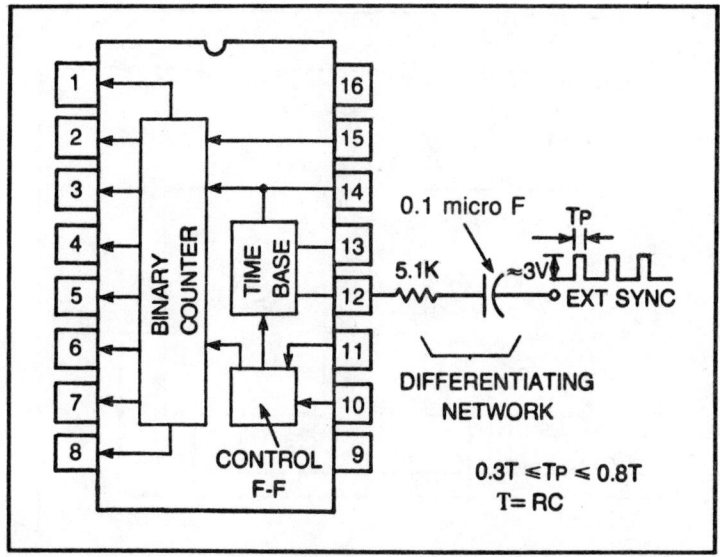

Fig. 3-7. Synchronizing the internal clock to an external timing source.

is connected. The use of the XR-2240 IC timer will all but eliminate such problems because you are allowed to use a higher clock frequency to tame lower-value components to satisfy Eq. 3.11. It is usually easier to specify and obtain quality, precision components in the lower values. This both makes the period initially more accurate and results in less drift-with-temperature.

An example of a long-duration timer is shown in Fig. 3-8; it consists of two XR-2240 IC timers cascaded to increase the time duration. In this circuit you see the time-base output of IC2 (pin 14) used as an input for an external time base. Timer IC1, also an XR-2240, is used as a time base for the second. The next most significant bit of IC1 (pin 8, weighted 128 t) is connected to pin 14 of IC2. This pin will remain low from the time IC1 is triggered until t0 = 128 R1C1 and will then go high and trigger IC2.

The binary counter in IC2 will remain increment once every 128 t. Timer IC1 is essentially operating in the astable mode since its reset pin is tied to the rest of IC2, and that point does not go high until the programmed count for IC2 forces its output to go high.

The total time duration for this circuit under the conditions shown (IC2 input from pin 8 on IC1, and with all IC2 outputs connected into the wired-OR configuration) is 256^2, or 65,536 t. You can, however, custom-program the timer to your own application by manipulating three factors: time-base period (R1C1), the output pin on IC1 used to trigger IC2, and the strapping configuration on IC2. In other words

$$t0 = R1\ C1 \times t'0 \times t''0 \qquad (3.13)$$

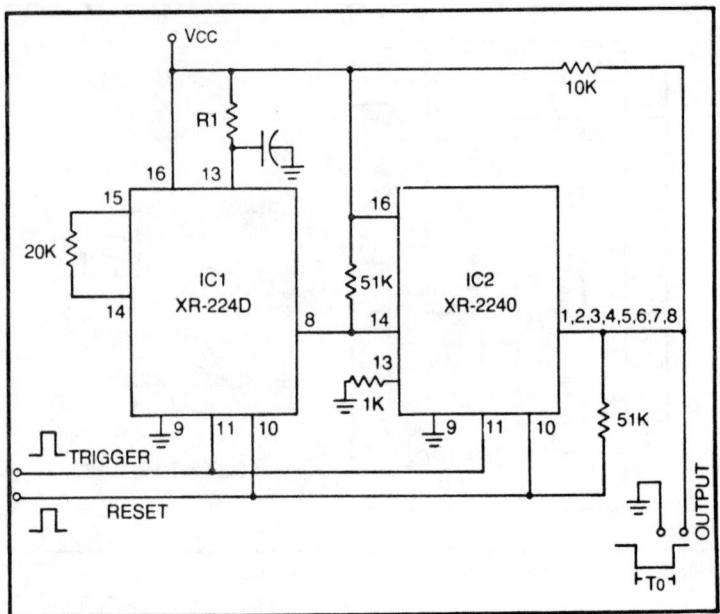

Fig. 3-8. Use of two XR-2240 timers in cascade will result in a total time of over 65,000t.

where:

    $t0$ is the total period in seconds that the output is low

    $t'0$ is the total time that the selected output of IC1 is low

    $t''0$ is the total time-weighting of the IC2 output.

As an example, let us assume that the product R1C1 is one second (i.e., R1 = 1 megohm, C1 = 1 μF). If the output remains low for a period of t0 = 65,536 (R1C1) = 65,536 t, it will remain low for 65,536 seconds, or about 18 hours! Of course, it is almost impossible to accurately generate a time delay this long using any of the other techniques. It is, for example, relatively common to find high-value electrolytic capacitors (necessary in long-duration, RC timers) rated with a -20 to $+100$ percent tolerance in capacitance. This is incompatible with the goal of making a precision time base of long duration. A further complication is that high-value electrolytic capacitors, even the tantalum types, will tend to change value while in service. For most bypassing or filtering applications this is tolerable, but in timing or other cases where the value is important it is a nightmare. The use of cascaded, XR-2240 timers allows us to pick values for R1 and C1 that are in the lower, more easily managed ranges.

An alternative method for accomplishing the same trick with an XR-2240 is the use of an external time-base oscillator. This is used to drive the TBO input of IC1. The time duration of the output-low condition is then 65,536 t' is the period of the external reference pulse. You will see additional examples of the XR-2240 in other chapters.

Any of the timers discussed thus far can be operated in a delayed, monostable mode simply by using the "outside world" pulse to trigger one timer and the output of that timer to trigger a second timer. The output of the second timer is then fed to whatever circuitry is being controlled. It is delayed from the initial pulse by a factor equal to the time duration of the first.

It is also possible to use unijunction-transistor, relaxation, relaxation oscillators or regular TTL or CMOS, digital counters in timing-circuit applications. However, these are not as versatile as single IC timers or are more difficult to implement.

ANALOG IC MULTIPLIERS

One of the most useful IC devices introduced by modern semiconductor technology, outside of the operational amplifier, is the analog multiplier. It is, perhaps, unfortunate that both IC manufacturers and engineers usually designate this device by only one of its functions (e.g., "multiplier"), for it is actually a uniquely versatile device.

Before taking a look at the IC analog multiplier let us examine the various methods for performing this function. Some of these techniques are used for making IC multipliers, while others were previously implemented using discrete operational amplifiers and transistors.

An analog multiplier is any electrical circuit or IC device which is governed by a transfer function of the form

$$E_{out} = k\,V_x\,V_y \tag{3.13A}$$

$$I_{out} = k\,V_x\,V_y \tag{3.13B}$$

or,

$$Z = k\,V_x\,V_y \tag{3.13C}$$

where
V_x is the voltage applied to the X input
V_y is the voltage applied to the Y input
k   is a constant ranging from 1/25 to 10, depending upon the particular IC or circuit configuration chosen. Typical values are 1/25, 1/10, and 10.

It should be pointed out that the multiplier transfer function is not linear and that it actually may describe a saddle surface instead of a simply X-Y cartesian plane. It is best though, to use the notion of cartesian quadrants when discussing multiplier properties.

One method for classifying analog multipliers is with respect to the number of cartesian quadrants in which any particular multiplier will operate. Many multiplier circuits, even some that are an IC form, are limited to one or two quadrants. The transfer function will not be valid in any other quadrant. Some are single-quadrant multipliers, and these require that both V_x and V_y be of the same polarity, usually positive. This will place the operation of the multiplier in what is usually called the first quadrant of the cartesian plane, 0° to 90°.

A two-quadrant analog multiplier is one that accepts voltages of either polarity at one input but limits the permissible voltages at the other input to positive values. An example of a two-quadrant multiplier is one that accepts signals over the range $-10\text{ V} \leqslant V_x \leqslant +10\text{ V}$ at the X input and $0 \leqslant V_y \leqslant +10\text{ V}$ at the Y input.

In this respect one can easily see that a four-quadrant multiplier is the most versatile, but this is not for free because the four-quadrant multiplier is also more costly. It can accept signals over the range $-10\text{ V} \leqslant V_{xy} \leqslant +10\text{ V}$ at *both* inputs. As you can see from this discussion, you can completely specify analog multipliers from the properties of their input terminals.

Another method for defining analog multipliers is by the process used to obtain multiplication. There are several techniques for actually performing this job. Those that have become popular over the years include quarter-square, log-antilog, variable-transconductance, and the Gilbert transconductance cell.

The quarter-square technique takes advantage of the fact that

$$XY = \frac{(X + Y)^2 - (X - Y)^2}{4} \qquad (3.14)$$

This class of multiplier is implemented using operational amplifiers for the summation and diode-piece wise-squaring modules to perform the squaring of the parenthetical expressions in EQ. 3.14. Inputs V_x and V_y are summed in appropriate operational-amplifier circuitry to form $V_1 = (V_x + V_y)$ and $V_2 = (V_x - V_y)$. Examples of summing circuits can be found in Chapter 5 (Fig. 5-4) and in *Op-Amp Circuit Design & Applications* (TAB book No. 787). Diode-squaring modules (covered in the referenced book) will produce the outputs $(V_1)^2$ and $(V_2)^2$ which are then summed in another operational-amplifier circuit to produce the difference-of-squares term. The constant, 4, in the denominator of Eq. 3.14 can be either designed into the circuit as an actual value or made implicit. In most cases, though, some constant will be required to rationalize the output voltage; so we must designate the output as

$$E_{out} = k\, V_x V_y \qquad (3.15)$$

Operational amplifiers, through the mechanism of feedback can be used to create output potentials that are proportional to either the logarithm or antilogarithm of the input voltage. Various ways of doing these neat tricks are known, but only the most common will be treated, and those in another chapter. In the meantime accept the fact that amplifiers exist that have transfer functions of the form

$$E_0 = -\ln(E_{in}) \tag{3.16}$$

and

$$E_0 = -\ln^{-1}(E_{in}) \tag{3.17}$$

These can be combined in a circuit to compute the function

$$E_{out} = k (\ln V_x + \ln V_y) \tag{3.18A}$$

$$= k \ln (V_x V_y) \tag{3.18B}$$

which is the multiplication transfer function and

$$E_{out} = k (\ln V_x - \ln V_y) \tag{3.19A}$$

$$= k \ln (V_x/V_y) \tag{3.19B}$$

which is the analog-division transfer function.

To actually implement these functions requires two log amplifiers, an operational-amplifier summing circuit, and an antilog amplifier. The two input voltages, V_x and V_y, are first fed through their respective log amplifiers where they are converted to logarithmic form. The outputs of these amplifiers are either summed or differenced in an operational-amplifier summer of appropriate design. The output of the summer is then passed through an antilog amplifier to return the voltage levels to real-world values.

The log-antilog multiplier looks appealingly simple on first glance, but problems appear almost immediately. For one thing, it is difficult, albeit not impossible, to obtain a nicely linear log or antilog amplifier that covers more than a few decades of input voltage range. Since most of these circuits require a few hundred microamperes of feedback current (in fact, they are not logarithmic at currents greater than that!) it becomes rather difficult to have large, V_x and V_y excursions. On the upper end of the range, the circuit is limited by the maximum feedback current that is consistent with the creation of the log function, and on the lower end by noise, offset currents and voltages, and the other errors and problems that seem associated with electronic circuitry in general or operational amplifiers in particular. This brings to mind Farquhar's Principle: *The perversity of the universe tends to a local maximum whenever and wherever it can do the most damage*—named after Barg Farquhar, legendary circuit designer who usually takes transistors and makes them into silicon-to-carbon converters, regardless of what he does.

Another problem is that most log amplifiers are usually unipolar, and that limits analog multipliers using these circuits to one-quadrant operation. Probably the most serious limitation, especially if applications can tolerate the small-dynamic-range problem, is thermal errors. Most log amplifiers operate using the logarithmic properties of certain semiconductor junctions, and this is highly dependent upon temperature. In fact, in a later section I will discuss at least one thermometer circuit that operates using this very principle.

The thermal problem can be alleviated somewhat by using a dual diode or transistor to form the log-feedback elements of both log and antilog amplifiers, thus making them track together even though with error—they have the *same* error.

Also, good temperature-coefficient design, or the use of a special IC log amplifier can also help. These "fixes," though, add complexity and cost to the circuit, and that destroys the appealing simplicity, leading us to consider other designs. The log-antilog amplifier is used where errors on the order of two to eight percent can be tolerated and where the temperature of the environment will not vary over a wide range. The sole possible exception is the function-module type of log-antilog multiplier, but these are expensive.

Variable-transconductance multipliers use a differential pair of transistors such as shown in Fig. 3-9. If the emitters of the pair are driven from a common non-constant-current source, this amplifier will have stage gain and is variable. If the current driving the emitters is generated proportional to V_y, a differential output voltage is created such that

$$E_{out} = k2 \ V_x \ V_y \qquad (3.20)$$

The problem with this lies in the requirement that both transistors be perfectly matched and that they track with each other over wide variations in temperature. When individual transistors were used this was an almost impossible goal to achieve since the physical separation of the components gave them a different thermal environment. In an IC package, however, the transistors are built on a common substrate; so the thermal tracking is almost perfect. The initial match of the transistors will also be very good, and differences can easily be trimmed out using external bias components. The variable-transconductance multiplier forms the basis for many IC analog multipliers and at least one multiplier application of a non-multiplier IC—the transconductance operational amplifier. Multipliers manufactured by Analog Devices, Inc. (Norwood, Massachusetts) use a similar system based on the Gilbert transconductance cell introduced by them in their AD-500-series multipliers in 1970.

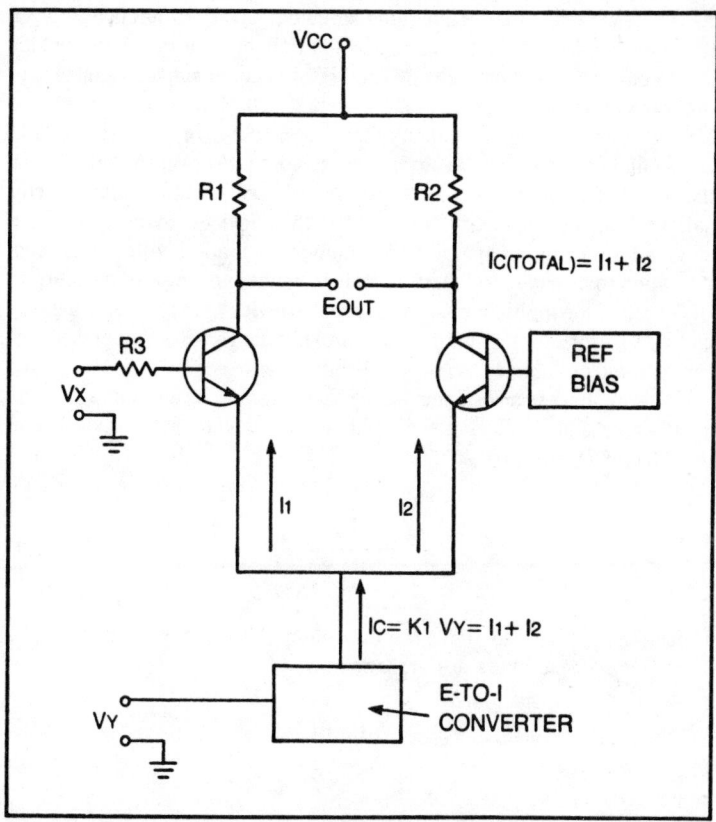

Fig. 3-9. Basic circuit for a transconductance multiplier.

Some Actual IC Multiplier Devices

Although there are a respectable number of analog multipliers to be found in manufacturers' catalogues, it is necessary and advisable to limit this discussion to those that give reasonable performance at relatively low cost. If an application requires a better device you will find that the information presented here is still generally valid and that various manufacturers offer versions of these chips with tighter specifications. In fact, the specifications of many multiplier families are similar enough among the various grades to make one believe that the real difference is merely a selection process whereby the better ICs are labeled with the premium-type number and the lesser given a low-cost-type number. Consult the manufacturers' catalogues and data sheets for appropriate type numbers.

The first analog-multiplier IC that I will consider is the type MC1495 /MC1595 (Fig. 3-10) by Motorola Semiconductor Products, Inc. and others.

As is true in many types of integrated circuits, the only difference between two devices of similar type number is the temperature range over which the specifications are guaranteed. The MC1495, for example, operates over the commercial temperature range specified as $0°$ C to $+70°$ C, while the MC1595 operates over the military temperature range of $-55°$ C to $+125°$ C. Needless to say that the latter is more costly. As you saw in Fig. 3-9 the output of a transconductance multiplier is usually a differential-collector pair, and this also holds true for the 1495/1595 devices. In my specific examples I am using an operational amplifier (A1) in a dc-differential configuration to provide level-shifting and conversion to single-ended output. Of course, if the multiplier is used to drive a device or instrument that normally has a differential input, it will provide the required suppression of common-mode voltages; so no level-shifting is needed.

This multiplier can be trimmed to provide various scale factors by varying the values of R_a, R_b, R_x, and R_y in addition to currents I1 and I2. The scale factor is given by

$$k = \frac{2 R_L}{I1 R_x R_y} \qquad (3.21)$$

These parameters also affect the maximum value of the input voltages, V_x and V_y. These limits are prescribed by

$$V_{x(max)} < (I2)(R_x) \qquad (3.22)$$

and,

$$V_{y(max)} < (I1)(R_y) \qquad (3.23)$$

where
$$0.5 \text{ mA} \leq I1 \approx I2 \leq 2.0 \text{ mA}$$
Sample Calculation:

I want to set $V_{x(max)}$ to $+10$ volts and I2 to 0.5 mA. What is the appropriate value of R_x? Of course, this is an Ohm's law problem; so I express current I2 as 0.0005 amperes.

$$R_x = \frac{V_{x(max)}}{I2} \qquad (3.24)$$

$$= 10.0 \text{ volts}/0.0005 \text{ amps}$$
$$= 20,000 \text{ ohms} = 20 \text{ kilohms}$$

Typically, though, it has become somewhat standard practice to set I1 = I2 = 1 milliampere, and $R_x = R_y$ at either 10 kilohms, or 15 kilohms if

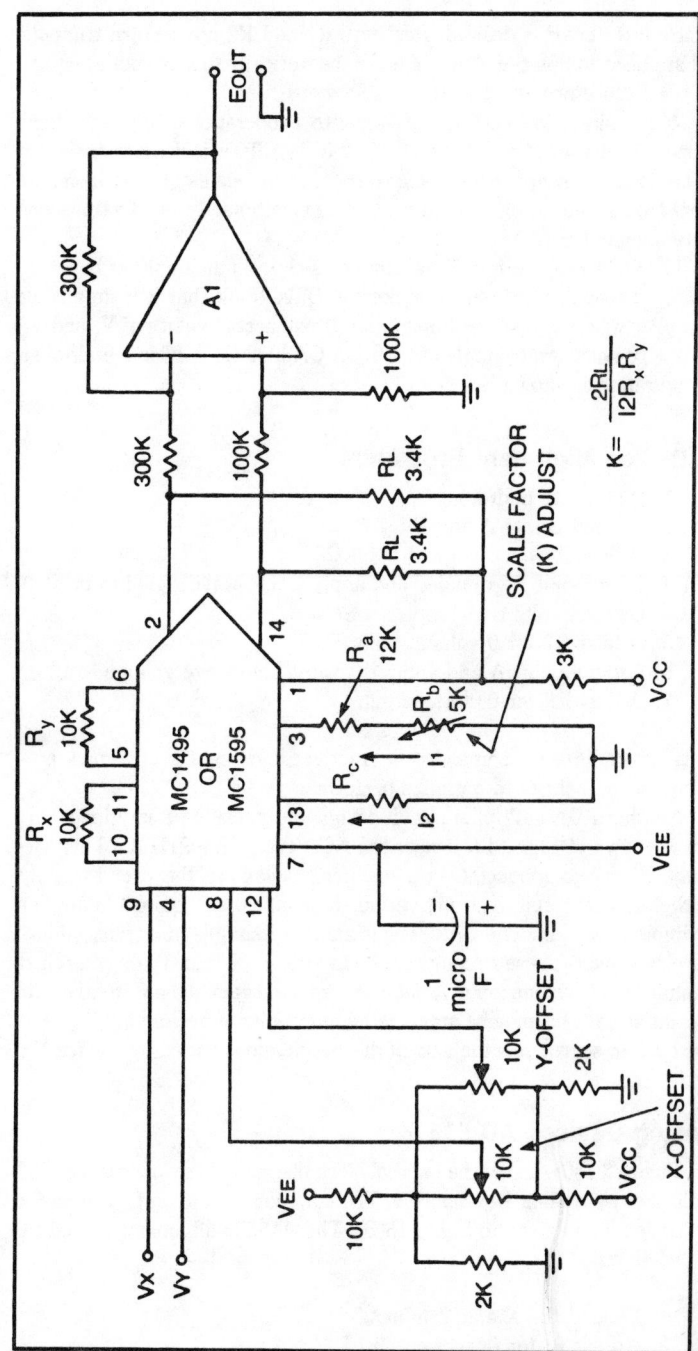

Fig. 3-10. Circuit for using the MC1495 multiplier.

$$K = \frac{2R_L}{12R_x R_y}$$

63

a margin for error is desired. Resistors R1 and R2 are 10-turn trimpots and are used to null the X and Y error theoretically to zero, but "negligible" is a more prudent and reasonable word.

RCA makes a low-cost device similar to an operational amplifier called an operational transconductance amplifier (OTA). This device is very similar to the op amp except that the transfer function relates change in *output current* to changes in *input voltage*; so we can legitimately call it a transconductance amplifier.

The OTA has a terminal that can be used to set an amplifier bias current (I_{abc}) that can be used to make the OTA think that it is an analog multiplier with four-quadrant operation. It will accept values of V_x and V_y up to ± 10 volts. An example of the RCA CA3080 OTA IC in a multiplier configuration is shown in Fig. 3-11.

Multiplier Alignment Procedure

1. Set R9 to midrange.
2. Ground points A and B.
3. Adjust R6 for 0 volts at point C.
4. Keep point B grounded and apply a sine wave signal to point A (several volts p-p).
5. Adjust R7 for 0 volts at point C.
6. Ground point A and apply the same sine wave voltage to B.
7. Adjust R9 for 0 volts output.

These procedures are somewhat interactive; so repeat the alignment procedure several times to optimize performance.

Another RCA multiplier circuit designed especially for multiplier service is shown in Fig. 3-12. It uses the RCA type CA3091D IC. This particular device is connected in a manner that allows the user to easily reprogram the circuit to permit various functions that require a multiplier for implementation. The CA3091D is another example of a transconductance multiplier. These circuits operate as little more than a gain-controlled amplifier in which one input variable controls the gain of a differential pair of transistors. This may be appreciated more fully by restudying Fig. 3-9. The current source sets the gain of the circuit and is varied by voltage V_y.

Analog Devices AD-533

Figure 3-13(A) shows the internal block diagram of the low-cost AD-533 multiplier by Analog Devices, Inc. The common circuit configuration for multiplication is given in Fig. 3-13(B). The AD-533 alignment procedure is as follows

1. Ground the X and Y inputs.
2. Adjust R_z for 0-volts output.

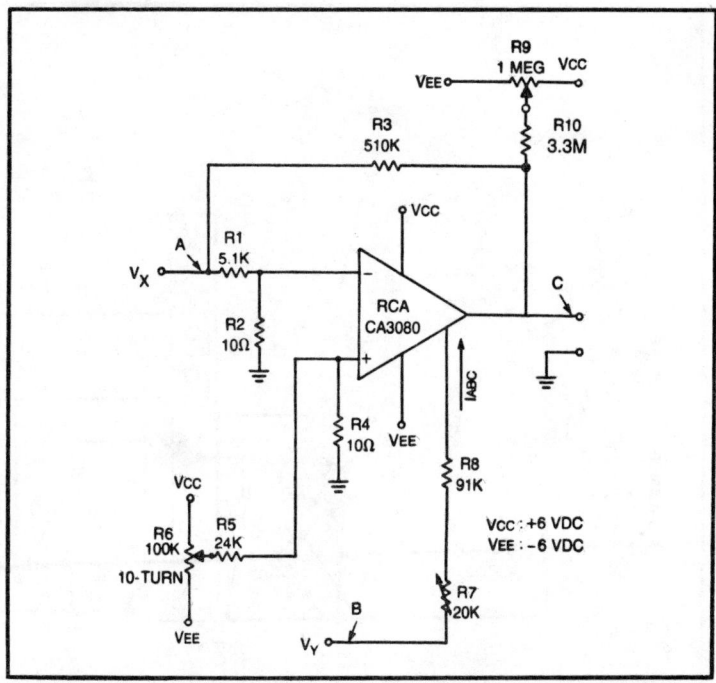

Fig. 3-11. Analog multiplier made from an operational transconductance amplifier (OTA) by RCA.

3. Keep the X input grounded and apply a 20-volts p-p, 50-hertz signal to the Y input.
4. Adjust R_x for minimum output.
5. Ground the Y input and apply a 20-volt p-p, 50-hertz signal to the X input.
6. Adjust R_y for minmium output.
7. Again adjust R_z for 0-volts output.
8. Set $V_x = +10$ volts dc and apply a 20-volts p-p, 50-hertz signal to the Y input. Adjust the gain potentiometer for $E_{out} = V_y$.

Additional applications for multipliers which justify our concern over the misfortunately limiting label given this IC are to be given in a later chapter.

COMPARATORS

One electronics author once referred to a comparator as an amplifier with too much gain, and that is probably the best definition that I have heard to date. Most comparators are operational amplifiers with the feedback resistors removed. This means the circuit gain will essentially be the

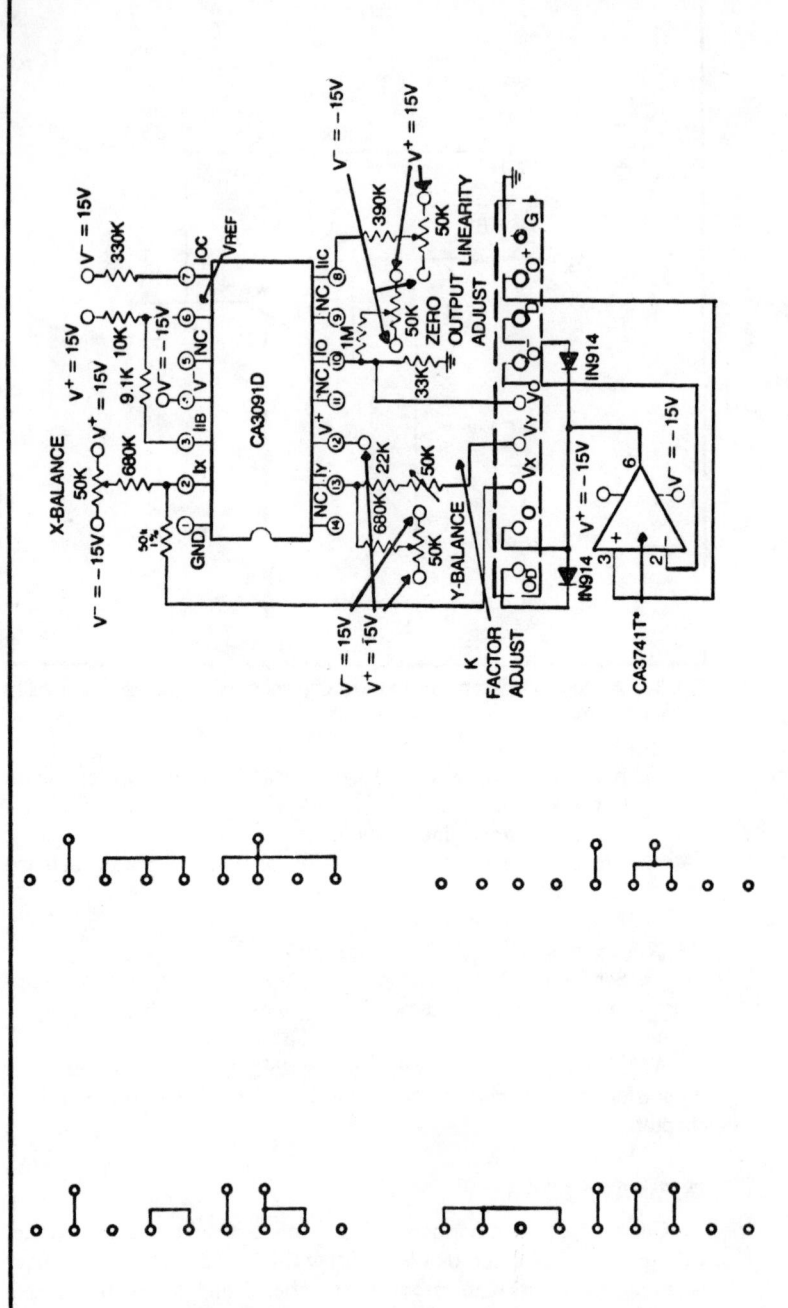

AC Alignment Procedures For CA3091D, Four-Quadrant Multiplier

Step No.	Voltage Setting		Control Adjust	Test Equipment Used	Measure	Notes
	Vx	Vy				
1	-	-	-	-	-	Set all potentiometers to center of range.
2	0	VM	X Balance	ac VM	VO	Adjust for a minimum reading.
3	0	VM	Linearity	ac VM	VO	Adjust for a minimum reading.
4	-	-	-	-	-	Repeat Steps 1 and 2 until no further improvement is noted.
5	VM	0	Y Balance	ac VM	VO	Adjust for a minimum reading.
6	0	0	Zero	dc VM	VO	Adjust for zero output.
7	VMID	VMID	Output Rk	ac/dc VM	VO	Adjust for $\sqrt{2}\,V_{MID}/10$ at the output
8	-	-	-	-	-	Check multiplier for alignment in all four quadrants

VM—Is the maximum ac swing of the sine wave that will be applied to the multiplier. A 20-volt p-p value is the nominal maximum swing of the ac sine wave with input resistors of 50 kilohms.

VMID—An ac or dc voltage that approximately satisfies the equation $V_{MID} = V_{IM} / \sqrt{2}$. For example, if a 50-kilohm resistor is used with a 7-volt input, then Rk should be adjusted for a 9-volt output

Divider Alignment Procedure

Step No.	Set		Measure	Output Coupling	Test Equipment Used	Adjust	Notes
	Vz V	Vy V					
1	-	-	-	-	-	-	Set all potentiometers to center of range
2	0	Vs	VO	ac	ac—VM	OZERO	Adjust for minimum reading
3	0	10V dc	VO	dc	dc—VM	XBALANCE	Adjust for 0V dc output
4	Vs	Vs	VO	ac	ac—VM	YBALANCE	Adjust for minimum reading
5	5V dc	5V dc	VO	dc	dc—VM	KADJUST	Adjust for 10V dc output.

Fig. 3-12. Four quadrant analog multiplier/divider using the RCA CA3080. (Courtesy RCA Corporation.)

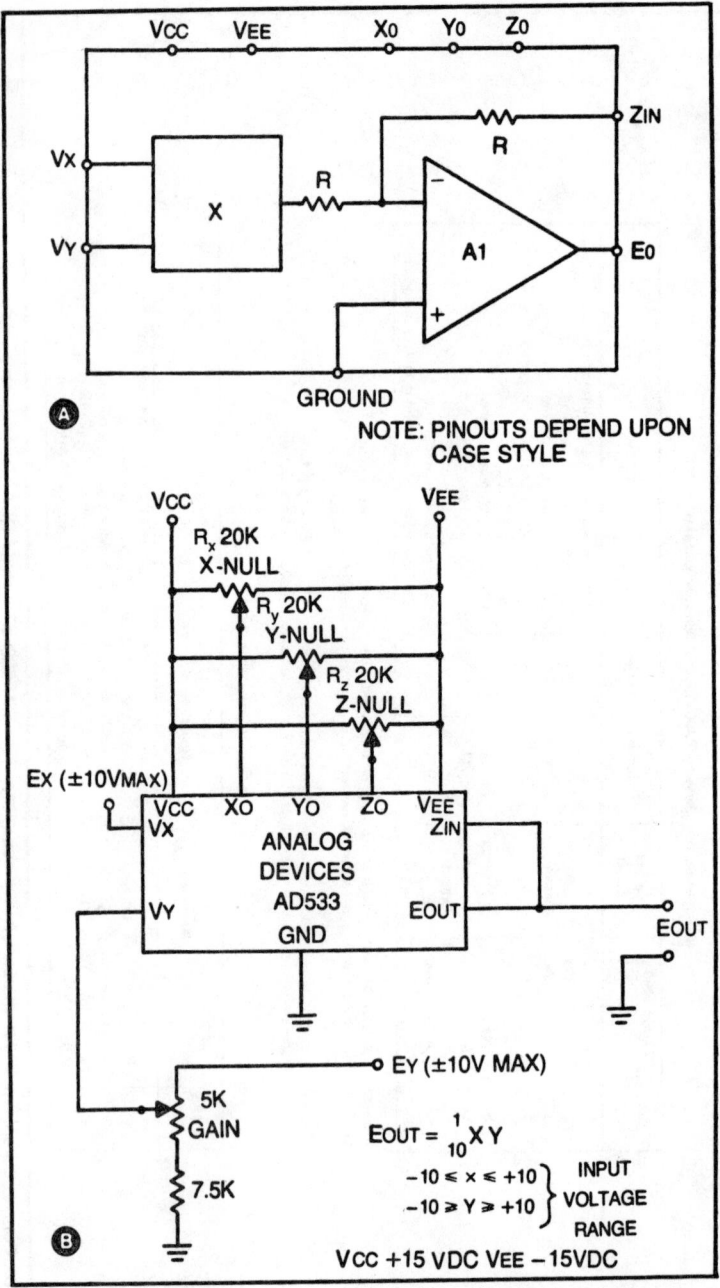

Fig. 3-13. Block diagram and schematic of the Analog Devices AD533 analog multiplier IC.

open-loop, voltage gain of the operational amplifier. In most IC op amps the value of the open-loop gain is typically greater than 20,000 and may exceed 1,000,000. Even the lowly 741 family of devices offers a respectable gain that is typically in the range of 50,000 to 100,000. The result of this is that almost any signal applied to the inputs will cause the output to saturate. In most linear circuits that would be a decidedly bad feature, but in comparator circuits it allows us to tell whether two voltages are in agreement with each other.

Consider Fig. 3-14. Amplifier A1 is an operational amplifier with a high-value, open-loop, voltage gain. When E_1 equals E_2 the differential voltage applied to A1 is zero. Let us assume that A1 has a voltage gain, without feedback, of 100,000. If the negative and positive, saturation voltages for this device are 10 volts (not unreasonable because V_{CC} and V_{EE} are each 12 volts), the signal would be:

$$E_{in} = \frac{E_{out}}{A_{vol}}$$

$$= \frac{10 \text{ volts}}{100,000} = 0.0001 \text{ volts}$$

(3.25)

In other words, as soon as the differential voltage ($E1$-E_2) reaches 100 microvolts, the operational amplifier is saturated. Much-lower voltages, though, can cause a meaningful output that can easily be put to work. For example, in the situation above, a 10-microvolt signal will cause an output voltage of 1 volt, which is certainly large enough to be detected as an indication that E_1 and E_2 are no longer equal. For all practical purposes, then, the operational-amplifier comparator will detect non-equality of two voltages applied to its two inputs.

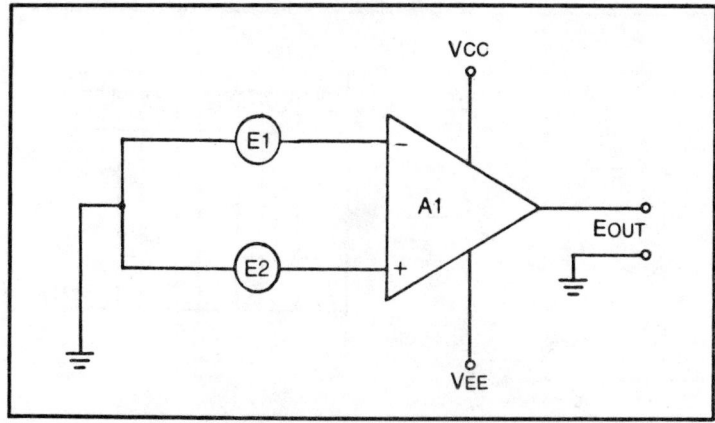

Fig. 3-14. Open-loop gain on an operational amplifier is high.

If the voltages thus applied are such that E_1 is less than E_2, it is equivalent to applying a negative voltage to the inverting input, and the output will be high positive ($+10$ volts in the example). On the other hand, a situation in which E_1 is greater than E_2 causes the operational amplifier to think a positive voltage is being applied to the inverting input, and that causes the output terminal to swing negative. This allows you to tell not only whether the two input potentials are equal but also which of them is greater. Although operational amplifiers are often used for this type of circuit, many designers prefer to use comparator integrated circuits such as the LM311 and LM339.

The LM311 is a single comparator available from almost all semiconductor manufacturers featuring a generalized line (instead of all special products). It may be purchased in 14-pin DIP, 8-pin mini-DIP, and 8-pin metal-can packages, but for most purposes I recommend the mini-DIP package—it is much easier to work yet does not take up the space on a circuit board required of the 14-pin variety.

The LM339 is a quad comparator in a single 14-pin DIP package. It has four independent comparators sharing a common power supply and ground-connection terminals.

In the LM311 and LM339 IC comparators the output terminal will remain high anytime $E_1 < E_2$. This IC can be used in many ways, but it is usually the case that one input will be tied to a reference voltage or ground. The reference voltage (including ground state) may be set manually or be tied to some circuit function. The remaining input sees a variable voltage from an analog input. An example is shown in Fig. 3-15. In this example, the comparator issues an output every time $V_{sig} \geq V_{ref}$. This could be useful in many ways, so is a very popular configuration.

In one medical alarm circuit (designed by C.E. McCullough), for example, V_{sig} is an electrical output from a respiration transducer. As long as V_{sig} has sufficient amplitude to generate output pulses from the com-

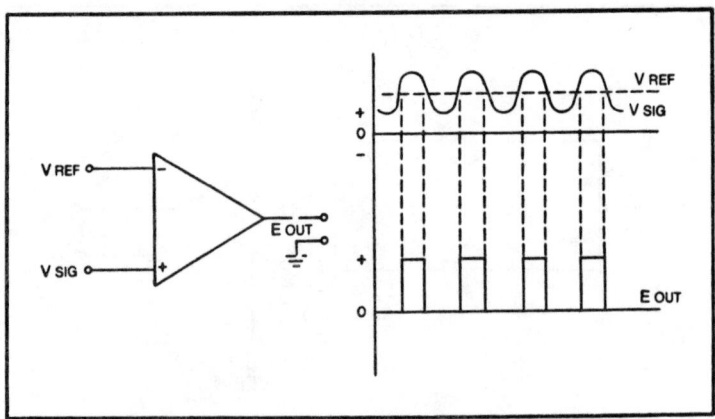

Fig. 3-15. An open-loop operational amplifier makes a voltage comparator.

parator, which are used to reset the alarm, nothing happens. If, however, V_{sig} stops or reduces in amplitude, no pulse will be issued by the comparator; so the alarm times out, and a warning beeper turns on. Another use will be seen in the chapter on analog-to-digital converters where V_{sig} is a linear ramp and E_{out} is used to stop the internal A/D clock.

SPECIAL-FUNCTION MODULES

Several companies offer special-function modules that combine operational amplifiers, discrete components, and other integrated circuits in a circuit that is designed to perform a particular analog-circuit function. These are usually packaged in a black epoxy or ceramic, potted block and have a minimum of user-access pinouts. They can be used with only a minimum of effort and are usually treated as a component in their own right rather than as a circuit.

Generally speaking, special-function modules are quite expensive, but their cost may be economically justified if convenience or tight specifications are required. In another chapter we will take a detailed look at some of these modules.

Chapter 4

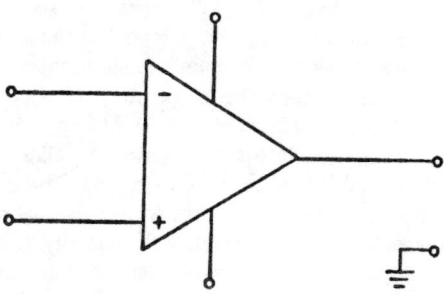

Power Supply Design

MOST ELECTRONIC CIRCUITS REQUIRE DIRECT CURRENT (DC) TO OPerate properly. The function of a power supply is to convert the alternating current delivered by the power mains to the correct dc voltage and current levels demanded by the circuit being powered.

There are several desirable properties in any good power supply. The output voltage should be precisely that which is required, and there should be an ample amount of current available for the purpose at hand. The voltage should not vary as the load (amount of current drawn) varies. Furthermore, there should be no noise or *ripple* (remnant of the ac power mains waveform) on the dc output. Also, the ac impedance across the output terminals at the lowest ac frequency that the circuitry will process should be very low. Of course, since we deal in reality, we know that not all power supplies will meet these requirements precisely, especially those of economy designs. An attempt should always be made, though, to ascertain the quality of power supply required for the general purpose.

POWER-SUPPLY BASICS

Regardless of the final form taken by any specific power-supply circuit, there are certain common characteristics that can be recognized. There are, for example, several parts to most power supply circuits

1. Voltage and current level conversion
2. Rectification
3. Filtering
4. Regulation

The conversion of voltage and current levels is accomplished through the use of a transformer. For most instrumentation applications using semiconductor electronic circuits you will want to step down the voltage and, possibly, step up the current. This is because most such circuits operate from dc levels less than the 115-volt potential of the ac mains.

Various voltage levels are commonly found in this area of electronic technology. Many transistor circuits operate from a positive-to-ground power supply in the range of 24 to 28 volts dc. Most operational amplifiers, on the other hand, prefer dual-polarity power supplies in which V_{CC} is positive to ground and V_{EE} is negative to ground. These will most often be ± 12 volts dc to ± 15 volts dc. A few circuits exist that require voltages outside that range. These are, however, in the minority. If you contemplate using TTL digital logic, and you will if you spend any amount of effort in this business, then have a +5-volt dc regulated source. The current levels required will be dependent upon just what you want to do, but in general a 500-mA or better rating for the ± 12-to-± 15-volt supplies is adequate, while at 5 volts one will want not less than 1 ampere. In the 5-volt supply it is almost a case of "the more the merrier." I personally like to see at least 1.5 amps at 5 volts, and 3 to 5 amperes really delights me.

A transformer is a passive device and cannot create *power*. The power drawn from the secondary can never exceed the power drawn from the mains by the primary. In fact, it will always be a little less since nothing is ever 100 percent efficient. You might be surprised to know that a transformer is one of the *most* efficient devices around, something that ought to delight everybody in this energy conscious age. Since the efficiency is on the order of 95 percent or better, we are justified in using the relationship

$$(E_{pri}) (I_{pri}) = (E_{sec}) (I_{sec}) \qquad (4.1A)$$

$$\frac{E_{pri}}{E_{sec}} = \frac{I_{sec}}{I_{pri}} \qquad (4.1B)$$

Of course, purists would question the use of an equals sign for Eqs. 4.1, but I won't quibble over a few percent error here. As you can tell from scrutinizing Eq. 4-1(B) it is a reciprocal arrangement; if the secondary voltage goes down, the current draw that is possible for any given wattage ($E \times I$) goes up. Clever things these transformers!

Rectification is the conversion of ac into dc or something like dc. A *rectifier* does this neat trick by allowing current to flow in only one direction, but ac flows in opposite directions on alternate half cycles. The rectifier, then, converts ac into a dc that pulsates.

Filter circuits are used to smooth out the pulsations left by rectification. They will convert the pulsating dc from the output of the rectifier to an almost-smooth dc usable by most electronic circuits—not perfect, but often usable.

Regulator circuits are used in some, but not all, power supplies to stabilize the dc output voltage under varying-load conditions. In most unregulated supplies we can expect the voltage at the output to vary somewhat with load changes and fluctuations of ac power-mains voltage. The electronic voltage regulator prevents these variations. You will see more use of regulators especially now that there are easy-to-apply IC regulators available at low cost, and they are highly recommended.

RECTIFIER TYPES

Several different types of rectifier devices are commonly used in electronic equipment. Almost all forms of rectifiers are electron devices called diodes. In older equipment you might find either vacuum-tube rectifiers or an obsolete semiconductor type made of selenium. This consisted of a series of thin metal plates coated with selenium material. The area of the plates determined the approximate current rating, while the total number of plates set the voltage level. The plates were connected in series.

Modern rectifiers are made of silicon, another semiconductor material. This element allows the construction of rectifier types that are much smaller than previous types. Typical silicon diodes may be only 1/8 inch by 1/4 inch, yet have a current rating of over an ampere. Some are so small that the axial connecting wires all but dwarf the diode.

RECTIFIER CIRCUITS

Figure 4-1 shows several different rectifier circuits. In each case power is supplied to the diodes from a transformer. This is done for two main reasons: scaling the voltage and current levels to those needed and isolation from the ac power mains. Safety considerations make it *imperative* to use a transformer even if you want 115 volts (a ratio of 1:1). Some consumer electronic products are transformerless, but these are extremely hazardous when used in a grounded environment. This is why TV and other appliance makers often pack a "do not use outdoors" label with their sets. DO NOT EVER make or press into service any transformerless power supply; this can be lethal.

The circuit in Fig. 4-1(A) is a *half-wave rectifier*. It gets its name from the fact that it utilizes only half of the ac input waveform. This, unfortunately, creates several problems that tend to limit the usefulness of the half-wave rectifier. Its single saving grace is the doubtful economy of using a single diode in the rectification process—doubtful because diodes are very cheap.

To understand the operation of this power supply circuit it is necessary to recall that diode D1 can pass current in only one direction; that is when its anode is positive with respect to its cathode, the diode will be reverse biased so no current will flow.

As a result of rectifier action the voltage and current waveforms across the output load R_L, will exhibit only the positive halves of the ac cycles.

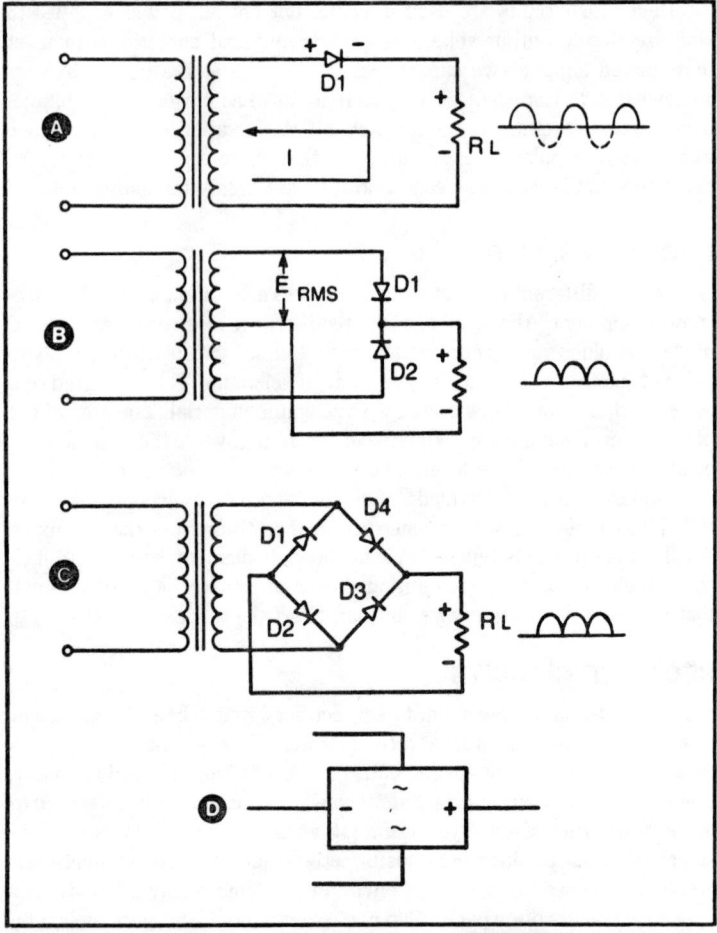

Fig. 4-1. At (A), a half-wave rectifier; (B) full-wave rectifier; (C) full-wave bridge rectifier; and (D), the symbol for a full-wave bridge rectifier.

The missing halves, shown here as dotted lines, are blocked by D1 so will not appear across the load resistor. The pulse-repetition rate of the pulsating dc is exactly the same as the ac line frequency.

The second type of rectifier that I shall consider is the conventional full-wave rectifier shown in Fig. 4-1(B). This type of circuit depends for its operation on a center-tapped power-transformer secondary winding. Diodes D1 and D2 are connected such that one will be forward biased on each half of the ac cycle. The center tap is used as the power-supply common and is often grounded. This makes one end of the secondary winding positive with respect to the center tap while the other end is negative with respect to the center tap at any given point on the ac cycle.

76

The result of this configuration is that both halves of the ac cycle are used. This produces a pulsating dc waveform in which the missing halves of the waveform are inverted and appear as half cycles between the normal positive cycles. This is not magic but simply a matter of diode steering.

The frequency of the pulsating dc is exactly twice the frequency of the ac input waveform. This fact makes the full-wave supply more useful than the half-wave type. It seems that the pulsating dc output from this type of supply is much easier to filter economically than that from the half-wave supply.

The final type of common circuit is the full-wave bridge rectifier of Fig. 4-1(C). Although this circuit seems uneconomical at first glance, it is the type usually preferred for instrumentation power supplies. It preserves the advantage of full-wave operation, yet also offers an output voltage twice that of the conventional full-wave circuit. Recall that the former full-wave circuit used a center-tapped transformer. For any given power transformer, then, the bridge circuit will produce twice the output voltage, but only half the output current, compared to the conventional full-wave circuit that uses a center-tapped transformer.

This type of circuit uses a four-arm bridge circuit made of rectifier diodes. On any given half cycle only two diodes will be conducting, the remaining two being reverse biased. On the alternate half cycle roles reverse; and the diodes which had been conducting become cut off, and those that had been off begin to conduct.

Consider the situation in which point X is positive with respect to point Y. In that case diodes D4 and D2 will be forward biased and will conduct current. Diodes D1 and D3, on the other hand, will be reverse biased and will not conduct current. On this half cycle current flows from point Y through diode D2, the load resistance, and diode D4 to point X.

On the alternate half cycle the opposite situation obtains. Diodes D2 and D4 are cut off, and D1 and D3 are conducting. Now point X is negative with respect to point Y. Current will flow from point X through diode D1, the load resistance, and diode D3 to point Y. The important thing to note is that the direction of current flow *through the load resistance* is the same for both half cycles. This produces the full-wave pulsating dc waveform.

Most semiconductor manufacturers offer four-diode bridge-rectifier stacks in a sealed unit molded of black epoxy or some similar material. A typical schematic representation for such a rectifier stack is shown in Fig. 4-1(D). The points where the transformer wires are attached are marked by either the letters AC or the symbol for an alternating current as shown in the diagram.

FILTERING

A typical electronic filtering circuit is shown in Fig. 4-2(A). This is called an RC filter because its principal constituents are a resistor and two capacitors. Although we have shown this filter connected to the output of

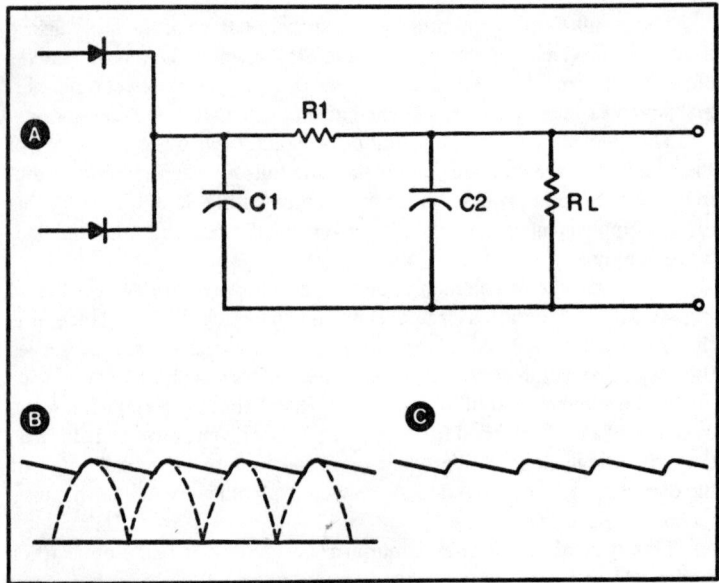

Fig. 4-2. At (A), an RC filter circuit; (B) composite waveform; and (C) an output waveform alone.

a conventional full-wave bridge rectifier, you may assume that the same type of circuit will also suffice for the half-wave and bridge rectifiers.

The output waveforms associated with this circuit also are shown in Fig. 4-2. The waveform in Fig. 4-2(B) is the composite showing the various components; and that in Fig. 4-2(C) is the way it will appear on an oscilloscope.

Capacitors C1 and C2 will become charged during each loop of the pulsating dc waveform. As the pulse fades away they will begin to discharge. This dumps current into the circuit; so it fills the spaces between the humps. The dashed lines in Fig. 4-2(B) represent the pulsating dc that you would find in a full-wave, unfiltered power supply. The solid line represents the capacitor discharge. Soon after turn-on an equilibrium is established such that the capacitor never completely discharges before the next pulse begins to recharge it. The result is the quasi-sawtooth of Fig. 4-2(C).

Although it somewhat tortures the truth, we can look upon the combination R1 and C2 as a low-pass filter designed to reduce the 120-hertz component of the waveform. In fact, the 120-hertz reactance of C2 is usually set such that it is one-tenth of the resistance of R1, further supporting our contention that it is a low-pass filter.

There are several variations on the circuit of Fig. 4-2(A). In some instances resistor R1 is replaced by an inductor. This is a common tactic found in older, high-voltage, high-power designs. In low-voltage, high-current applications, on the other hand, we may not use R1 and C2 at all.

In that case C1 is used alone, and it has a much higher value, on the order of 1,000 μF to 10,000 μF.

SOME NUMBERS

Thus far I have used descriptive techniques in treating the power supply, but now I should quantify the process and give you what is needed to actually build working supplies. I will discuss terms such as *percentage ripple, peak reverse voltage, percentage of regulation*, as well as the appropriate output calculations.

The ripple component in a half-wave circuit is over twice that in full-wave circuits, which is why the full-wave circuit is much easier to filter. The full-wave ripple component is on the order of 48 percent for a perfectly symmetrical sine wave input.

Another critical difference between full- and half-wave types is that less energy is delivered by the latter. In this respect we find the full-wave circuit more economical in the long run. We must have approximately 40 percent greater volt-ampere capability in the transformer primary for any given amount of work required at the output circuit. The transformer for the half-wave case, then, must have a volt-ampere rating 1.4 times that of a full-wave of the same rating.

In either type of rectifier circuit the output voltage, which is the charge across the capacitor, is not the rms value of the ac, as you might expect, but the *peak* value. Funny enough, it is usually the rms value that is quoted in transformer catalogues and spec sheets. The output voltage, as measured across the first filter capacitor, is 1.41 times the rms value delivered by the transformer secondary winding.

A difference in output voltage is apparent in unfiltered cases. A normal dc voltmeter across the load will read the average value, not the peak. This is 0.45 $\times$ rms for the half-wave and 0.9 $\times$ rms for the full-wave.

Diode rectifiers have a rating called the *peak reverse voltage* (PRV), also called *peak inverse voltage* (PIV). This is defined as the maximum reverse-bias voltage that the diode can withstand without blowing out. This is the voltage rating usually quoted in diode-specification sheets. You will, for example, see a diode rated as 1000 volts at 1 ampere. This means that the PIV is 1000 volts and the forward current is 1 ampere. The maximum allowable rms voltage that is used to forward-bias the diode and cause current to flow is somewhat less than the peak value.

The PIV must be considered when selecting an appropriate diode. In a filtered power supply the capacitor will charge to a level of 1.41 $\times$ rms. Add this voltage to the peak voltage applied by the transformer (also 1.41 $\times$ rms), and you have a potential of 2.82 $\times$ rms applied to the diode at least once every half cycle. As a rule of thumb, select a diode that has a PIV rating at least 2.82 times the rms voltage that will be applied. For added safety use a factor of three times rms. This matter, incidentally, is probably the single most-frequent mistake made by people servicing or constructing power supplies. Most will assume that any diode with a PIV greater than 1.41 × rms is adequate; it isn't.

VOLTAGE REGULATION

Most power supplies will exhibit some degree of variation of output voltage as the load is changed. In general, an increased current drain will reduce the output voltage. The mechanism by which this happens is shown in Fig. 4-3. All power supplies have some amount of internal resistance even though it *may* be quite low. Such power supplies, the real kind that we see in the lab instead of on a blackboard, can be represented (as in Fig. 4-3) as a perfect battery and a series resistance. These are E and R_i in the figure. The value of R_i is

$$R_i = \frac{E}{I} \qquad (4.2)$$

where

E is the *open-circuit* voltage across points A and B with R_i disconnected

I is the current that would flow if A and B were shorted together.

The output voltage will depend upon the value of the load resistance (R_L) so may be said to vary with current. The actual value of the output potential (E_o) will be

$$E_o = \frac{E\ R_L}{R_L + R_i} \qquad (4.3)$$

A voltage regulator is a circuit that keeps the output voltage constant despite changes in load resistance or input voltage. Although there seems to be a large variety of regulator circuits in the literature, there are actually only a few generic types.

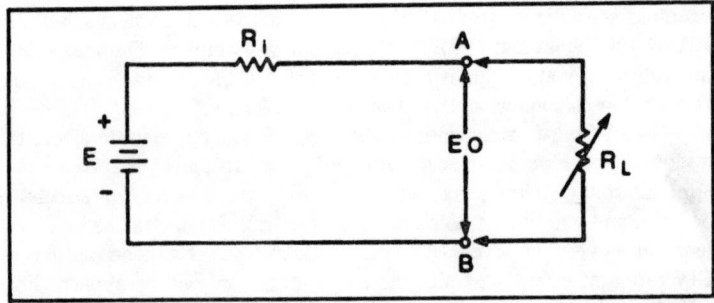

Fig. 4-3. Output voltage is divided by internal resistance of the power supply and the load resistance.

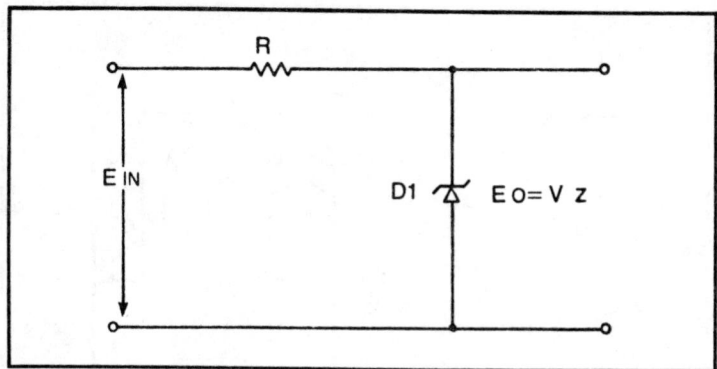

Fig. 4-4. Zener diode regulator.

Perhaps the simplest voltage regulator is the zener diode. This is a special type of silicon diode which breaks down if a certain predetermined reverse voltage is applied. If the zener current is limited to a safe value the diode will regulate at its zener voltage (V_Z).

Figure 4-4 shows an example of a simple, zener-diode voltage regulator. The resistor R is used to limit current in zener diode D1 to a value less than would cause destruction. The output voltage will equal V_Z as long as E_{in} is above a minimum threshold defined as ($E_r + V_z$). Above the limit, input voltage E_{in} may vary substantially with but with small changes in output voltage.

On the disadvantage side, we find that zener diodes also have some substantial problems. For example, the actual value of V_z may vary quite a bit with temperature. Also, one generally finds that V_z is not well calibrated in ordinary off-the-shelf zener diodes. Unless precision is required, though, the zener potential is usually close enough.

Another disadvantage—and this is more serious in most equipment designs—is the relatively-small dynamic range. Zener regulation is usually not good enough under those circumstances. Related to this are the current limitations imposed by the series resistor. Both of these problems can be remedied by using a zener as a voltage *reference*, which is actually its intended purpose, to control a series-pass transistor. Such a circuit is shown in Fig. 4-5.

Series-pass transistor Q1 is used as an electronically-variable resistor. Its base is kept at a constant voltage by zener diode D1. The emitter voltage is dependent upon the current drain and the base-emitter voltage (V_{be}). It will have a value of

$$E_o = (V_z - V_{be}) \tag{4.4}$$

If the output current goes up, the output voltage will go down. To maintain the equality of Eq. 4.4, V_{be} must go up. This turns transistor Q1 on

81

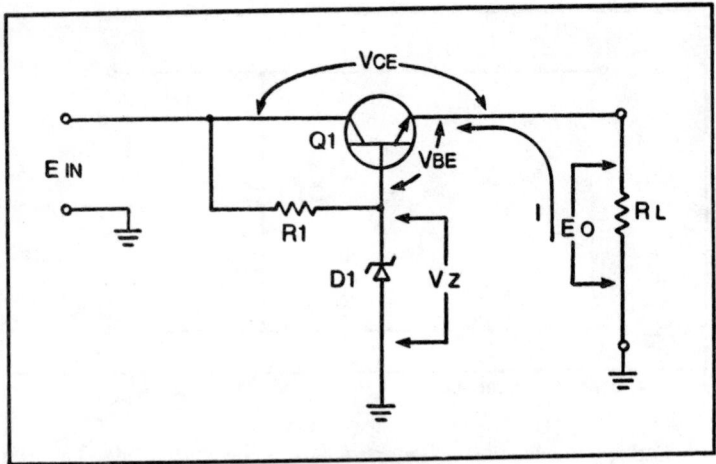

Fig. 4-5. Zener diode controlling a series pass transistor.

harder which effectively lowers its series resistance. This will cause E_o to rise back to its equilibrium value. Although this circuit is also affected somewhat by temperature drift it is far superior where a large dynamic range or a heavy current drain is anticipated.

A modern version of the series-pass idea is manifest in the three-terminal IC voltage regulator. This device has all but supplanted regulators in the 100-mA-to-3-amperes current range because they have only three terminals, so are very easy to apply. For most designs requiring modest current levels the three-terminal IC regulator is probably the most sensible approach.

Figure 4-6(A) shows the circuit for a typical three-terminal IC voltage regulator. Voltage E_{in} is the unfiltered output of the power supply rectifiers, while E_o is the *regulated* output voltage.

Capacitors C1 and C2 are not necessary but do tend to improve performance. Capacitor C1 is used to provide stability to the system, and C2 is used to improve transient response. This is used because a sudden increase in load may not be tracked by the regulator; so C2 dumps charge into the circuit.

Figures 4-6(B) through 4-6(D) show the basing of several standard IC regulator types. Generally, those built into cases such as Fig. 4-6(B) are rated at 100 milliamperes. The current rating for the other two is usually 1 ampere, although Lambda makes one in a TO-3 case rated at 3 amperes, and one reportedly exists that will deliver 10 amperes.

The three-terminal regulator will offer very respectable regulation when used in the configuration of Fig. 4-6(A). The value of the output voltage is fixed by the manufacturer and may be determined through the type number.

A popular 5-volt regulator is the LM309 series. The LM309H is rated at 100 mA and comes in a TO-5 package. The LM309K is a 1-ampere device in a TO-3 package. Most, incidentally, can be overrated to approximately 1.5 amperes by placing them on a large, finned heatsink.

The LM340 and LM320 series devices are available in both of the 1-ampere package types, shown in Figs. 4-6(C) and 4-6(D). They are available in output voltage levels of 5, 6, 12, 15, 18, and 22 volts dc. The rating may be determined by a hyphenated suffix following the type number. For example, an LM340-12 is a 12-volt, 1-ampere regulator.

The LM340 is a positive-to-ground device, while the LM320 is the same thing but accepts a negative-to-ground input voltage. THE PINOUTS ARE NOT THE SAME! Check the spec sheet before using!

Similar, and often considered equivalent, to the LM340/LM320 are the Motorola 7800/7900 series. The 7800 is a positive regulator, while the 7900 is negative. The last two digits of the type number tell you the output voltage rating; for example,

Type No.	
7805	5 volts
7812	12 volts
7815	15 volts
7818	18 volts

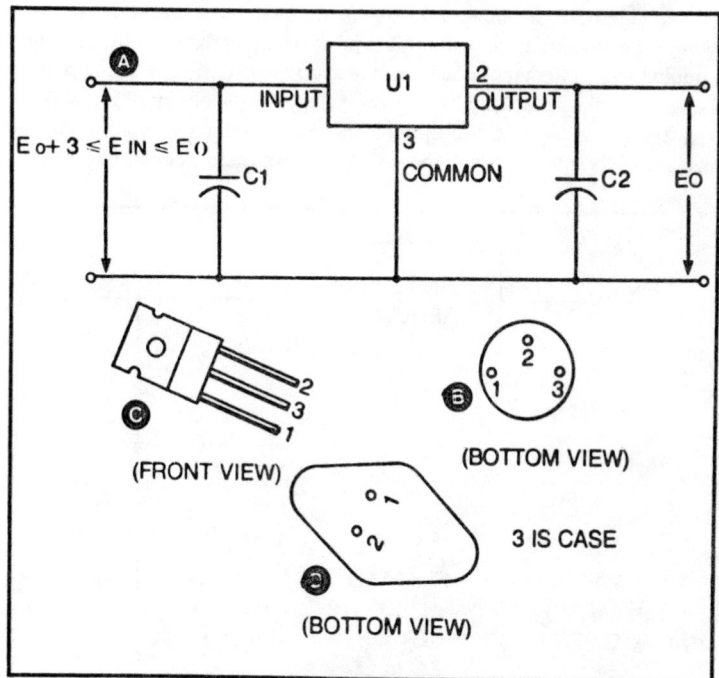

Fig. 4-6. Three-terminal IC voltage regulator and some case styles.

(The same convention applies to the 7900 series.)

All of the these regulators are better off and will prove more reliable if operated on a heatsink. You may buy a commercial or surplus finned heatsink, or you may bolt the TO-3/TO-220 to an aluminum chassis.

When operated in a circuit such as Fig. 4-6(A) the rated voltage and the output voltage will be very close to each other. Oddball voltages can be accommodated by such a circuit as in Fig. 4-7 although at some cost in regulation. The output voltage in Fig. 4-7 is given by

$$E_{out} = E_o\left[1 + \frac{R2}{R1}\right] + (I)(R2) \qquad (4.5)$$

where

E_{out} is the nominal rated output voltage in the three-terminal mode, per Fig. 4-6(A).

It must be realized, however, that in most three-terminal regulators, this will deteriorate the regulation an amount roughly proportional to the change in output level.

One application for a circuit such as Fig. 4-7 is to trim the output voltage to a precise level. Few, if any, three-terminal devices produce exactly the rated voltage. An assortment of 5-volt regulators I tested offered output voltages over a range of 4.92 to 5.12 volts, and *none* were exactly 5 volts. The indication for using this type of circuit is the need for precision coupled with a demand that is lightly and consistently loaded. If regulation, rather than precision, is important then use the circuit of Fig. 4-6(A).

Another application for Fig. 4-7 is the use of the three-terminal regulator in place of the zener diode in the series-pass, transistor regulator discussed

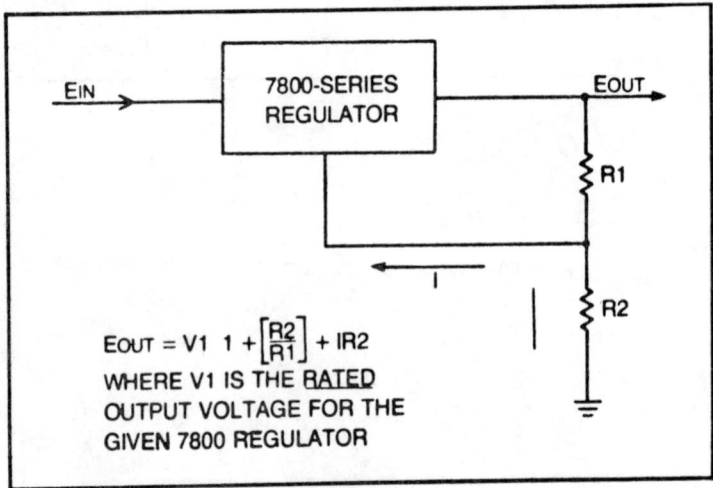

Fig. 4-7: Changing the output voltage.

earlier. The base-emitter voltage of silicon series-pass elements such as the HEP S7000 or the 2N3055 (good up to several amperes and dirt cheap!) is around 0.6 to 0.7 volts. I found that a resistor of 60 to 80 ohms would raise the output of an LM340-5 to approximately that amount. When this combination was used to keep the base of a series-pass transistor stable it resulted in a relatively well-regulated, 5-volt dc output.

CURRENT REGULATORS

There are basically two types of current regulators. One allows current drain to vary at will up to a threshold point, after which no additional current may be drawn. This is a current-limiting regulator and is useful as a protection circuit. The other type of current regulator will keep current constant despite large changes in load resistance. This is called a *constant current source* (CCS).

The current-limiting regulator is usually based on an operational amplifier or a transistor amplifier that senses the voltage across a resistor in series with the output line. This resistor must be very low in value (as low as several *milliohms* in high-current applications) or it will seriously deteriorate voltage regulation by increasing the series resistance. The current-sensing amplifier is triggered if the voltage across the series resistor exceeds the threshold programmed by the designer.

Applications for the current limiter are mostly in the area of protection. You might use one on a power supply so that output short-circuits will not damage the power supply. This becomes especially important if the supply is subject to abuse or is used as a power source when designing electronic circuits (substantially the same thing). Another idea is to use it when troubleshooting electronic equipment. A number of bench power-supplies used by designers and service technicians have a current-limit circuit that is adjustable. It is a good idea to set the current limit so the circuit will have adequate operating current but will not be able to draw enough to cause further damage. This becomes especially desirable when the equipment uses high-cost integrated circuits.

There are several approaches to making a constant-current source. A pseudo-CCS can be made by series-connecting a voltage source and a resistor that has a value greater by two orders of magnitude than any possible variation in load resistance. The current will be

$$I_{CCS} = \frac{E}{R_L - R_{CCS}} \qquad (4.6)$$

To provide an example let us assume a load resistance of 1 kilohm that changes to 10 kilohms (a 10:1 change). Also, the CCS resistance is 1 megohm. The total change in current is related to the total change in resistance, and that is less than one percent.

$$\frac{1,010,000}{1,001,000} = 1.009:1 \qquad (4.7)$$

The problem with this approach is that it requires very high values of input voltage if the current level is high, and "high" in this context is ridiculously low. For example, if the current required is 1 microampere and R is still 1 megohm, the voltage is only 1 volt. But what if a 1-milliampere current is required?

$$E = IR \quad \text{(Ohm's law)} \qquad (4.8)$$
$$E = (0.001)(1,000,000) \qquad (4.9)$$
$$E = 1,000 \text{ volts} \qquad (4.10)$$

That may be a little hard to come by in low-voltage circuits. Besides that, it can be very dangerous! Fortunately, there are safer and easier ways to implement alternatives to the pseudo-CCS that work even better.

Junction field effect transistors (JFETs) can be made to act as a CCS by virtue of their knee current near saturation. Figures 4-8(A) and 4-8(B) show two common methods for making a CCS using a JFET.

In Fig. 4-8(A) we see the simplest JFET CCS. It will provide an output current approximately equal to I_{dss}, an ordinary JFET parameter.

Siliconix, Inc., and perhaps others, offer CCS diodes—symbol shown in Fig. 4-8(C). These can be purchased at low cost, in ratings from a few microamperes to a few milliamperes.

In some applications we find it wise to use the types of CCS shown in Fig. 4-8 because they help us keep the design as simple as possible. One of the great principles of electronic design is the so-called kiss principle: keep it simple, stupid!

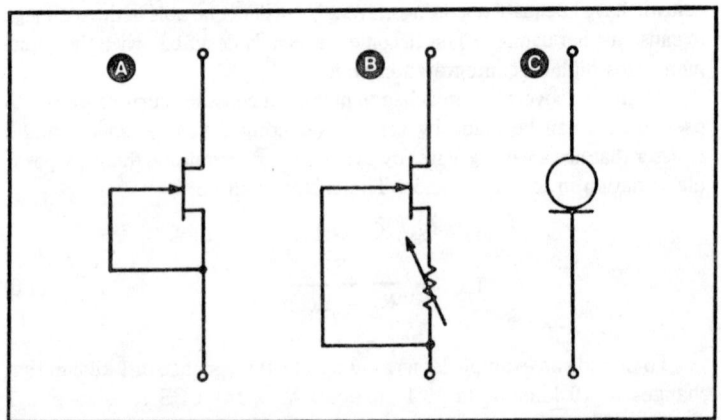

Fig. 4-8. Constant current circuits: (A) simple JFET; (B) with resistor; and (C) a CCS diode.

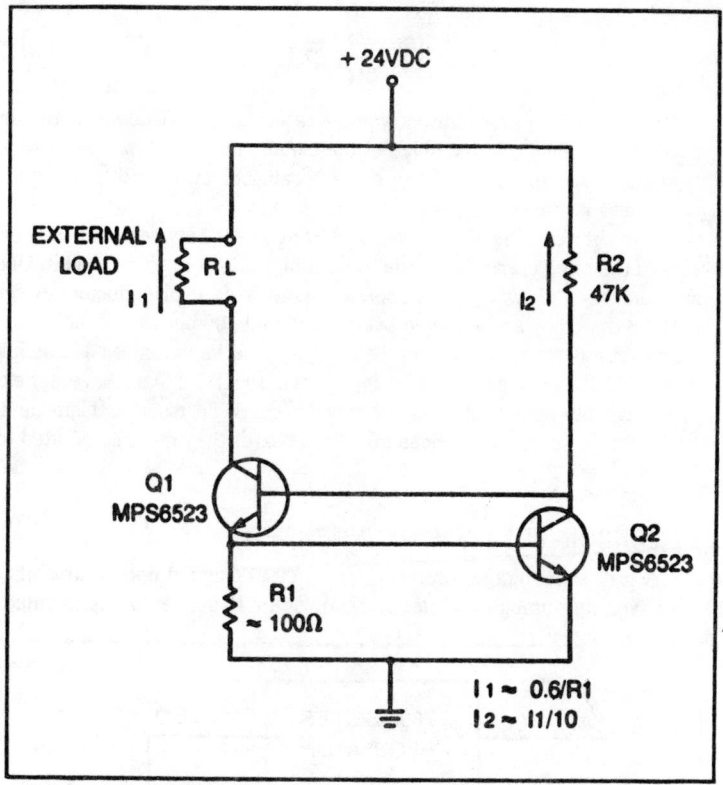

Fig. 4-9. Active constant current source.

As nice as that idea sounds, it is not always possible because in some cases elegant simplicity must be given second billing to precision, in which case more complexity is required. It is unfortunate that *simple* and *precision* are not words that prove universally compatible.

Several approaches to more complex CCS designs can be taken. One such circuit is the simple bipolar CCS shown in Fig. 4-9.

This circuit is different from the others in that it is used to *sink* a constant current rather than to *source* a constant current like the others. The transistors shown are discrete devices, but superior tracking should result if a dual transistor is used instead. A good example would be the Precision Monolithics MAT-01 series described in the transducers chapter.

The actual current-sink transistor is Q1, and it is controlled by Q2. The constant current is I1, and it has a value approximately equal to 0.61/R1. Current I2 is the reference, and it is set to be approximately equal to I1/10.

For somewhat-larger currents you may use a three-terminal, *voltage* regulator in a circuit such as Fig. 4-10. This circuit will give us an output current of approximately

$$I = \frac{E_o}{R1} \qquad (4.11)$$

where E_o is the nominal output voltage rated for the particular regulator used at U1. Since this is stable but not usually precise, it may be wise to trim the value of R1 until the correct value of current is produced—cleverness is a 10-turn potentiometer.

The circuit of Fig. 4-10 works well for current between 5 and 50 percent of the rated current. Beyond those limits constancy suffers a bit. One should always be aware in any application involving semiconductor devices that anytime they are operated between 50 and 100 percent of their maximum ratings reliability will probably suffer markedly unless a heatsink is provided. The three-terminal regulators in the TO-220 package, for example, can be operated at 1 ampere only if properly heatsinked. If left standing in free air or on a non-heatsinked PC board, they must be derated to 750 mA.

PRECISION-VOLTAGE REFERENCES

A precision-voltage reference is a circuit designed not for its ability to deliver huge amounts of electrical power but to deliver precision values

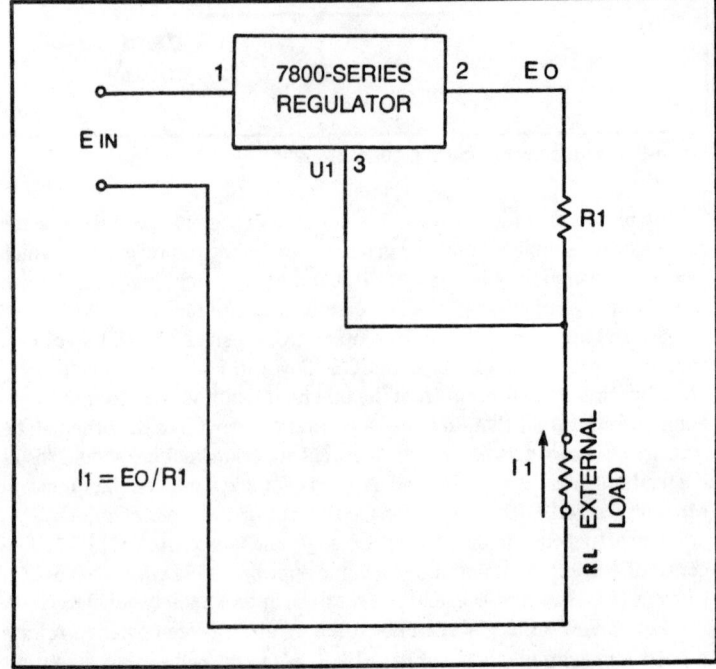

Fig. 4-10. Three-terminal voltage regulator used as a current source.

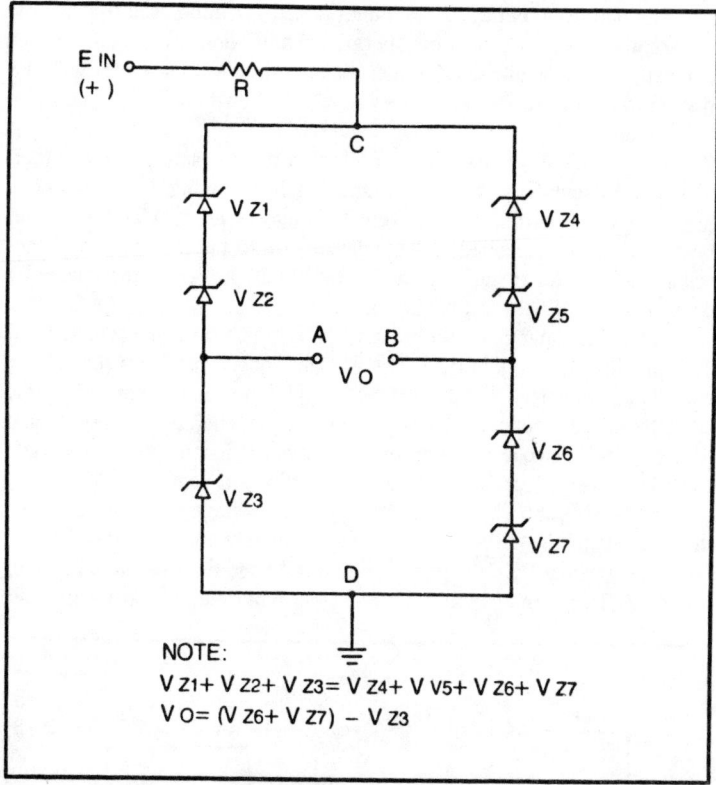

Fig. 4-11. Multiple zener diode reference voltage source.

of voltage. Amplifiers following the reference can produce any power needed. References may be used as an external standard for instrument calibration or internally as an integral part of instrument design.

A crude sort of reference is the precision zener diode. These devices are optimized (which usually means *selected* for certain properties from a batch of regular diodes) for superior performance. These diodes, at least from most sources, still suffer enough from the ordinary, zener diode troubles to make them useful only where the word "precision" is used loosely or where all concerned realize that taffy is being distributed as the spec sheet is read aloud.

The main problem seems to be thermal tracking of the zener-point voltage V_Z. The zener stack shown in Fig. 4-11 is an impromptu solution to this problem. Several zener diodes, preferably all of the same brand, are arranged in two, series-connected stacks. It is important that the sum of the zener voltages in each stack be equal or problems will result. If point D is considered common, the differential output voltage across points A and B will be $(V_{Z6} + V_{Z7}) - (V_{Z3})$.

The output potential across points A and B can be used to drive the differential inputs of a low-drift operational amplifier. If the gain is scaled correctly, a reasonable quality operational amplifier (not a 741) will produce a relatively stable reference voltage, KV_o, where K is the op-amp stage gain.

This circuit only works nicely if all zener diodes share a common thermal environment. This can be accomplished by clamping the zener diodes to a common heat sink. Alternatively, they may be encapsulated in a good heat-conducting compound such as silicone grease, but that is messy. Modern semiconductor manufacturers have virtually leaped to the rescue by providing us with several elegant alternatives.

At least one manufacturer offers an IC in which the zener is embedded in an amplifier circuit that runs class-A. One property of a class-A amplifier with its input shorted is that the heat dissipation is almost constant. This sets a thermal environment that is constant for the zener. It is a four-terminal device in which two are for the zener and two are for the power terminals to an amplifier which is shown schematically as a heater.

Precision Monolithics, Inc., offers two IC voltage reference devices, the REF-01 and the REF-02. These offer output potentials of 10 and 5 volts, respectively. The REF-01, and by implication the REF-02, are shown in Fig. 4-12. They are available in the TO-99 8-pin metal IC package, and

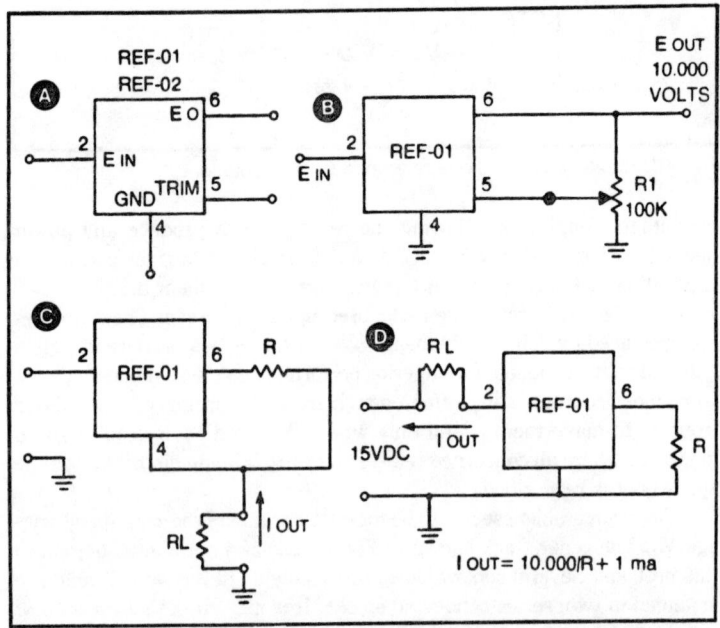

Fig. 4-12. At (A), the REF-01/REF-05 precision voltage reference; (B) basic circuit; (C) REF-01 as a current source; and (D) the REF-01 as a floating current source.

both commercial and military temperature ranges are provided.

The basic voltage reference circuit is shown in Fig. 4-12(B). This circuit seems elegantly simple and is much like the three-terminal devices discussed earlier. Potentiometer R1 allows adjustment of output voltage over a range of approximately ±300 millivolts. The range of normal operation, then, is 9.7 to 10.3 volts. This range may be too large for some applications, but the trimming resolution may be improved by making R1 a combination of 10-turn potentiometers or a single 10-turn pot and a few precision resistors of fixed value. It may seem utterly stupid to use precision resistors in a circuit with a potentiometer, but the reasoning becomes more sound when you consider the fact that precision resistors are predictably good performers under varying temperature conditions.

Battery operation of a voltage calibrator is easy using the REF-01 because it draws only a few milliamperes. At least 12 volts are required for the REF-01 and 7.5 volts for the REF-02.

Constant-current source and sink circuits are shown in Figs. 4-12(C) and 4-12(D), respectively. In both cases, the output current is

$$I_{out} \text{ (ma)} = \left[\frac{10.000}{R} \right] + 1 \qquad (4.12)$$

MISCELLANEOUS TOPICS

All voltage-regulator circuits have a certain voltage drop between the unregulated input and the unregulated output. For some circuits, such as the three-terminal IC regulators, this drop may be considerable. The popular 5-volt LM309 regulator, for example, will operate normally with input voltages from 7 to 35 volts, although heatsinking is recommended if the input voltage exceeds about 10 volts, maybe 12 volts at the outside.

SCR Crowbar Protection

If the series-pass element inside the regulator shorts out or if the internal reference goes haywire, the output voltage will rise to a level almost equal to the input voltage, and this is potentially damaging to the circuitry being supplied. A very common occurrence in regulators where a series-pass-transistor base is regulated by a zener diode is to lose control when the zener open-circuits. This raises the base voltage to almost the same as the collector, turning the transistor on hard. Disaster inevitably follows if the designer was not clever enough to incorporate a method for preventing overvoltage conditions.

Some IC regulators provide a shutdown terminal that will kill the output when overvoltage is present. But what if it is the regulator that fails? This brings about the ludicrous situation in which the failure of a two-dollar component will destroy thousands of dollars worth of circuitry attached to the power supply—clearly unacceptable since the "fix" is so simple!

Two types of circuits are used: zener diode and SCR crowbar. In the zener-diode type, useful only in low-current applications where a slow-response time is not disastrous, a zener with a voltage a little above the nominal voltage of the output line is shunted across the line. If the voltage increases, the diode will either clamp the voltage to the zener potential or blow a series-connected fuse.

A better and more popular technique is the SCR crowbar, shown in Fig. 4-13. It is not elegant, as befits its name "crowbar," but it is effective in most cases. It is a line of defense and is not intended to appeal to the person who revels in complex-circuit descriptions. One is reminded of the action appropriate when a thug with a gun decides to kill you—hit him hard with anything you can get your hands on, including a cannon. The "cannon" in this case is a silicon-controlled rectifier (SCR).

An SCR is a power-rectifier diode that remains turned off unless a current is injected into a gate terminal. It will then turn on and remain turned on until the forward current is dropped below a critical threshold value.

In Fig. 4-13 the SCR is normally turned off but can be turned on if a sufficient current (I1) is created. If that current flows, the SCR will conduct a large current (I2) that is sufficient to blow fuse F1. This is crude and decidedly inelegant but is appropriate to protect high-cost equipment.

If current I1 remains below the minimum current level needed to gate-on the SCR, nothing happens, and I2 is zero. In this circumstance the output voltage is normal, and the protected circuitry is allowed to operate normally.

Resistors R1, R2, and R3 are selected to keep I1 below the minimum SCR-firing threshold unless an excessive output voltage appears across the output terminals. If that occurs, I1 will exceed the minimum required SCR-

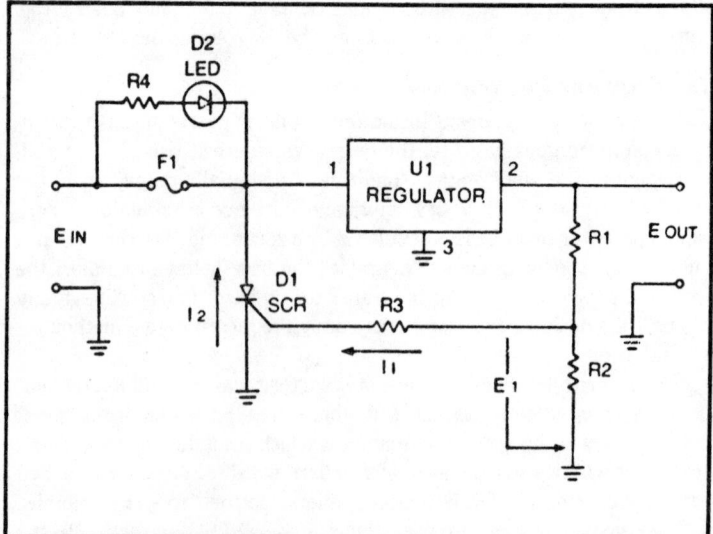

Fig. 4-13. SCR crowbar protection circuit.

gate current, causing D1 to conduct, resulting in high current that blows fuse F1.

In some circuits R1 is replaced by a zener diode. Its V_z value is selected so that it breaks down at whatever maximum output voltage is permitted. In some circuits only the zener and R3 will be used.

Blown-Fuse Indicator

Troubleshooting a system can be made much easier if certain indicators are built into the circuit. You want, for example, an LED across each power-supply line so that you can tell at a glance whether there is voltage across the line. If the LED is not lit the power supply is dead.

Another type of indicator is the blown-fuse indicator of Fig. 4-13. Fuse F1 normally keeps the R4/D2 combination shorted out, so the LED will *not* be lighted. If F1 opens, however, the voltage across R4/D2 will be sufficient to turn on D2. The current through D2 is only a few milliamperes; so no damage to protected circuits is possible. In 115-volt ac circuits, replace D2 with a small, neon lamp (NE-2, NE-51, etc.) and make R4 equal to something over 100 kilohms. The blown-fuse indicator is especially useful on panels with more than one fuse.

Ac/Dc Power Supplies: a Warning Repeated for Emphasis

Some designers with an eye for "economy" try to save the cost of a power transformer by using a resistor to drop rectified, ac-mains voltage to the level required by the circuitry. In these ac/dc power supplies the ac-mains neutral wire is, we hope, connected to chassis ground or a counterpoise ground. The ac hot wire is connected to the rectifier.

THIS PRACTICE SHOULD BE BRANDED STRAIGHTAWAY AS AN *EXTREMELY* DANGEROUS AND STUPID PRACTICE!

There are NO circumstances where it should be allowed. Forget ac/dc power supplies altogether, and roundly condemn those who *do* use them. Be suspicious of the motives of anybody who suggests that *you* use one!

Purchased Power Supplies

There are quite a number of companies who make as their only product power supplies to go into other people's equipment. These may be a good alternative to building your own, especially if the current requirement is over the 1-ampere rating of three-terminal regulators. What they cost in cash outlay is often repaid by saved assembly and parts cost.

Three basic types of purchased power supplies usually available are: close-frame, open-frame, and encapsulated. The first of these is a power supply inside its own enclosure and can be bolted into your assembly in a few moments. It is, however, the most expensive in any given voltage/current rating.

Open-frame power supplies actually include two different basic designs. Some are only a printed circuit board that includes the transformer, filters, and regulators, as needed. Most of these are found only in current ratings up to a few hundred milliamperes or perhaps 1 ampere. The other type of open-frame power supply is built onto a metal, half-shell frame that is large enough to support the weight of the larger power transformers needed for increased current loads.

Encapsulated power supplies are light duty (0.1 to about 2 amps) and are entirely enclosed inside a black epoxy block. Most are not repairable since the potting compound cannot be removed. Most encapsulated power supplies have only four or five pins protruding from the bottom. Two are for ac (115 volts), and two for the dc output—positive and negative. A fifth pin may be provided to allow you to trim the output voltage to an exact value. For small instruments with modest power requirements and limited-space requirements these are often considered the best deal. In most cases they can be mounted directly onto the printed circuit board.

Batteries

When simplicity is a factor, what could be simpler than a battery? All that is required is a holder and a few sprigs of wire; and the holder can be homemade if necessary.

Batteries have an undeserved reputation for inelegance, but they are sometimes the best alternative. In some experimental instrumentation packages the electronics draw only a few milliamps, and a high-current battery will last a year or more; then it will die, due to expired shelf life rather than current drain!

When circuit isolation is a factor there is often no better alternative than batteries. There is, for example, no power supply that is 100 percent free of ripple, especially where low cost is desired. Some very-high-gain electronic circuits will respond to even a small ripple or a tiny magnetic field. In that type of application battery power is not merely a matter of convenience but is the indicated design.

A popular method to produce a battery power supply is to use nickel-cadmium batteries affectionately known as *ni-cads*. A charger is used to trickle-charge the ni-cad when it's not being used, but the ac power to the charger is turned off when the battery is in use.

Be aware that ni-cad batteries must be charged to 140 percent of their rating at a current equal to one-tenth of their nominal ampere-hour rating. One would ordinarily expect to charge them for only 10 hours at the one-tenth rate, but 14 hours is required.

Also, do not let the charger exceed the one-tenth-ampere-hour-rating current level lest damage be experienced. Ni-cad batteries may blow up if charged too rapidly.

One final topic is a warning that the ni-cad batteries sold in blister packs at the local, parts wholesalers are not usually rated to the same ampere-hour capacity as those found in commercial power packs. If you find an instrument with a bad battery, do not expect blister-pack ni-cads to be good replacements. A ni-cad C cell, for example, is normally rated at 2 ampere-hours, but the blister-pack replacement types are rated at only 1.2 ampere-hours. Similarly, the D size industrial ni-cad cell is rated at 4 ampere-hours, while the replacement cells are rated at only 2 ampere-hours.

BUILDING FIXED AND ADJUSTABLE DC POWER SUPPLIES

The dc power supply is one of the most important parts of any electronic construction project; yet it is often the least-considered portion of the design. Dc power supplies include both integral power supplies mounted inside a project cabinet and "universal," bench power supplies used for testing, adjusting, troubleshooting, or otherwise messing with an electronic circuit. I will discuss the methods for designing simple dc regulated power supplies; both fixed-voltage and variable-voltage models will be considered.

Fixed-Voltage Types

The problem of designing a fixed-output voltage regulator is made much simpler these days by the three-terminal, integrated-circuit, voltage regulators that are now available. There are quite a few of these devices on the market, but they all share certain characteristics. For one thing, the output voltage is fixed at some standard value; the actual value is usually identified from the type number. For example, a 7805 is a 5-volt regulator, while the 7812 is a 12-volt model. There are exceptions to the numbering rule, but in general these are summarized as follows:

1. LM-309 is a 5-volt regulator @ 100 mA or 1 ampere, depending upon case style;
2. LM-323 is a 5-volt, 3-ampere regulator in a "K" package; i.e, "same as TO-3" transistor package;
3. LM-340n-xx is a positive output voltage regulator; the "n" in the type number denotes the package type, while "xx" is the voltage rating; for example, the LM-340-12 is a 12-volt regulator in a "K" package so it will pass 1 ampere;
4. 78xx is a family of regulators that are similar to the LM-340n-xx series; the letters "xx" denote output voltage (7812 is a 12-volt regulator); package style determines current capacity;
5. LM-320n-xx is a negative output version of the LM-340n-xx, while 79xx is a negative output version of the 78xx (note: the input and ground terminals on the LM-320 and 79xx are reversed from the pinouts of the LM-340 and 78xx; failure to observe this convention will result in destruction of the regulator!).
6. The package designations are as follows: "H" indicates a TO-5 case, or its plastic equivalent, and a current of 100 mA; "K" indicates a TO-3 case and a current of 1-ampere (1.5 amperes if prop-

erly heatsinked); "T" indicates a TO-220 plastic power-transistor-type case, and a current rating of 750 mA in free air or 1 ampere if properly heatsinked.

Regarding the current ratings, it doesn't make sense to use the maximum rating if accompanied by the admonition ". . . if properly heatsinked." Heat and high current are the twins that destroy electronic circuits. If these devices are routinely operated at or near maximums, they will experience a higher than usual failure rate.

Figure 4-14 shows the basic circuit for a three-terminal, fixed-voltage, IC regulator. The transformer T1, rectifier BR1 and filter capacitor C1 are selected according to the usual rules for any dc power supply. The transformer steps the 115-volt ac line voltage down to a level required for the input of the voltage regulator. Normally there will be a 2.5-volt difference between the rated, regulator output and the minimum-allowable, input voltage. For a +5-volt regulator, therefore, a minimum of +7.5 volts is required.

The rule for selecting a transformer is to provide at least the minimum voltage required for proper operation of the regulator. Keep in mind that the voltage across the regulator input, which is also the voltage across the filter capacitor, will be approximately 0.9 times the peak ac voltage across the transformer secondary (under load). Since the secondary voltage is specified in rms values, you must multiply the rated value by 1.414 in order to find the peak voltage. If all terms are accounted for the output voltage will be approximately $0.9 \times 1.414 \times V_{rms}$. The minimum rms value of the secondary voltage should be the minimum value of dc input required to the regulator divided by 1.26. For the +7.5 volts required for a +5-volt regulator you need an rms rating of 7.5/1.26, or 5.95 volts. Because 6.3 volts rms is the next higher standard value, you would select a 6.3 volt ac transformer for this application.

The current rating of the transformer should be the highest expected dc value plus a margin for safety. In addition, you should be aware that most transformers that have a center-tapped secondary are rated for regulator full-wave rectification, not full-wave *bridge* rectification. The current available when the bridge circuit is used is one-half that rated value. This is true because the voltage is twice, and you still do not want to exceed the volts-x-amperes rating of the transformer. Some transformers will "take it" when a higher current is drawn, but it is not good practice to make the thing work so hard. The current rating must be at least the current rating of the regulator, and preferably more. There is a general design rule that requires us to use only about 75 percent of capacity on the average.

There are two ratings on the rectifier that need attention: the forward current and the peak inverse voltage, "PIV." The forward current is simply the amount of current that the bridge rectifier will normally pass in the forward direction without suffering failure. In most cases the rating

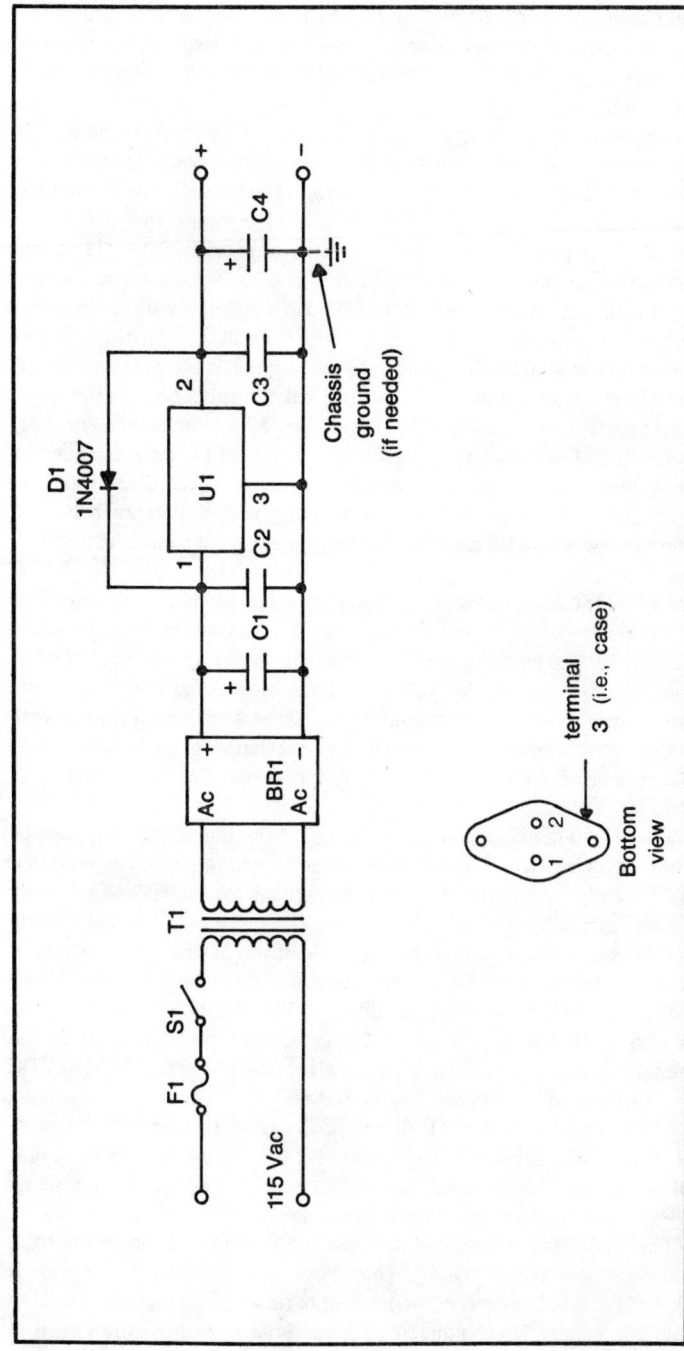

Fig. 4-14. Three-terminal regulator power supply.

97

should be equal to or greater than the regulator's forward-current rating. Again, having some excess capacity, so the rectifier never operates for long at its maximum rating, is good design practice and will make the circuit more reliable.

The peak-inverse-voltage rating, PIV, is the maximum reverse-bias voltage that the rectifier will withstand without breaking down. If this voltage is exceeded, the rectifier will be destroyed. The normal rule of thumb is to use a minimum PIV rating that is 2.83 times the applied rms. The reason for this rule is that the normal PIV seen by the rectifier is $1.414 \times$ rms plus the voltage on the capacitor (C1 in Fig. 4-14), which is also $1.414 \times$ rms; thus the total voltage is $2 \times 1.414 \times$ rms, which is $2.83 \times$ rms. This rule does not mean much when dealing with 6.3-volt transformers because the minimum available PIV rating is 25 volts, which is above the maximum reverse voltage generated in the circuit, but the rule becomes more and more important as the voltage increases. I recall a famous amateur radio transceiver that had a habit of popping the bridge-rectifier diodes frequently. The problem was that the 2.83 rule was violated. When three 1000-volt PIV diodes were connected in series (along with 470-kohm balancing resistors in parallel with each diode), the problem went away; the 2.83 rule is *not trivial*.

The filter capacitor C1 is selected to provide enough ripple reduction to make the regulator happy but does not have to provide all of the ripple reduction needed by the external circuitry powered by the regulator (the regulator adds considerable ripple reduction). Most authorities recommend a capacitance for C1 equal to 1000 μF per ampere of current drawn, with some recommending 2000 μF per ampere. Another rule requires us to keep a capacitance of at least 500 μF in the circuit, even when the forward current is less than 500 mA.

The working voltage rating, (WVDC), of the filter capacitor must be somewhat higher than the maximum expected voltage. Keep in mind that most electrolytic capacitors have a 20 percent tolerance, and normally voltages vary ± 15 percent. Thus, you will require a 35 percent margin for error on the WVDC rating. For example, if you have a 12-volt regulator that inputs 18 volts, the filter capacitor will normally see 18 volts. Using the 35-percent rule, you would specify 18 volts $\times 1.35$, or 24.3 (or more). Since 25 WVDC is a "standard" rating, you can use that value as the minimum value for the WVDC rating of C1. I prefer to use 35 or 50 WVDC if practical; again, safety is a consideration.

The capacitors marked "C2" and "C3" in Fig. 4-14 are used for noise reduction. These capacitors have a value of 0.1 μF to 1 μF. They are normally mounted as close as possible to the regulator device; in fact, many designers mount them on the regulator itself.

The output capacitor C4 is optional and is used to improve the transient response of the regulator. When external current demand increases very rapidly, it will take a certain amount of time for the regulator to catch up (microseconds). During this time, the external circuit will draw current from the capacitor, thereby preventing a "glitch" in the power supply voltage. The value of C4 is 100 μF per ampere of current drawn, with a WVDC rating of not less than 1.35 times the rated output voltage of the regulator.

If capacitor C4 is used it is also advisable to use diode D1. The purpose of this diode is to dump the charge in C4 when the circuit is turned off; otherwise, the charge can be dumped back into the circuit through the regulator, which can cause damage. Any diode in the 1N4002-through-1N4007 series will suffice.

Adjustable-Voltage Types

Adjustable-voltage regulators used to be somewhat harder to design than fixed-voltage types, but today there are several three- and four-terminal devices on the market that will serve nicely. I will limit my discussion to the LM-317 and LM-338 devices since those two are readily available in the hobby and mail-order markets. The LM-317 and LM-338 devices are similar to each other in function, except that the LM-317 handles 1.5 amperes, while the LM-338 device is rated at 5 amperes. Information given below for the LM-338 is generally usable for the LM-317 also.

Figure 4-15(A) shows a circuit based on the LM-338K device, while Fig. 4-15(B) shows the bottom view of the LM-338K package. Note in Fig. 4-15(B) that the output terminal is the case, which is exactly the opposite from fixed-voltage, three-terminal devices. It also means that the builder must insulate the case and/or the heatsink it rests on from the chassis, especially if the chassis is used as common ground.

The transfomrer, rectifier, and filter capacitor are selected from the same criteria as mentioned above, so will not be discussed again here.

The LM-338K can accept an input voltage up to 35 volts and will produce a maximum output voltage of several volts less than that figure.

The exact output voltage is set by the ratio of two resistors, R1 and R2. In Fig. 4-15(A), resistor R2 is a variable resistor made from a potentiometer with the wiper and one end terminal shorted together. The output voltage will be approximately

$$V_o = 1.25v \left(\frac{R2}{R1} + 1 \right) + (R2 \times I_{adj}) \qquad (4.13)$$

values given in Fig. 4-15(A), the output voltage can be varied from 1.25 volts to over 35 volts (if the input would allow it). In some cases, where the voltage is set and then forgotten, the potentiometer R2 will be a trimmer pot and will be mounted on the power-supply printed-circuit board. In other cases, it will be a front-panel control so that the operator can adjust it.

Diodes D1 and D2 serve the same protection function in this circuit as in the previous circuits. Once again, any diode in the series 1N4002 through 1N4007 will suffice.

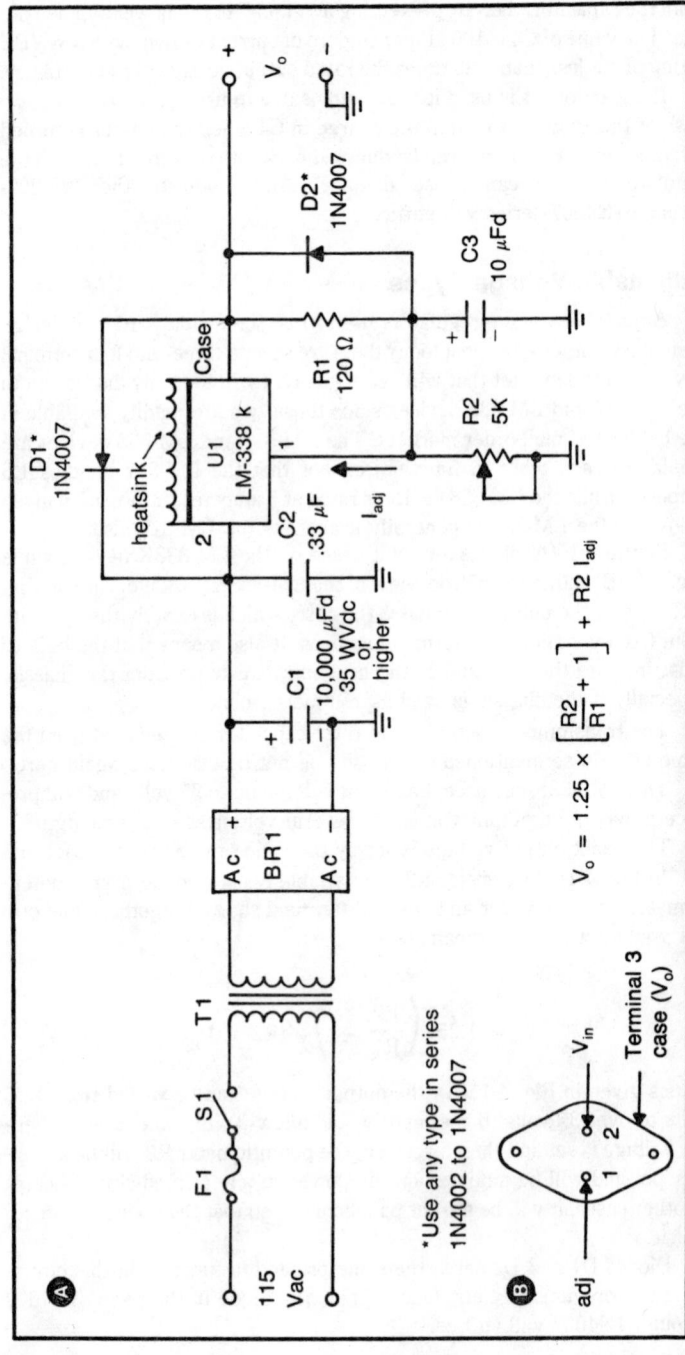

$$V_o = 1.25 \times \left[\frac{R2}{R1} + 1 \right] + R2 \, I_{adj}$$

*Use any type in series
1N4002 to 1N4007

Fig. 4-15. (A) Circuit for adjustable 5-ampere power supply. (B) LM-338 pin-outs.

If you want to make the LM-338K or the LM-317 into a fixed-voltage regulator, adopt one of the two circuit modifications shown in Fig. 4-16. In the one case, two fixed resistors are used for R1 and R2. Normally, R1 will be 120 ohms, and R2 will be selected to set the output voltage at the required level. In the other case, R2 is broken into two components, R2A and R2B. The value of R2B is roughly 10 to 20 percent of R2A and is set to trim the output to a precise value. This arrangement has the advantage of fixed operation while allowing trimming of the output voltage to the exact value required.

The next circuit is shown in Fig. 4-17. Here there are two LM-338K devices connected together to form a 1.2-to-16-volt dc, 10-ampere, regulated power supply. Because the cases of the LM-338K devices are connected together, you can use the same heatsink for both. Potentiometer R2 sets the output voltage and is adjustable to just over 16 volts.

The input voltage for this circuit can be acquired by rectifying the output of a 12.6-volt-rms transformer and filtering it with 15,000 μF, or more, of capacitance.

USING ZENER DIODES

The output voltage of ordinary rectifier/filter dc power supplies is not stable but varies considerably over time. There are two main sources of variation in the output of this type of power supply. First, there is always a certain amount of variation of the ac input voltage. Ordinary commercial power lines normally vary from 105 to 120 volts rms, and may drop to less than 100 volts during the infamous, summer "brownouts."

The second source of variation is created by load variation (see Fig. 4-18). The problem is caused by the fact that real dc power supplies are

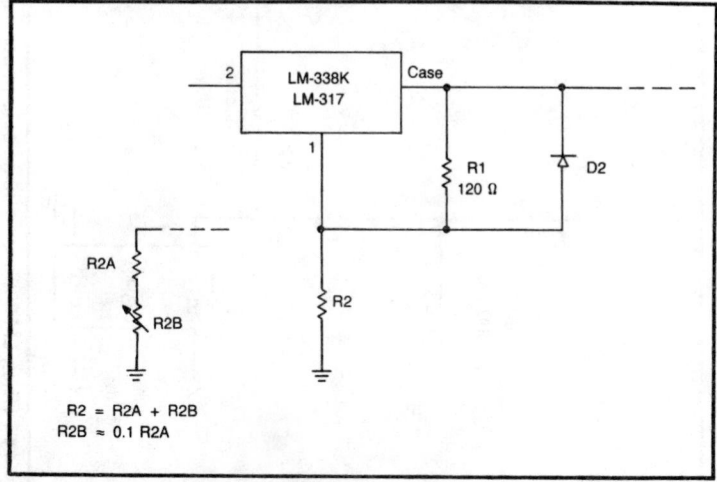

Fig. 4-16. Alternate circuit for LM-338.

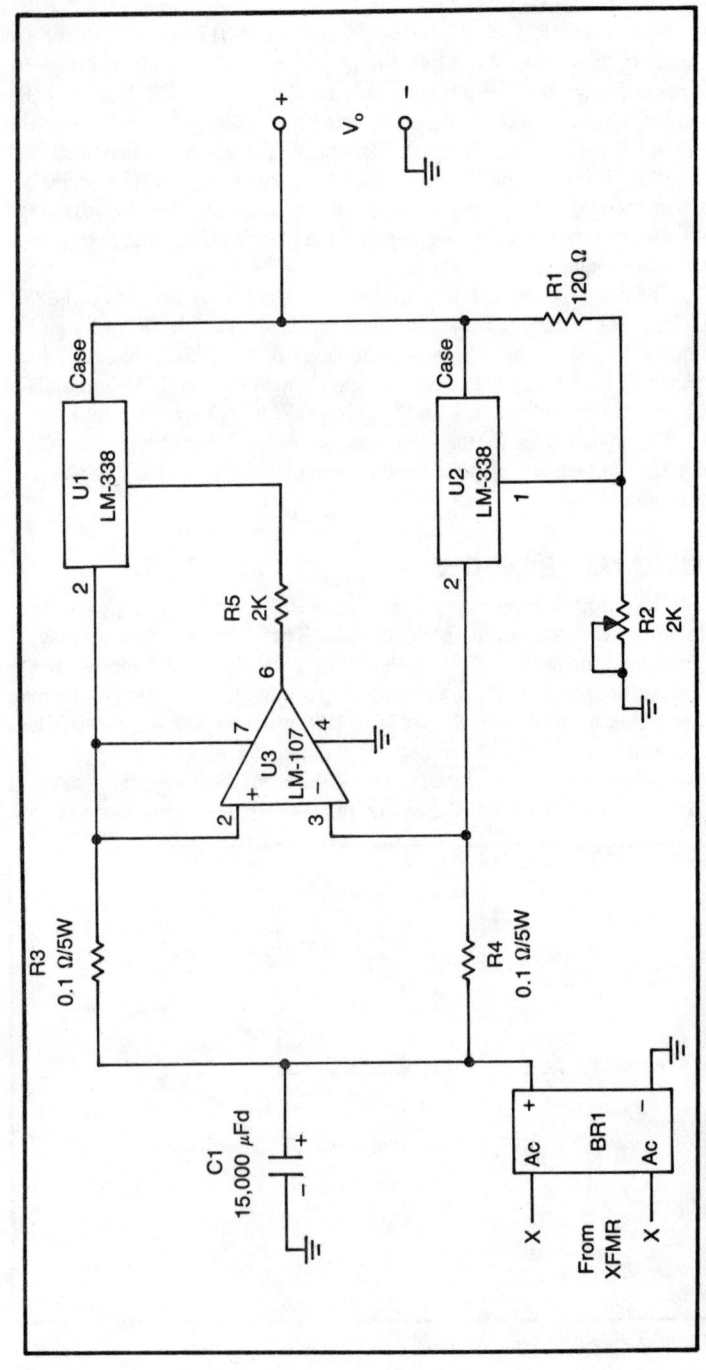

Fig. 4-17. Using LM-338's for a 10-ampere power supply.

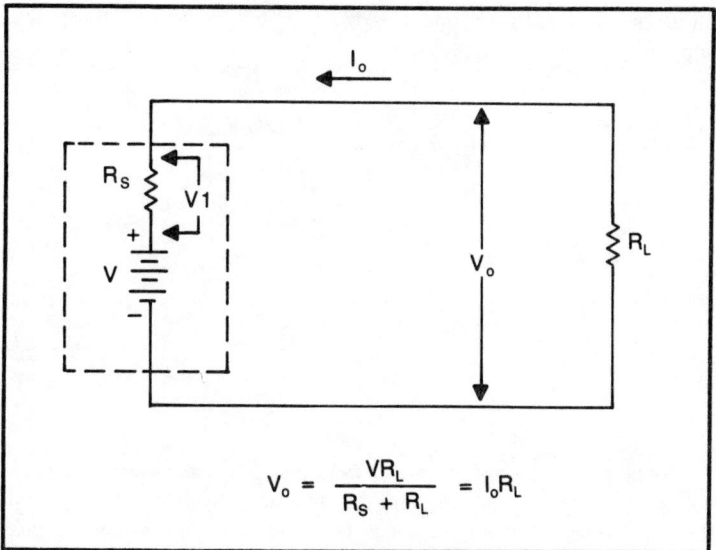

$$V_o = \frac{VR_L}{R_S + R_L} = I_o R_L$$

Fig. 4-18. Equivalent circuit for a dc power supply.

not ideal. The ideal textbook power supply has zero internal resistance, while real power supplies have a certain amount of internal resistance (represented by R_s in Fig. 4-18). When current is drawn from the power supply there is a voltage drop (V1) across the internal resistance, and this voltage is subtracted from the available voltage (V).

In an ideal power supply, output voltage V_o is the same as V, but in real supplies V_o is equal to (V − V1). Since V1 varies with changes in the load current I_o, the output voltage also will vary with changes in current demand.

The "goodness" or "badness" of a power supply can be defined in terms of its percentage of regulation. This specification is a measure of how badly the voltage changes with changes in load current and is found from

$$\% \text{ REG} = \frac{(V_o)}{V} (100) \qquad (4.14)$$

where
V is the open-terminal (no current) output voltage
V_o is the output voltage under full-load current (I_o = max)
% REG is the percentage of regulation.

Example

What is the percentage of regulation if a dc power-supply output voltage drops from 15 to 13 volts as the output current is raised from 0 to 2

103

amperes, which is the maximum allowable output current for this power supply?

$$\%REG = \frac{(V - V_o)\,(100)}{V}$$

$$= \frac{(15 - 13)\,(100)}{15}$$

$$= \frac{(2)\,(100)}{15}$$

$$= 13.3$$

Many electronic circuits do not work properly under varying supply-voltage conditions. Oscillators, for example, tend to change frequency if the dc power-supply voltage changes. Obviously, some means must be provided to stabilize the dc voltage. The zener diode is perhaps the simplest such regulator device.

Zener Diodes

The zener ("zee-ner") diode is a special case of the old-fashioned, PN junction diode; Fig. 4-19 shows both the circuit symbol and the I-vs-V curve for a zener diode.

In the forward-bias region operation is the same as for other PN junction diodes. For this case, the anode is positive with respect to the cathode; so a forward-bias current ($+I$) flows. For voltages greater than V_g, which is approximately 0.7 volts, the current flow increases approximately linearly with increasing voltage. At potentials less than V_g, the current increases from a small reverse current (I_L) at $V = 0$, to a small forward current at V_g. The diode is, like all other PN junction diodes, nonlinear in this low-voltage region.

The zener diode also acts like any other PN junction diode in the reverse-bias region between $V = 0$ volts and the zener potential V_Z. In this region only the small reverse-leakage current flows.

At an applied potential of V_Z or less the zener diode breaks down and allows a large reverse current to flow. Note in Fig. 4-19 that further increases in negative potential do not cause an increased voltage drop across the diode. Thus, the zener diode regulates the voltage to its zener potential. Another way of saying the same thing is that the zener diode clamps the applied voltage to the value of V_Z.

ZENER-DIODE-REGULATOR CIRCUIT

A zener diode is essentially a parallel regulator because it is connected

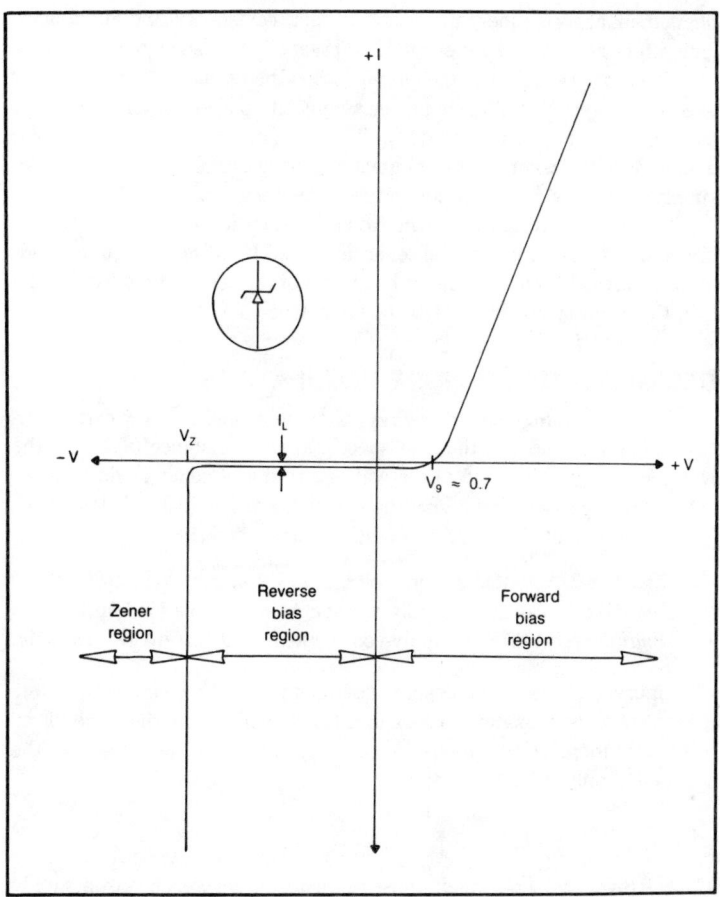

Fig. 4-19. I-vs-V curve for zener diode.

in parallel with the load. It regulates by clamping the output voltage across the load to the zener potential. Figure 4-20 shows a typical zener-diode regulator circuit that takes advantage of these attributes.

In Fig. 4-20 resistor R_L represents the load placed across the power supply; i.e., the circuits that draw current from the supply. The value of the load resistor is merely V_Z/I_o. Resistor R1 is used as a series current limiter to protect the diode. Refer back to Fig. 4-19, and you will see why it is needed. Note that $-I$ increases sharply when $-V$ reaches the zener potential. If R1 were not used this current would destroy the diode. One of the tasks in designing a zener-diode regulator circuit is picking the resistance value and wattage rating of resistor R1.

Capacitor C1 is optional and is used to suppress the hash noise generated by the zener diode. The zener process is essentially an avalanche phenomenon, so is inherently noisy. In fact, certain RF and audio noise generators often use a zener diode to create the noise signal.

The other capacitor in the circuit, C2, is the regulator-filter capacitor used in any rectifier/filter dc power supply. Its purpose is to smooth the pulsating dc into nearly-pure dc (well . . . sort of!). It doesn't really serve a function in the zener-regulator circuit except as a reminder that the power supply should be filtered prior to the regulator circuit.

The main current I3 drawn from the rectifier is broken into two branches: I1 flows through the zener diode and I2 flows through the load. In most cases I1 is approximately 10 percent of I2. According to Kirchhoff's law, the relationship I3 = I1 + I2 holds true.

DESIGNING ZENER-DIODE REGULATORS

When designing zener-diode regulators you need to know certain circuit conditions and use them to specify a) the resistance of R1, b) the wattage rating of R1, and c) the wattage rating of zener diode D1.

There are three circuit conditions which are designated I, II and III. The properties of these three conditions are as follows:

Condition I. Variable supply voltage with constant load current;
Condition II. Constant supply voltage with variable load current;
Condition III. Variable supply voltage with load current also variable.

Figure 4-21 shows the design equations for all three conditions. Note that the power-dissipation expressions for R1 and D1 are the same for all three conditions. Of course, V_{in} and V_{in} (max) are the same for the constant-supply voltage instance.

Example

A 6.8-volt-dc zener diode must regulate in the presence of a supply voltage that varies from +9 to +12 volts. The current remains constant

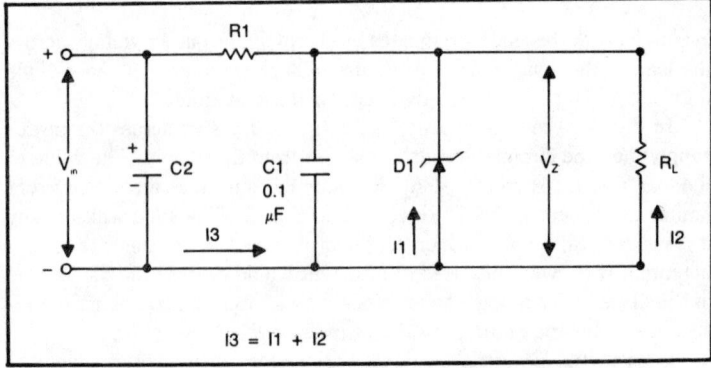

Fig. 4-20. Zener diode regulator.

106

Condition I.   Variable V_{in}, Constant I_o.

$$R1 = \frac{V_{in \, (min)} - V_Z}{1.1 \, I2}$$

Condition II.   Constant V_{in}, Variable I_o.

$$R1 = \frac{V_{in \, (max)} - V_Z}{1.1 \, I2_{(max)}}$$

Condition III.   Both V_{in} & I_o Variable.

$$R1 = \frac{V_{in \, (min)} - V_Z}{1.1 \, I2_{(max)}}$$

For all three conditions

Diode dissipation

$$P_{d1} = \frac{(V_{in \, (max)} - V_Z)^2}{R1} - (I2 \times V_Z)$$

Resistor dissipation

$$P_{R1} = P_{d1} + (I2 \times V_Z)$$

Fig. 4-21. Zener diode design equations.

at 120 milliamperes (0.12 amperes). In this case, V_{in} (min) is 9 volts dc and I2 is 0.120 amperes.

$$R1 = \frac{V_{in} \, (min) - V_Z}{1.1 \, (I2)}$$

$$= \frac{(9 - 6.8)}{1.1 \, (0.12)}$$

$$= 2.2/0.132 = 16.7 \text{ ohms}$$

Because 16.7 ohms is not a standard value, you will either series-parallel two or more resistors to make 16.7 ohms or use the nearest standard value (perhaps 15 or 18 ohms). The power dissipated by D1 is given by

$$P_{d1} = \frac{[V_{in}\,(max) - V_Z]^2}{R1} - [I2 \times V_Z]$$

$$= \frac{[12 - 6.8]^2}{16.7} - (0.12)(6.8)$$

$$= \frac{[5.2]^2}{16.7} - (0.816)$$

$$= \frac{27.04}{16.7} - (0.816)$$

$$= 1.62 - 0.816 = 0.803 \text{ watts}$$

A power dissipation of 0.803 watts means that a 1-watt or greater rating is required for the zener diode D1. For the sake of improved reliability, I prefer at least a 2:1 ratio between the rated and dissipated wattage, so would recommend a 1.6-watt (which means 2-watt standard value) or more. For most intermittent, hobby applications, however, a 1-watt zener diode should suffice.

The power rating of R1 is found from calculating the actual power dissipation of R1 and then picking a higher standard wattage rating. The power dissipation of R1 is

$$P_{R1} = P_{d1} + (I2 \times V_Z)$$

$$= (0.803 \text{ watts}) + (0.012)(6.8)$$

$$= 0.803 + 0.816 \text{ watts}$$

$$= 1.62 \text{ watts}$$

Since P_{R1} is 1.62 watts, use at least 2 watts for R1; a 3-watt or 5-watt resistor would yield even greater reliability.

The values for this example are 16.7 ohms @ 2 watts (or more) for R1 and 6.8 volts @ 1 watt (or more) for D1.

Solving the other conditions is similar, so will not be repeated here. Select the correct value for the supply voltage and current (with due regard for maximum and minimums as called for in the equations of Fig. 4-21) and plug them into your calculator. Alternatively for those who own a personal computer, use the BASIC program shown in Fig. 4-22.

```
100 REM   The name of this program is ZENER
110 REM   This program allows you to design simple Zener diode
120 REM   voltage regulator circuits
130 GOSUB 770
140 PRINT "ZENER is a design aid for zener diode voltage"
150 PRINT "regulator circuits."
160 PRINT
170 PRINT "There are three conditions:"
180 PRINT
190 PRINT
200 PRINT TAB(5);"1.  Variable Supply Voltage, Constant Load Current"
210 PRINT
220 PRINT TAB(5);"2.  Constant Supply Voltage, Variable Load Current"
230 PRINT
240 PRINT TAB(5);"3.  Variable Supply Voltage, Variable Load Current"
250 PRINT
260 PRINT "SELECT one (1) from menu above:";
270 INPUT N
280 GOSUB 770
290 PRINT "ENTER REGULATED OUTPUT VOLTAGE:";
300 INPUT VZ
310 GOSUB 770
320 IF N = 1, THEN GOSUB 350
330 IF N = 2, THEN GOSUB 490
340 IF N = 3, THEN GOSUB 600
350 PRINT "ENTER MINIMUM VALUE OF INPUT VOLTAGE:";
360 INPUT VMIN
370 PRINT
380 PRINT "ENTER MAXIMUM VALUE OF INPUT VOLTAGE:";
390 INPUT VMAX
400 PRINT
410 PRINT "ENTER LOAD CURRENT:";
420 INPUT I
430 R = (VMIN - VZ)/(1.1*I)
440 PD = ((VMAX - VZ)^2)/R
450 PRMIN = PD
460 PD = PD - (I*VZ)
470 GOSUB 840
480 GOTO 950
490 PRINT "ENTER VALUE OF CONSTANT INPUT VOLTAGE:";
500 INPUT V
510 PRINT
520 PRINT "ENTER MAXIMUM VALUE OF LOAD CURRENT:";
530 INPUT I
540 R = (V - VZ)/(1.1*I)
550 PD = ((V - VZ)^2)/R
560 PRMIN = PD
570 PRMIN = PD + (I*VZ)
580 GOSUB 840
590 GOTO 950
600 PRINT "ENTER minimum value of input voltage:";
610 INPUT VMIN
620 PRINT
630 PRINT "ENTER maximum value of input voltage:";
640 INPUT VMAX
650 PRINT
660 PRINT "ENTER maximum value of load current:";
670 INPUT IMAX
680 PRINT
690 PRINT
700 R = (VMIN - VZ)/(1.1*IMAX)
710 PD = (,VMAX - VZ)^2)/R
720 PRMIN = PD
730 PD = PD - (IMAX*VZ)
740 GOSUB 840
750 GOTO 950
760 GOTO 1050
770 FOR I = 1 TO 30
780 PRINT
790 NEXT I
800 RETURN
810 PRINT "Press ANY key to continue:"
820 A$=INKEY$: IF A$="" THEN 820
830 RETURN
840 PRINT
850 R = R*100
```

Fig. 4-22. Zener design program in BASIC.

```
860 R = INT(R)
870 R = R/100
880 PRINT
890 PRINT "Resistor R1:";R;" Ohms"
900 PRINT
910 PRINT "Resistor dissapation:";PRMIN;" Watts"
920 PRINT
930 PRINT "Zener Diode Dissapation:";PD;" WATTS"
940 RETURN
950 PRINT "FINISHED?"
960 PRINT TAB(3);"1. Yes"
970 PRINT TAB(3);"2. No"
980 PRINT
990 INPUT H
1000 IF H > 2, THEN GOTO 950
1010 IF H = 2, THEN GOTO 100
1020 PRINT
1030 PRINT
1040 PRINT "PROGRAM ENDED"
1050 END
```

Fig. 4-22. (Continued.)

Increasing Output Current

The output current that can be supplied by a zener-diode regulator is somewhat limited. In cases where a large output current is needed, it is possible to amplify the effect of the zener diode by using it to control the base of a series-pass transistor (Q1 in Fig. 4-23). The output voltage produced by this circuit is approximately 0.6 to 0.7 volts less than the zener potential. This reduction is accounted for by the base-emitter potential V_{be} of the transistor. The current rating is the collector-current rating of the transistor with due regard for the collector dissipation. The collector will dissipate a power of $(V_{in} - V_o) \times I_o$. If there is a large difference between V_{in} and V_o, it is possible to exceed the maximum collector-dissipation rating of Q1 if the maximum collector current is drawn. You must limit the output current to a value less than that required to exceed the collector dissipation.

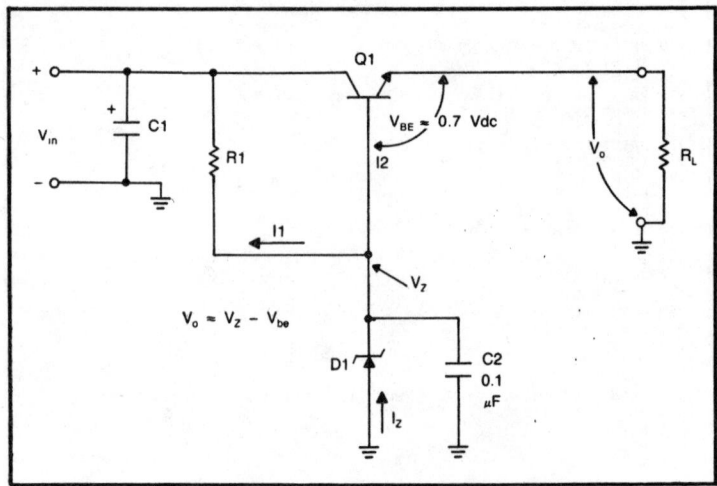

Fig. 4-23. Series-pass transistor increases current capacity.

110

In any event the series-pass transistor will survive much longer if it is properly heatsinked and cooled. When I built my first microcomputer, a Digital Group, Inc. Z80 system, I constructed a 15-ampere, +5-volt dc power supply using a HEP S-7000 as the regulator. The transistor's TO-3 case ran too hot to touch without losing skin, even though a Motorola HEP heatsink was used. Adding a 40-c.f.m. "muffin fan" cured the heat problem and extended the life of that power supply (and hence, the life of my computer also).

For some reason that escapes me, the design of zener-diode regulators is regarded as a deep, dark secret by too many electronics hobbyists who would otherwise undertake such an effort. There is little reason why this should be true; I hope this discussion burns off a little of the fog.

BUILD AN ELECTROPLATER/ELECTROFORMER

(A small voltage-regulator dc power supply that can also be used on your electronics workbench.)

I once took up jewelry-making as a hobby. After learning the arts of bending, sawing, and soldering silver (no mean trick, by the way!), I looked into electroplating and electroforming. When I found that the electrical-current source that electroplaters call the "rectifier" (only partially correct, by the way) is nothing more than an ordinary, variable dc power supply, I decided to build my own. With all the parts in my junk box it was easy to beat the $150 price of a 0-to-5-ampere, electroplating "rectifier."

Most readers are familiar with elementary electroplating in which a silver or gold anode supplies metal to be electro-deposited on the surface of another metal: copper, brass, etc. Using electroplating you can make a low cost, copper or brass piece look like sterling silver or gold. There are also electronic applications of electroplating. For example, silver is used to make high-"Q" inductors and capacitors for UHF applications, and gold plating makes a contact or surface corrosion-resistant.

Electroforming is a little different but still involves electro-deposition of metal onto a surface. In electroforming a non-conductive model is made of the form desired (electronic part, jewelry piece, etc.) and coated with a conductive paint. The model is then treated almost as if being electroplated so that metal is deposited onto the surface. When the job is finished the model is destroyed, and its form is retained by the metal overlay.

Several publishers, among them Crown Books, have texts on electroplating and electroforming. Since some of the chemicals used are either acids or extremely poisonous non-acids, it is ESSENTIAL that you read a text first before attempting to plate anything. When I asked a professional plater what the caution note on the poisonous electroplating-solution bottle meant by "small quantities dangerous . . .," he said, "The first thing I tell a new apprentice is 'Don't lick your fingers, bite your nails or pick your nose when working here.' "

The current requirements are different for electroplating and electroforming. Because of this difference, two devices were built, a Main Controller and a Fine Controller. The Main Controller is a 0-to-5-ampere unit used for electroplating, while the Fine Controller is a 0-to-1-ampere unit used for electroforming. They could have been built into a single cabinet, but I wanted to donate the units to a recreation department community center where the Fine Controller might be used with either the Main Controller or another dc power supply.

ELECTROPLATER/MAIN CONTROLLER

The electroplater set consists of two separate units: Main Controller and Fine Controller. The Main Controller is a variable source capable of delivering up to 14 volts dc at currents up to 5 amperes. This controller is voltage-regulated and filtered to produce nearly-pure direct current. The nature of the controller is such that it does not do well in both the high- and low-current ranges, hence the need for the fine controller. The Fine Controller is a 0-to-1-ampere controller.

Main Controller

The Main Controller is a filtered and voltage-regulated, 5-ampere dc power supply. The front-panel layout is shown in Fig. 4-24. The functions of the front-panel elements are as described below:

1. **On-Off Switch.** This switch turns the unit on and off.
2. **"On" Lamp.** The "ON" lamp is a red, light-emitting diode and indicates when the unit is turned on. Two conditions must be satisfied: 1) the unit must be plugged in, and 2) the ON-OFF switch must be in the UP position. Don't laugh at this simplistic description—I once ran a service call 25 miles away for a TV set that was not plugged into the wall outlet.
3. **Power Controller.** This control is a vernier knob that sets the output level.
4. **Dc Ammeter.** This current meter indicates the level of output current in amperes. It is a 0-to-5-ampere model. For lower current levels (0 to 1 amperes) used in small jewelry electroplating or electroforming, use the add-on Fine Controller.
5. **Outputs.** The output connectors are five-way, binding posts. The BLACK is the NEGATIVE (–) output, and the RED is the positive (+) output.
6. **Fuse.** The fuse is not shown in Fig. 4-24 but is located on the back panel next to the ac power cord. Use a 1- or 2-ampere fuse for this application.

The main controller is a 1.2-to-14-volt, regulated dc power supply. The circuit is shown in Fig. 4-25. Transformer T1 steps down the 115-volt ac

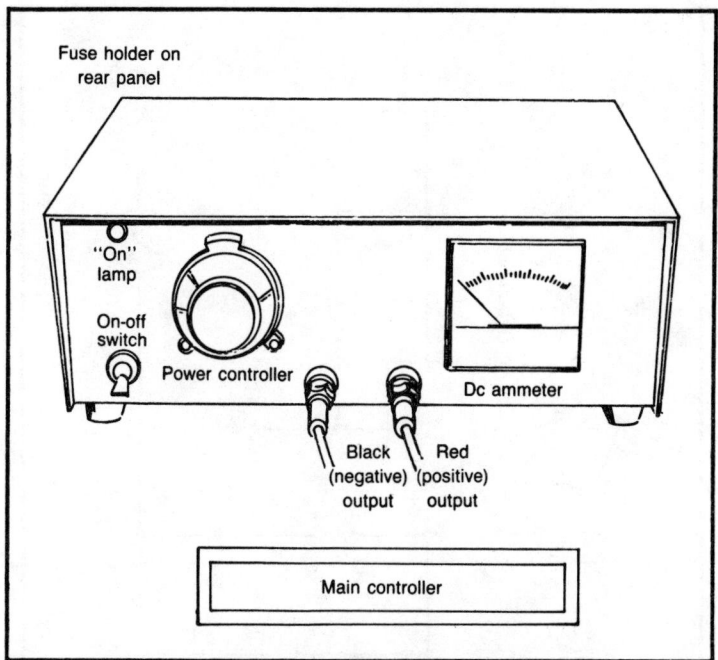

Fig. 4-24. Electroplater power supply (0-13 volts, 5-amperes).

line voltage to 12.6 volts rms for use by the supply. The 12.6 volts ac is rectified by full-wave bridge rectifier BR1 and filtered by C1. The value of C1 was selected according to the $1000\text{-}\mu\text{F}$-per-ampere rule.

The ON indicator is shown in Fig. 4-25 as D1. The 1.2-kohm resistor in series with D1 limits the current to the required 15 milliamperes.

The output meter is a 0-to-5-ampere, taut-band model and is placed in the negative leg of the output line. As such, it will measure only the current in the electroplating solution or other external load connected between the RED and BLACK outputs.

The regulator is a National Semiconductor LM-338K or equivalent. This three-terminal, adjustable, IC regulator will provide outputs up to 5 amperes, at voltages up to 35 volts; although only 14 volts are possible with the transformer chosen here for T1.

Adjustment of the regulator output is provided by the combination of R1 and R2; potentiometer R1 is ganged to the Power-Controller, vernier knob on the front panel. The output voltage (hence current) is determined by

$$V = 1.25 \frac{R1}{R2} + 1 \tag{4.15}$$

The regulator is a voltage regulator; yet the output required is a cur-

113

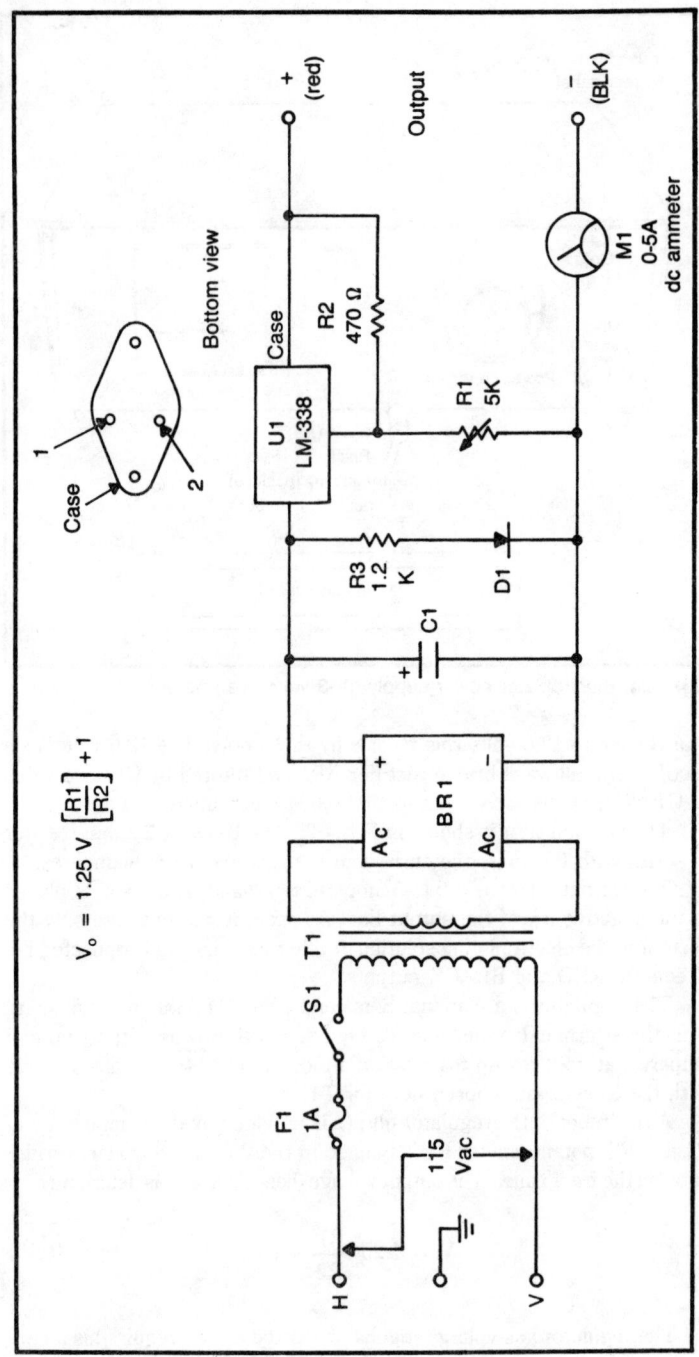

Fig. 4-25. Electroplater power supply schematic.

114

rent (0 to 5 amperes). You can determine the output current from the known voltage by judicious applications of Ohm's law

$$I = V/R$$

where

V = is the output voltage from the equation above
R = is the resistance of the plating solution or other load.

Construction

The construction of this project is simple and straightforward. The cabinet is a standard, black & white, LMB chassis cabinet costing less than $40. Figure 4-26 shows the mounting of the transformer, rectifier, filter capacitor, and regulator/heatsink assembly. The latter must be insulated from chassis ground because the case of the TO-3 regulator package is "hot." I used a small piece of *Vector* "perf-board" left over from one of a hundred earlier projects (never throw away perf-board!) as the insulator. The heatsink was bolted onto the perf-board, and the perf-board in turn was mounted to the chassis on metal, stand-off posts. Alternatively, you can use nylon or ceramic, stand-off posts to insulate the heatsink from the chassis.

The main-adjustment potentiometer (R1 in Fig. 4-25) is mounted on a small, homemade L-bracket and then connected to the vernier knob.

The Main Controller is designed to work best in the 1-to-5-ampere range. Operating instructions are given below.

Operation

Read instructions completely before beginning operation!

1. Set ON-OFF switch to "off" (down position);
2. Set POWER CONTROLLER to the zero ("0") position on the vernier dial;
3. Connect the black, electrode wire to the black output and the red, output wire to the red output;
4. Connect the black and the red alligator clips (on the ends of the electrode wires) to the "anode" and the "work" per instructions given in an electroplating book (note: a copy of *Electroplating and Electroforming* by L.S. and J.H. Newman is supplied with this equipment). Remember: BLACK = (−) and RED = (+);
5. Turn Main Controller "on" by placing ON-OFF switch in up position;
6. Adjust Power-Controller, vernier dial for the correct current (Note: YOU are responsible for determining the correct current level);
7. When you are finished electroplating, turn off the ON-OFF switch and *then* disconnect the electrode, alligator clips. *This sequence is very important to the continued health of the main controller!*.

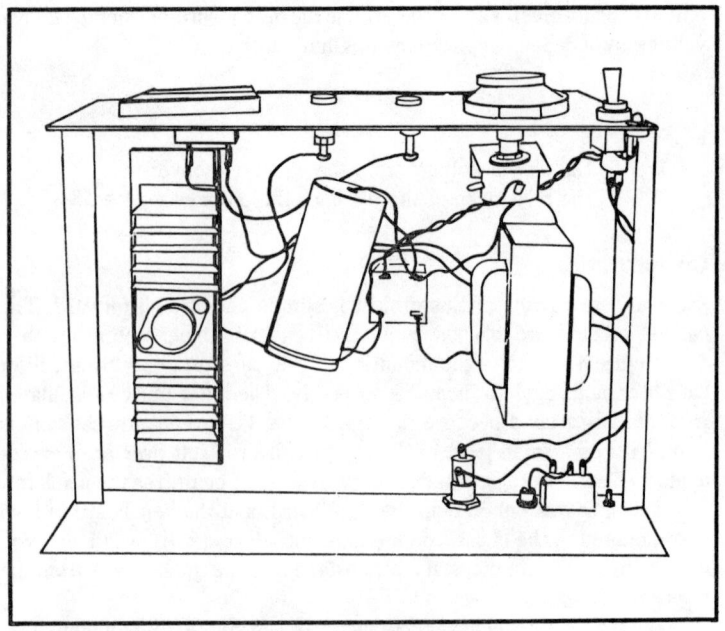

Fig. 4-26. Internal parts layout for electroplater supply.

Caution!!!!! Read This Notice!!!!

Electrical-current meters are very delicate instruments. If you short the electrode leads together it is possible to damage the EXPENSIVE meter movement. This is a problem on all electroplating equipment. Be careful to prevent the leads from coming together accidentally! This caution should be followed when using any metered power supply; failure to observe this simple rule is what sends most dc power supplies to the repair bench.

Parts List for Main Controller

- F1 1-ampere Littlefuse
- S1 SPST ac toggle switch
- T1 12.6-volt, 5-ampere (or higher) transformer
- BR1 10-ampere (or higher), 50-volt (or higher) PIV full-wave bridge rectifier assembly (boltdown mounting type)
- C1 5000-μF, 50-WVDC electrolytic tubular
- R1 5000-ohms linear-taper potentiometer with 1/4-inch, round shaft
- R2 470-ohm, 1/2 watt carbon or film resistor
- R3 1200-ohm, 1/2-watt carbon or film resistor
- U1 LM-338K

- M1 0-to-5-ampere dc ammeter
- D1 15-mA red, light-emitting diode

Miscellaneous: perf-board to insulate U1, RED and BLACK five-way binding posts, 6-32 and 4-40 machine screws, #18 or #20 hookup wire, 11 × 4 × 8-inch LMB cabinet, vernier knob for 1/4-inch round shaft (prefer 270-degree model instead of 180-degree model used).

Fine Controller

The Fine Controller (Fig. 4-27) is an adapter for the main controller. The circuit is shown in Fig. 4-28. The main controlling element is a National Semiconductor LM-337, negative, three-terminal, adjustable regulator connected in the current-source configuration. The output current is set either by varying the input voltage (with R1 set to minimum resistance) or by keeping the input voltage fixed and adjusting R1.

The meter is a 0-to-1 milliammeter with a current-shunt resistor to in-

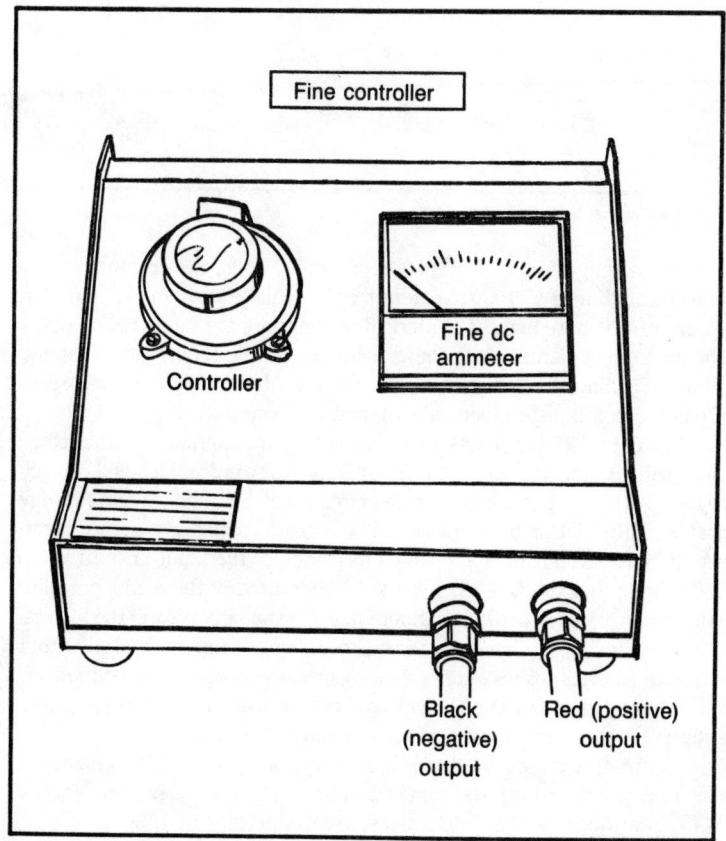

Fig. 4-27. Fine controller adapter.

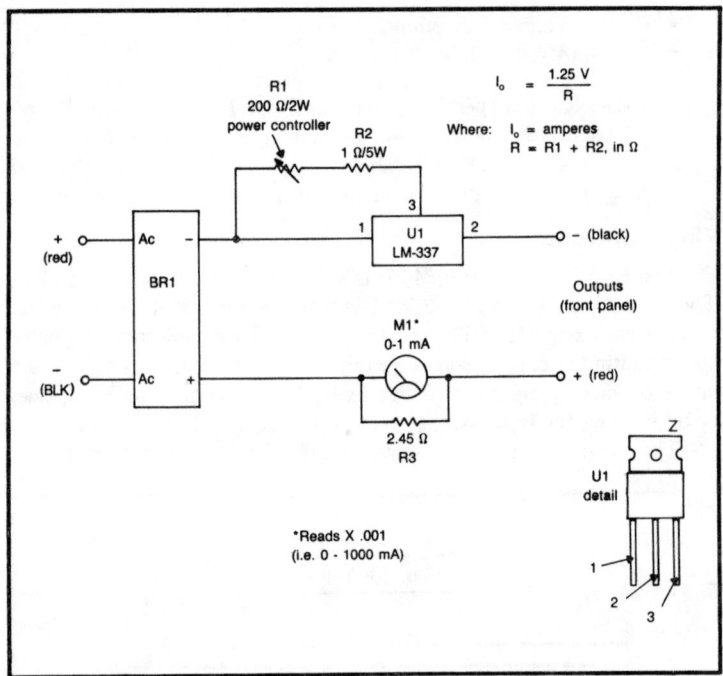

Fig. 4-28. Fine controller schematic.

crease the range to 0 to 1 ampere (1000 mA). Thus, the current reading must be multiplied by 1000 to determine the actual current in milliamperes. Alternatively, note that the numerical readings are the same, but the units change from milliamperes to amperes because of the 1000:1 current shunt. Thus, a reading of 0.2 mA means 0.2 amperes. In replacing this meter for repairs, use a 0-to-1-ampere model, and disconnect R3.

Rectifier BR1 is used as a safety feature. It is anticipated that some Bozo will attempt to connect the input lines backwards which will destroy regulator U1. By placing BR1 in the circuit we obtain a switching bridge that is indifferent to input polarity. The regulator will always see the correct polarity even if Bozo the clown misconnects the input lines. It is too bad that neither the LM-337 nor the LM-338 used in the main controller is amenable to a current-knee output limiter to similarly protect the meters.

The Fine Controller is designed to provide a fine control of 0 to 1 amperes for small-project electroplating or electroforming applications. It can be used either with the Main Controller or with a fixed-voltage dc power supply; both modes of operation are described below.

Figure 4-27 shows the Fine-Controller front panel. The elements of the front panel perform the same function as the equivalent ones on the Main Controller; so the descriptions will not be repeated here.

NOTE: The meter on the Fine Controller indicates 0 to 1000

118

milliamperes (0 to 1 ampere) but is marked "0 to 1 milliamperes." Multiply the indicated reading by 1000 to determine the actual current. Example: If the reading is 0.2 mA, the actual current is 0.2 mA × 1000, or 200 milliamperes (0.2 amperes). The following table serves as a rough guide:

Reading (mA)	=	Current (amperes)	=	Current (mA)
0.2		0.2		200
0.4		0.4		400
0.6		0.6		600
0.8		0.8		800
1.0		1.0		1000

Read the instructions for the Main Controller first, including all cautions.

Operation

There are two modes of operation for the Fine Controller. First described is operation in conjunction with the Main Controller. The Fine Controller is not a dc power supply in its own right, but rather, it is an adapter that permits fine control of a larger power supply.

Operation with Main Controller. The outputs of the Main Controller are connected to the corresponding inputs on the **rear panel** of the Fine Controller. The Fine-Controller outputs (located on the front panel) are then connected to the work in the manner described in the book on electroplating (see citation under Main Controller above). Prior to turning on the ON-OFF switch of the Main Controller, perform the following:

1. Connect the black wire from the Main-Controller output to the Fine-Controller, black, input connector (see Fig. 4-29);
2. Connect the red wire from the Main-Controller output to the Fine Controller, red, input connector (see Fig. 4-29);
3. Set the Fine-Controller dial to "0" (note: when operated with the Main Controller, the dial on the Fine Controller is left sitting on "0" and is not changed; current levels are set by the vernier, "Power Controller" dial on the Main Controller);
4. Connect the red and the black output wires from the Fine Controller to the anode and the cathode, respectively, of the plating-solution electrodes (see plating textbook);
5. Turn on the ON-OFF switch on the Main Controller;
6. Adjust the Power-Controller, vernier knob on the Main Controller for the desired current level (note: you are responsible for determining the proper current level);
7. Observe ALL cautions given for the Main Controller above.

Operation with Fixed-Voltage Dc Power Supply. The Fine Controller is designed to work also in conjunction with sources of dc other than the Main Controller described above. Examples of these other dc sources include any dc supply rated at 1 ampere or more with a voltage of 9 to 16 volts dc; e.g., batteries, radio repairmen's "bench supplies," laboratory power supplies, or the sort of "battery eliminator" used by Citizens Band enthusiasts to operate their mobile units in the house. Instructions are given below:

1. Turn power supply off (or disconnect one terminal of battery);
2. Set the output vernier Controller knob to "100;"
3. Connect the red input to the positive (+) output of the power supply (as in Fig. 4-29);
4. Connect the black input to the negative (−) output of the power supply (as in Fig. 4-29);
5. Connect the red and the black outputs per the instructions given in the electroplating textbook (RED = + = anode, BLACK = − = cathode);
6. Turn on the dc power supply and adjust the vernier Controller knob on the Fine Controller for the desired current level (note: you are responsible for determining the correct current level);
7. Follow instructions for turn-off given for Main Controller Operation above.

Parts List for Fine Controller

BR1   1.5-ampere (or higher), 50-volt PIV (or higher), full-wave bridge rectifier assembly

R1   200-ohm, 2-watt, linear-taper potentiometer with 1/4-inch shaft

R2   1-ohm, 5-watt, wirewound resistor

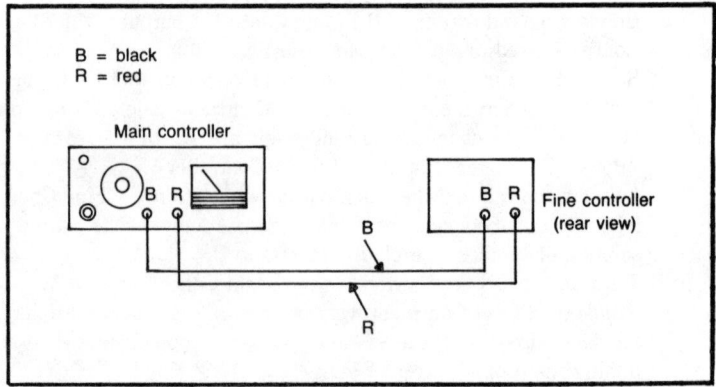

Fig. 4-29. Interconnection between main controller and fine controller.

R3	2.45-ohm meter shunt (four 10-ohm & one 100-ohm: all connected in parallel)
M1	0-to-1-mA, dc milliammeter
U1	L M - 3 3 7 T

Same miscellaneous parts as Main Controller, except the cabinet is smaller.

Calculating a Meter Shunt Resistor

The meter shunt R3 is used to increase the range of the meter movement. In this section I will show you a method for calculating the value of R3 for any given meter. Look at Fig. 4-30. In this case, there is meter M1 with an internal dc resistance of R_m connected in parallel with shunt resistor R_s.

You must know the actual dc resistance of the meter. If the resistance is not specified, either in the papers that come with the meter or on the face of the meter, you must measure the resistance. There are a couple of ways to find this value. If you have a low-current digital voltmeter, measure the resistance using the voltmeter. **DO NOT USE THE "DIODE" POSITION OF THE OHMMETER!!!!** Otherwise you might construct a circuit with a 1.5-volt battery and a 2000-ohm resistor and connect the meter in series with the battery, the resistor, and a milliammeter. Note the current reading on the meter (about 0.75 mA or 0.00075 amperes). Next, take a voltmeter and measure the voltage drop across the meter terminals. From Ohm's law, find the resistance ($R_m = V/I_{reading}$).

We define the currents as follows:

1. I1 is the desired full-scale reading (1 ampere)
2. I2 is the meter full-scale current (0.001 ampere)
3. I3 is the current in the shunt resistor

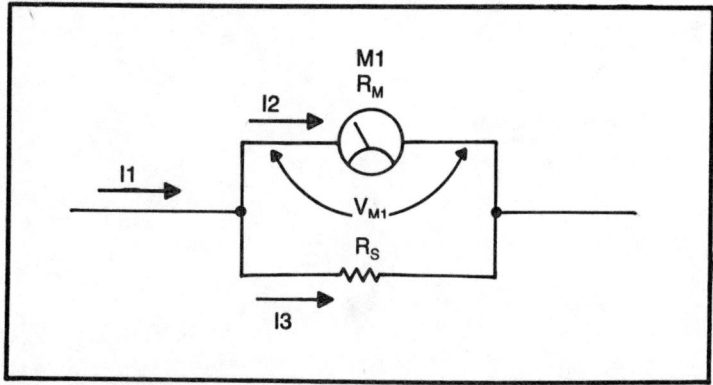

Fig. 4-30. Making a meter shunt.

Using these values, we can find the shunt resistance from

$$R_s = (I2 \times R_m)/I3 \qquad (4.16)$$

Example

In building the electroplater Fine Controller I found, using my digital multimeter, that the internal resistance of the 0-to-1-mA meter was 2.448 kilohms (2,448 ohms). In this case, $I1 = 1$ ampere, $I2 = 0.001$ amperes, and $I3 = 0.999$ amperes. Using the equation

$$
\begin{aligned}
R_s &= (I2 \times R_m)/I3 \\
&= (0.001 \times 2448 \text{ ohms})/(0.999) \\
&= 2.448/0.999 = 2.45 \text{ ohms}
\end{aligned}
$$

Chapter 5

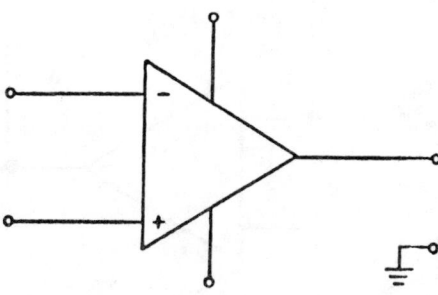

Single-Ended Dc Amplifiers

MANY INSTRUMENTATION CIRCUITS REQUIRE DC AMPLIFIER STAGES that must have good response characteristics at very low frequencies. This chapter will delve into these configurations.

INVERTING FOLLOWER CIRCUITS

This class of operational amplifier circuits is probably the most widely used of all op-amp configurations. An example of the inverting follower is shown in Fig. 5-1. In Chapter 1 (Eq. 1.12) I developed the fact that these circuits obey the relationship

$$E_{out} = -E_{in} \left[\frac{R_f}{R_{in}} \right] \tag{5.1}$$

To set the gain of such a stage we need only choose a pair of resistors, R_f and R_{in} that have a ratio equal to the required voltage gain. For example, let us say that we need a voltage gain of 100. If that is true,

$$(R_f/R_{in}) = 100 \tag{5.2}$$

We need only pick two resistors which satisfy Eq. 5.2. As a general rule, you will want R_{in} to be greater than 500 ohms but both resistors to be as low as practical for reasons that will become apparent. Paradoxically, there are some situations in which it is best to keep R_{in} very high. A standard rule in almost all electronic design is that amplifier input impedances should

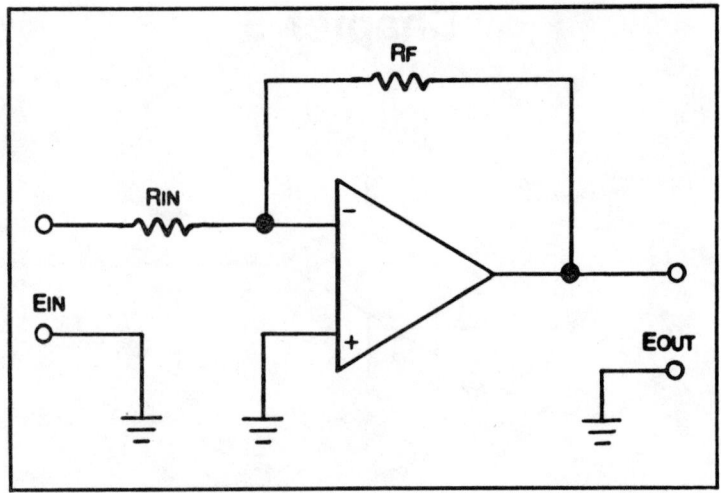

Fig. 5-1. Simple inverting follower.

be at least 10 times as large as the impedance of the driving source. In other words

$$R_{in} \leqslant 10R_{source} \qquad (5.3)$$

where

    R_{in} is the input resistor

    R_{source} is the source resistance of the input circuit (external).

In the case where input source resistances are large, as might happen in the case of bioelectric measurements or when pH or gas electrodes are the signal source, we find that R_{in} also must be very large, and that would make the feedback resistor excessively large for even relatively-low gain figures.

Variable or selectable gain can be accomplished by varying the feedback resistor (R_f). Examples are shown in Fig. 5-2. The first example, Fig. 5-2(A), shows a method by which gain can be made continuously variable by placing a potentiometer in the feedback loop. In this case R_f = R1 + R2; so the voltage gain is given by

$$A_v = \frac{R1 + R2}{R_{in}} \qquad (5.4)$$

It is standard practice to make R1 equal to R_{in} if the minimum gain is to be unity. When potentiometer R2 is at its minimum-resistance setting (zero ohms), the stage gain will be unity. If R1 were equal to zero ohms (there is no R1 in that case), the minimum gain would be zero.

This same circuit is also used to trim the stage gain to some specific value. Errors in voltage gain can occur because of errors in the operational amplifier, or (more commonly) with normal tolerances in the resistors. Where the potentiometer is used to fine tune the stage gain, it might be wise to make R1 something like ten times greater than R2 so that good resolution is obtained. Also, it is prudent to use a 10-turn potentiometer at R2.

A switch-selectable method for adjusting gain is shown in Fig. 5-2(B). In this system gain is varied in steps by using switch S1 to change feed-

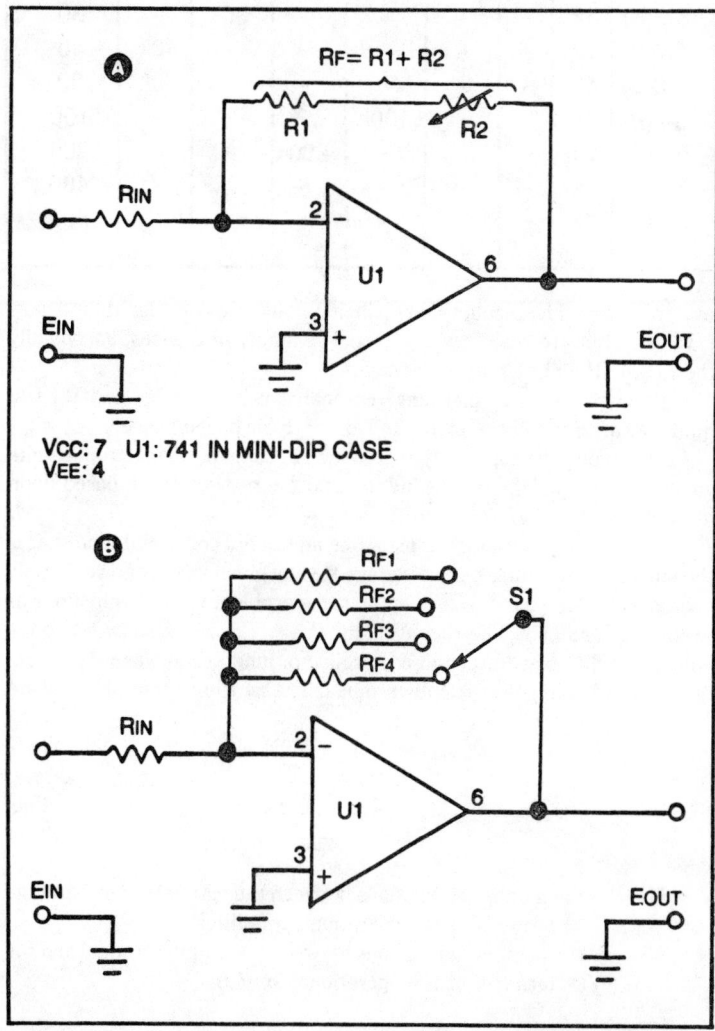

Fig. 5-2. At (A), an adjustable-gain inverting follower; and at (B), a selectable-gain inverting follower.

Table 5-1. Resistor Value/Voltage Gain Chart.

Fig.	R$_{IN}$	R1	R2	R$_{F1}$	R$_{F2}$	R$_{F3}$	R$_{F4}$	A$_V$
5-2(A)	10K	10k	100k	-	-	-	-	1– 11
5-2(A)	10k	-	100k	-	-	-	-	0– 10
5-2(B)	1k	-	-	10k	-	-	-	10
5-2(B)	1k	-	-	-	20k	-	-	20
5-2(B)	1k	-	-	-	-	30k	-	30
5-2(B)	1k	-	-	-	-	-	40k	40
5-2(B)	1k	-	-	-	-	-	80k	80
5-2(B)	1k	-	-	100k	-	-	-	100
5-2(B)	1k	-	-	-	200k	-	-	200
5-2(B)	1k	-	-	-	-	400k	-	400
5-2(B)	1k	-	-	-	-	-	800k	800

back resistors. The pinouts shown, incidentally, are considered "industry standard" but are really those of the 741 family of devices, specifically the mini-DIP and metal-can packages.

Although you may pick any reasonable values for R$_{in}$ and R$_f$, you might want to use those given in Table 5-1 for convenience.

If the input resistor values are too low for some particular application, go ahead and vary them, keeping in mind the restrictions imposed upon the ratio R$_f$/R$_{in}$.

You must also consider at least one additional constraint, and that is the maximum allowable output voltage for any given level of power-supply voltage and gain factor. We have the same problem with PN junctions as we did in Chapter 1. The rule of thumb is also the same: allow 0.7 to 0.9 volts for each base-emitter and base-collector junction between the output terminal and each power-supply terminal. The two limits thus obtained are

$$E_{max(+)} = V_{CC} - 0.9N + \tag{5.5}$$

$$E_{max(-)} = V_{EE} - 0.9N - \tag{5.6}$$

where

N +   is the number of junctions between the output terminal and the V$_{CC}$ terminal of the operational amplifier.

N –   is the number of junctions between the output terminal and the V$_{EE}$ terminal of the operational amplifier.

It is necessary to consider these limits in the light of maximum expected input voltage and the voltage gain. For example, let us suppose that the

126

maximum positive output voltage ($E_{max(+)}$) is 10 volts and that the voltage gain is 1000. What is the maximum negative input voltage that will *not* saturate the output?

$$E_{out} = - A_v E_{in} \qquad (5.7)$$

; so

$$E_{in} = - E_{out}/A_v \qquad (5.8)$$

Substituting the values given,

$$\begin{aligned} E_{in(max)} &= -(10)/1000 \qquad (5.9)\\ &= -0.010 \text{ volts}\\ &= -10 \text{ millivolts} \end{aligned}$$

As you can see, this amplifier will saturate with only 10 millivolts applied to the input!

MULTI-INPUT CIRCUITS

Consider the circuit of Fig. 5-3. Here we have an inverting follower with more than one input. By Kirchhoff's current law

$$I_f = I1 + I2 + I3 \qquad (5.10)$$

The correct expression for the output voltage will be

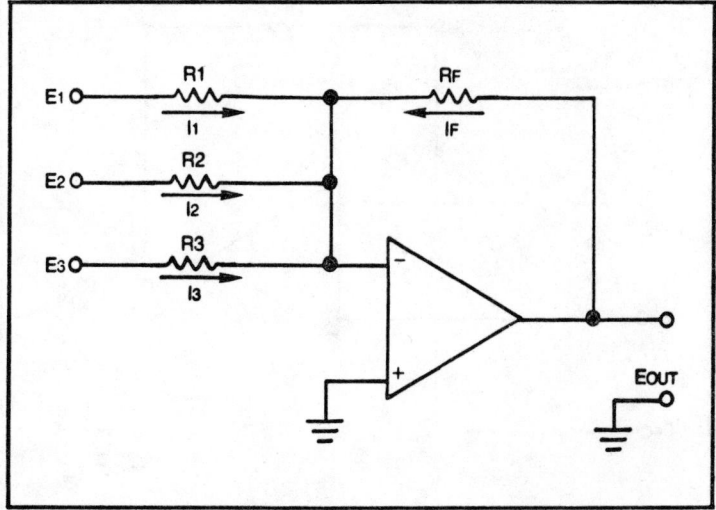

Fig. 5-3. Inverting summer.

$$E_{out} = -\left[\frac{E_1}{R1} + \frac{E_2}{R2} + \frac{E_3}{R3}\right] \times R_f \qquad (5.11)$$

but, by Ohm's law,

$$I1 = E_1/R1 \qquad (5.12)$$
$$I2 = E_2/R2 \qquad (5.13)$$
$$I3 = E_3/R3 \qquad (5.14)$$

; so Eq. 5.11 can also be written as

$$E_{out} = -R_f (I1 + I2 + I3) \qquad (5.15)$$

In some cases you will also use the noninverting input. In that type of circuit it is necessary to write the overall equation reflecting all currents. This requires that the signs be kept correct. Consider Fig. 5-4. This circuit has a transfer equation such as Eq. 5.16.

$$E_{out} = -R_f \left[-\frac{E_1}{R1} - \frac{E_2}{R2} + \frac{E_3}{R3} + \frac{E_4}{R4}\right] \qquad (5.16)$$

Circuits such as those in Figs. 5-3 and 5-4 have many uses in electronic instrumentation. They are sometimes used to sum voltages or currents that may represent quantities or numbers. In fact, they are a type of analog com-

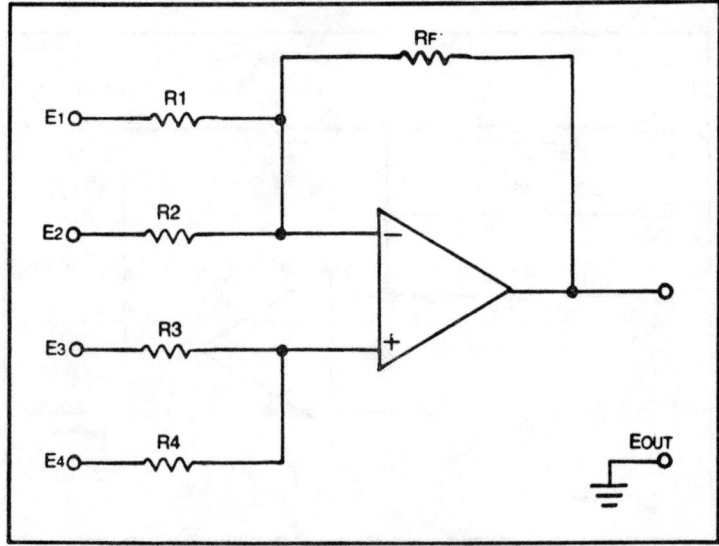

Fig. 5-4. Dual-polarity summer.

puter circuit. It is often said that most of the measurement instruments used in science and engineering are merely analog computers programmed to solve a specific function. Now that microprocessors are cheap, it is likely that future instruments will use *digital* computers in the same way.

NONINVERTING FOLLOWERS

This type of circuit is configured so that the input signal is applied directly to the noninverting input of the operational amplifier. The feedback, though, is still applied to the *inverting* input. If it were also applied to the noninverting input, feedback situations could easily arise that would generate oscillations, not beneficial in an amplifier!

The single most important feature of the noninverting follower is greatly increased input impedance. The input circuit of the inverting follower uses a resistor that has one end connected to a *virtual ground*. This limits input impedance to the value of the input resistor. The noninverting follower, on the other hand, applies input signal directly to the noninverting input which, you should recall, has a high impedance (Z_{in} is infinite!). In actual practice the input impedance has a value equal to the product of the voltage gain and the input impedance for that particular device—a figure published in the spec sheet. Although not actually infinite, it is very high.

Figure 5-5 shows the common unity-gain, noninverting follower. It is probably the simplest linear operational-amplifier circuit you will see anywhere. Note that 100 percent of the output signal is fed back to the inverting input. This makes the B term of Eq. 1.22 equal to unity; so we may rewrite it as

$$A_v = \frac{A_{vol}}{1 + A_{vol}(1)} \tag{5.17}$$

But for reasons already given, Eq. 5.17 is essentially equal to unity; so we may conclude that the closed-loop voltage gain of the circuit in Fig. 5-5 is also unity. That statement is true to a very good approximation, especially for high-gain operational amplifiers. Unity-gain followers can be made easily using regular operational amplifiers, or they may be bought as a special op-amp device that has the output and the inverting input terminals strapped together inside the case.

There are two main reasons for using unity-gain follower circuits. One of these is buffering without either the loss or amplification of the input signal. In many designs you may want to feed a signal or voltage level to several different points but do not want any interaction between them. An example might be control circuits in which the signal at a particular point is to be fed to both the outside world via panel jacks and to another part of the circuit which is actually being controlled. This is often done to allow simultaneous monitoring and use of a parameter in a circuit. Unless the monitor output is isolated from the rest of the circuit, a short in the out-

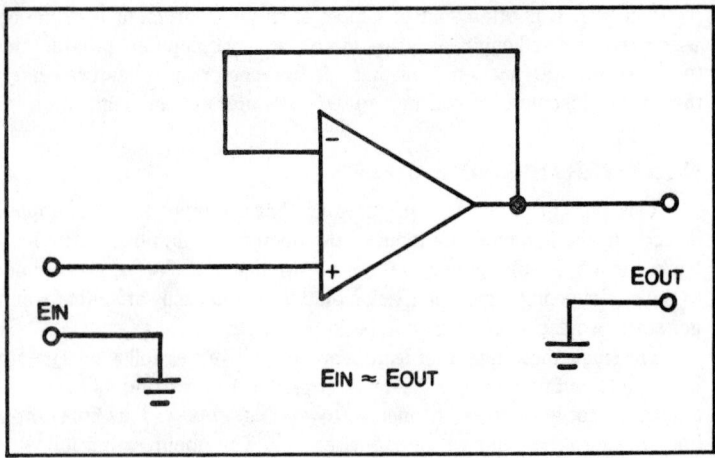

Fig. 5-5. Unity gain, noninverting follower.

side world connecting cables (a likely possibility) will cause the control circuit to lose its input signal. This could be catastrophic in some instruments and will definitely be at least annoying even if the results are harmless. A unit-gain, noninverting follower between the output jack and the signal path will prevent this from happening.

The second major application in which we find the unity-gain, noninverting follower is impedance transformation. Recall from Chapter 1 that the output impedance of an operational amplifier is typically very low and that the input impedance is very high. These two criteria in a single stage allow us to use it as an impedance transformer.

If a driving circuit for a stage has a source impedance far too high to comfortably drive the input, then such a follower is needed. There are, for example, many electrodes and transducers used in scientific instrumentation that have relatively high source impedance values. These devices must look into a high impedance to prevent loading. In fact, with some of the electrodes and transducers used in the life sciences, one might have to use a JFET, super beta, or BiMOS-type operational amplifier at the input to obtain the impedance needed. Good possibilities for this service are the RCA CA3140 and CA3160 devices. These have an input impedance on the order of 1.5 teraohms (10^{12}). In addition, these BiMOS op amps have superior drift and offset characteristics when operated from a low V_{CC}—V_{EE} supply (i.e., ±5 volts dc), especially if a top-hat heatsink is used on the case. The CA3140/3160 types are available in different grades covering a 10-to-1 price range. The low cost devices are basically super 741s and have superior performance for low cost. Premium grades of the devices will offer even better performance but at a somewhat higher cost. In any event, they are worth trying because the next economic step is in the $50 range! In making some of the projects for Chapter 22, I found that these op amps frequently performed better than devices that cost much more.

A noninverting follower *with* gain is shown in Fig. 5-6. In this circuit, a feedback-voltage divider is used in a manner similar to that used in the inverting circuit. The difference is that the "free" end of the input resistor is connected to ground. Some writers erroneously give the equation for gain in this circuit as the same R_f/R_{in} used for inverting circuits. This is true *only* at high gains; the correct expression for all gain levels is

$$A_v = \left[1 + \frac{R_f}{R_{in}}\right] \qquad (5.18)$$

As you can ascertain from an inspection of Eq. 5.18, use of the simpler version is only appropriate where the gain is high. When the voltage gain is 100 or more, deletion of the +1 term costs you an error of one percent or less. At lower gains, on the other hand, the error becomes significant. If, for example, the gain is to be two, and R_f/R_{in} is set equal to two, deleting the +1 factor results in an error of 50 percent. It is good advice to always use Eq. 5.18 for noninverting follower circuits—and that holds especially true for gains under 100.

FREQUENCY RESPONSE

Since the title of this chapter is dc amplifiers, you may well wonder why a section on frequency response is included—after all, frequency implies ac circuitry. The dc amplifier is given that name because it responds to signals at all frequencies down to dc. It will also respond to ac signals up to their upper cut off limits. For low cost, unconditionally-stable (i.e., frequency-compensated) operational amplifiers such as those in the 741 class, this upper limit is on the order of 10 kHz or less. Premium-grade

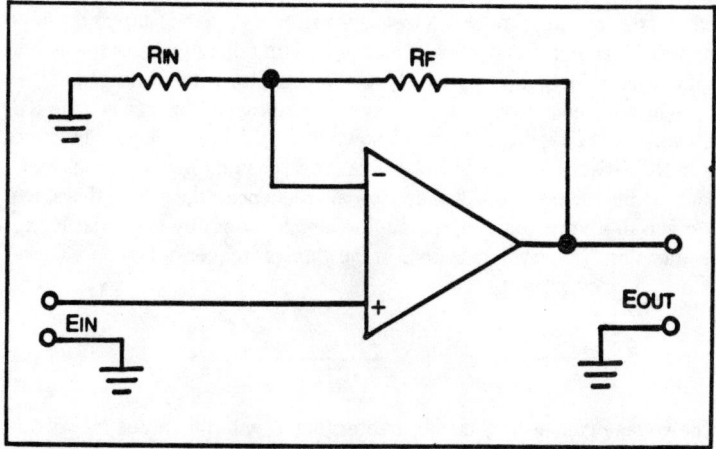

Fig. 5-6. Noninverting follower with gain.

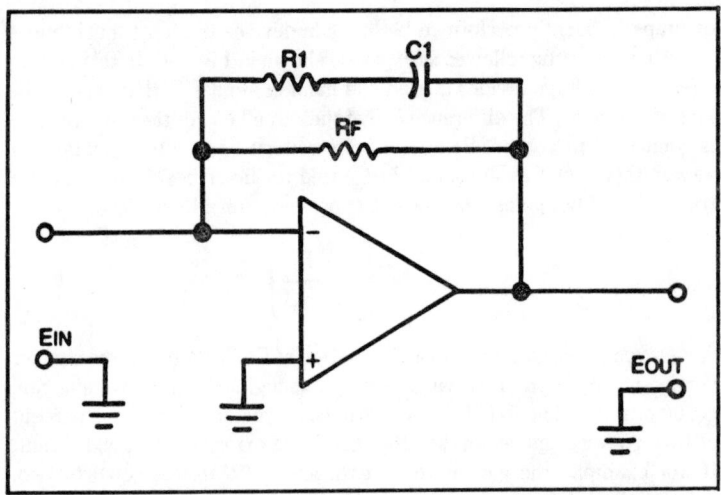

Fig. 5-7. Custom tailoring the ac frequency response without affecting the dc response.

operational amplifiers may have frequency responses into the megaHertz region. At least one moderate-cost family of BiMOS devices has a gain-bandwidth product over 5 MHz.

Do not be too quick to leap at the fancy, high cost, op amp that offers fast slew rates and high frequency response. Such properties can cause more problems than they cure unless there is a critical need for the excess frequency response. Most scientific instruments, probably all devices used in the life sciences, use low frequency circuits because the waveforms and signals that are expected do not vary rapidly. Performance achievable in the modest 741 operational amplifier is usually quite satisfactory. I will leave the problem of too much frequency response to the chapter on solving problems, but for now I will be content with telling you how to custom tailor operational amplifier frequency response to your own needs.

The frequency response of a stage may be shaped by means of an RC network in the feedback loop as shown in Fig. 5-7. In this circuit there is an RC network with a particular time constant shunting the regular feedback element, a resistor. At dc and low ac frequencies the gain of this stage is given to a very good approximation (*exactly* at dc) by the usual R_f/R_{in} relationship. There will be a knee in the gain-vs-frequency response curve at a frequency given by

$$f_0 = \frac{1}{2\pi R1C1} \qquad (5.19)$$

The voltage gain at frequencies greater than f_0 will roll off at a rate of approximately 6 dB/octave. This type of circuit is used to reduce artifacts in scientific, engineering, and medical instruments and to custom tailor the passband in audio circuits. It is also used in servomechanisms to custom tailor the damping properties of the circuit's feedback control loop. In some instances you will see simple RC integrators performing similar functions, but this is not always feasible because size and component values become too large for comfortable use as frequencies become low. It seems that capacitors at low frequencies become just too large.

Circuits such as Fig. 5-7, though, use the feedback loop, and this allows us to use more reasonable values of resistance and capacitance. Such circuits work because frequencies above the knee described through Eq. 5.19 see a feedback resistance that is approximately equal to the parallel combination of R_f and R1. The voltage gain is given approximately by

$$A_v = \frac{R1 + R_f}{R_{in}} \tag{5.20}$$

This value for voltage gain can be considerably lower than the voltage gain at dc, provided R1 and C1 are given appropriate values. In some cases two or more RC networks are provided to create two or more knee frequencies. This is the case in phonograph equalization stages, usually the preamp.

Zero-offset control is provided by the circuit shown in Fig. 5-8. This

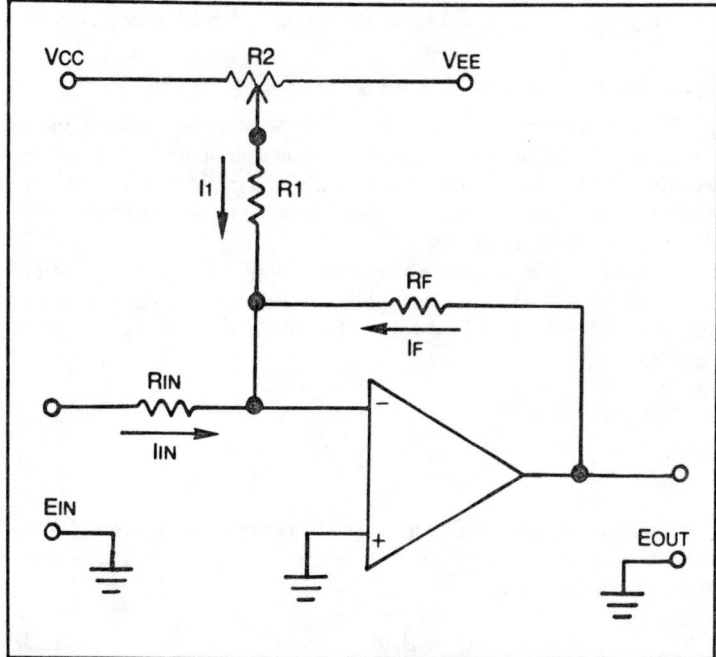

Fig. 5-8. Offset control.

circuit can be used to shift the baseline of the operational-amplifier output terminal. Normally the output will be zero volts when the input potential is also zero. Current I1, however, will always flow, regardless of the input voltage, and this means that a current (I_f) must flow and be equal to I1. The output voltage when the input voltage is zero will be equal to

$$E_{out} = -I1\ R_f \qquad (5.21)$$

This type of circuit is also used to null certain operational-amplifier errors (of which, more later). In that case, the potentiometer will be a 10-turn trimmer type that is screwdriver adjusted. In other cases the potentiometer might be a zero-suppression control or a position control to drive the vertical channel of an oscilloscope or strip chart recorder. In most applications R2 should be a 10-turn potentiometer, if only to improve resolution or "adjustability." Normally it is some value between 5 kilohms and 100 kilohms, with both 10 kilohms and 20 kilohms being very common. Resistor R1 will almost always be 10 kilohms, but anything within the same range as R2 may be found.

Superior resolution is also obtainable by making R2 a combination of several resistors. If two equal resistors are connected between the respective ends of R2 and the VCC/VEE terminals, the resistance of R2 will be proportionally less than the total resistance which sets I1. This allows a 10-turn potentiometer at R2 to set the output potential very precisely.

SUPERGAIN DC AMPLIFIERS

There are several reasons why it might be bad practice to try to obtain large gain figures from a single operational amplifier if the inverting follower circuit is to be used. In those cases you might have to use very-high-value feedback resistors and very-low-value input resistors, both of which are potential problems.

If high gain is required of a single stage (note that it is still preferable to cascade several stages to obtain the gain required), you can use the circuit of Fig. 5-9. Current I_{in} is caused by the input signal voltage and is equal to

$$I_{in} = \frac{E_{in}}{R_{in}} \qquad (5.22)$$

In order to satisfy Kirchhoff's current law there must exist a current I equal to I_{in}. This means that a voltage must be created at point A great enough to produce a current I_A

$$I_A = E_A/R_f \qquad (5.23)$$

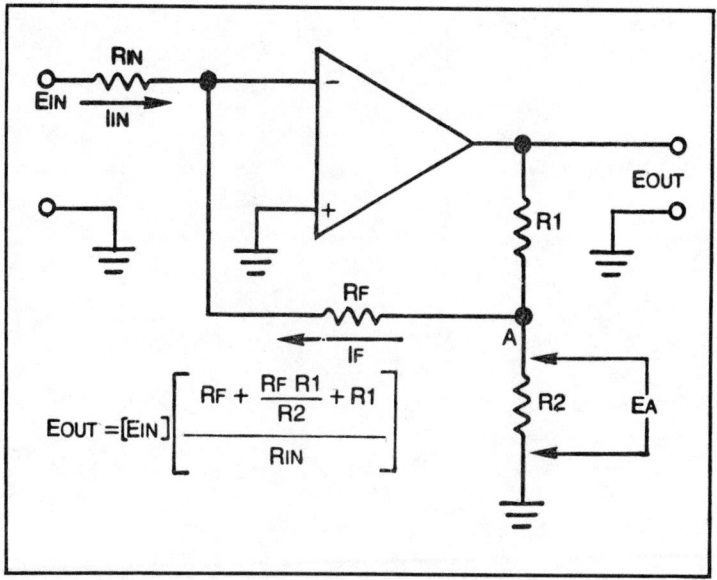

Fig. 5-9. High-gain dc amplifier using a single op amp.

The voltage at point A is derived from the output voltage E_{out} through a resistor voltage divider consisting of R1 and R2. The appropriate transfer function is

$$E_{out} = (E_{in}) \left[\frac{R_f + \dfrac{R_f R1 + R1}{R2}}{R_{in}} \right] \qquad (5.24)$$

Be forewarned, though, that offset currents and voltages, and layout difficulties accompany high gain from a single stage. You will find, for example, a 1-mA-input offset current able to saturate the output terminal if there are high gains (i.e., 5000 to 10,000). It is usually easier to tame these circuits if the gain per stage is low and an offset null is provided for each stage or every two stages.

Chapter 6

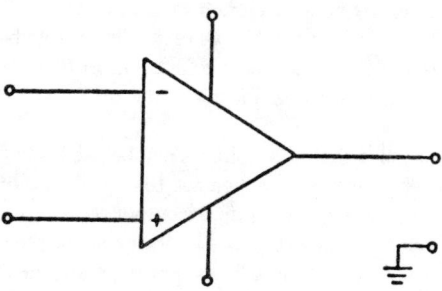

Ac Amplifiers

V ERY SIMPLY DEFINED AN AC AMPLIFIER IS ONE THAT IS SENSITIVE to ac signals, but not dc signals. This does not imply, however, that the dc amplifiers discussed in other chapters will not also respond to ac signals. Most dc amplifiers, in fact, have a response from dc to some specified ac frequency. An ac amplifier, on the other hand, will only respond to ac, or let us say "varying," signals even though the frequency response on the lower end is only a fraction of a hertz. In fact,many medical and scientific instruments have ac responses down to 0.05 hertz, but since they will not respond to a steady dc signal they are classified as ac instruments.

WHY AC AMPLIFIERS?

Of what possible use, then, is an ac amplifier if most dc amplifiers will also handle ac signals out to quite a few kilohertz? Although this may appear to be one of those questions that is asked merely for the sake of asking, it must be answered in a satisfactory manner. In many cases there will be no reason to prefer the ac amplifier over a dc amplifier; so one might just as well opt to build the dc version because it is simpler. In other cases you may be able to find any number of good reasons why an ac amplifier is to be preferred. Most common among them, perhaps, is the need to combat dc offset components in the signal voltage. Take as an example the common physician's electrocardiograph machine used to record electrical activity of the heart. The American Hospital Association's recommended bandwidth for a diagnostic quality ECG machine is 0.5 to 100 hertz. Note that the lower response limit is very nearly dc, so why not use a simple dc amplifier? The reason is that the electrodes applied to the patient's skin to pick up the tiny ECG potentials also form one plate of a voltaic cell that forms a dc offset that may typically have a value several times the amplitude of the ECG signal's main features. In addition to this high amplitude, we find that the dc offset is not well behaved in that it tends to drift badly over a period of a few minutes. If it were stable it would be a simple matter to build into the ECG machine a counter-offset that would reduce the offset artifact, but in practical machines the only result is a wandering baseline.

Another reason is that ac amplifiers can be used as a type of filter to get rid of certain unwanted frequencies that tend to interfere with the desired signal and thereby clutter up the output waveform. In this application high and low frequency response limits are set so the unwanted components are not amplified nearly as much as the desired frequency components. Care must be exercised, however, when setting these limits because all periodic functions may be represented as a Fourier series of sine and cosine products. Waveform features that change rapidly represent the high frequency Fourier components, while features that change slowly in time are represented by the low frequency Fourier components. If you set the limits incorrectly the output waveform will be distorted by the frequency response. It is, therefore, necessary to appreciate what frequencies can be attenuated without wreaking mental havoc with your results.

Also, the judicious use of feedback can make the ac amplifier more stable and freer of distortion. The former property, when used in certain types of ac amplifiers, becomes extremely important. For example, in the ac amplifier used in a chopped dc system we can amplify very low-level signals that would ordinarily be close in frequency and amplitude to the drift component of a dc amplifier. A chopper switch can be used to chop the dc signal so that it may pass through the ac amplifier, which is feedback-controlled in order to be far more stable.

AC AMPLIFIER CIRCUITS

Figure 6-1 shows what is, perhaps, the easiest method for building an ac amplifier, although optimum performance is not usually achieved. All three circuits shown in this figure are based on the single-ended dc amplifier circuits discussed in a previous chapter.

The circuit in Fig. 6-1(A) uses the noninverting input follower configuration. It is converted from dc to ac service by transformer T1. The transformer may have a voltage step-up factor of its own that depends upon the turns ratio. This must be accounted for in the gain equations. The ac-voltage gain is given by

$$A_v = \left[\frac{N_s}{N_p} \right] \left[1 + \frac{R_f}{R_{in}} \right] \qquad (6.1)$$

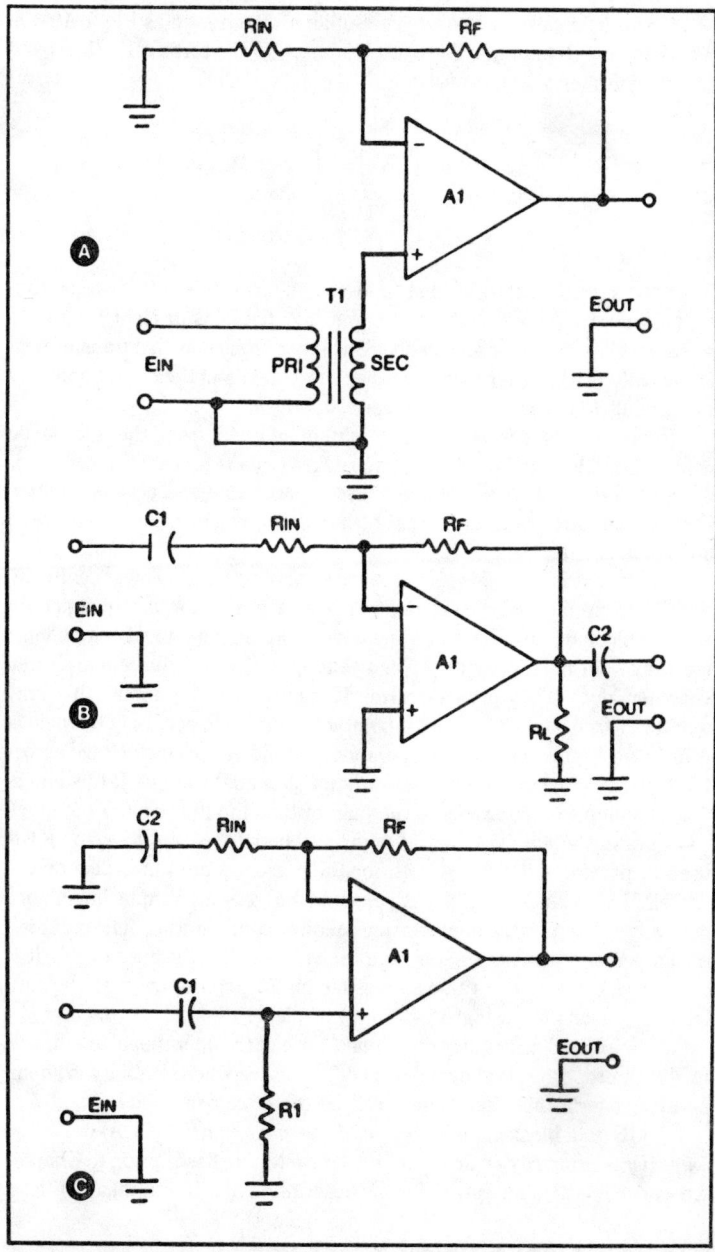

Fig. 6-1. At (A), a transformer coupled, noninverting ac amplifier; (B) shows a capacitor coupled inverting ac amplifier, and (C) shows a capacitor coupled, noninverting ac amplifier.

Let us calculate the voltage gain of an ac stage such as Fig. 6-1(A) in which R_f = 100 kilohms, R_{in} = 10 kilohms and the turns ratio (N_s/N_p) of the transformer is 3:1

$$A_v = \left[\frac{3}{1}\right] \times \left[1 + \frac{100}{10}\right]$$
$$= (3)\,(11)$$
$$= 33$$

There are two problems that may plague this circuit. One is that transformers are imperfect devices and may have insufficient frequency response for a particular application. The other problem is that transformers are usually made larger than the other components in the circuit and may be difficult to accommodate in some situations.

Figure 6-1(B) shows an ac amplifier adapted from the simple dc inverting-follower circuit. In this particular version of the circuit capacitor C1 is used to block dc yet will allow ac signals to pass. There is, though, some attentuation of ac signals at frequencies for which the reactance of C1 is high.

Another version of this idea uses the dc noninverting-follower configuration. This circuit is shown in Fig.6-1(C). Again, the purpose of capacitor C1 is to block dc but to pass ac. In an ideal circuit using ideal components we would not need resistor R1, but real operational amplifiers are often very non-ideal. They produce a tiny bias current at the inputs that may tend to charge capacitor C1 to a point where trouble occurs. The mechanism of the problem is that the operational amplifier sees the voltage across C1, caused by these bias currents, as another signal voltage. If this potential builds up to a point where the gain of the amplifier causes the stage to saturate, the amplifier would appear to be latched up. Resistor R1 is used to provide a discharge path for the current stored in capacitor C1. This is, unfortunately, at the expense of the very high input impedance normally offered by the noninverting follower configuration. The input impedance reduces to no greater than the value of R1. Because of this it is usual to select as high a value as possible for R1—consistent with the anti-latchup reasons for using R1 in the first place.

An improved ac amplifier that makes use of the operational amplifier's feedback properties is shown in Fig. 6-2. In cases where stability is an important criteria this circuit may well be the circuit of choice.

An IC gain block called a *current difference amplifier* (CDA) or *Norton amplifier* is primarily an ac amplifier. I have left its description to Chapter 7 because it is also necessary for the description of differential amplifiers.

TAILORING FREQUENCY RESPONSE

The capacitors and certain RC time constants determine the frequency response characteristics of ac amplifiers. Bypass, decoupling, or those

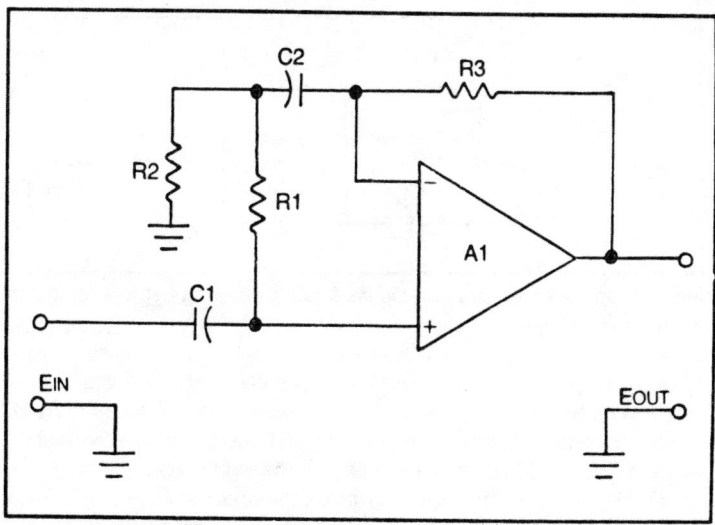

Fig. 6-2. Bootstrapped ac amplifier.

capacitors doing jobs such as C2 (dc blocking) in Fig. 6-1(C), follow a rule of thumb stating that the capacitive reactance of the capacitor selected must not be greater than one-tenth the value of the associated circuit resistance. The reactance value for the capacitor is calculated at the lowest frequency where normal operation is desired, usually the lower 3 dB point. This presents a problem in amplifiers used to process signals with very low frequency components though. Such exist often, incidentally, in preamplifiers used to process signals in medical and scientific instrumentation. The trouble results because capacitive reactance is inversely proportional to frequency. We would, therefore, require a rather large-value capacitor at those low frequencies. This makes them physically large even when electrolytics are employed. Of course, the electrolytics are much smaller in, say, 100 μF sizes but also suffer from the fact that they are polarity sensitive. That property is hardly useful in an ac amplifier.

The problem of size is alleviated somewhat by the use of electrolytic capacitors, but the problem of polarity sensitivity must be handled using a circuit such as in Fig. 6-3. This is a method for overcoming polarity sensitivity while retaining the high capacitance. Do not make the mistake of looking at this circuit too quickly, assuming that it is a simple series circuit employing two capacitors. In that case you would assume that the normal formula for series capacitors would apply. If that were true the total capacitance would be one-half the capacitance of a single unit, provided that C1 = C2. In reality, though, the total capacitance of the combination shown in Fig. 6-3 is equal to the value of either C1 or C2, again assuming their equality. This occurs because of diodes D1 and D2.

Consider what happens on positive-going excursions of an input signal.

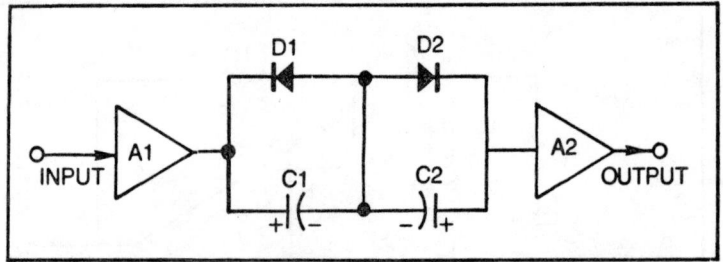

Fig. 6-3. Use of capacitors and diodes forms a coupling network for low frequency operation.

In this case diode D1 is reverse biased; so it does not affect capacitor C1 in any way. Signals will pass through C1 to the junction of C1-C2-D1-D2. Diode D2, on the other hand, will be forward biased; so it effectively shorts across C2, and directly passes signals to the next stage.

On the negative alternation, however, the situation is reversed, so that diode D1 is forward biased and D2 is reverse biased. Since diode D2 is now cut off, it has no effect on capacitor C2. The signal will pass through C2 to the following stages.

From this we can see that each capacitor is effectively in the circuit for only half of the input signal. The polarities are arranged such that each capacitor sees only the signals of the proper polarity. Modern electrolytic capacitors, in both tantalum and aluminum, can be made relatively small in the voltage ratings usually associated with operational amplifiers. You will see this circuit idea in many biological and physiological amplifiers.

Chapter 7

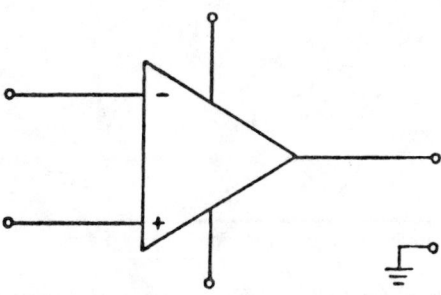

Dc Differential Amplifiers

MOST OF THE AMPLIFIER TYPES THAT I HAVE CONSIDERED THUS far have used a single-ended input circuit configuration. That is to say there is only one input terminal with respect to ground. Most scientific instruments and measurement circuits, however, require a differential input. Of course, the standard IC operational amplifier is almost uniquely suited to filling this requirement because it has a differential input pair.

COMMON-MODE REJECTION

Recall from Chapter 1 the operational amplifier rules governing the circuit of Fig. 7-1. Voltages E_1 and E_2 each affects only one input, but voltage E_3 is applied equally to *both* inputs. Thus, E_3 is known as a *common-mode voltage*. An ideal operational amplifier in the differential mode will respond only to

$$E_{in} = E_1 - E_2 \tag{7.1}$$

and will generate an output voltage equal to

$$E_{out} = A_v E_{in} \tag{7.2A}$$

$$E_{out} = A_v (E_1 - E_2) \tag{7.2B}$$

Common-mode voltage E_3 will cause the respective input terminals to have equal, but opposite, effects on the output terminal; so the output voltage is theoretically nulled to zero. *Common-mode rejection ratio* (CMRR) is a

143

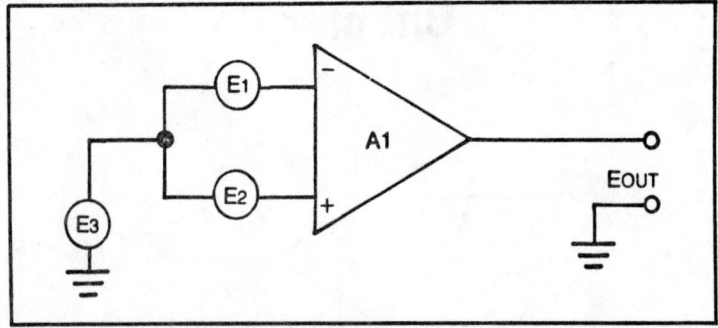

Fig. 7-1. Differential input configuration of the op amp.

measure of the differential amplifier's ability to reject common-mode signals and is expressed in decibels

$$CMRR_{(dB)} = 20 \log_{10} (A_{vol}/A_{v(cm)}) \qquad (7.3)$$

where

A_v is the differential voltage gain

$A_{v(cm)}$ is the common-mode voltage gain by the test circuit of Fig. 7-1 in which $(E_1 - E_2) = 0$ and E_{out} is solely due to E_3

$$A_{v(cm)} = \left[\frac{E_{out}}{E_3} \right] (E_1 - E_2) = 0 \qquad (7.4)$$

Most operational amplifiers can offer inherent CMRR figures up to around 120 dB, and even "el cheapo" types frequently have CMRR figures exceeding 80 dB. This means that, at 120 dB, they have a differential circuit gain up to one million times the circuit gain for common-mode signals.

Clearly, this will mask the effects of common-mode signals in all but the most critical of applications. Or will it? Unfortunately, CMRR will deteriorate almost immediately as soon as the operational amplifier is connected into a real circuit because other factors tend to unbalance the differential inputs of the amplifier device.

SINGLE-OPERATIONAL-AMPLIFIER CIRCUITS

The simplest operational amplifier differential circuit is shown in Fig. 7-2. The differential gain of the circuit is given by

$$A_v = R3/R1 \qquad (7.5)$$

If

R1 = R2
R3 = R4 + R5

There are two main problems usually associated with this circuit, and they may or may not affect a particular application. One problem is that input impedance is limited by the values of the resistors R1 and R2, not by the actual Z_{in} of the IC.

If input impedance is raised by increasing the value of these resistors, it might become difficult to obtain high gain. This is a situation where two possible requirements have mutually exclusive solutions; so some trade-offs are necessary. For example, it is usually allowable to reduce R1 and R2 to a value that is 10 times the output impedance of the driving source. All practical input sources (i.e., other electronic circuits or a transducer) will have some specific output impedance. In most cases this will be in the range of 100 to 1000 ohms. The exceptions, and these can be very important, are some of the electrodes (e.g., pH) used in biophysical or chemical experiments. If it is the case that the source impedances are under 1000 ohms, then a value of 10 kilohms for R1 and R2 is appropriate. The maintenance of a high CMRR requires that R1 = R2 and R3 = R4 + R5. Potentiometer R5 is usually labeled "CMR Adjust" because it helps optimize CMRR by cancelling out the effects of resistor mismatch. Of course, we must still see to it that the resistors used for R1 and R2 are matched, but it is usually easier just to use 1-percent-tolerance resistors. The alternative is to hand-sort from a pile of resistors using a precision ohmmeter or Wheatstone bridge.

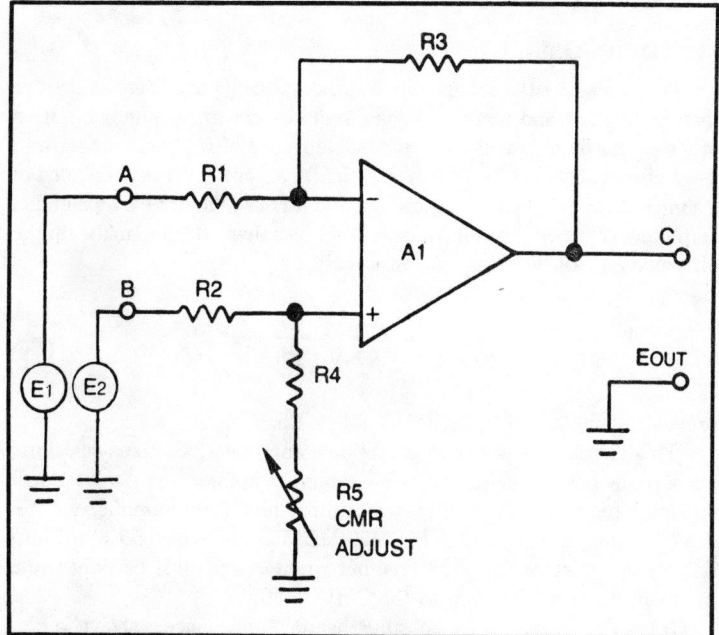

Fig. 7-2. Simple, single op-amp differential amplifier.

It is generally considered to be good practice to arrange your circuit, if possible, such that R3 = R4 + R5 when potentiometer R5 is set somewhere close to the middle of its range. Also, R5 should have a full range value of approximately 10 percent of R4 and should be a 10-turn type. It must be realized that the tolerance of the resistor selected for R4 has some bearing on the permissible full-range value of R5. What we are trying to do is create a situation in which R5 causes as small a change in the quantity (R4 + R5) per turn of R5 as possible. This will aid in obtaining the highest possible resolution in the CMR adjustment.

CMRR Adjustment Procedure

1. Connect a sensitive oscilloscope to terminal C of E_{out}.
2. Short together points A and B.
3. Apply a 100 Hz (or so) sine wave to shorted terminals A and B. Adjust the sine wave amplitude for a level of several volts rms or until a point is reached just below clipping.
4. Adjust R5 for minimum output voltage at point C by monitoring E_{out} on an oscilloscope.
5. Repeat step 4, using ever more sensitive positions of the oscilloscope vertical input selector, until no more reduction of E_{out} is noted. This should prove to be somewhere close to the noise level and results in the best CMRR.

Increasing Gain

In cases where the source impedance requires R1 and R2 to be greater than 10 kilohms and there remains a requirement for voltage gain of an order not easily obtained with reasonable values of R3, you may wish to use a circuit such as Fig. 7-3. In this circuit, amplifier A1 is connected in a simple dc differential configuration, and it is used to drive single-ended amplifier A2. Keep the voltage gain of A1 less than 100, and make up the difference in A2. For this type of circuit

$$A_V = \frac{R3}{R1} \times \frac{R7}{R6} \tag{7.6}$$

provided that R1 = R2 and R3 = R4 + R5.

This circuit uses just two operational amplifiers, six fixed resistors, and a single potentiometer. If the operational amplifiers are part of a dual op-amp integrated circuit, a very small, but high-gain, amplifier will result. Types such as the 747 or MC1458/MC1558 (also called 5558) will suffice for low tolerance applications, but where specs must be tight use a premium dual op amp such as the PMI/OP-10.

A low cost technique for boosting the input impedance is shown in Fig. 7-4. It also allows us to use values for R1 and R2 as low as 1000 ohms

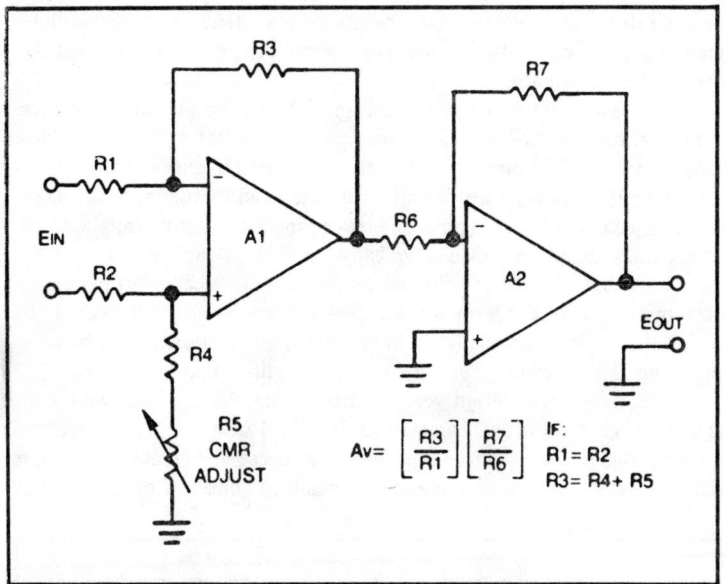

Fig. 7-3. Adding a post amplifier increases gain simply.

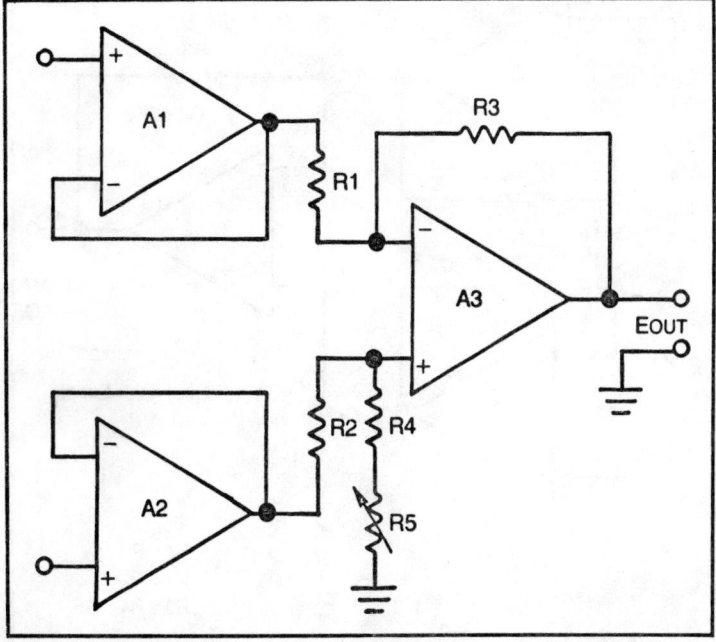

Fig. 7-4. Unity gain input amplifiers raise the effective input impedance of the previous circuit.

147

or so. In this circuit there are two operational amplifiers connected as unity gain, noninverting followers, and they are used to provide the needed impedance transformation.

Since even low cost operational amplifiers have output impedance values on the order of 50 to 200 ohms, we are allowed to use low values for R1 and R2. The immediate advantage conferred by this is that it allows us to obtain relatively high voltage-gain figures with little increase in circuit complexity. This configuration proves especially useful in applications where the source impedance is typically very high. Amplifiers A1 and A2 may be either regular operational amplifiers with the inverting inputs strapped to the output terminals, or they may be of the LM302 variety in which this is done internally. In very critical applications, though, some premium JFET or superbeta operational amplifier may be needed.

Figure 7-5 is a different version of the same idea in which we obtain a voltage gain from operational amplifiers A1 and A2. This circuit is generally called an *instrumentation amplifier* because it combines the high gain and very high input impedance usually required of dc differential

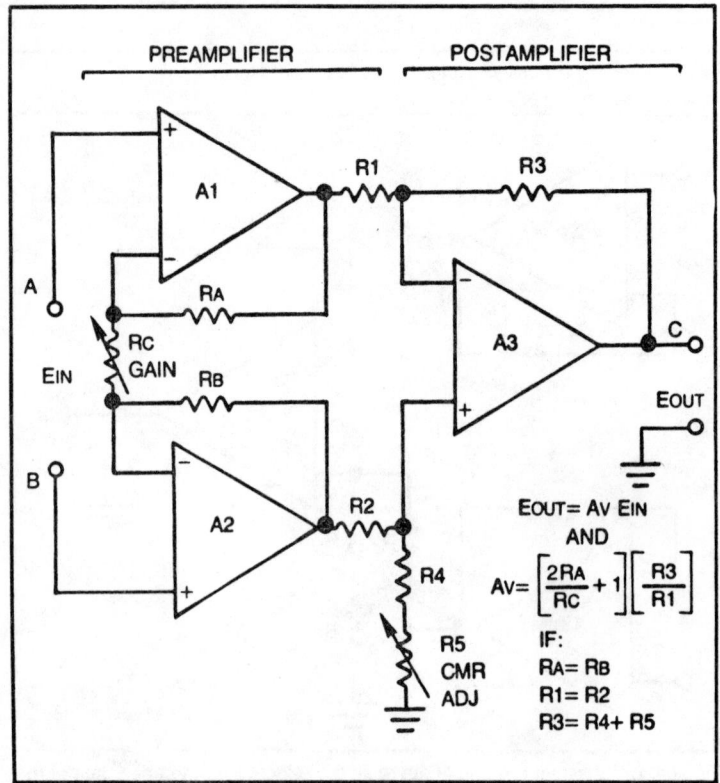

$$E_{OUT} = A_V E_{IN}$$
$$\text{AND}$$
$$A_V = \left[\frac{2R_A}{R_C} + 1 \right] \left[\frac{R3}{R1} \right]$$

IF:
$R_A = R_B$
$R1 = R2$
$R3 = R4 + R5$

Fig. 7-5. High-gain instrumentation amplifier.

in scientific instrumentation. This circuit is aligned for optimum **CMRR** in exactly the same manner as the circuit of Fig. 7-2. The gain of the classic instrumentation amplifier of Fig. 7-5 is given by

$$A_v = \left[\frac{2R_a}{R_c} + 1\right]\left[\frac{R3}{R1}\right] \tag{7.7}$$

preamplifier gain ⟋   ⎿post amplifier gain

provided that

$$R_a = R_b, R1 = R2, R3 = R4 + R5$$

Resistor R_c will typically have a value of 500 to 2000 ohms and can be made to perform the function of gain control if desired. This can result in a problem unless R_c is made up of a fixed resistor and a potentiometer in series. Note that R_c appears in the denominator of Eq. 7.7. If it is allowed to go to values in the neighborhood of zero ohms, the value of A_v tries to go to infinity. Since this is not possible we find the output of the amplifier in saturation.

Here is a typical voltage gain calculation.
Set:

$$R_a = R_b = 50 \text{ kilohms}$$
$$R_c = 1 \text{ kilohm}$$
$$R1 = R2 = 2 \text{ kilohms}$$
$$R3 = R4 + R5 = 20 \text{ kilohms}$$

$$A_v = \left[\frac{(2)\,(50)}{1} + 1\right] \times \left[\frac{20}{2}\right]$$

$$= (101)\,(10)$$

$$= 1010 \text{ (better than 60 dB)}$$

OTHER DC DIFFERENTIAL AMPLIFIERS

The IC operational amplifier is not the only linear integrated circuit that may be used as a dc differential amplifier. Although it is probably the easiest to apply in practical situations, there are competitors that have advantages of their own.

These other devices fall into two categories. Some are like op amps, while others are IC versions of the three- (or more) transistor discrete differential amplifiers. In the first class we find devices such as the current-difference amplifier (CDA), also called the Norton amplifier, and the *operational transconductance amplifier* (OTA).

Current-Difference Amplifier Devices

The CDA, or Norton, amplifier is represented graphically by the modified op-amp symbol shown in Fig. 7-6. It must be kept in mind that these are predominantly ac devices although dc signals can be accommodated if an output offset voltage can be tolerated, if the input is balanced, as in the differential case. The equation (R_f/R_{in}) yields only an approximation of the voltage gain, but it is sufficiently accurate in most cases to allow us to forget about some error terms. The transfer equation for quiescent conditions is

$$E_{out} = \left[\frac{V_{ref}R_f}{R1}\right] + 0.7 \left[1 + \frac{R_f}{R_{in}}\right] \tag{7.8}$$

The value for the reference voltage is less than, or equal to, VCC. Resistor R1 is selected, keeping the reference voltage in mind, such that $5\ \mu A \leq I1 \leq \mu A$. This sets the static output operating point around which the signal must swing. For the inverting circuit of Fig. 7-6.

$$A_v \approx R_f/R_{in} \tag{7.9}$$

The noninverting CDA configuration is shown in Fig. 7-7. The only essential different aspect of this circuit is that the input resistor has been moved from the inverting to the noninverting input. The gain is given by

$$A_v = \left[\frac{R_f}{R_{in}}\right] \times \left[\frac{I1}{26}\right] \tag{7.10}$$

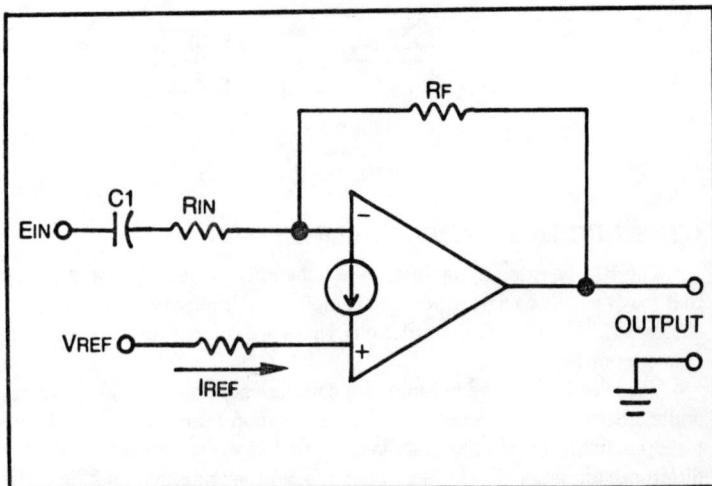

Fig. 7-6. Basic circuit of a Norton, or current-difference amplifier (CDA).

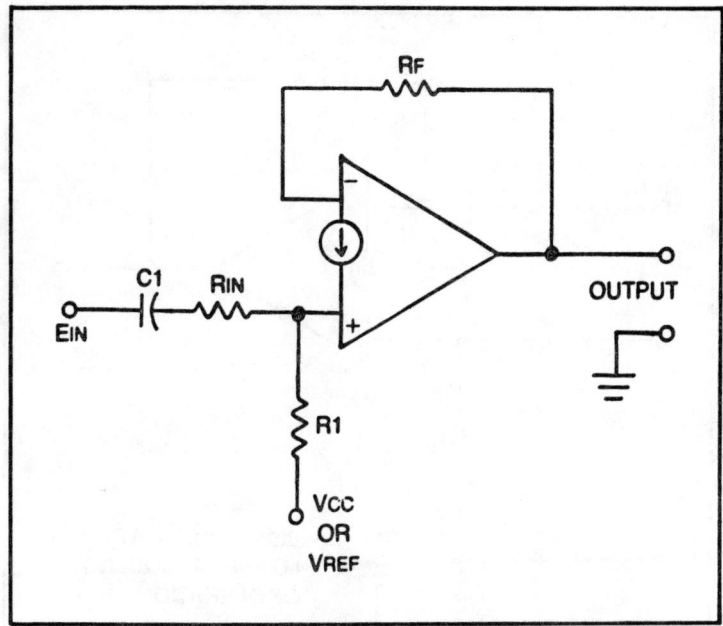

Fig. 7-7. Noninverting CDA amplifier.

In all cases we are bound to the rule given earlier for the permissible values of I1. The circuit for the CDA differential amplifier is given in Fig. 7-8.

Operational Transconductance Amplifiers (OTAs)

The OTA (see Fig. 7-9) is very much like an operational amplifier but with an important difference. The normal op amp is a device with a transfer function that relates an output *voltage* change to a change in the input voltage, or

$$A_v = \frac{\Delta E_{out}}{\Delta E_{in}} \tag{7.11}$$

In the OTA family of devices, however, the transfer function relates an output *current* change to changes in input voltage, or

$$A_v = \frac{\Delta I_{out}}{\Delta E_{in}} \tag{7.12}$$

But Eq. 7.12 is also the relationship that defines the property called *transconductance*; so we may safely refer to the device as a transconductance

151

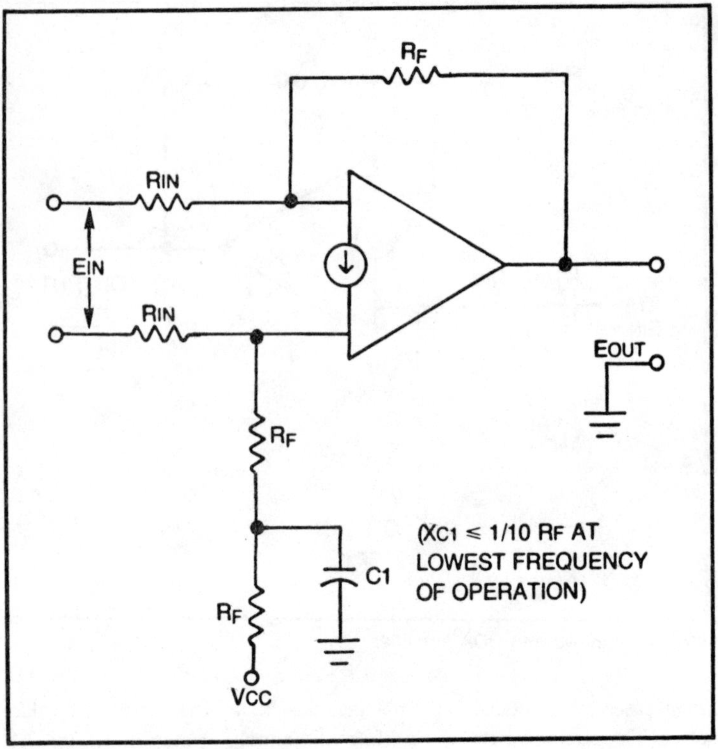

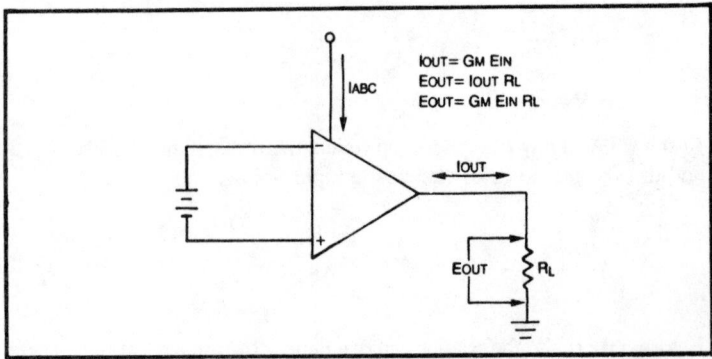

Fig. 7-8. Differential CDA amplifier.

amplifier. Equation 7.12, then, may also be written:

$$gm = \frac{I_{out}}{E_{in}} \qquad (7.13)$$

IOUT= GM EIN
EOUT= IOUT RL
EOUT= GM EIN RL

Fig. 7-9. Circuit of a programmable operational transconductance amplifier (OTA).

We may also rearrange Eq. 7.13 to find an expression for I_{out}

$$I_{out} = (gm) (E_{in}) \tag{7.14}$$

One noteworthy difference between the operational amplifiers and the OTA is output voltage dependency upon load resistance. In the classical operational amplifier we find that the output voltage is relatively independent of external load resistance up until the point is reached where the op amp is saturated. After that point, of course, a new situation occurs, and none of the normal equations gives valid results. In the OTA family, however, we must realize that the output voltage is Ohm's-law dependent and is given by

$$E_{out} = I_{out}R_L \tag{7.15}$$

The transconductance values of even modestly-priced OTA devices such as the RCA CA3080 typically approach 0.01 siemens. It is interesting to note that the exact value of the transconductance may be manipulated by varying I_{abc}, the amplifier bias current.

The operational transconductance amplifier may be used in almost all of the same circuit applications as the normal operational amplifier. It may also be used in those circuit applications where an analog multiplier is required, even though the accuracy and linearity of actual multiplier integrated circuits aren't important.

Besides ordinary analog multiplication of voltages, we see its use as an amplitude modulator or a heterodyne mixer (which are both multiplication processes, although not always recognized as such).

One should not overlook the variations in circuit design that are possible due to the availability of the amplifier bias-current terminal. This can be used, for example, to turn on and off the amplifier from a remote location, or in a switching type multiplexer circuit.

Chapter 8

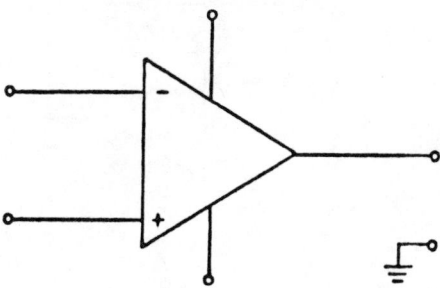

Beyond Operational Amplifiers

T HE OPERATIONAL AMPLIFIER REVOLUTIONIZED ANALOG INSTRU-
mentation design about a million years ago, but early models were
not well-suited to general use in analog circuits. In the mid-sixties when
the μA-709 IC operational amplifier was developed, it became possible to
use this wonderful invention for a wide range of applications where it had
been hopeless before.

Figure 8-1 shows the four basic forms of the op-amp circuit: inverting
follower (Fig. 8-1(A)), noninverting follower with gain (Fig. 8-1(B)), unity-
gain noninverting follower (Fig. 8-1(C)), and dc differential amplifier (Fig.
8-1(D)).

The inverting follower circuit shown in Fig. 8-1(A) produces an output signal that is 180° out of phase with the input signal—which is where
it gets its name, "inverting" follower. The gain of the inverting follower
is simple: $-R2/R1$, where the minus sign indicates the inversion. Thus,
when a circuit like Fig. 8-1(A) has an input resistor (R1) of 10 kilohms and
a feedback resistor (R2) of 100 kilohms, the gain is -100 k/10 k, or -10.
The input impedance of this circuit is equal to R1.

The noninverting follower with gain (Fig. 8-1(B)) provides no phase
reversal between input and output. The input impedance of this circuit is
very high, and is not related to the value of any resistors in the circuit.
The gain of the circuit is $(R2/R1) + 1$; so in the example above, the gain
when R1 equals 10 kilohms and R2 equals 100 kilohms is 101.

The unity-gain noninverting follower (Fig. 8-1(C)) is a special case of
Fig. 8-1(B) in which the feedback resistor is zero ohms; the output is connected directly to the inverting input. This circuit provides a voltage gain
of one (unity gain). There are two main purposes for the unity-gain noninver-

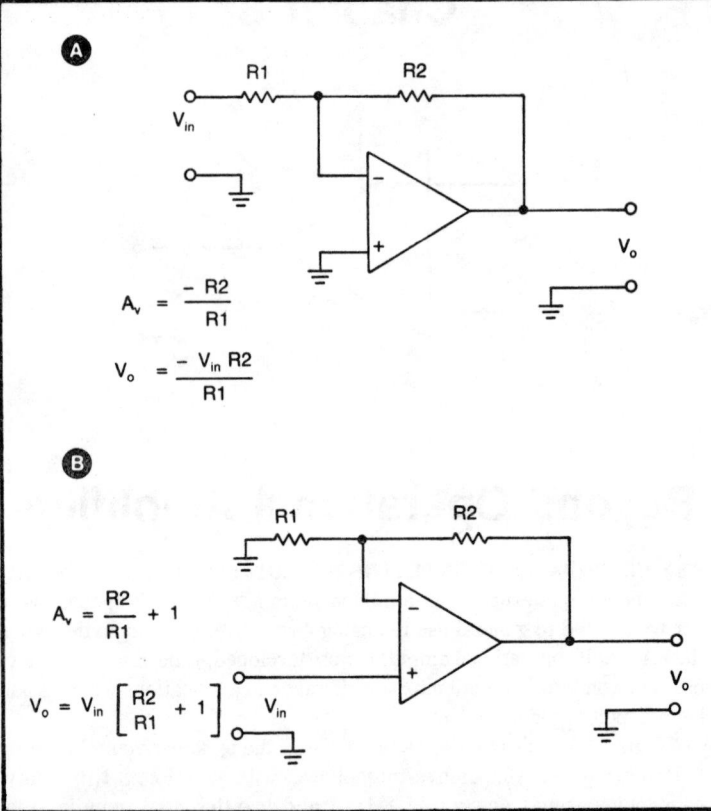

Fig. 8-1. Operational amplifier configurations (A) inverting follower, (B) noninverting follower with gain, (C) unity gain inverting follower, (D) Dc differential amplifier.

ting follower: buffering and impedance transformation. Buffering means that the circuit provides isolation between input and output. Thus, variations in load will not affect the input circuit—an ideal situation where oscillators and other sensitive circuits must drive unstable loads. Impedance transformation is high-to-low: a high input impedance is reduced to a low output impedance. Since the voltage remains the same at the output and the impedance drops at the output, you can see that the unity-gain noninverting amplifier provides a power gain greater than unity while providing a voltage gain of one.

Finally we have the dc differential amplifier (Fig. 8-1(D)). This type of amplifier is used to provide a single-ended (i.e., unbalanced) output from a differential-(i.e., balanced-) input-signal source. The output voltage V_0 is proportional to the differential-voltage gain (A_{vd}) and the difference between input voltage V_1 and V_2 (i.e., $V_1 - V_2$)

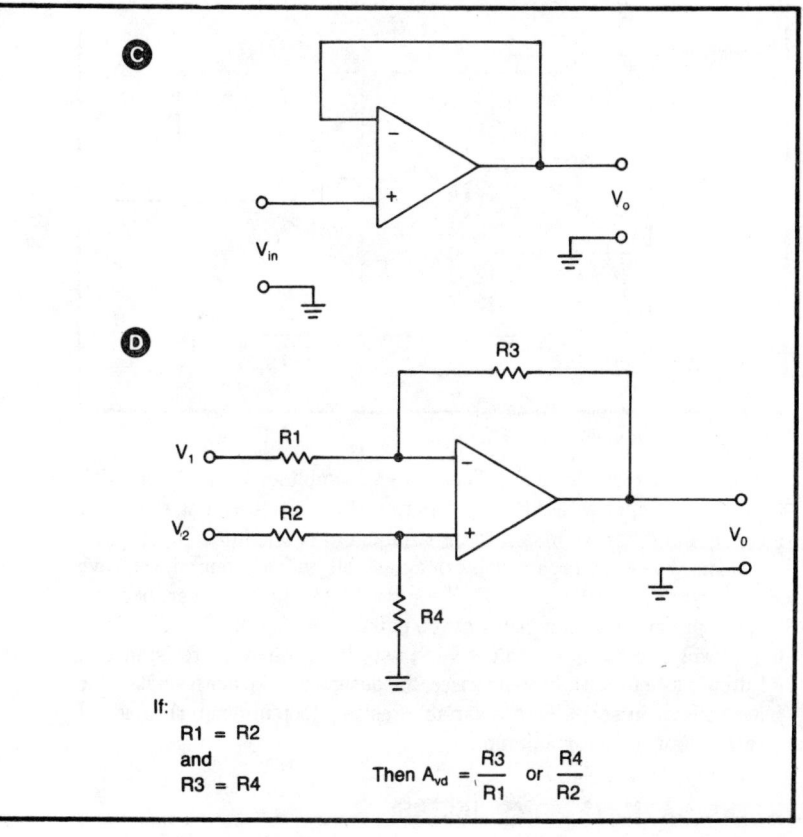

If:
R1 = R2
and
R3 = R4

Then $A_{vd} = \dfrac{R3}{R1}$ or $\dfrac{R4}{R2}$

$$V_0 = a_{vd} \times (V_1 - V_2) \qquad (8.1)$$

The gain is calculated in much the way of the inverting amplifier and is equal to either R3/R1 or R4/R2, provided that R1 equals R2 and R3 equals R4 (These equalities are important!).

The circuits above are based on single op amps. When we allow two or more op amps, however, even more complex circuits are possible. In the remainder of this chapter we will deal with IC versions of the Instrumentation Amplifier (I.A.) circuit shown in Fig.8-2. The I.A. provides an extremely high input impedance (similar to the noninverting follower circuits), a high possible gain, and easy design. The gain equation of this circuit is shown in Fig. 8-2.

The I.A. circuit shown in Fig. 8-2 consists of two sections: A1—A2 and A3. Amplifier A3 forms a simple dc differential amplifier such as Fig.8-2

157

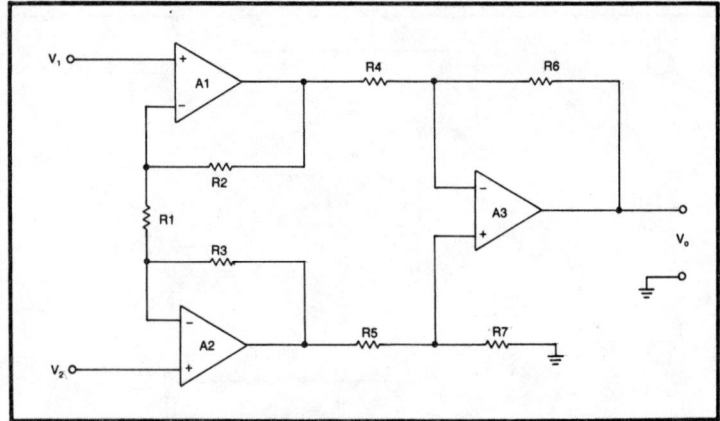

Fig. 8-2. Instrumentation amplifier.

and obeys the same rules. The A1—A2 amplifier is a differential-noninverting-input-with-differential-output stage. By cascading these two forms of amplifier we obtain the instrumentation amplifier.

In many cases, where variable or adjustable gain is required, we leave all resistors constant except R1. We must be careful, however, because R1 appears in the denominator of the equation in Fig. 8-1(D). This location means that the gain can get very, very large when the resistance of R1 drops close to zero. In some cases, the designer will place a small-value fixed resistor in series with a variable resistor (potentiometer) to adjust gain but limit it to a maximum.

IC OPERATIONAL AMPLIFIERS

The operational amplifier truly revolutionized analog-circuit design. For a long time, the only additional advances were that op amps became better and better (They came nearer the ideal op amp of textbooks!). While that was an exciting development, it was not really a new device. The next big breakthrough came when the analog-device designers made an IC version of Fig. 8-2, the integrated circuit instrumentation amplifier (ICIA). Today, manufacturers are offering better and better ICIA devices; we can truly say with an early op-amp textbook "the contriving of contrivances is a game for all."

Figure 8-3 shows a popular ICIA, the AMP-01 by Precision Monolithics, Inc. The AMP-01 is housed in an 18-pin DIP package (Fig. 8-3(A)).

The basic circuit for the AMP-01 is shown in Fig. 8-3(B). Notice how simple the circuit is! There are few connections: differential inputs, dc power supplies (V – and V +), output, ground, and two gain-setting resistors. The voltage gain of this circuit is given by

$$A_{vd} = 20 \, R_s/R_g \qquad (8.2)$$

Suppose you want to make a differential-voltage amplifier with a gain of 1000. You need to make a resistor ratio of 1000/20, or 50:1. Thus, if R_s is set to 100 kilohms and R_g is 2 kilohms, you will have the required gain of 1000. The permissible-gain range is 0.1 to 10,000.

The dc power-supply voltages are up to ± 18 volts. Notice in Fig. 8-3(B) that the dc power-supply lines are heavily bypassed. The 0.1-μF units are used to bypass high frequencies, while the 1-μF units are for low frequencies. The 0.1-μF units must be mounted as close as possible to the body of the amplifier.

The maximum operating frequency depends upon the gain. At a gain of one, the maximum small-signal input frequency is 570 kHz, while at a gain of 1000 it drops to 26 kHz.

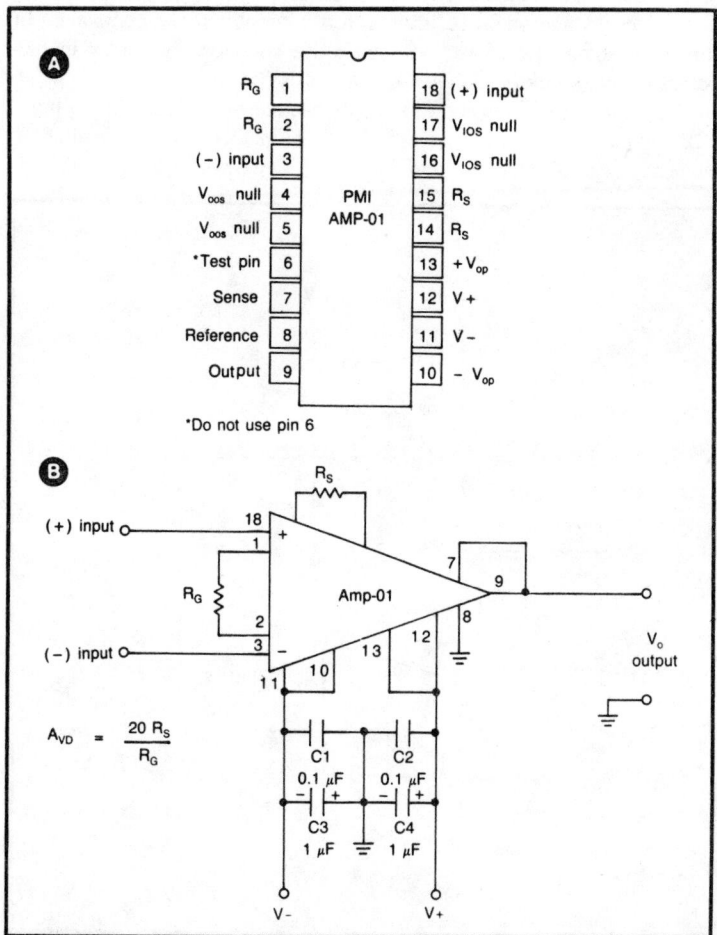

Fig. 8-3. Precision Monolithics, Inc. AMP-01 ICIA: (A) pinouts, (B) typical circuit.

159

The Burr-Brown INA-101 is another new ICIA device. This amplifier is similarly easy to connect. There are only dc power connections, differential-input connections, offset-adjust connections, ground, and an output. The gain of the circuit is set by

$$A_{vd} = 40 \text{ k}/R_g + 1$$

The INA-101 is basically a low-noise, low input-bias-current, integrated-circuit version of the I.A. of Fig. 8-2. The resistors labeled R2 and R3 in Fig. 8-2 are 20 kilohms, hence the "40 k" term in the equation of Fig. 8-4.

Potentiometer R1 in Fig. 8-4 is used to null the offset voltages appearing at the output. An offset voltage is a voltage that exists on the output at a time when it should be zero (i.e., when $V_1 = V_2$, so that $V_1 - V_2 \cdot 0$). The offset voltage might be internal to the amplifier or a component of the input signal. Dc offsets in signals are common, especially in biopotentials amplifiers such as in ECGs and EEGs.

Still another ICIA is the LM-363 device shown in Fig. 8-5; the mini-DIP version is shown in Fig. 8-5(A) (an 8-pin metal-can is also available), while a typical circuit is shown in Fig. 8-5(B). The LM-363 device is a fixed-

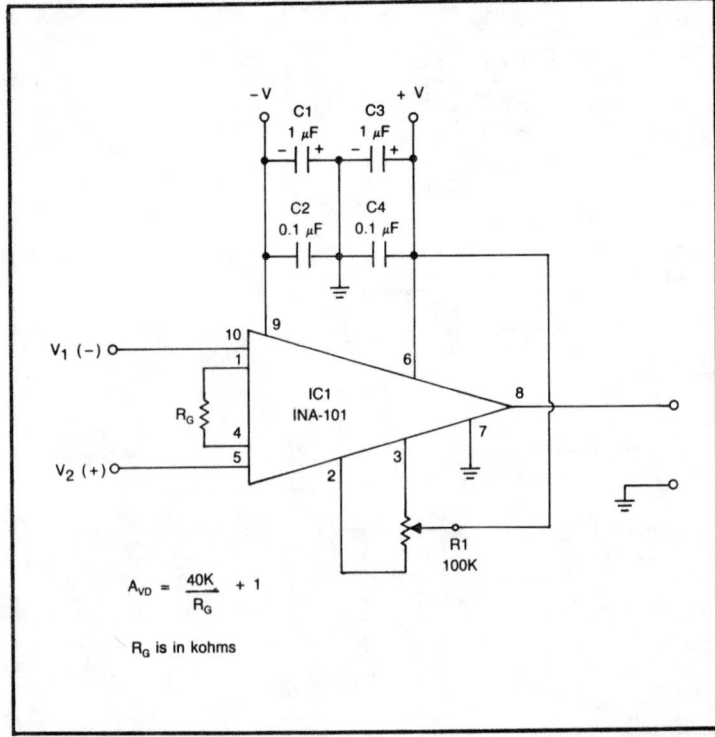

Fig. 8-4. Burr-Brown INA-101 ICIA.

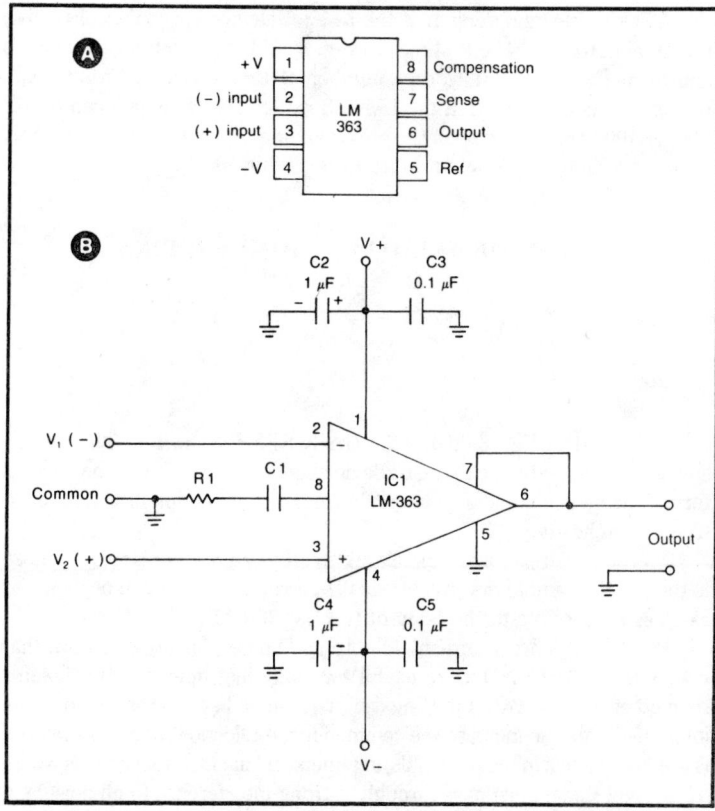

Fig. 8-5. National Semiconductor LM-363; (A) pinouts, (B) typical circuit.

gain ICIA. There are three versions:

DESIGNATION	GAIN
LM-363-10	10
LM-363-100	100
LM-363-500	500

The LM-363-xx is useful in places where one of the standard gains is required and there is minimum space available. Two examples spring to mind. You could use the LM-363-xx as a transducer preamplifier, especially in noisy signal areas; the LM-363-xx can be built onto (or into) the transducer to build up its signal before sending it to the main instrument or signal-acquisition computer. The other example is in bio-amplifiers. The bio-potentials are typically very small, especially in lab animals. The LM-363-xx can be mounted on the subject and a higher-level signal sent to the main instrument—a little exotic, but none the less useful.

161

A selectable-gain version of the LM-363 device is shown in Fig. 8-6; the 16-pin DIP package is shown in Fig. 8-6(A), while a typical circuit is shown in Fig. 8-6(B). The type number of this device is LM-363-AD, which distinguishes it from the LM-363-xx devices. The gain can be 10, 100, or 1000 depending upon the programming of the gain-setting pins (2, 3, and 4). The programming protocol is as follows:

GAIN DESIRED	JUMPER PINS
X10	All Open
X100	3 & 4
X1000	2 & 4

Switch S1 in Fig. 8-6(B) is the GAIN SELECT switch. This switch should be mounted close to the IC device but is quite flexible in mechanical form. The switch could also be made from a combination of CMOS electronic switches (e.g., 4066).

The dc-power-supply terminals are treated in a manner similar to those in the previous amplifiers. Again, the $0.1-\mu F$ capacitors need to be mounted as close as possible to the body of the LM-363-AD.

Pins 8 and 9 are guard-shield outputs. These pins are a feature that makes the LM-363-AD more useful for many instrumentation problems than other models. By outputting a signal sample back to the shield of the input lines, you can increase the common-mode-rejection ratio. This feature is used very much in bio-potentials amplifiers and in other applications where a low-level signal must pass through a strong-interference (high-noise) environment.

The LM-363 devices will operate with dc supply voltages of ± 5 to ± 18 volts dc, with a common-mode-rejection ratio (CMRR) of 130 dB. The 7-nanovolt-per-SQR(Hz) noise figure makes the device useful for low-noise applications; 0.5-nanovolt model is available at premium cost.

ISOLATION AMPLIFIERS

There are many applications for instrumentation amplifiers that are dangerous to either the circuit or the user. In bio-medical applications the issue is patient safety. There are numerous signal-acquisition situations in bio medical instrumentation in which the patient is at risk. Even the simple ECG machine, which measures and records the heart's electrical activity, was once implicated in patient-safety problems. Another problem area in bio medical applications is catherization instruments. There are several procedures in which the doctors insert an electrode or transducer into the body and measure the resulting signal; the intracardiac ECG places an electrode inside the heart by way of a blood vein, the cardiac-output computer uses a signal from a thermistor inside a catheter placed in the

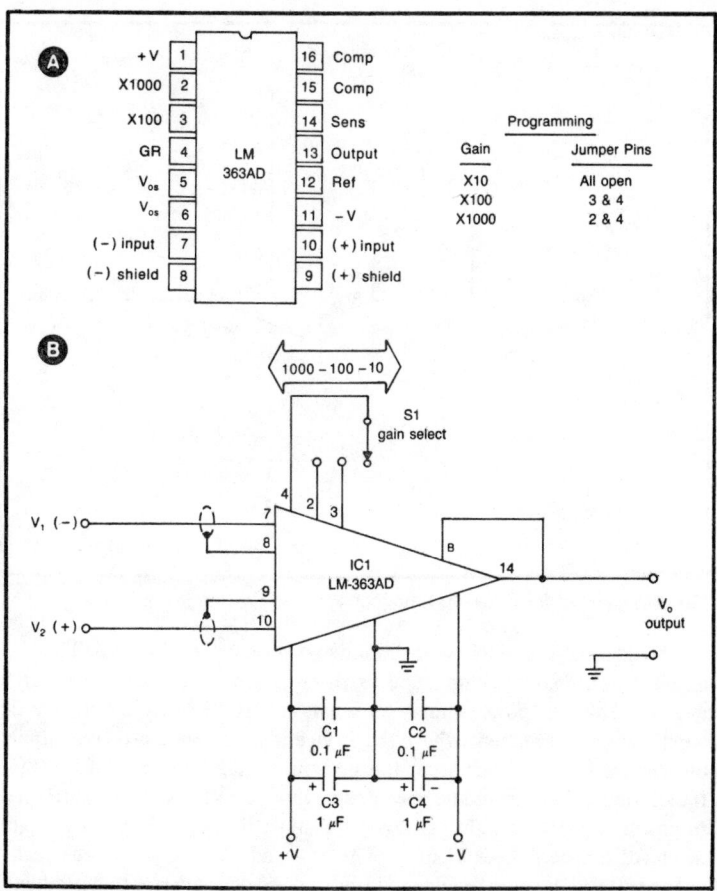

Fig. 8-6. National Semiconductor LM-363AD; (A) pinouts and gain options, (B) typical circuit.

heart (also through a vein), and simple electronic blood-pressure monitors use a transducer that connects to an artery. In all of these cases we do not want the patient exposed to small differences of potential due to current leakage from the 60-Hz ac-power lines. The solution is the use of an isolation amplifier.

Another application is signal acquisition in high-voltage circuits. We do not want to mix high-voltage sources with low-voltage electronics because we don't want the low-voltage circuits to blow out. Again, the solution is the isolation amplifier.

Figure 8-7 shows the basic symbol for the isolation amplifier. The break in the triangle used to represent any amplifier denotes the fact that there is an extremely high impedance (typically 10^{12} ohms) between the inputs and the output terminal of the isolation amplifier.

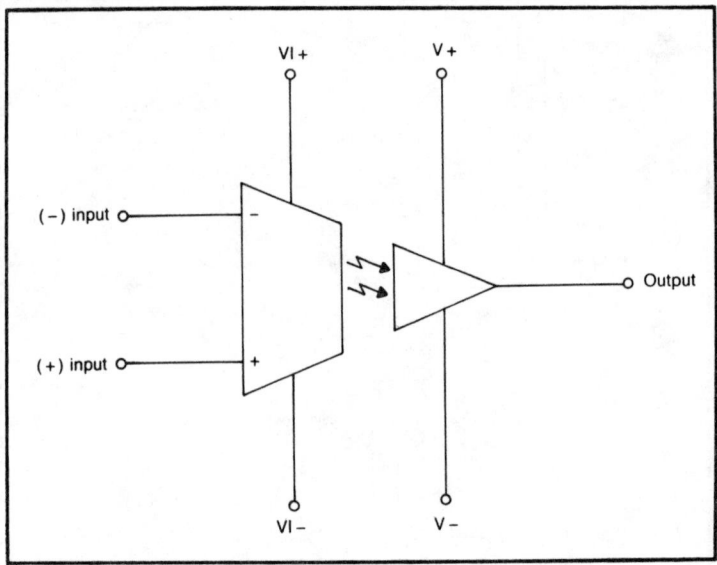

Fig. 8-7. Symbol for isolation amplifier.

Notice that there are two sets of dc-power-supply terminals. The V − and V + terminals are the same as those found on all ICIA or op-amp devices. These dc-power-supply terminals are connected to the regular dc supply of the equipment where the device is used. Such a power supply derives its dc potentials from the ac power mains by way of a 60-Hz transformer. The isolated dc power supply inputs (VI − and VI +) are used to power the input-amplifier stages and must be isolated from the main dc power supply of the equipment. The VI − and VI + terminals are usually either battery-powered or powered from a dc-to-dc converter that produces a dc output from the main power supply by using a high frequency (50 kHz to 500 kHz) oscillator. The high-frequency "power supply" transformer does not pass 60-Hz signals well; so the isolation is maintained.

Figure 8-8 shows the circuit of an isolation amplifier based on the Burr-Brown 3652 device. This isolation amplifier is not generally available to hobbyists but would be used even in small, "one of a kind" professional labs.

The dc power for both the isolated and nonisolated sections of the 3652 is provided by the 722 dual dc-to-dc converter. This device produces two independent, ± 15-volt-dc power supplies both of which are isolated from the 60-Hz ac power mains and from each other. The 722 device is powered from a + 12-volt-dc source that is derived from the ac power mains. In some cases, the nonisolated section (which is connected to the output terminal) is powered from a bipolar dc power supply that is derived from the 60-Hz ac mains, such as a ± 12-volt-dc or ± 15-volt-dc power supply. In no instance, however, should the isolated dc power supplies be derived from the ac power mains.

164

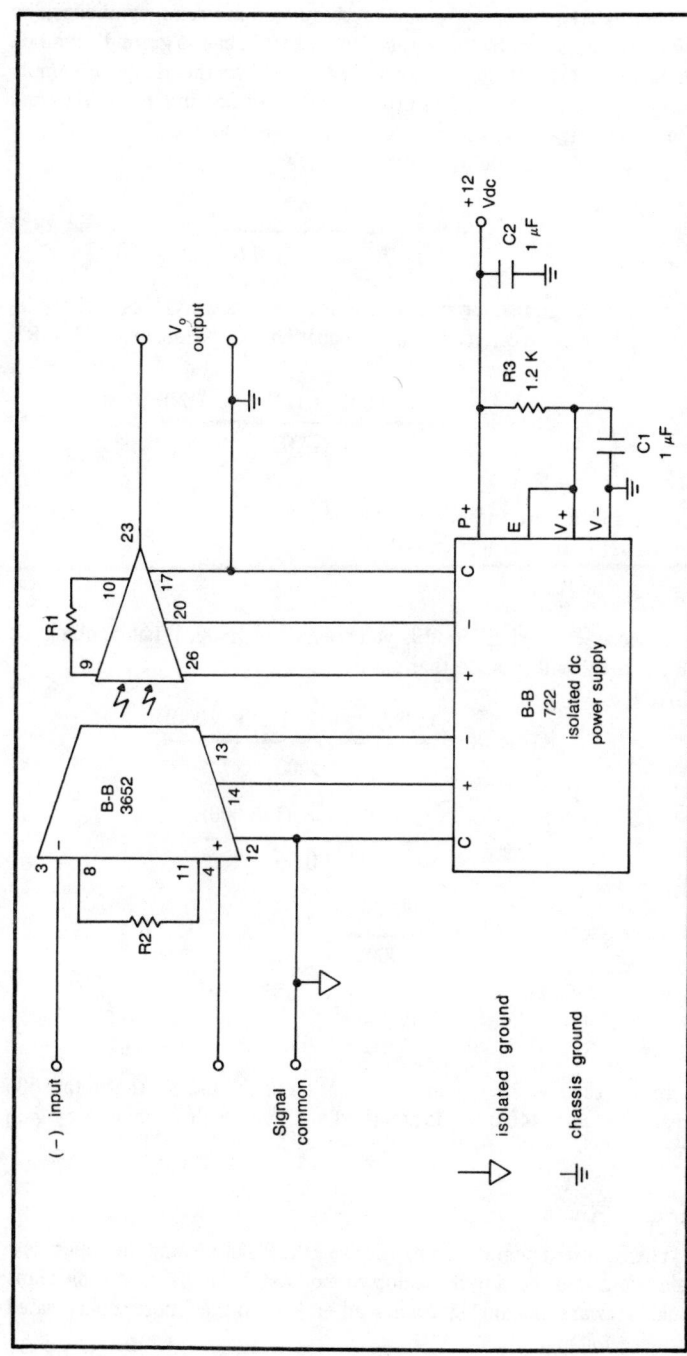

Fig. 8-8. Circuit for the Burr-Brown 3652 instrumentation amplifier.

165

There are two separate ground systems in this circuit, symbolized by the small triangle and by the regular, three-bar, "chassis ground" symbol. The isolated ground is not connected to either the dc-power-supply ground/common or the chassis ground. It is kept floating at all times and becomes the signal common for the input-signal source.

The gain of the circuit is approximately:

$$GAIN = \frac{1,000,000}{R1 + R2 + 115} \qquad (8.3)$$

In most design instances, the issue is the unknown values of the gain-setting resistors. We can rearrange the equation above to solve for R1 + R2:

$$R1 + R2 = \frac{1,000,000 - (115 \times GAIN)}{GAIN} \qquad (8.4)$$

where

- R1 and R2 are in ohms
- GAIN is the voltage gain deired.

Example

Suppose we need differential-voltage gain of 1,000. What combination of R1 and R2 will provide that gain?

If GAIN = 1000

$$R1 + R2 = \frac{1,000,000 - (115 \times 1000)}{1000}$$

$$= \frac{1,000,000 - (115,000)}{1000}$$

$$= \frac{885,000}{1000}$$

$$= 885 \text{ ohms}$$

In this case we need some combination of R1 and R2 that totals 885 ohms. The value 440 ohms is standard and will result in only a tiny gain error.

Conclusion

The IC instrumentation amplifier and the isolation amplifier open new applications that the simple op amp cannot match. Digital electronics fans should be aware that analog is not dead—it lives in even more sophisticated manifestations.

Chapter 9

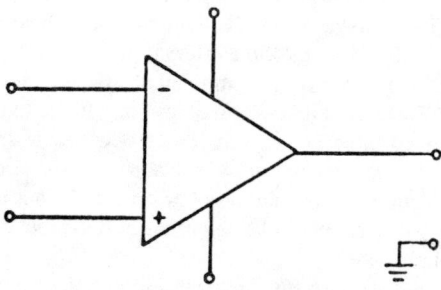

Miscellaneous Useful Circuits—Mostly Analog

T HIS CHAPTER WILL TREAT CERTAIN CIRCUIT CONFIGURATIONS THAT do not seem to fit nicely into the structure of the book, even though they are interesting and immensely useful. Although most of these circuits do not rate a chapter of their own, they are particularly useful for instrumentation design.

INTEGRATION

In mathematics the term *integration* refers to a process for finding the area under the curve of function. In many areas of engineering and those physical sciences where mathematics is a tool rather than an entity in its own right, integration over time is frequently required. This allows us to find the time average of the function, which is occasionally somewhat more interesting than the function itself.

Certain electronic circuits can be made to approximate the process of time-averaging. In the case of an electronic integrator

$$E_{out} = K \int_{t1}^{t2} E_{in} \, dt + E_1 \tag{9.1}$$

where:

E_{in} is a voltage function of time
E_1 is a constant of integration, or "initial condition"
K is a constant
t is time

A PASSIVE INTEGRATOR

Figure 9-1 shows the basic form of the passive RC integrator. In this circuit current I is generated by the input voltage, and it is used to charge capacitor C1. The voltage across C1, then, is proportional to the integral described by Eq. 9.1 over the time interval (t2 – t1).

The circuit of Fig. 9-1 may seem familiar to you under the title "RC low-pass filter," as indeed it is a form of low-pass filter. In fact, wherever the term "low-pass filter" appears in certain scientific-instrument-design descriptions you may suspect that the actual use for the circuit is integration over time. This dictum is not absolute, of course, but it is seen often enough to make it desirable to be cognizant of this practice of industrial technical-manual writers.

The knee frequency of the circuit in Fig. 9-1 is given by

$$f_o = \frac{1}{2\pi RC} \qquad (9.2)$$

You can probably deduce from this that the integration properties of this circuit depend somewhat on the frequency of operation. At frequencies above f_o we find that integration is minimal, if existent at all, and reduces sharply as frequency increases. Integration is possible only if the frequency of the applied waveform is well below f_o. Another way of saying this is that the period of the input signal (1/f) must be less than the time constant of the RC network.

One disadvantage of the simple RC integrator is that it is lossy and attenuates the signal quite a bit, especially if the circuits which follow place a substantial load on resistor R1, a problem if the load resistance seen by the integrator is less than 100 R1, or in some cases 1000 R1. One advantage of the circuit is that it will provide integrator action while requiring no active components or power-supply connection. You may want to use this type of integrator as a "fix" for certain noise problems that might exist in an instrument system. There might be, for example, a high-frequency artifact riding on the waveform of interest, or alternatively, perhaps small, high frequency variations are of no interest in the present experiment, but the time average over a short period is of prime concern. The integrator of Fig. 9-1 can be built inside a shielded, metal box that is fitted with connectors that mate with existing instruments and then be connected into the system with little effort. As one widely-experienced systems engineer claimed, "Let me have my box of capacitors and resistors, and I can fix almost any (artifact) problem." That may actually be a little grandiose, but there is a ring of solid truth in it!

The only real constraint on the use of this technique is that it should be driven from a low-impedance source, such as an amplifier output, and should in turn drive a high-impedance load, such as the noninverting,

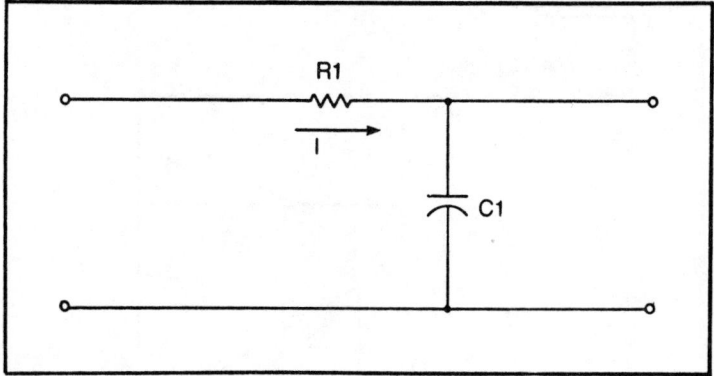

Fig. 9-1. RC low-pass filter/integrator.

operational-amplifier follower. It is also important to try the technique only where the circuit will work with low values of R and C. The integration effect is, after all, dependent upon the RC time constant, and high values of either can cause problems. Keep R much less than 1 megohm, and C less than 1 μF. In any event, use only high-quality capacitors.

ACTIVE INTEGRATORS

An active integrator can be made by sandwiching the circuit of Fig. 9-1 between a pair of operational amplifiers. A unity-gain follower is used to drive the input of the integrator, while another follower is used to provide output. The capacitor of the integrator is connected between the noninverting input of the second operational amplifier and ground. The first operational amplifier can be almost any device, provided that it has either a low inherent offset present at its output, or appropriate circuitry for nulling the offset. This will create an excessive value of error, corresponding to an unwanted initial condition term (E_1 in Eq. 9.1). The second operational amplifier, though, must have an extremely high input impedance, or it will tend to discharge the capacitor even as it is integrating the input waveform. Use a JFET or MOSFET (BiMOS) input operational amplifier for this one. Also, be aware that connecting a capacitor between an operational-amplifier input and ground can give rise to the generation of high-frequency noise. Therefore, this technique is only recommended where the input waveform has a relatively large amplitude.

Figure 9-2 shows an active integrator that overcomes some of the disadvantages of previous circuits, even at the expense of a certain headache to be described shortly.

In this integrator circuit, which most texts list as the classic op-amp integrator, the capacitor used to store the charge proportional to the integral is in the feedback loop of the operational amplifier. The current used to charge this capacitor is equal to the current generated by the input volt-

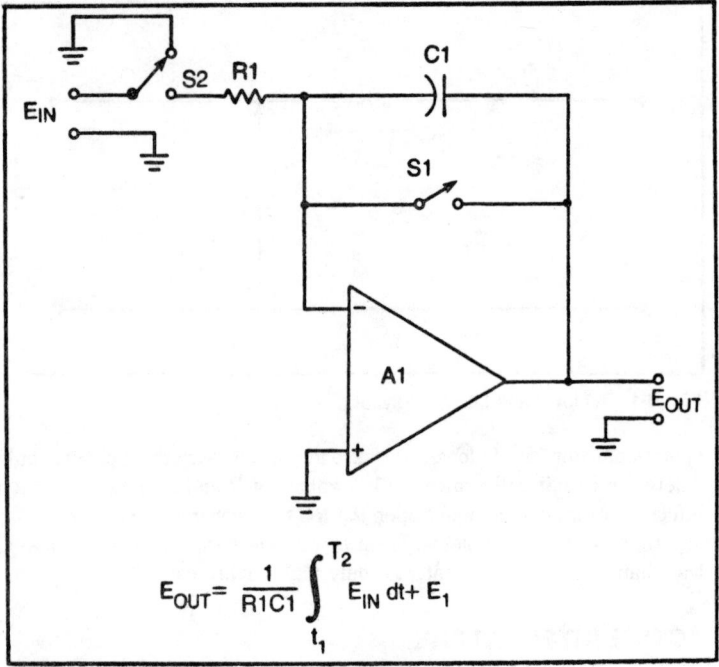

$$E_{OUT} = \frac{1}{R1C1} \int_{t_1}^{T_2} E_{IN}\, dt + E_1$$

Fig. 9-2. Operational-amplifier active integrator.

age in resistor R1. The active integrator obeys the equation

$$E_{out} = \frac{1}{R1\ C1} \int_{t1}^{t2} E_{in}\, dt + E_1 \qquad (9.3)$$

As was also true in Eq. 9.1, the term E_1 represents an initial condition. It will take the form of a voltage already present in the output when the time interval (t2 − t1) commences. It may be due to errors or offset artifacts or may be intentionally preloaded into the integrator prior to performing the integration. This term is necessary, outside of the mathematical purity of providing an integration constant for all integrals, because many instrumented processes either preload a constant or will sum a voltage representing some signal process that occurred prior to integration. In some instruments, for example, the integrator input might be switched from one part of the circuit to another at some predetermined point in the process, so the output result will actually be a double integration, or, if you prefer rigor, the summation of two integrations. In that case the result of the first integration might be treated as an E_1 term by the second, allowing the economy of using a single integrator to perform the allotted task.

170

The integrator will theoretically retain its final voltage at time t2. In practice, though, we find that the perversity of the universe takes over and the output diminishes, seemingly without cause, other then perhaps the occult. What we are seeing here is the result of using non-ideal components in the circuit. Both capacitor leakage and the input impedance of the operational amplifier (its input bias currents, mostly) conspire to cause a little bit of output-voltage droop, that is, a decreasing of the output voltage without a corresponding change from the input signal.

Switch S1 is used to reset the integrator to zero, which should be done prior to integration so that spurious E_1 voltages might be eliminated. Input switch S2 is used to set the time interval over which integration takes place. It will be connected such that the input is grounded prior to, and after, the time interval, but will be connected to the integrator during the time interval.

Almost any operational amplifier will theoretically work for A1, especially if a little sloppiness can be tolerated. This might be the case, for example, where the integrator is being used primarily as a low-pass filter, or in certain other applications. In most cases, though, it is necessary to use a premium-grade operational amplifier for A1. This will eliminate some of the problems which are discussed in the article on the practical integrator in Chapter 22. It becomes especially desirable if the circuit is being used to process a slowly-changing signal or if the output level either must be retained over a long period following integration or if the integration must be performed over a long time—long time being measured in seconds. Again, the problem is that bias currents from the operational amplifier tend to charge capacitor C1. Also, in some few cases, if the open-loop voltage gain of the device is too low, other error terms become apparent.

Another problem in this type of circuit is output-voltage droop due to leakage currents through the capacitor. This phenomenon results because all capacitors have a certain amount of parallel resistance. Fortunately, most of the droop problems associated with electronic integrators can be all but eliminated through the simple expedient of using only high-quality components. For example, don't *ever* use an electrolytic capacitor for C1. All electrolytics have a rather high leakage and are suited for only the most gross applications where their small size for relatively large values of capacitance is the most important consideration. The rest of the time we must consider the lowly electrolytic to be among the great unwashed—doomed to lowly estate.

In moderate applications you can get away with using mylar, mica, ceramic, metal film oxide, or Pacer Filmite capacitors. In critical applications—where the time constant might be long or the hold time lengthy or little droop can be tolerated—it is necessary to use high-grade capacitors with glass, polycarbonate, or polystyrene dielectrics.

The operational amplifier also must be of very high input impedance. The type 741 should not be considered for any but the sloppiest of applications, and even there it will require some method for counteracting the input bias currents. The best thing to do is specify an operational amplifier with an extremely high input impedance, such as those with FET or superbeta (Darlington) input stages, or the newer diode-protected RCA BiMOS series such as the CA3130, CA3140, and the CA3160.

It must also be realized that it is easy to absolutely ruin your best efforts by picking the wrong values for R1 and C1. Examine Eq. 9.3 and note that the product R1C1 appears in the *denominator*. Real values of R1 will be on the order of 10^7 ohms or less while real capacitors are available with values on the order of 10^{-12} farads up to around 10^{-6} farads in the types of capacitor that are actually suitable for integrator service. If you use the wrong values, the integrator will have a fantastic gain and will remain saturated most of the time. Consider an example in which a 100-kilohm resistor and a 1-μF capacitor is used. The constant outside the integral symbol in Eq. 9.3 becomes

$$\frac{1}{(10^5)(10^{-6})} = 10^1$$

Such an RC network will multiply the output by a factor of 10; therefore with most operational amplifiers the input range is limited to something considerably less than 1 volt. The situation is even worse when lower-value capacitors are used. If, for instance, a 1000-pF capacitor had been used instead, the multiplication factor of the constant would be

$$\frac{1}{(10^5)(10^{-9})} = 10^4 = 10,000$$

In that case the integral of the input voltage would be multiplied by

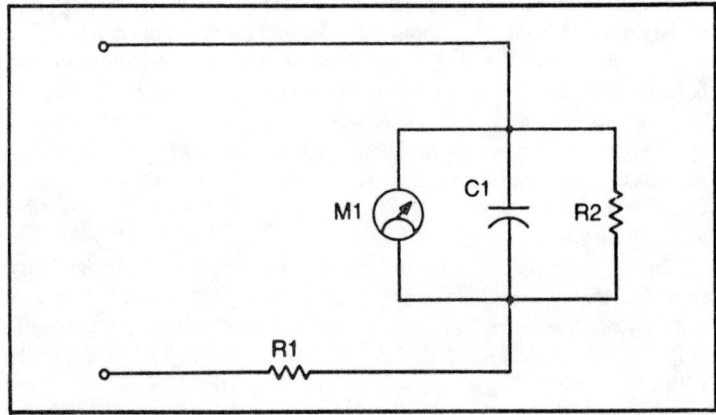

Fig. 9-3. Electromechanical integrator.

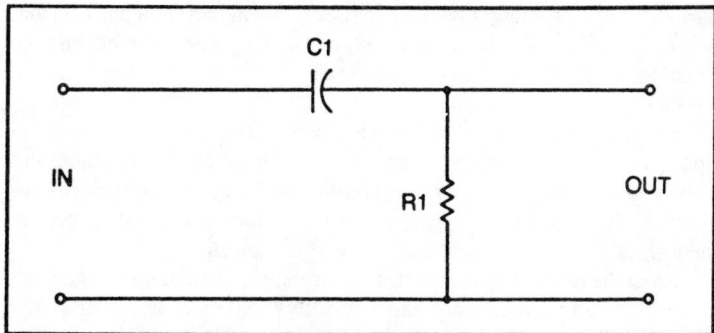

Fig. 9-4. RC high-pass filter/differentiator.

a factor of 10,000! This means that as soon as an input current greater than a few microamps accumulates, the output voltage will rise against the upper, or power supply, limit, saturating the stage. Hardly useful. It is also true that in such a case input bias currents will cause the output to rise as a ramp with a short rise time!

ELECTROMECHANICAL INTEGRATION

Before leaving the subject of integrators let us consider a crude but effective integrator used in many instruments, especially, but not necessarily, those of low cost. The circuit shown in Fig. 9-3 is basically an RC/mechanical integrator that depends upon the meter movement's own inertia and bearing friction, plus the actual electronic integrating effect of the RC time constant. This consists of capacitor C1 and the sum of resistances, including the meter's coil resistance. Capacitor C1, incidentally, may be an electrolytic in direct contradiction to our earlier advice because it must have a value of several dozen to several hundred microfarads, and it is not the sole source of integration. In fact, it's used mainly for damping.

DIFFERENTIATORS

A differentiator is a circuit that gives an output voltage that is proportional to the input signal's *rate of change*. For dc there is no output from a differentiator because the input is not *changing*. The output of a differentiator, then, is the derivative of the input function. A simple differentiator is shown in Fig. 9-4.

A differentiator is also useful as a high-pass filter with a knee frequency that is given by Eq. 9.2. Again we see here the implication that properties of a circuit, in this case differentiation, are frequency dependent. The RC time constant of R1C1 must be very short relative to the period of the input signal. In some cases you would use the actual period defined by the relationship $T = 1/f$, but in others you would have to use some other time period. Such a case might be the rise time of a long-duration pulse, if that property is of interest. In either event, it is usually wise to make the time constant of the differentiator approximately one-tenth of the time period selected.

An active differentiator is shown in Fig. 9-5. This circuit used an operational amplifier as an active current-source element. Notice the placement of R1 and C1 is reversed from the placement seen in the active integrator circuit in Fig. 9-2. This may lead you to understand a fact that proves immensely useful in electronic instrumentation design.

From the calculus we know that integration of the derivative of a function will yield that function. Similarly, differentiation of the integral of a function will also yield the function. Therefore, we may conclude that integration and differentiation are complementary processes. If you pass a signal through first an integrator and then a differentiator, it should theoretically emerge unchanged. As you will find out in the next chapter (on transducers) this can be very useful when the actual information desired is not easy to obtain but its derivative or integral is.

PEAK FOLLOWERS

A peak follower is a circuit that will produce an output voltage equal to, or at least proportional to, the highest peak voltage on the input waveform.

An example of a peak holder, or follower as it is often called, is shown in Fig. 9-6. Operational amplifier A1 is basically a buffer and is connected in the unity-gain, noninverting-follower configuration. The output of operational amplifier A1 is used to charge capacitor C1. Because the output stage is another unity-gain amplifier, the output voltage will be approximately equal to the voltage across C1.

The purpose of diode D1 is to prevent discharge of capacitor C1 through the low-impedance output circuit of amplifier A1. This diode *must* be a

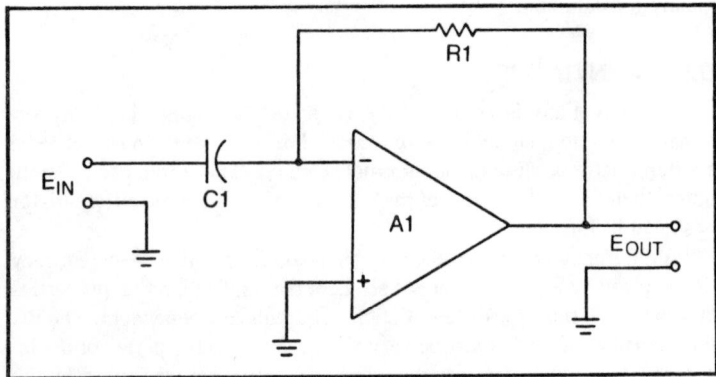

Fig. 9-5. Operational-amplifier active differentiator.

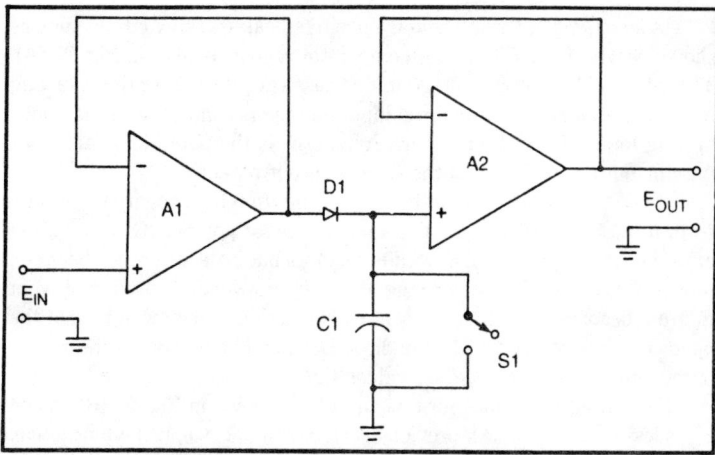

Fig. 9-6. Operational-amplifier peak follower/holder.

silicon type, and it must be selected, either by type-number specification or by manual means, to have an exceptionally high reverse impedance. In some instances we find amplifier A1, diode D1, and a second high-grade diode similar to D1 are formed into an active precision, or "ideal," rectifier circuit.

One primary problem with this type of circuit is output-voltage droop, not particularly surprising since the circuit bears a striking resemblance to the integrator discussed earlier. Droop occurs because of partial discharge of capacitor C1 through at least three separate paths: D1 leakage, A2 input impedance, and the parallel resistance of C1. The "fixes" for these are the same as for the active-integrator circuit's droop. These were presented earlier in this chapter.

The peak follower of Fig. 9-6 will follow only the peaks of a positive input signal. This is due to the direction allowed to current passing through diode D1. Negative peaks can be tracked by making either of two circuit changes: reverse diode D1, or make A1 into an inverting stage. Alternatively, you may also precede amplifier A1 with an inverting-follower stage.

The potential across capacitor C1 is actually a little less than the actual input voltage. This is due to the normal voltage drop across the silicon diode, approximately 0.7 volts, and this creates a slight error. This can be overcome by making the input stage an ideal or precision rectifier, or by giving A1 a slight dc offset to compensate for the error. Of the two, the former yields the better circuit.

In some instrumentation applications there is need for generation of a logic level every time a new peak is established. This can be done simply through the use of an operational-amplifier comparator (see Chapter 3). One input of the comparator is connected to E_{out}, while the other is connected to E_{in}.

As an example, let us assume a positive-peak-tracking circuit such as shown previously in Fig. 9-6. A comparator is connected as in Fig. 9-7(A). The peak-holder output voltage in this case is connected to the inverting input of the comparator, and the input voltage is connected to the noninverting input of the comparator. Voltage E_a is the potential at the comparator output, while E_b is the circuit output potential.

The output of the comparator may swing from high negative, through zero, to high positive, depending upon the input polarity. Of course, the extremely high gain of such circuits requires but little difference between input voltages to effect the change. Diode D1, however, is used to prevent E_b from becoming negative. In some other design, it may well be that the opposite is desired; in which case diode D1 would be reversed. The overall output, then, can only take on values that are zero or positive.

The output conditions for this circuit are shown in Fig. 9-7(B). When E_{in} is less than the highest previous peak, voltage E_a will be high negative. This reverse biases D1, making E_b zero. Similarly, when the input voltage rises to precisely the same value as the highest previous peak ($E_{out} = E_{in}$), voltage E_a will be zero, as will be E_b. When a new peak is

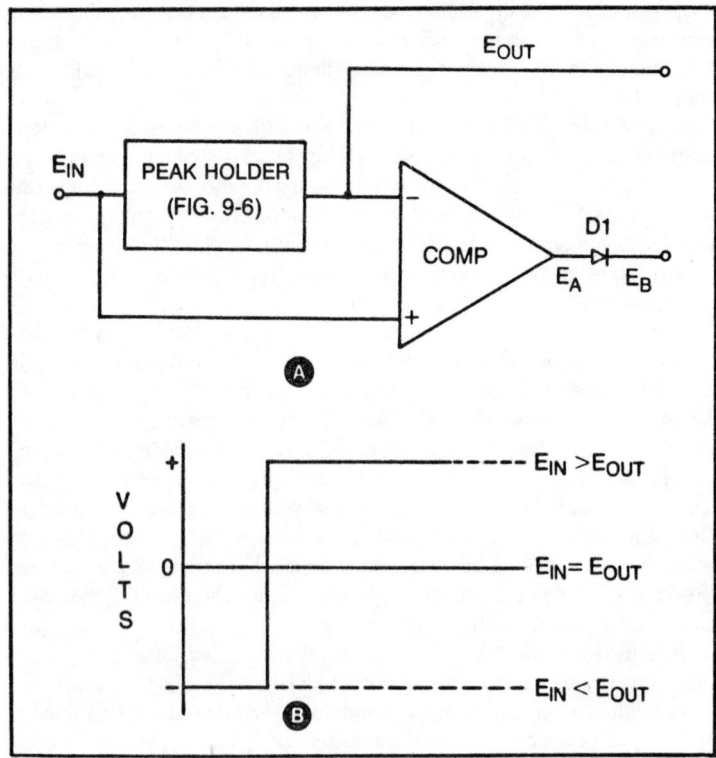

Fig. 9-7. Peak detector with digital output.

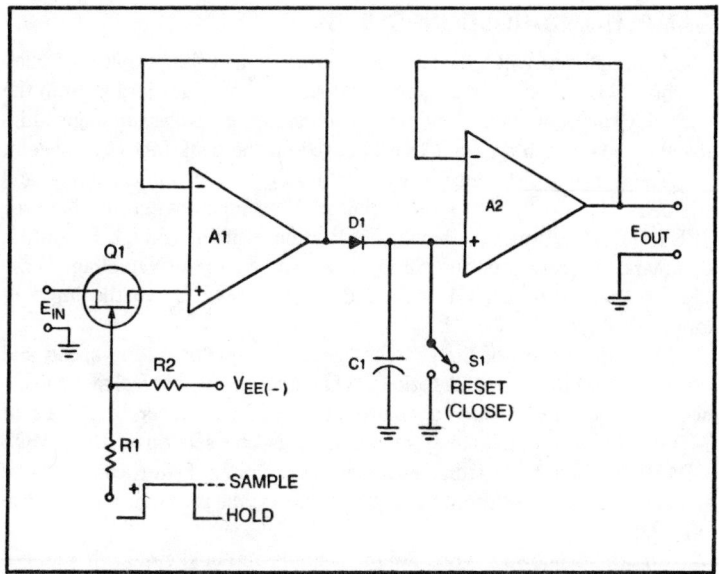

Fig. 9-8. Sample-and-hold.

established, however, E_{in} will be greater than E_{out} for the short period of time required to charge the holding capacitor. This causes the output of the comparator to snap positive for a brief time whenever a new peak is established. The width of this pulse is determined by the rate at which C1 (see Fig. 9-6) charges to full value. If this capacitor is too small, it will charge very rapidly, making for an extremely short pulse, perhaps too short for practical use. In any event, variation in pulse widths and other problems causes most designers to follow the comparator with a monostable multivibrator (one-shot) capable of responding to a short trigger pulse.

Such gated peak followers are sometimes seen in instrumentation or computer data acquisition systems, where only peak values are needed and only new peaks are of any further interest. A pulse from the E_b terminal or a subsequent one-shot stage could, for example, tell a computer to digitize the new data. This requires the use of much less, valuable computer memory, a blessing in microcomputer circuits. Only relevant data is stored; the rest is rejected.

Some manufacturers offer special peak-holder or peak-follower, analog ICs (function modules). Most of these will include the two required operational amplifiers, the isolation diode, internal-power-supply regulation, and internal-temperature compensation on a single chip or ceramic substrate. Some will operate from a monopolar power supply (Vcc-to-ground), but most require the same sort of power supply as an operational amplifier. These devices are preferred where either component count or thermal tracking is a critical specification.

177

SAMPLE-AND-HOLD CIRCUITS

A circuit that will measure and hold an input voltage that is applied but briefly is called a *sample-and-hold* circuit. In an older jargon from the early days of radar, such circuits were sometimes called *boxcar* stages. Figure 9-8 shows such a circuit that is based on the peak-follower concept.

The principal difference between an analog sample-and-hold circuit and the peak-follower circuit is the inclusion of an input switch, in this case JFET Q1. This transistor blocks application of input voltage E_{in} until a command is received on the control input. When a positive voltage is applied to that terminal, Q1 will conduct and apply E_{in} to the input of amplifier A1.

The actual form of the switches depends upon operating speed and certain other circuit considerations. They may be mechanical switches, a set of relay contacts (in very-slow-speed systems), discrete JFETs or MOSFETs, or an IC analog switch such as one section of the CMOS CD4016. Although electronic switches are definitely faster, care must be taken that the type selected has a very high reverse resistance in the turn-off mode.

Figure 9-9 shows a type of digital sample-and-hold circuit that all but eliminates output droop, albeit at the expense of increased circuit complexity. This makes the digital sample-and-hold especially useful for long-duration holds, or applications where little droop can be tolerated. The heart of the digital sample-and-hold is the IC digital-to-analog converter (DAC).

A DAC is, in this context, a circuit that generates an output current (I_o) that is proportional to the value of a digital word present at its inputs. Most IC DACs have a built-in, binary, digital counter, and the R2R resistor network needed to generate the current (see Chapter 14). The only external circuitry required is a source of clock pulses and whatever control logic is needed for the particular purpose at hand.

Current I_o is applied to the inverting input of amplifier A1. The voltage output, E_{out}, is equal to I_oR. This voltage is fed to both the output terminal and to the inverting input of amplifier A2, a comparator-connected operational amplifier.

The input voltage being sampled, E_{in}, is applied to the noninverting input of comparator A2. The output of the comparator will remain high so long as E_{in} is greater than E_{out}. Under the *hold* condition both inputs to the control-logic circuit are high; so its output is low. This tells the control-logic circuit to cease sending clock pulses to the DAC.

Let us now consider the sequence of events in a typical sampling cycle. A reset pulse is applied to the reset input. This resets the binary counter in the DAC to, say, 00000000 and makes sure that the control-logic circuit is inhibiting the DAC's clock input. Next, the sample *pulse* is applied to the control-logic circuit. This pulse is used to unlock the control-logic circuit and allow clock pulses to enter the DAC's binary-counter input.

Current I_o will be zero when the counter is in the zero state but in-

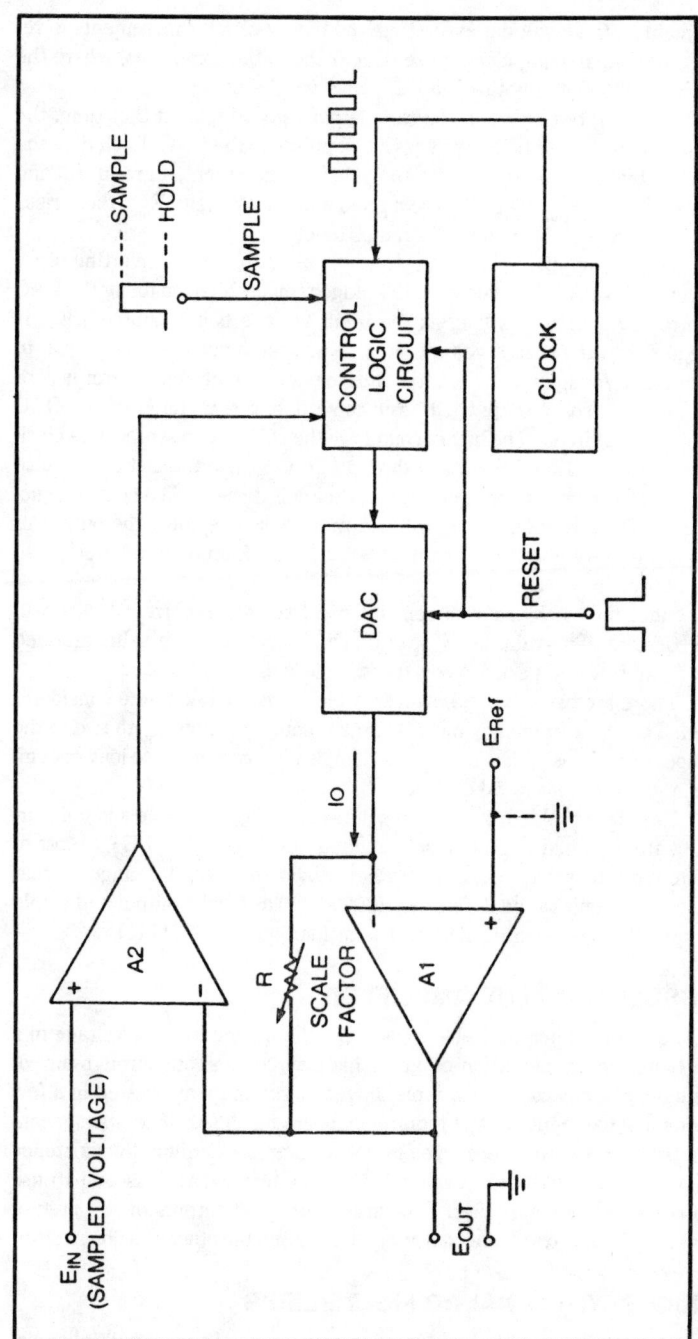

Fig. 9-9. Digital sample-and-hold.

179

crements a fixed amount every time the binary counter increments in response to an arriving clock pulse. Under the initial conditions, where the count is 00000000, both I_o and E_{out} also will be zero.

Sampling begins when the sample *pulse* goes high. At that time, the control-logic circuit begins to pass clock pulses to the DAC. Each time the DAC's binary-counter outputs (not shown) increment, current I_o, and hence voltage E_{out}, will increase a fixed amount. Voltage E_{out}, then, rises after the fashion of a slope, or ramp, function.

The voltage being sampled, E_{in}, is applied to the noninverting input of comparator A2. The output of this stage remains high, enabling the DAC control-logic circuit as long as the input voltage being sampled, E_{in}, is greater than the output voltage, E_{out}. When the input voltage is equal to the output voltage the differential input voltage to the comparator is zero ($E_{in} = E_{out}$); so its output will drop to zero, and this shuts off the DAC control-logic circuit. The binary counter in the DAC no longer receives clock pulses; so its output remains in the state it had when the last clock pulse had been received, prior to control-logic-circuit turn-off. The control-logic-circuit design is such that it will not turn back on even if the output of comparator A2 goes high again. It will turn back on only in response to another sample *pulse*.

Since the binary counter remains in a fixed state, current I_o also will be constant. The value of E_{out}, then, will be equal to the value attained by E_{in} during the period when the sample *pulse* was high.

There are two factors, however, which must be taken into consideration. The clock frequency must be great enough to allow E_{in} to rise to the proper level. This means that t0, the sampled interval, must be long enough for a good sample to take place.

The other problem is circuit resolution. All digital sample-and-hold circuits are designed to recognize maximum levels of E_{in} and E_{out}. Most of them typically work from 0 to 10 volts or over some similar range. In that case, our circuit would define zero (00000000) as 0 volts output and would supply 10 volts when the counter is at maximum (11111111).

ABSOLUTE-VALUE AMPLIFIERS

It is sometimes desirable to use only the magnitude of a voltage in a particular instrumentation circuit. This may come about through any of a number of means. For example, in vector summation (covered in a few moments) the input-voltage requirements are for the *absolute value*, meaning positive. In other cases, you might be interested only in the existence of a voltage level, but not the polarity. In either event, it is best to use an absolute-value amplifier. This circuit sums the outputs of two precise rectifier circuits, one connected as an inverter and the other as a noninverter.

TRICKS WITH ANALOG MULTIPLIERS

In an earlier chapter I discussed various forms of analog multiplier circuits. In most cases you will want to use the IC type that obeys the relationship $E_{out} = KV_xV_y$.

It is unfortunate that most of these ICs are called "multipliers" by their manufacturers, because they have an immense usefulness beyond simple analog multiplication and multiplication processes such as amplitude modulation. In this section I will cover four other arithmetic functions using the IC analog multiplier as the heart.

Squarer

It takes almost no cleverness to realize that a circuit with the transfer function

$$E_{out} = KV_xV_y \tag{9.4}$$

can also yield

$$E_{out} = K(V_x)^2 \tag{9.5}$$

if the X and Y inputs are tied together to form a single input. A *squarer*, then, is an analog circuit in which a multiplier has its two inputs together.

Analog Dividers

Division is the inverse of multiplication. One of the factors that makes operational amplifiers so useful is that *the transfer function of the overall circuit is the inverse of the transfer function of the feedback-network circuit*. Recall that a simple dc amplifier has a resistor voltage divider (R_f/R_{in}) in the feedback loop. This makes the stage a voltage multiplier in which the ratio of those two resistors sets the multiplication constant. Similarly, we can form an analog divider by placing an analog multiplier in the feedback loop of an operational amplifier. Incidentally, please remember that fact about the transfer function of an operational amplifier. It is a very useful bit of knowledge because many times you can easily produce a passive or active circuit in which the inverse of our desired function is achieved. By placing that circuit in the feedback loop of an operational amplifier you get the desired circuit action.

An example of a simple multiplier-based, analog divider is shown in Fig. 9-10. Recall from an earlier chapter that an analog multiplier will obey the rule

$$E_o = KV_xV_y \tag{9.6}$$

in the circuit of Fig. 9-10, where V2 is E_o and E_{out} is V_y; so

$$V_2 = KV_xE_{out} \tag{9.7}$$

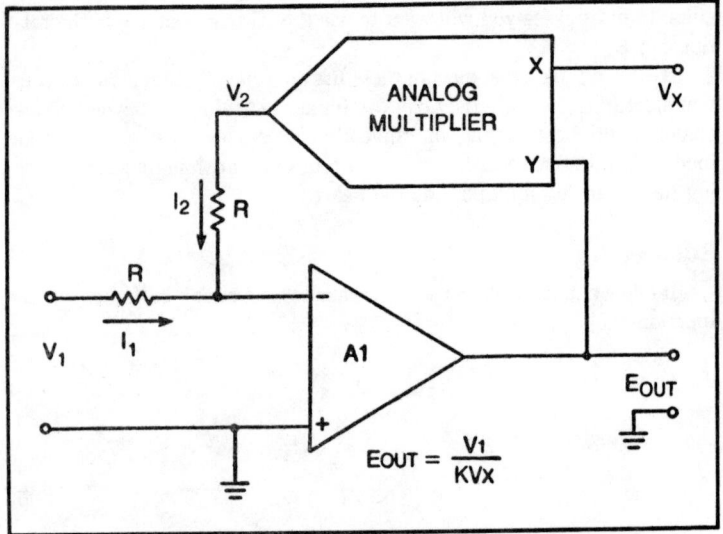

Fig. 9-10. Analog divider featuring analog multiplier in a negative feedback loop.

From my discussion of operational amplifiers in Chapter 1 you should recall that currents I1 and I2 will remain equal to keep Mr. Kirchhoff from spinning in his grave. Therefore, you can write the following

$$I1 = \frac{V_1}{R} \qquad (9.8A)$$

$$I_2 = \frac{V_2}{R} \qquad (9.8B)$$

and

$$I1 = I2 \qquad (9.8C)$$

;therefore

$$\frac{V_1}{R} = \frac{V_2}{R} \qquad (9.8D)$$

$$V_1 = V_2 \qquad (9.8E)$$

By substituting 9.7 into 9.8E, you find

$$V_1 = KV_x E_{out} \qquad (9.9)$$

by algebraic manipulation of Eq. 9.9

$$E_{out} = \frac{V_1}{KV_x} \qquad (9.10)$$

;so you see proof that an analog multiplier can function as a divider if it is placed in the feedback loop of an operational amplifier.

Square-Rooter

Another circuit that makes use of inverse, transfer functions is the square-rooter. Squaring and square-rooting are inverse processes. We may, therefore, place a squarer-connected, analog multiplier in the negative feedback loop of an operational amplifier. This generates the square-root function, as shown in Fig. 9-11.

The process of Fig. 9-11 is called the *inverse method* for reasons which should be, by now, clear to you. An alternative method, called the *implicit* technique, makes use of an IC that is specially designed for analog division. Consider the following process. If you connect the output back to the input, it will create a circuit with the transfer function

$$Y = \frac{X}{Y} \qquad (9.11)$$

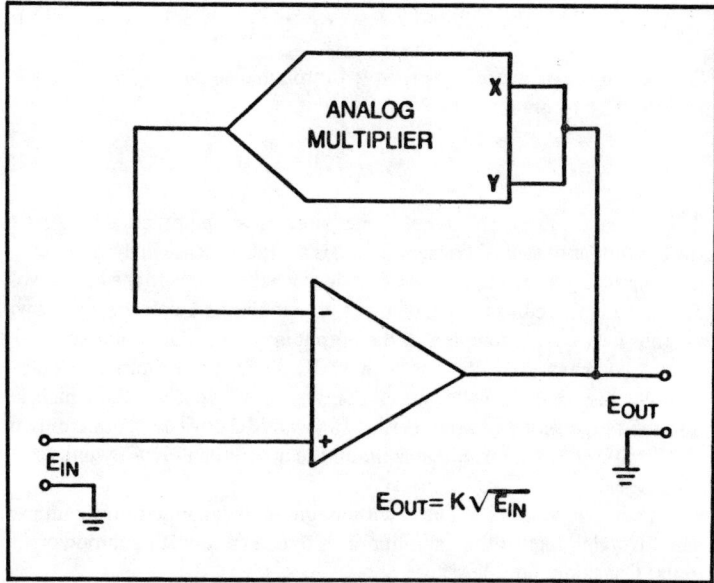

Fig. 9-11. Analog square-rooter using analog multiplier in the negative-feedback loop.

which yields

$$Y^2 = X \qquad (9.12A)$$

or

$$Y = (X)^{1/2} \qquad (9.12B)$$

Vector Summation

The vector sum of a set of voltages is defined as the square root of the sum of the squares of the individual terms. In other words, the vector sum of voltages V_1, V_2, V_3 . . ., V_n is

$$\overline{V} = [\ (V_1)^2 + (V_2)^2 + (V_3)^2 + \ldots + (V_n)^2\ ]^{1/2} \qquad (9.13)$$

There are two methods for computing the vector sum: direct and implicit. The direct method is simpler to implement in most cases but is somewhat limited in accuracy if wide-dynamic-range signals are being processed.

In the direct method it is necessary to design a circuit in which there is one squarer for each input voltage. The respective outputs of these squarers are fed to an operational-amplifier summer. The output of this stage represents the voltage

$$V = V_1^2 + V_2^2 + V_3^2 + \ldots + V_n^2 \qquad (9.14)$$

This voltage, nonvector V, is then fed through a square-rooter stage such as Fig. 9-11 to obtain

$$V = (V)^{1/2} \qquad (9.15)$$

This process falls down almost immediately in actual practice because of the limitations of real IC devices. There is a large increase in dynamic range inherent in any squaring process. Consider what happens to the output voltage of such a circuit for various input potentials. At 1 volt the output will be only 1 volt. At 2-volts input, the output is 4 volts. At 5-volts input, the output becomes 25 volts. Since most IC devices are limited to output voltages less than 15 volts, we can only process those signals which fall into the range root-15 to minus root-15. Many IC devices operate only up to 10 volts; so the input range is limited to approximately -3 volts to $+3$ volts.

A solution to this difficulty, without going to dynamic-range compression involving logarithmic amplifiers, is to use an implicit solution or formula. Consider the following

$$V = (V_1^2 + V_2^2)^{1/2} \qquad (9.16)$$

(Square both sides.)

$$V^2 = V_1^2 + V_2^2 \tag{9.17}$$
$$V^2 - V_1^2 = V_2^2 \tag{9.18}$$

(Factor out a difference of squares)

$$(V + V_1)(\overline{V} - V_1) = (V_2)^2 \tag{9.19}$$

$$V + V_1 = \frac{V_2^2}{V - V_1} \tag{9.20}$$

$$V = \frac{V_2^2}{V - V_1} - V_1 \tag{9.21}$$

This technique also works for multiple-term vector summation by adding the terms required over the $(V + V_1)$ denominator.

It is, of course, necessary to design an analog circuit which will compute this sum—not always a non-trivial exercise. The effort can, however, be reduced considerably in the case of a two-term vector summation of the form

$$V = (V_a^2 + V_b^2)^{1/2} \tag{9.22}$$

by using an analog IC device with the transfer function

$$V_{out} = \frac{Y\,Z}{X} \tag{9.23}$$

Examples of this type of chip, which is the universal analog multiplier/divider, include the Analog Devices AD533 family, a low-cost alternative. A typical circuit is shown in Fig. 9-12. The operational amplifier in this circuit can be any of the devices of the 741 family, but performance is predictably superior if one of the premium 741 operational amplifiers is used. Analog Devices recommends its AD741.

The definition of a vector involves *both* the magnitude computed by the circuit of Fig. 9-12 and the *direction*. If the direction information is desired in your application, a circuit must be provided that yields this data. In my system, direction can be specified by an angular notation such that

$$\theta = \arctan \frac{V_b}{V_a} \tag{9.24A}$$

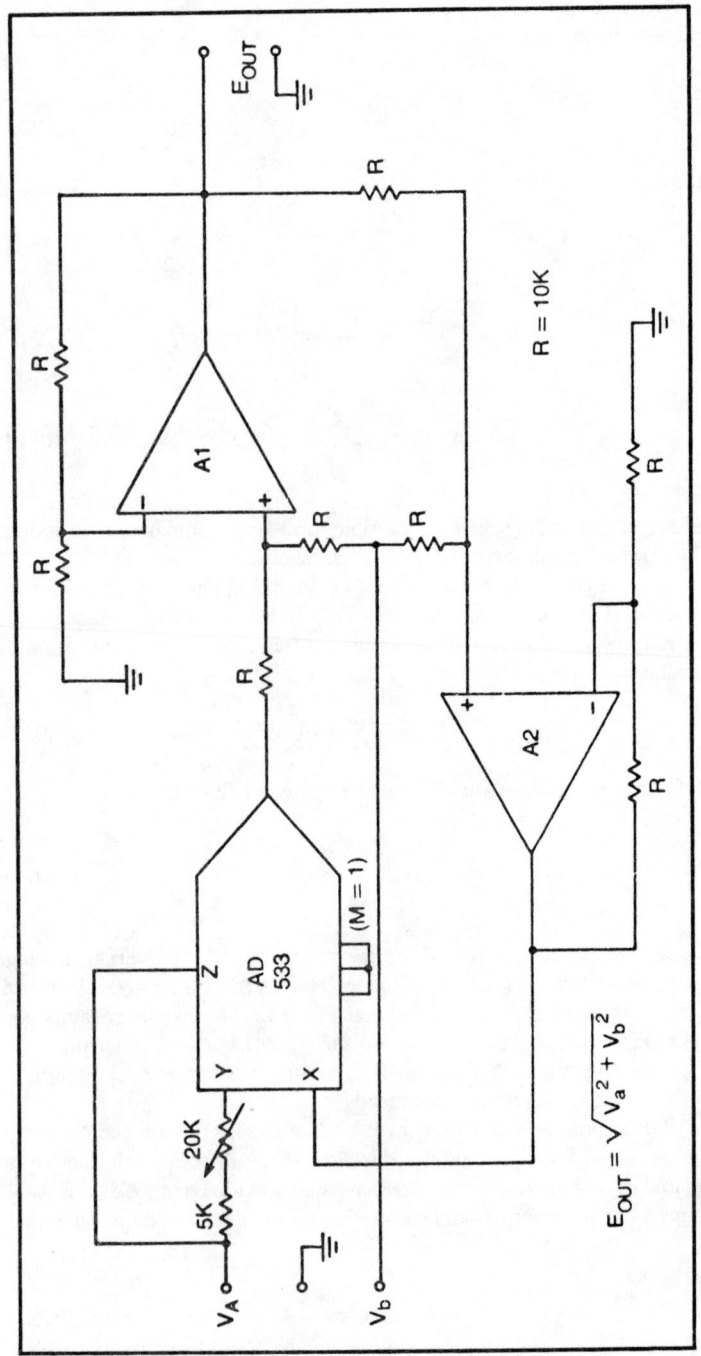

Fig. 9-12. Vector summer.

186

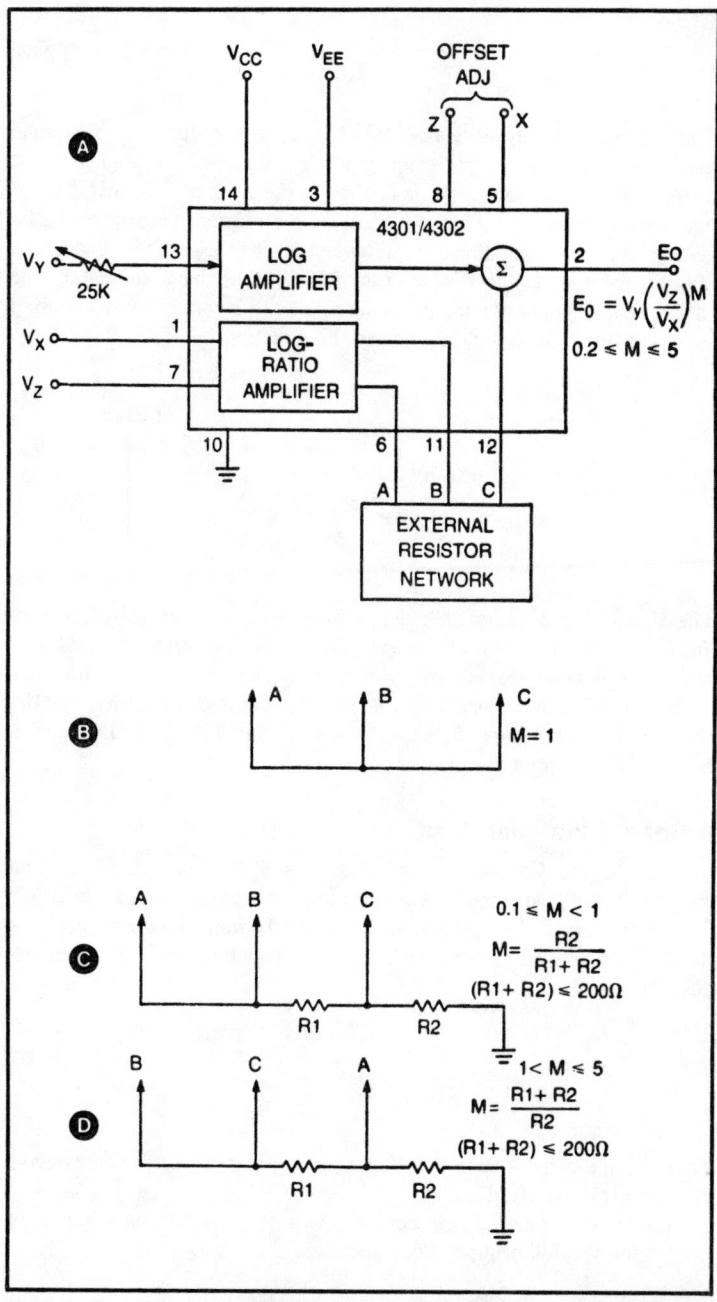

Fig. 9-13. Burr-Brown 4301/4302 function module: (A) block diagram; (B) connection for m = 1; (C) for m between 0.1 and 1.0; and (D) for m between 1 and 5.

187

$$\theta = \tan^{-1} \frac{V_b}{V_a} \qquad (9.24B)$$

This can be realized by using an AD533 IC in an inverse-function role around the feedback loop of an operational amplifier. Voltage V_a is applied to the chip's Z input, and voltage V_b is applied to the X input. The AD533 output is summed with a constant current from a reference source at the inverting input of an operational-amplifier unity-gain follower (A2). The output of this stage is fed to the Y input of the AD533 through an attenuator. The value of the exponent factor (m) is set to 1.2125 by judicious selection of resistor values. Such a circuit solves the equation

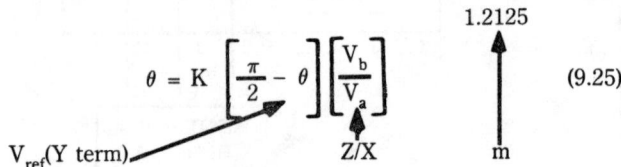

$$\theta = K \left[\frac{\pi}{2} - \theta \right] \left[\frac{V_b}{V_a} \right]^{1.2125} \qquad (9.25)$$

The V_a and V_b inputs of this circuit must always be greater than zero (must be positive); so it is only a one-quadrant design. All four quandrants can be accommodated, however, by inclusion of a sign-detector comparator (an inverter that switches when the sign is of the wrong polarity) and the appropriate control logic. It is usually the practice to connect the respective V_a and V_b inputs in parallel.

A Special-Function Module

Burr-Brown Corporation manufactures a line of special-function modules that should come to your attention if you design analog, electronic instrumentation. Figure 9-13 shows the block diagram and pin connections for the Burr-Brown 4301 and 4302 modules. This function module has the general transfer function

$$E_o = V_y \left[\frac{V_z}{V_x} \right]^m \qquad (9.26)$$

where

m is an exponent such that $(0.2 \leqslant m \leqslant 5)$, and is set by an external resistor network, R1/R2.

Depending upon circuit configuration this module is capable of generating the following analog functions:

- Multiplication
- Division
- Roots
- Sine

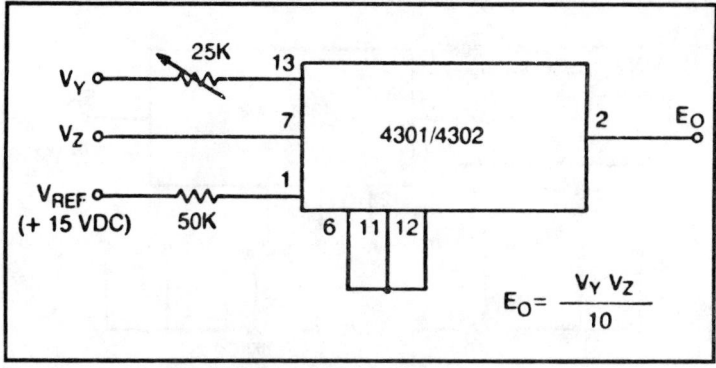

Fig. 9-14. Analog multiplier connection of the 4301/4302.

- Squaring
- Squarerooting
- Exponentiation
- Cosine
- Arctan (Y/X)
- Vector summation

This module is programmed through manipulation of exponent m. This is done by setting the values of resistors R1 and R2. With all three terminals shorted together, m = 1; so the transfer function becomes (A, B & C)

$$E_o = \frac{V_y V_z}{V_x} \qquad (9.27)$$

This configuration is used to form analog multipliers (Fig. 9-14) and dividers (Fig. 9-15). The analog-multiplier configuration, shown in Fig. 9-14, uses the Y and Z inputs, and X is set to a constant reference voltage such

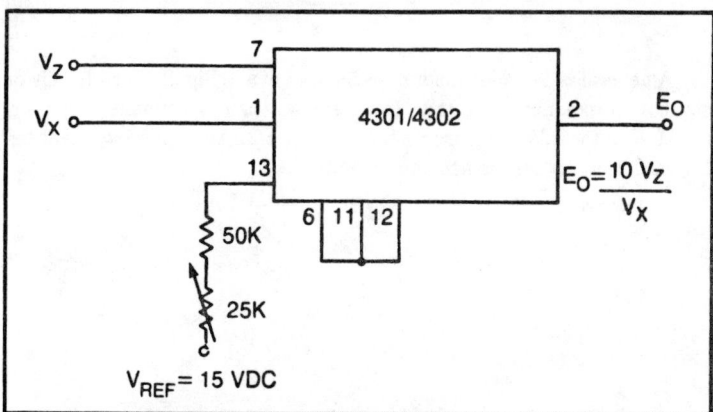

Fig. 9-15. Analog-divider connection of the 4301/32.

189

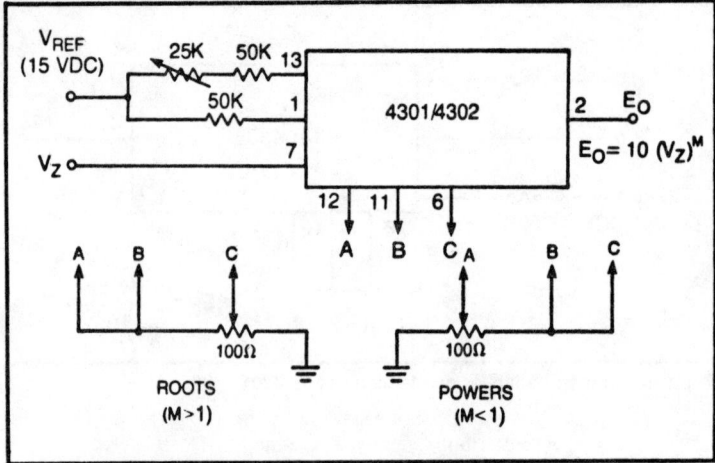

Fig. 9-16. Exponentiation connection of the 4301/02.

that $V_x = 10$. This yields a transfer function of

$$E_o = \frac{V_y V_z}{10} \qquad (9.28)$$

The same idea is used to form an analog divider, an example of which is shown in Fig. 9-15. Note that it is basically the same, except that the X input is active, and the Y input is tied to the constant reference. The transfer function is

$$E_o = \frac{10 V_z}{V_x} \qquad (9.29)$$

A generalized exponentiation circuit is shown in Fig. 9-16, with resistor circuits for exponents less than 1 and greater than 1. To make a squarer, set $M = 2$; to make a square-rooter set $m = 1/2$. Any other exponent between 0.2 and 5 can be accommodated.

Chapter 10

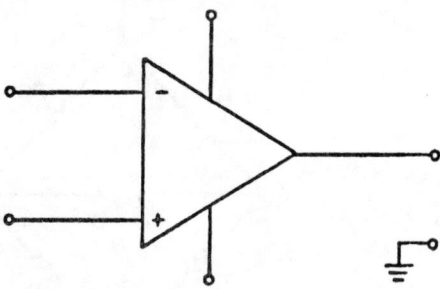

Transducers

$\mathbf{A}$ *TRANSDUCER* IS A DEVICE THAT CONVERTS ENERGY FROM ONE form to another for purposes of instrumentation, control, or measurement. For this discussion I will further stipulate that the form of energy *converted to* is electrical.

THE WHEATSTONE BRIDGE

Before becoming involved with different types of transducers let us first review the fundamentals of the classic Wheatstone bridge. The basic circuit for this form of bridge is shown in Fig. 10-1. You must realize that this circuit is basic to most types of transducers used in measurement. I won't go so far as to say it is universal, there are too many exceptions for that, but it at least seems ubiquitous.

The voltage drop appearing between points A and B is the output voltage from the bridge. This voltage is equal to the difference between the voltage drops across resistors R2 and R4. From elementary circuit theory we know that

$$E_{R2} = (E_1)\left[\frac{R2}{R1 + R2}\right] \qquad (10.1)$$

$$E_{R4} = (E_1)\left[\frac{R4}{R3 + R4}\right] \qquad (10.2)$$

Combining Eq. 10.1 and 10.2, we may also write

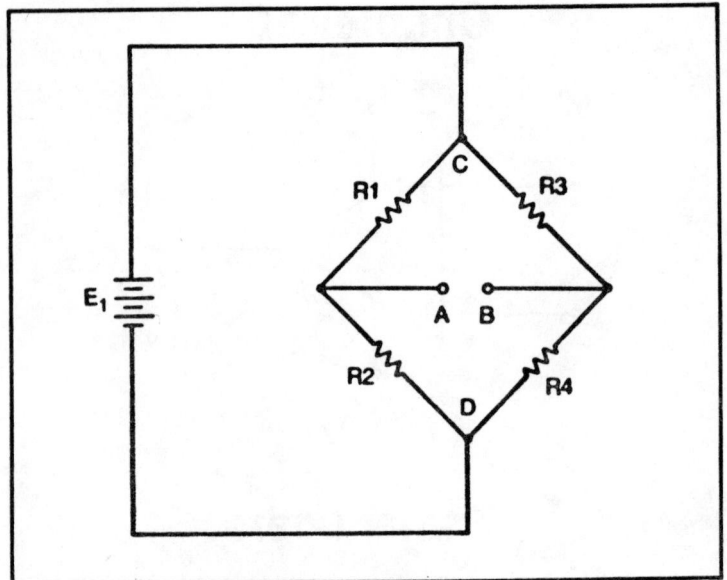

Fig. 10-1. The classic Wheatstone bridge circuit.

$$E_{AB} = \left[\frac{E_1\, R2}{R1 + R2} \right] - \left[\frac{E_1\, R4}{R3 + R4} \right] \tag{10.3}$$

$$= (E_1) \left[\frac{R2}{R1 + R2} - \frac{R4}{R3 + R4} \right] \tag{10.4}$$

The null condition, in which $E_{AB} = 0$, occurs when

$$\frac{R2}{R1 + R2} = \frac{R4}{R3 + R4} \tag{10.5}$$

$$R2 = \frac{R1R4}{R3} \tag{10.6}$$

and this relationship holds true

$$\frac{R2}{R1} = \frac{R4}{R3} \tag{10.7}$$

It is not necessary that these resistances be equal, only that the ratios of Eq. 10.7 be equal. In most transducers, however, these resistors will be made equal to each other for reasons of manufacturing simplicity. In those transducers R1 = R2 = R3 = R4.

The term ΔR in Fig. 10-1 represents a small change in the value of R4. This would occur in response to the type of external stimulus the transducer is designed to measure. If the bridge had been in the null condition, this change would perturb the bridge and produce an output voltage with a value

$$E_{AB} = (E_1)\left[\frac{R}{2R} - \frac{R}{2R + \Delta R}\right] \tag{10.8}$$

—assuming that all resistors are equal in the null condition. Changes in R4 (i.e., ΔR) occur because it is actually a transducer element. It is to some of these elements that I shall now turn.

POSITION TRANSDUCERS

A simple linear potentiometer may be used to provide an analog indicator of position. A slide potentiometer is used for the direct indication of rectilinear motion while a rotary potentiometer is used for angular motion.

In some cases mechanical linkage may be required between the stimulus and the actual potentiometer-adjustment shaft or arm. In some situations you would want the linkage to convert rectilinear motion to angular, or vice versa, but in other cases the purpose of the linkage is to reduce the length of the displacement to something that is compatible with the mechanical properties of the potentiometer.

In other situations, the motion might be very complex, and the purpose of the linkage would be to drive several potentiometers in proportion to the motion in a certain direction. In one such application, the so-called *joystick,* a shaft is designed so that it can move anywhere within a square area. A vertical-component potentiometer and a separate horizontal-component potentiometer are used to indicate exact position as functions of X and Y coordinates on a graph.

An example is shown in Fig. 10-2(A). In this case, we want to be able to know the position of an object that is ganged to potentiometers R_x and R_y, anywhere within the X-Y Cartesian system defined by Fig. 10-2(B).

The two potentiometers are oriented and the linkage designed such that the X component of the motion vector controls R_x and the Y component controls R_y.

One end of each potentiometer is connected to the positive voltage source, while the other is connected to the negative supply. The purpose of this is to create a four-quadrant system. When the object is located in the center of the square (point (0,0), the Cartesian origin), both potentiometers are centered and the condition $E_x = E_y = 0$ volts dc exists. The rules are:

- Movement to the left along the X axis produces a negative voltage at E_x.

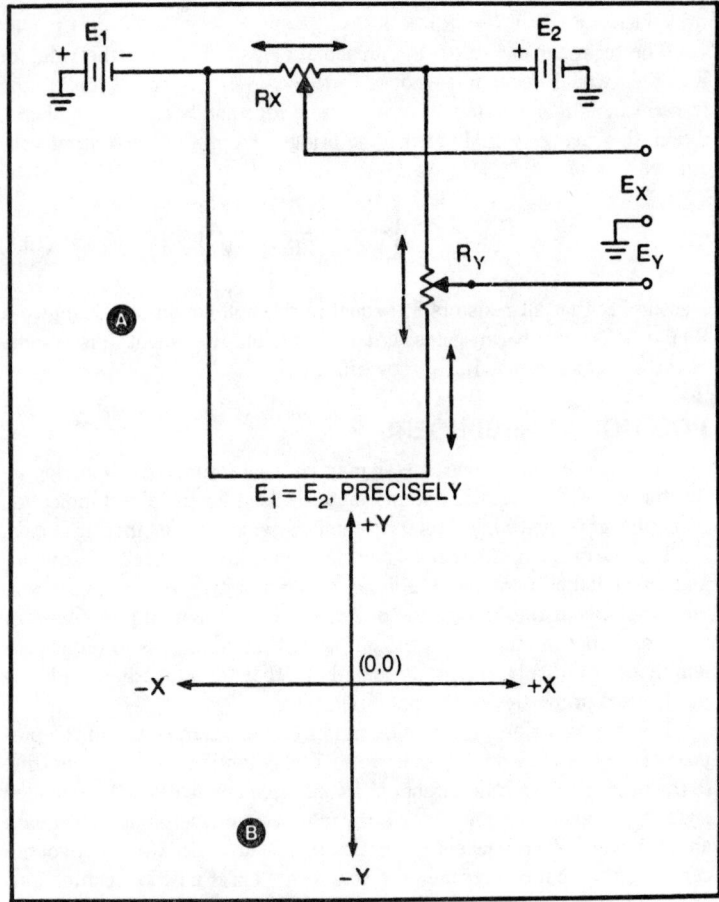

Fig. 10-2. X-Y position transducers.

- Movement to the right along the X axis produces a positive voltage at E_x.
- Movement upward along the Y axis produces a positive voltage at E_y.
- Movement downward along the Y axis produces a negative voltage at E_y.

In many cases we may also want to press into service joystick devices used originally in TV games. These are available through several of the nationally-advertised electronic-surplus houses. These will have four 5-kilohm or 100-kilohm potentiometers arranged one each for X+, X−, Y+, and Y−.

Some instrumentation requirements dictate that we take a single-value-vector version of the X and Y signals. To create the vector sum we must compute

$$\overline{E} = (E_x^2 + E_y^2)^{1/2} \qquad (10.9)$$

In that case it will be necessary to process the E_x and E_y potentials in a vector-summation circuit, such as given in Fig. 9-12.

If the motion is purely rectilinear, use a linear-taper potentiometer as the sensor. Most manufacturers of these components offer precision, highly-linear potentiometers in several possible values, tapers, sizes, and configurations. Also available are certain special potentiometers that offer sine, cosine, sine and cosine (in one unit), and several logarithmic tapers.

DERIVED DATA

It is often true that position data alone is insufficient, but it may be used to derive the desired information. Position transducers are simple and easy to implement. Furthermore, they tend to be low-cost. Velocity and acceleration transducers, on the other hand, tend to be expensive and tricky to use. We know from high school physics, however, that velocity is the first derivative of position with respect to time. This is expressed

$$V_x = \frac{dx}{dt} \qquad (10.10)$$

Similarly, acceleration is the first time derivative of velocity and, therefore, the second time derivative of position, or

$$a_x = \frac{dV}{dt} = \frac{d^2x}{dt^2} \qquad (10.11)$$

We may, therefore, generate a position signal in a transducer and then use electronic differentiation fo find V_x and a_x.

Position might also, by clever contrivance, give us certain other types of parameters. Figure 10-3 shows a simple example in which position, of some part of the apparatus, is proportional to some variable which we actually want to measure. This illustration shows a syringe pump that was custom made for a biophysicist who wanted to do certain studies on the eyes of animals. It consists of a syringe filled with a liquid and a pumping mechanism to infuse it into the subject. Although schematically simplified, this drawing represents a system that is in actual use.

The pumping chamber is shown here as a piston tank, which is filled through the upper valve when the piston is fully withdrawn and is then sealed. Potentiometer R1 is a 100-kilohm, linear-taper, slider type. It is ganged to the traveling rack through linkage in such a way that $R = 0$

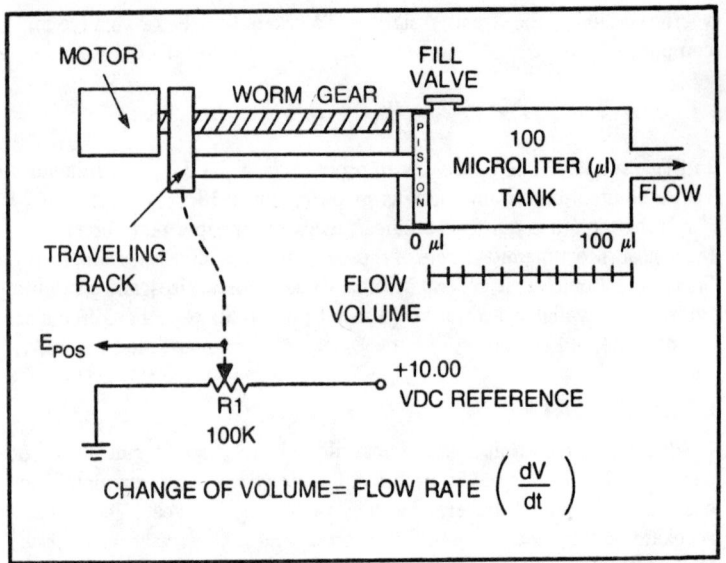

Fig. 10-3. Application for a position transducer to show pumped volume in a syringe-pump system.

ohms when the piston is in its totally-withdrawn position, and R = 100 kilohms when the piston is fully inserted into the tank.

The electrical configuration which you would expect for R1 will depend upon the information sought. As shown in Fig. 10-3 E_{pos} = 10 volts when the piston is fully inserted into the chamber. This gives us the voltage analog of volume pumped, where the scale is 0.1 volts/microliter.

Alternatively, if the connections had been reversed, the output voltage would be proportional to the volume remaining in the syringe.

We may also use this signal to obtain a flow-rate signal. It seems that the flow rate is the first time derivative of the tank volume. Therefore the value $d(E_{pos})/dt$, as taken from the output of an electronic differentiator following the position signal, will be proportional to the flow rate.

There are many parameters in all of the physical sciences and engineering that can be derived by measuring an easier-to-obtain signal and then passing it through some sort of signal-processing circuitry. It is, of course, necessary that you understand the mathematics and physical situation before all of these possibilities in your own work become apparent. You may, for example, be required to differentiate, integrate, vector sum, amplify, multiply, or divide two or more signals or constants, or process just a single voltage waveform.

STRAIN-GAUGE TRANSDUCERS

A strain gauge is a resistance element which changes electrical resistance when mechanically deformed. This phenomenon is known as piezoresistance. The deformation may be length, cross-sectional area, or shape.

Most practical strain-gauge transducers, at least those offered commercially, have four such elements arranged into the circuit of a Wheatstone bridge. Before proceeding with my strain-gauge discussion, though, let me present some definitions and terms. Among these are *gauge factor* and *sensitivity factor*. These are represented and K and Ψ, respectively.

A strain-gauge element produces a change in electrical resistance for a given change in applied strain. This element will change dimensions, either shrinking or stretching, precisely as the applied compression or tension in its host carrier. Recall from elementary physics that the electrical resistance of a wire is given by

$$R = \varrho \frac{L}{A} \qquad (10.12)$$

where

> ϱ is the resistivity constant that is unique to the type of electrical conductor being used
> L is the length
> A is the cross-sectional area
> R is the electrical resistance in ohms

Clearly then, by Eq. 10-12 we should anticipate a change in electrical resistance in a conductor designed to allow the ratio L/A to change under strain.

By definition, the gauge factor is

$$K = \frac{\dfrac{\Delta R}{R}}{\dfrac{\Delta L}{L}} \qquad (10.13)$$

This is also sometimes written as

$$K = \frac{\dfrac{\Delta R}{R}}{\epsilon} \qquad (10.14)$$

where

> ϵ is the factor $\Delta L/L$ and represents the axial strain in μin/in.

The sensitivity factor of a strain-gauge transducer gives us a means for predicting the output voltage from consideration of the excitation voltage, E_1, and the applied physical parameter. The sensitivity factor is

$$\Psi = \frac{E_{out}}{E_1 X} \tag{10.15}$$

where

Ψ is the sensitivity factor in volts/volt/unit of X
E_{out} is the output potential
E_1 is the excitation potential
X   is the physical parameter the transducer is designed to measure.

In practical instrumentation transducers this factor is given in units of $\mu V/V/X$ in order to keep the numbers easy to handle.

Example

A pressure transducer is rated to have a sensitivity factor of 10 $\mu V/V/torr$. What is the output voltage if 400 torrs is applied and the excitation potential is 10 volts?
By rearranging Eq. 10-15

$$
\begin{aligned}
E_{out} &= (E_1)\,(X)\,(\Psi) \\
&= (10 \text{ volts}) (400 \text{ torr}) (10 \ \mu V/\text{torr-volt}) \\
&= 10 \times 400 \times 10 \ \mu V \\
&= 4 \times 10^4 \ \mu V, \text{ or } 0.04 \text{ volts}
\end{aligned} \tag{10.16}
$$

We may also use the gauge factor as part of a bridge calculation. For a bridge such as Fig. 10-4 we may claim that

$$\frac{dE_o}{dS} = \left[\frac{R_a R1}{R_a + R1} \right] (I1)\,(K) \tag{10.17}$$

where

dE_o is the rate of change of the output voltage
dS is the rate of change of strain gauge A
I1   is the current through strain gauge A
K   is the gauge factor

Strain gauges are used in transducers dsigned for the measurement of strain, force, load, torque, pressure, weight, vibration, flow, and many other mechanical and physical parameters.

Although every resistive strain gauge operates on the piezoresistive principle described earlier, there are significant differences in the type of physical construction. Both wire and bonded types are manufactured.

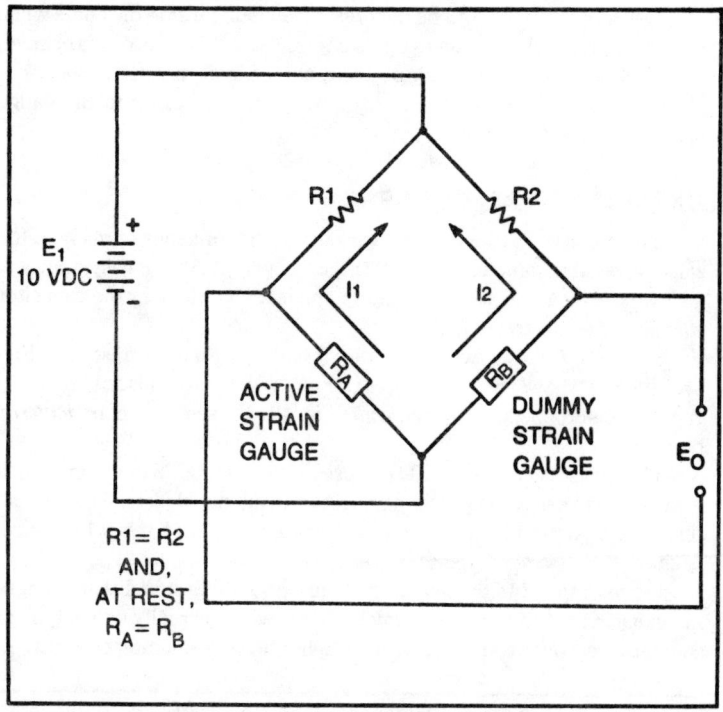

Fig. 10-4. Use of strain gauges in a Wheatstone bridge.

A wire-type strain gauge uses a short length of wire suspended taut between two supports. These supports act to stretch the wire under tension and let it shrink under compression.

Although single-element strain gauges exist, most use four elements connected in a Wheatstone-bridge configuration. The gauges are arranged so that two are in compression and the other two are in tension for any given applied strain. Electrically, this results in a somewhat more sensitive bridge circuit.

A bonded strain gauge uses a thin, metal foil, or semiconductor-resistor element, cemented to an insulator attached to a metallic diaphragm. The strains are applied to the substrate, and this flexes the elements. Again, there are four elements connected in the Wheatstone-bridge configuration.

Consider, for example, the fluid pressure transducer typically used by physicians and physiologists in both clinical and research applications. The transduction takes place on a thin, deformable, metal diaphragm. The four strain-gauge elements are cemented, or "bonded," to this structure.

Figure 10-5 shows how this fine diaphragm is mounted in the transducer assembly. It is sealed against liquid seepage. The opposite side of the diaphragm rests inside a plastic dome (clear for visibility) fitted with a pair of stopcock valves to admit liquid.

199

The zero reference will be the pressure existing inside the dome when one of the stopcocks is open to atmosphere. After this stopcock is opened, the amplifiers will be adjusted for zero output, and then the stopcock is closed. The strain gauge, then, will see a pressure equal to the gauge pressure in the fluid line.

CAPACITIVE TRANSDUCERS

Capacitances can be made to vary under the influence of many of the same physical parameters as affect the strain gauge. Additionally, they will also vary as position varies. In many applications, though, the capacitor transducer is superior.

Figure 10-6 shows two elementary forms of capacitor transducer. Figure 10-6(A) shows an axial, or *diaphragm,* type reminiscent of early capacitor-microphone elements, while Fig. 10-6(B) shows a rotary *butterfly* type.

The transducer in Fig. 10-6(A) uses a thin, metal diaphragm, which is flexible and easily deformed, parallel to a fixed, metal plate. These structures are separated by insulating supports and dry air; so they form a capacitor.

Forces applied to the diaphragm cause it to deform and that changes the capacitance. In a capacitor microphone, which is really only a type of sound-to-electrical transducer, the diaphragm will operate by itself. In

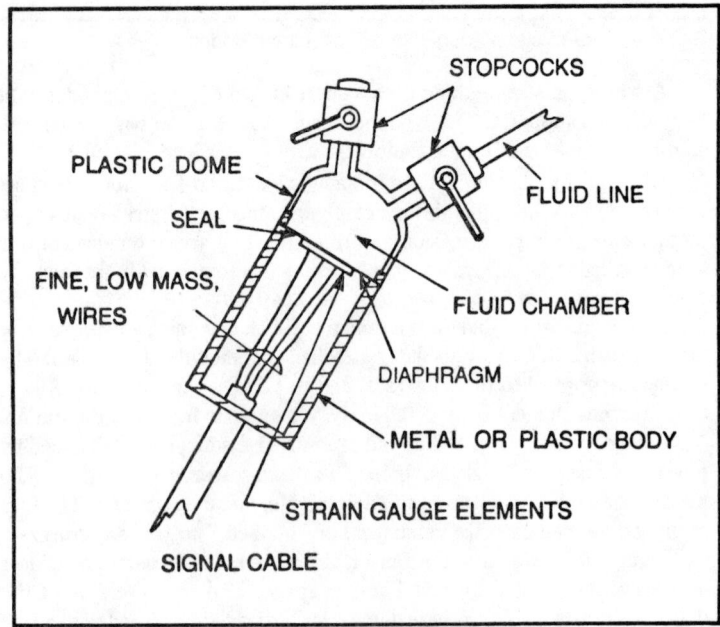

Fig. 10-5. Typical strain-gauge pressure transducer.

others, however, there may be a thin rod, or other linkage, coupling the forces to the diaphragm.

A second form of capacitor transducer is the rotary type shown in Fig. 10-6(B). In this configuration, there are two sets of plates separated by air. One set is stationary while the other rotates on a mechanical shaft. Angular displacement causes the amount of area on the fixed plate shaded by the rotary plate to vary. This, of course, changes the capacitance between the plates.

Capacitive transducers generally require an ac carrier signal for excitation unless they are operated in an FM oscillator circuit. An FM system uses an L.C. oscillator in which the capacitance of the transducer is part of the total capacitance of the frequency-determining tank circuit. When the transducer is strained, it changes the frequency of the oscillator a proportional amount.

A carrier system operates using a fixed-amplitude ac signal. It may be in a bridge circuit, but whatever the electrical configuration, the object is to create an amplitude variation that depends only on the incident stimulus.

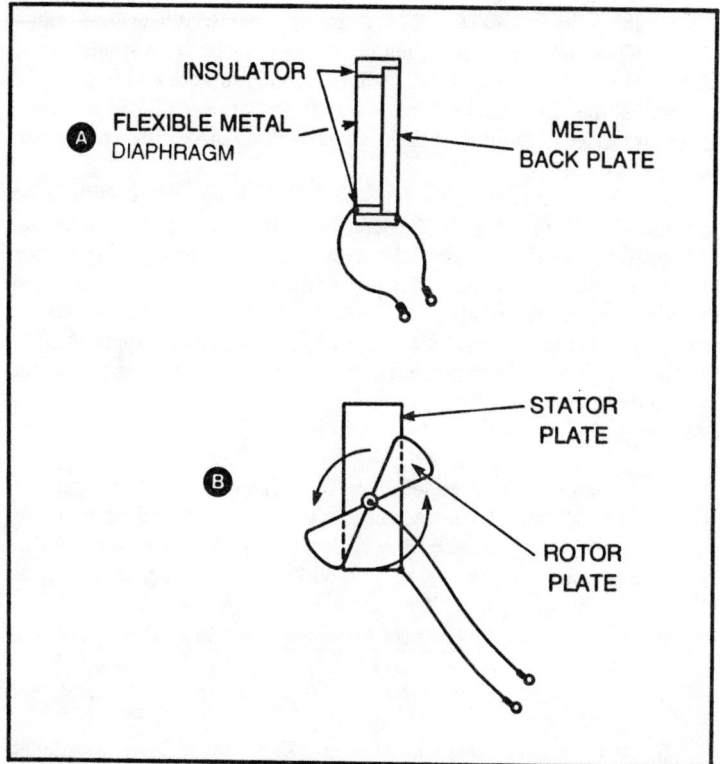

Fig. 10-6. A pair of capacitive transducers.

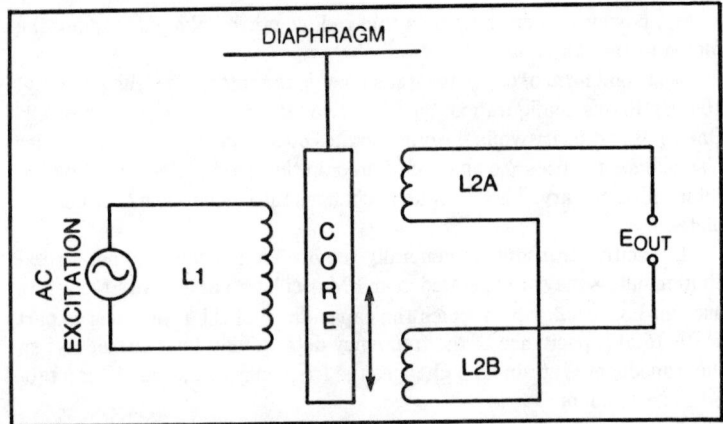

Fig. 10-7. The linear voltage-difference transducer (LVDT).

INDUCTIVE TRANSDUCERS

There are two basic types of inductive transducer. In one, there are two variable inductors and two fixed resistors in an RL, Wheatstone-bridge circuit. When the transducer is not under the influence of a stimulus, the mutual core of the two inductors will be equally in both coils. Applying a stimulus, though, causes the core to shift so that one coil has more and the other has less. This makes their inductive reactances different and unbalances the bridge.

The other form of inductive transducer is the linear-voltage-difference transducer (LVDT) of Fig. 10-7. Again, the core is equally inside L2A and L2B when the core is not being stimulated. These windings are connected to be series-opposing. An excitation carrier from L1, then, is cancelled when the core is in its rest position. Variation of the core position unbalances the null and creates an output voltage which has a phase proportional to the direction of core displacement and an amplitude that gives the extent of core displacement.

References

1. Simmons, J. and Soderquist, D.; "Temperature Measurement Method Based On Matched Transistor Pair Requires No Reference," Precision Monolithics, Inc. Applications Note AN-12.
2. Simmons and Soderquist used 59.73 μV/°K which corresponds to a current ratio f_o 2.06 rather than 2.0.
3. Oliver, Frank J.; *Practical Instrumentation Transducers,* Hayden Book Co., New York, 1971.
4. Carr, Joseph J.; *Op-Amp Circuit Design & Application,* Book No. 787, 1976.
5. Carr, Joseph J.; *Servicing Medical & Bioelectronic Equipment,* Book No. 930, 1977.

Chapter 11

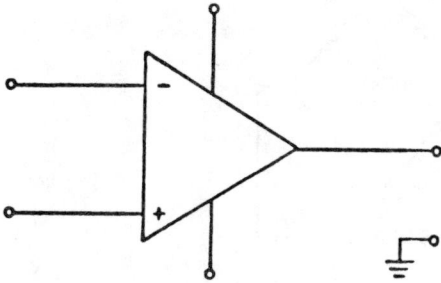

Bridge and Carrier Amplifiers

A LL WHEATSTONE-BRIDGE, STRAIN-GAUGE TRANSDUCERS, AS WELL as certain other types, require a special type of preamplifier—either a bridge amplifier or a carrier amplifier. The chief difference between bridge and carrier amplifiers is that the bridge amplifier responds to dc bridge excitation, while the carrier amplifier responds to ac excitation.

In many cases the two are totally interchangeable, but in others one must be very specific as which is required. The carrier amplifier is somewhat more universal because it will work with resistive, inductive, or capacitive transducers. The chief advantage of the bridge amplifier, which only works with resistive transducers, is that it is generally pretty inexpensive compared with the usual carrier amplifier.

Almost any amplifier can be used as a bridge amplifier, so one may simply pick from the various differential dc amplifiers of Chapter 7. The chief reason why an amplifier becomes a *bridge amplifier* is that the dc strain-gauge-excitation potential is also supplied.

DC-EXCITATION METHODS

Figure 11-1 shows three methods for providing a dc excitation potential to a resistive, strain-gauge transducer. The simplistic method of Fig. 11-1(A) uses an ordinary battery to power the bridge. Although it may appear ludicrous in this age of sophisticated electronics, it is actually used to good advantage now that some dry cells are available with high ampere-hour ratings. These are the lantern batteries one sees in drug and hardware stores at modest cost. The battery method fails only where quantitative measurements are made and calibration is difficult. The excitation

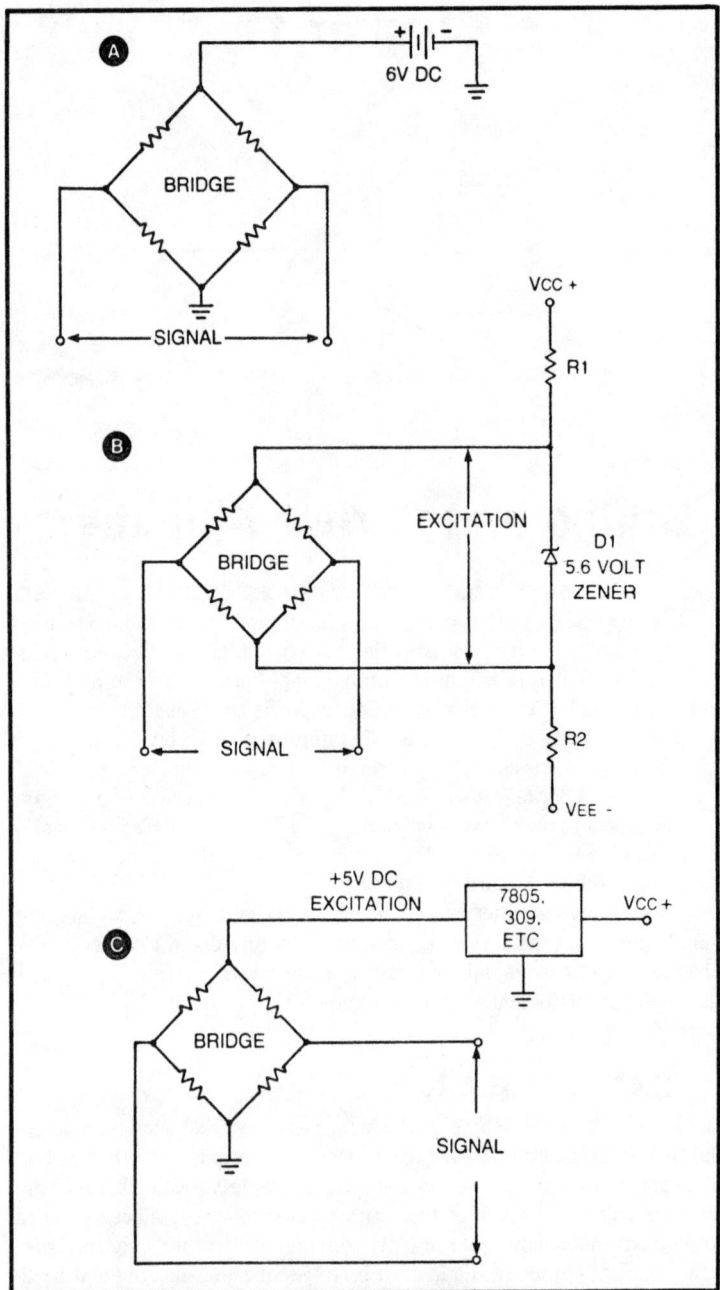

Fig. 11-1. At (A), battery excitation of a transducer; (B), zener diode excitation; and (C), three-terminal IC voltage regulator excitation.

voltage will not only lack precision but will tend to drop over the life of the battery. If the precision is low or if the system can be calibrated easily at each use, the battery method may be the way to go.

The next step in complexity is the technique of Fig. 11-1(B). Here we use a zener-diode regulator to stabilize the excitation potential derived from a dc power supply. This method can be used in moderately precise applications if the right diode is selected. Most ordinary zener diodes, though, have only a nominal zener potential, so may be used only if the requirements for precision are low or moderate. Reference-grade zeners, or those hand-selected from a large batch, can be used where precision is greater. One must be aware of thermal drift because most zener diodes will not maintain the same V_z over a wide temperature range. If the load is changing, internal heating often causes the potential to drift, even if ambient temperature is maintained reasonably constant.

A superior method is to use an electronically-regulated power supply designed to reduce the effects of thermal drift. This is shown in Fig. 11-1(C). You could use any of the precision voltage regulators found in the literature or one of the three-terminal IC regulators that are available. A 7805, LM340-5, or LM320-5 (for negative excitation potentials) may be used with good results. These devices will provide a potential that is close to 5 volts, and this is well within the "safe" range for most transducers. The actual excitation potential will not be precisely 5 volts dc; so some error will result. The usual device, however, produces a potential between 4.9 and 5.1 volts; so this error may well be negligible in your application. The important thing is that the potential will remain at its nominal value once the IC has reached thermal stability (a few minutes usually). This prevents artifacts that would result from changes in excitation voltage.

Of the several types of regulators available under each part number (differentiated by an alphabetic suffix), it is better to use those which have a current rating several times higher than the current drawn by the transducer. This can be determined by using Ohm's law, $I = E/R = 5/R$, where R is the resistance of *one arm* of the Wheatstone bridge. It is generally best to simply make it a rule to specify the 1-ampere or 750-milliampere versions, even when the current requirement is less than 100 milliamperes.

Overvoltage Protection

Most transducers will not survive long if the excitation potential exceeds the specified limits. Most will accept excitation between 0 volts and about 10 volts dc, but some are limited to less than 8 volts dc or ac rms. If potentials greater than these are applied, overheating will occur and this will either destroy the transducer (usually) or will change its properties so much that your data is in jeopardy.

Two methods are most often used for guarding against over-voltage. One is to place a large-wattage zener diode across the output of the excitation supply, and the other is to use an SCR crowbar. The latter technique is discussed more extensively in the chapter on power supplies; the former is used in a construction-project design for a transducer preamplifier shown in Fig. 22-9(A).

A Design Example

The purpose of a bridge amplifier is to produce a dc output voltage that is proportional to the value of the parameter being measured by the transducer. Furthermore, that output voltage should be easy to interpret, making it unnecessary to create a special meter scale for evaluation by humans.

Let us design a bridge amplifier with a digital readout to measure fluid pressures up to + 100 torr. To make things easier we will use an IC, three-terminal, voltage regulator to produce an excitation potential of + 5 volts dc. The particular transducer will be a Statham P23 with a looking-back resistance of 200 ohms and a scale factor (Ψ) of 50 μV/V/cm Hg, which translates to 5 μV/V/torr. Find the gain required of a dc differential amplifier that will produce a nicely-scaled output voltage for humans to read.

Since the scale is 0 to 100 torrs, we will want to use an output voltage of 0 to 100 millivolts, 0 to 1 volt, or 0 to 10 volts so the voltage can be read directly from a digital voltmeter attached to the output. This allows the pressure units to have the same digits as the output-voltage units, allowing humans to read it without much interpretation. We will select the correct range from the three candidates by examining available digital voltmeters and the properties of operational amplifiers. The 100-millivolt range is too small because only a few, digital, panel voltmeters are available for that range. This limits our selection of the readout device and may cause price increases. Furthermore, it would require us to use premium operational amplifiers so that low pressure readings would not be obscured by noise potentials and offsets from the operational amplifier itself. The 10-volt range is a little bit more appealing, but one would run into problems with the power supply again. The typical operational amplifier would require ± 15-volt supplies to output + 10 volts properly without linearity problems. Also, there are fewer, digital, panel meters available in that range, and it would require a great deal of amplification—which is harder to tame.

The 1-volt (1000-millivolt) range is best-suited to our purpose because there are many alternative, digital, panel voltmeters that will accommodate 0 to 999 millivolts in 2 1/2-, 3 1/2-, or 4 1/2-digit-precision models. These are also available at modest cost. Furthermore, the use of a 1000-millivolt output level allows gains in the hundreds rather than thousands. This last fact makes it easier to actually make the circuit work because under that condition circuit errors do not have as great an effect on the output. The scale factor of our pressure-measuring system is, therefore

$$\text{SF} = \frac{1000 \text{ mV}}{100 \text{ torr}} = \frac{10 \text{ mV}}{1 \text{ torr}} \qquad (11.1)$$

The full-scale output from the transducer, by Eq. 10.16, is

$$E_o = (5 \text{ volts}) (100 \text{ torr}) \; \frac{5 \; \mu V}{\text{torr-volt}} \qquad (11.2)$$

$$= 5 \times 100 \times 5 \; \mu V \qquad (11.3)$$
$$= 2500 \; \mu V = 2.5 \text{ mV} \qquad (11.4)$$

Our amplifier must provide enough gain to bring a 2.5-millivolt signal up to 1000 millivolts, the full-scale signal required at the amplifier output. The gain is, then

$$A_v = \frac{1000 \text{ mV}}{2.5 \text{ mV}} \qquad (11.5)$$

$$= 400 \qquad (11.6)$$

The digital voltmeter at the output of the amplifier will then read pressures at a scale factor of 10 mV/torr. If a low-cost, 2 1/2-digit meter is used, the decimal point can be blanked out, and the pressure reading will seem to make more sense. For example, 100 torr would read 1.00 with the decimal point and 100 without the decimal point. Most commercial, digital, panel meters have a provision for blanking the decimal point, and if not, you can always disconnect the LED used to form it.

Similarly, with the more-commonly-found, 3 1/2-digit meter, try relocating the decimal point so that it is always just to the left of the least-significant digit. This will make 1.000 read as 100.0. Some digital panel meters allow the user to set the decimal point; so look for this feature before buying. This is done usually by grounding the appropriate pin on the rear connector.

It cannot be overemphasized how easy things become if the correct scale factor is chosen. If you choose a voltage range that has the same digits as the range of the parameter being measured, ordinary meters can be used without the need for interpretation or for having to make a special meter scale. This method also allows the direct use of digital voltmeters. Creation of a special scale on a digital meter requires either an analog or digital circuit to translate the results for you. If this is not done, an oddball output reading will result, and this will require the *user* to make the translation. I can show you a servo-perfusion-pump system used by a prominent biphysicist in which the design engineer used a scale factor of 2 volts = 40 torr (50 mV/torr). The apparent reason for this was that the servoamplifiers following the pressure amplifier needed a signal of this magnitude. But this same signal was also fed to a digital voltmeter for readout; so some mental arithmetic is required of the user. A simple gain-of-2 operational-amplifier circuit would have eliminated this problem by changing the scale at the readout to 4 volts = 40 torr (100 mV/torr). In that case, the digits (i.e., 4 and 0) representing the full scale *pressure* would also represent the full scale *voltage*. I cannot buy the notion that a little horse sense is too much to expect from a circuit designer.

Ratiometric Measurements

Where high precision is required one cannot tolerate any change in the excitation potential because that would create a change in the output signal that was *not* due to variations in the stimulus parameter. Again consider my hypothetical pressure-measuring amplifier of the previous example. Calculate the error caused by variation of the excitation potential. At 100 torr the transducer output was

$$(5 \text{ volts}) (100 \text{ torr}) \times \frac{5 \ \mu V}{\text{torr-volt}} \tag{11.7}$$

$$= 2500 \ \mu V$$
$$= 0.0025 \text{ volts}$$

But if the excitation potential changed from 5 volts down to 4.5 volts, the output would change to

$$(4.5 \text{ volts}) (100 \text{ torr}) \frac{5 \ \mu V}{\text{torr-volt}} \tag{11.8}$$

$$= 2250 \ \mu V$$
$$= 0.00225 \text{ volts}$$

But the *transducer* scale factor is 25 microvolt/torr; so this change is interpreted at the readout as a 10-torr pressure change. The operator of the instrument has no way of knowing that the change in output reading is an artifact, not a real pressure change in the system being measured.

The diagram of Fig. 11-2 shows a ratiometric, bridge amplifier that will help to reduce this artifact considerably. In this circuit there is a strain-gauge output amplified a factor of 10 by A1. This voltage is applied to the X input of an analog divider. In this example the transfer function of the divider is of the form

$$E_{out} = \frac{10 \ X}{Y} \tag{11.9}$$

If we apply the output of amplifier A1 to the X input of the divider and the 5-volt excitation to the Y input, the divider output will be (using Eq. 11.9 and assuming 100 torr)

at $E_1 = 5$ volts

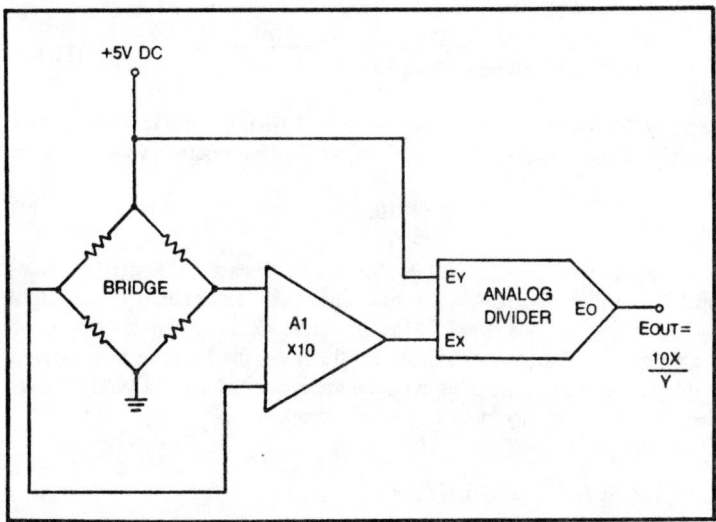

Fig. 11-2. Use of an analog divider to eliminate the drift artifact caused by variation in excitation potential.

$$E_{out} = \frac{10(0.0025)}{5} = 0.005 \text{ volts} \tag{11.10}$$

and at $E_1 = 4.5$ volts

$$E_{out} = \frac{10(0.00225)}{4.5} = 0.005 \text{ volts} \tag{11.11}$$

Note that with a 10-percent change in excitation potential, the output of the system remained constant. But before rushing into using the divider method, we must answer the question of whether a real change in stimulus would cause an output change. Of course, we know this is true intuitively by examining either Eq. 11.10 or Eq. 11.11, but to satisfy all critics let us actually compute the change in output if the divider input changed an equal amount as in Eq. 11.11, but the excitation remained at 5 volts. This situation would be the same as

$$E_{out} = \frac{10(0.00225)}{5} = 0.0045 \text{ volts} \tag{11.12}$$

The 10-torr change in real pressure resulted in a 500-microvolt change in the output voltage; so our system is valid. In this case we would have a scale factor of

$$\frac{500 \ \mu V}{10 \text{ torr}} = \frac{50 \ \mu V}{\text{torr}} = \frac{5 \times 10^{-5} \text{ volt}}{\text{torr}} \tag{11.13}$$

If we still wanted to let reason prevail and have a scale factor of 1 volt = 1 torr at the output of the system, we would now require a voltage gain of

$$A_v = 1000/5 = 200 \tag{11.14}$$

Amplifier A1 may not be necessary in some cases. It is used to scale-up the transducer-output voltage to a range that can be handled accurately by the analog divider circuit. Most of these will have a requirement for a gain greater than the 10 chosen for this example because most have a minimum-signal requirement in the neighborhood of 0.1 or 1 volt. The figure 10 was selected just to make an example.

CARRIER (AC) AMPLIFIERS

The carrier amplifier is designed to produce an ac excitation signal and then to process the resultant, ac, transducer-output signal. Although significant differences exist in the design of the individual stages, most carrier amplifiers on the market now more or less follow the block diagram of Fig. 11-3.

The transducer excitation will be a balanced, ground-referenced, ac signal. The two ends of the bridge are driven by signals that are 180° out of phase with each other. In some designs the transducer will feed a differential ac amplifier in the same manner as the dc bridge amplifier, but in Fig. 11-3 we have one port of the transducer grounded so that a single-ended ac amplifier can be used.

The ac-bandpass amplifier provides most of the system gain. It may have a frequency response down to a very low ac frequency, but not all the way to dc. This makes it able to take advantage of negative-feedback

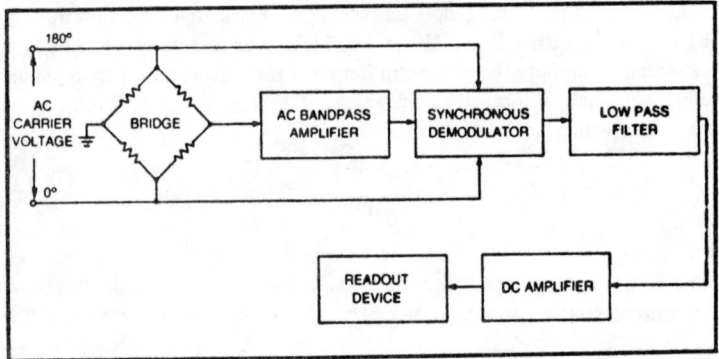

Fig. 11-3. Block diagram of carrier amplifier.

techniques which improve operating stability.

The signal applied to the ac-bandpass amplifier is fed to one input of a synchronous demodulator. This stage consists of a switching arrangement that converts the ac-amplitude variations into a dc level proportional to the transducer stimulus. The dc amplifier following the synchronous modulator may supply gain to scale the output to a reasonable level or it may have unity gain, in which case it acts as a buffer.

AC-EXCITATION TECHNIQUES

The ac carrier used to excite the transducer must have an rms amplitude less than the maximum dc potential that may be applied to the transducer. This carrier should have a stable amplitude lest artifacts sneak in to ruin the output. The same ratiometric techniques may be used, however, especially if the carrier frequency is low enough and the shape is a sine wave. The frequencies most often used for carrier amplifiers are between 400 hertz and 5 kilohertz, with 1 kilohertz and 2.5 kilohertz being very common. These are well within the frequency response of most analog dividers.

The actual ac signal used may be either a sine wave or a square wave. Methods for applying the signal are shown in Fig. 11-4. The signal may be originated by almost any technique, but selecting one known for amplitude stability is best. Examples are given in Chapter 13 (IC Waveform Generators).

In Fig. 11-4(A) we see the use of a balanced ac transformer to apply the excitation to the transducer. Although a proper transformer might be rather expensive, this is the method of choice where a sine wave is to be used.

If you want a good, clean waveform, with the proper 180° phase difference between the two outputs ($\phi1$ and $\phi2$) the transformer must be exceptionally well-balanced. This usually requires a high-cost, bifilar-wound transformer. Ninety dollars is not out of line.

The use of a power operational amplifier, or hybrid, power-amplifier module, is shown in Fig. 11-4(B). It is necessary that the amplifier have a differential (push-pull) output. This is not a popular technique, however, because most of the modules and power-amplifier ICs are either video amplifiers, which are costly, or produce too much power, so are wasteful. In either event, the amplifiers tend to be a little difficult to tame in real circuits.

Figure 11-4(C) shows a situation where a J-K flip-flop is used to produce the signals from a common chain of square waves. It does this by virtue of the complementary outputs of the flip-flop. A pair of operational amplifiers serve as buffers to provide isolation.

A biquadratic (i.e., 2 × 90°) phase-shift technique is shown in Fig. 11-4(D). The 0° reference is the signal from the source, passed through a unity-gain buffer amplifier, while the 180° signal is obtained by passing

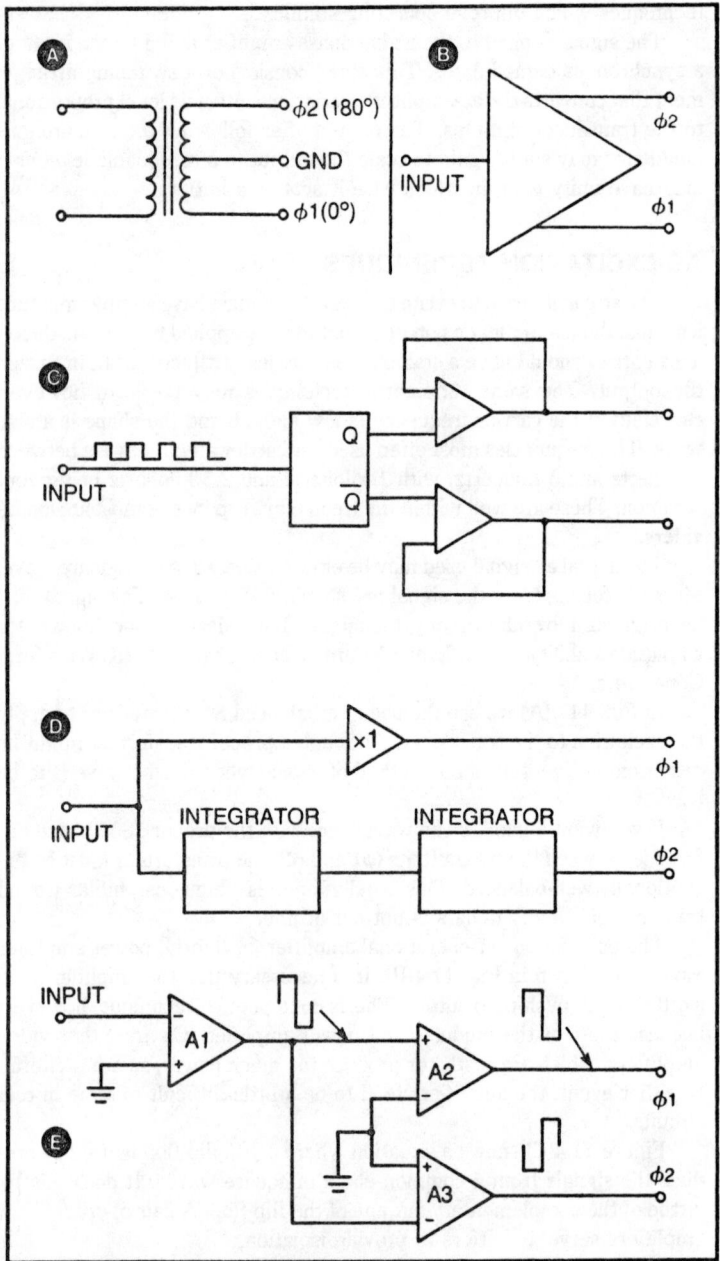

Fig. 11-4. Typical ac excitation techniques for carrier amplifiers: (A) transformer; (B) differential output amplifier; (C) flip-flop; (D) biquadratic; and (E) the comparator.

the original signal through two integrators in cascade. These are ordinary, operational-amplifier integrators and each will produce a phase shift of 90°; so the total phase shift is 180°. This method has achieved relative popularity because it allows sine-wave excitation without the use of costly and bulky transformers.

One last method is shown in Fig. 11-4(E). Here we use three operational amplifiers connected as ground-referenced comparators. Comparator A1 is used to do some signal shaping so that either sine- or square-wave source signals may be used.

The differential input to a comparator must have zero volts across it if a zero output is to be obtained. Since one input in each of these comparators is grounded, this occurs only when the input signal crosses zero axis. Once the signal has an amplitude greater than a few millivolts we find that the output is saturated at either Vcc- or Vee-supply rails. Op-amp A1 is connected as an inverter; so positive-input excursions result in negative output excursions and vice versa.

Comparators A2 and A3 also are ground-referenced but have opposite sense. Comparator A2 inverts the square wave; A3 does not. The resultant output is a pair of square waves that have the required push-pull relationship. Care must be taken, however, to limit the total-amplitude excursions to less than the maximum allowed to the transducer being used.

Typical Demodulators

Most demodulators are little more than electronic switches synchronized to the ac carrier signal. Two common types are shown in Fig. 11-5. The method of Fig. 11-5(A) uses two transistors and some push-pull transformers to achieve demodulation of the carrier. The respective base terminals of the transistors are driven on and off by the high-level signal produced by the carrier-oscillator reference signal. Both transformers must be well-balanced, and the transistors must be a matched pair.

A somewhat more recent design, shown in Fig. 11-5(B), uses a CMOS-integrated-circuit, electronic switch (called a *transmission gate*). This device, here the type 4016, has four transmission-gate switches, but only two are used in this application. When pin 5 is high, the connection between pins 4 and 3 is made, and bringing pin 13 high makes the connection between pins 1 and 2. In the circuit shown, only one switch is closed at a time.

Operational amplifiers A1 through A3 have functions much like those in Fig. 11-4(E). This is what causes each switch to toggle on and off in opposite phase to the other. The two switch-output terminals are fed to a differential amplifier (A4) where the respective outputs are summed. A low-pass filter following the amplifier will remove any residual carrier signal remaining on the dc output of A4.

Zero (Null) Circuits

Few transducers, or amplifiers for that matter, are perfect. Almost all

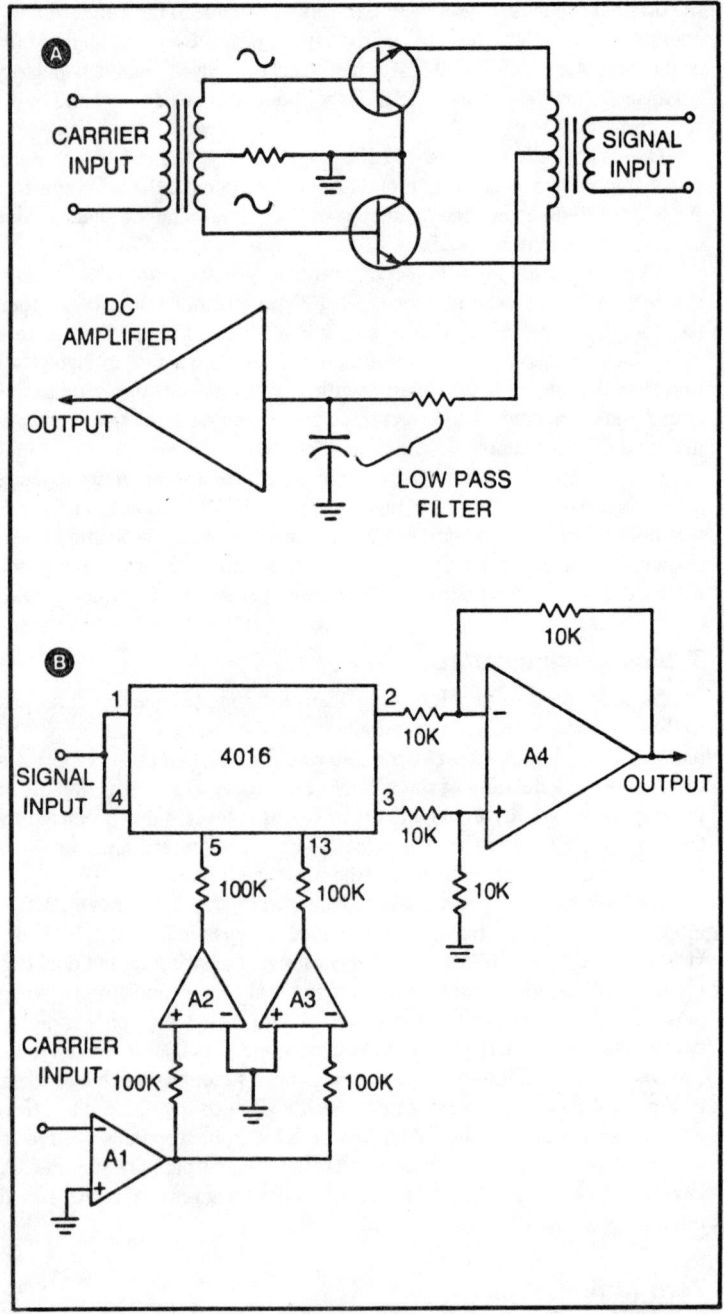

Fig. 11-5. Synchronous demodulator circuits: transformer (A) and digital (B).

will produce a dc offset potential that is not zero at a time when the input stimulus is zero. This can be due to a variety of faults, but regardless of the cause, all will bring about an error. A null circuit must be provided to cancel this error.

Most carrier and bridge amplifiers use manual methods to null out, or zero, the errors caused by offset problems. In a dc bridge amplifier one may use the ordinary offset null methods used for any operational-amplifier circuit. these methods produce a countercurrent that cancels out the offset. The operator adjusts the offset to zero when the transducer is in an unstimulated condition. For the pressure system described earlier, you would set zero with the transducer full of fluid but with its stopcocks open to air.

The ac carrier amplifiers usually use a potentiometer connected such that each end is at an opposite phase of the reference carrier. The wiper will then have a signal with a phase proportional to the resistance setting. This signal is connected to the bandpass amplifier or, possibly, the demodulator. At zero stimulus the phase and amplitude are adjusted so that the offset is cancelled.

Auto-zero circuits are much simpler to operate. If one of these is used, the operator need only put the transducer in a nonstimulated condition and then press the zero button.

Two different, yet similar, methods are used to produce an auto-zero feature. Both are actually variations of sample-and-hold techniques. A block diagram is shown in Fig. 11-6. If the zero button is pressed, the sample-and-hold circuit will sample the output of the amplifier, which at this point represents the offset potential and will store it in a memory circuit even after the button is released. The stored voltage will then be scaled, if necessary, and fed back to the input of the amplifier to cancel the offset.

One method for performing this job is to use a classical boxcar, sample-and-hold circuit, which stores the voltage in a capacitor. This is not very

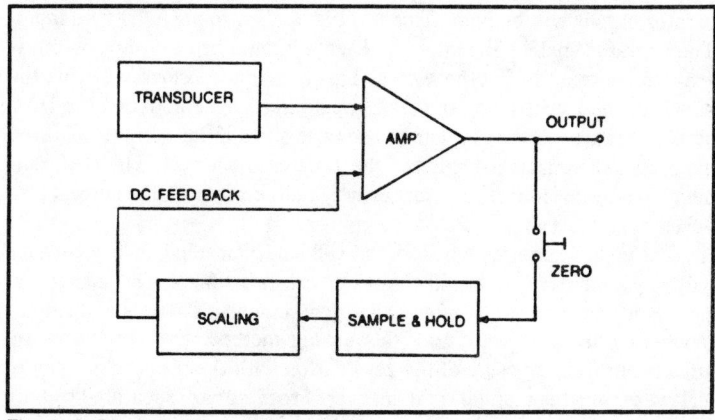

Fig. 11-6. Block diagram of an auto-zero circuit.

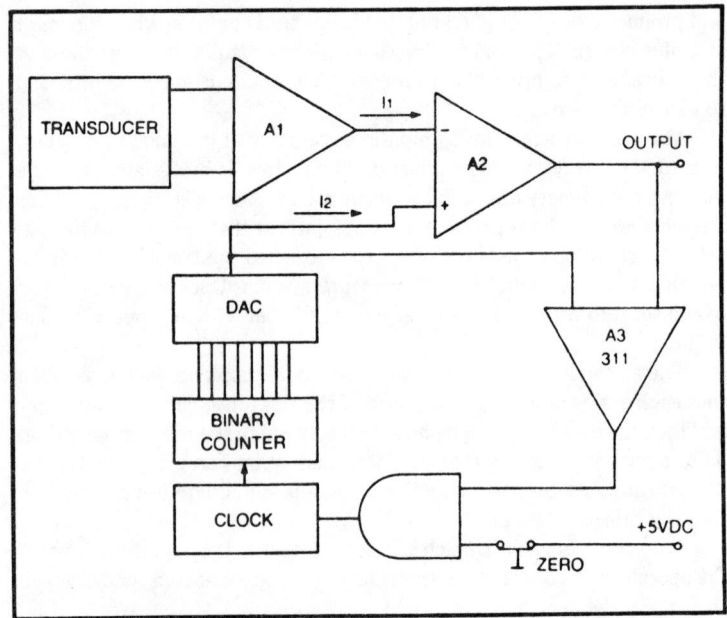

Fig. 11-7. Digital implementation of auto zero.

good, however, because high-cost capacitors and premium operational amplifiers will allow some droop of the stored value over time. This causes a shift in the claimed zero baseline that may not be detected during use. A superior method, now economically feasible because low-cost, digital/analog converter (DAC), IC devices are available, is shown in Fig. 11-7.

In the circuit of Fig. 11-7 a type 311 comparator (A3) is used to compare the output of a DAC (see Chapter 14) to the pressure-signal output of A2. If the pressure-signal output is larger than the DAC output, the comparator output will be high. This level is applied to an AND-gate input. The operator enables the other AND-gate input when the zero button is pressed, thereby closing the switch. This turns on the clock, allowing the binary counter controlling the DAC inputs to increment. When the DAC output equals the analog input (the pressure offset if the operator followed instructions), counting stops—but the counter *is not reset*. The DAC output generates current I2, and that exactly nulls current I1, the current produced by the offset.

The digital sample-and-hold circuit will hold almost indefinitely because only tiny amplifier offsets will creep in to create artifacts. The analog version, however, will droop after only a few minutes. There is an arterial-blood-pressure amplifier using this analog method, and it requires re-adjustment (i.e., pressing of the zero button with the transducer open to air) every time the reading is taken, even if readings are separated by only a few minutes.

Chapter 12

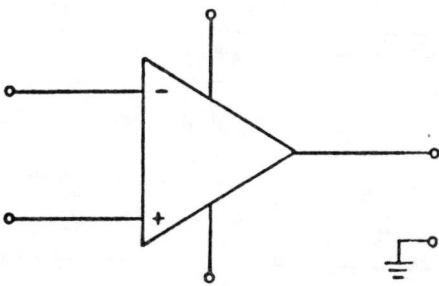

Temperature Measurements

ELECTRONICS HAS MADE TREMENDOUS STRIDES IN ALL AREAS OF physical measurement, but nowhere is it as commonplace as in temperature measurement. Using a variety of sensors (transducers) we can measure temperatures from near the liquid point of air to several thousands of degrees. Medicine has abandoned the mercury thermometer in all but a few specialized cases: the electronic clinical thermometer is faster, more accurate, and (over the long haul) less costly per use than old-fashioned models. Electronic thermometers are also available for consumer and household use (see Fig. 12-1). In this chapter I will discuss temperature transducers and the circuits used in temperature measurements. Let's deal first with the sensors.

TEMPERATURE TRANSDUCERS

There are several forms of temperature-sensitive devices that can be used as temperature transducers. These include thermal resistors (thermistors), thermocouples, and semiconductor junctions.

Of these, the simplest is probably the thermistor. This is a passive device which exhibits a change in electrical resistance in response to changes in temperature. It is known that all electrical conductors exhibit different resistivity figures at different temperatures. Furthermore, this change of resistivity is dependent upon the type of material. In general, though,

$$R_{t2} = R_{t1} (1 + \propto(\Delta t)) \tag{12.1}$$

where
R_{t2} is the resistance at the new temperature

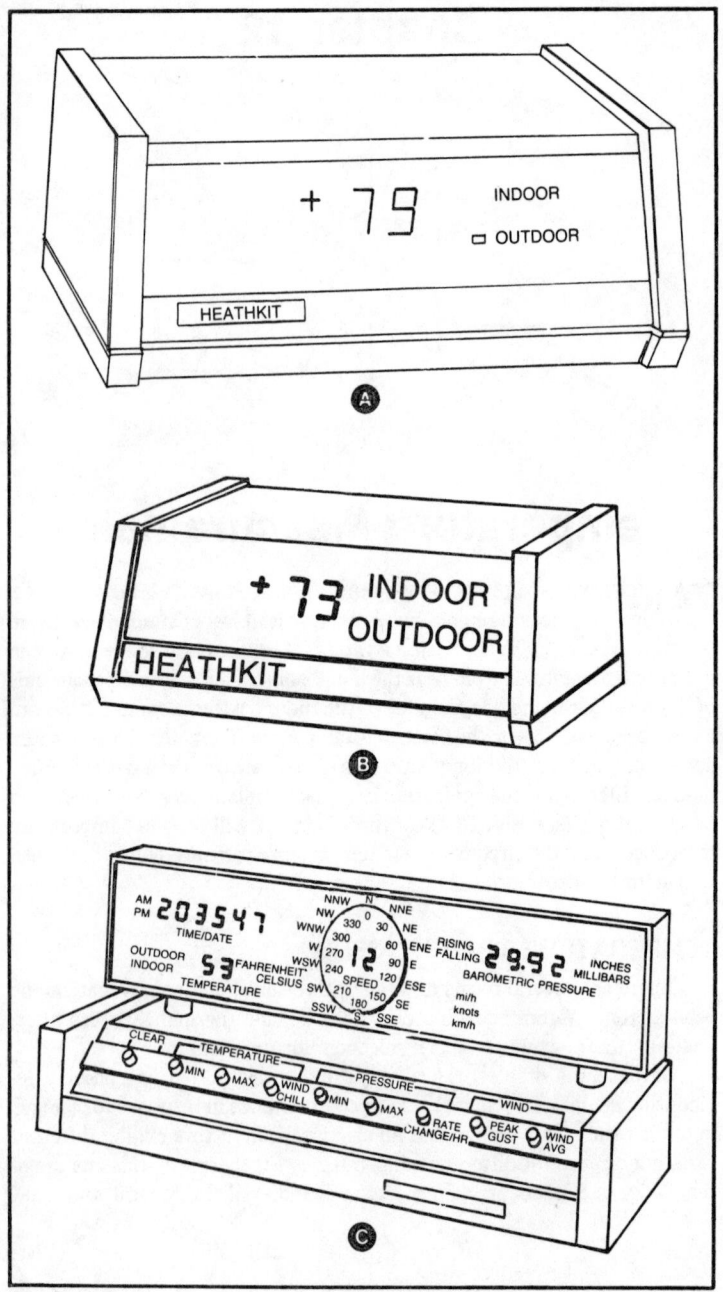

Fig. 12-1. (A) through (C) Electronic thermometers. (Courtesy of The Heath Company.)

R_{t1} is the resistance at the old temperature
Δt is the quantity (t2 − t1)
$\propto$ is the temperature coefficient (TC) of resistance specific to the type of material

The term $\propto$ may be positive (PTC) or negative (NTC), large or small. Copper, for example, is listed in the standard-resistivity tables as + 0.0039 ohms/ohm/° C, while graphite is listed as − 0.005 ohms/ohm/° C.

Example

What is the resistance of a copper wire at 100° C if the resistance at 25° C is 1 ohm?

$$
\begin{aligned}
R_{t2} &= R_{t1}(1 + \propto(\Delta t)) \\
&= (1)(1 + (0.0039)(100 - 25)) \\
&= 1.29 \text{ ohms}
\end{aligned}
$$

Although all electrical conductors possess the thermistor-like behavior described by Eq. 12.1, only a few, specially-designed types are used in electrical thermistor instruments. An instrumentation-type thermistor is a device in which this behavior is enhanced and optimized. (Most of these are nickel or manganese oxides.)

An appreciation of thermistor action can be realized through examination of the curves in Fig. 12-2. The curve at (A) shows the relationship between impressed voltage and thermistor current. The second curve relates current to the time after the application of voltage needed for the current to actually flow.

In Fig. 12-2(A) you see that current increases more or less linearly with increases in voltage, according to classical-Ohm's-law relationships, until voltage E_1 is reached. At this point the thermistor will begin to operate

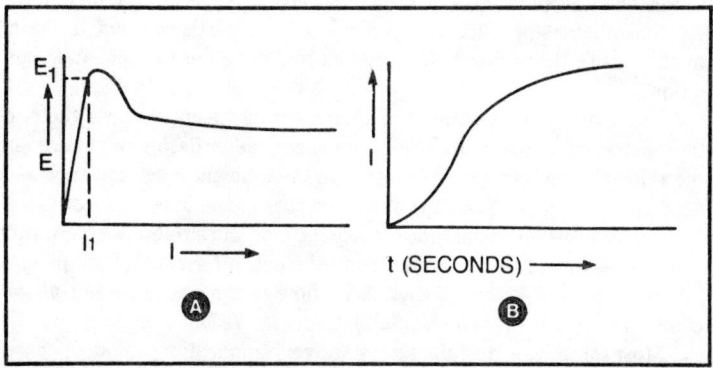

Fig. 12-2. Thermistor action: (A), impressed voltage vs. thermistor current; (B), current vs. time, showing the relative response time of a thermistor.

in a negative-resistance mode. In this region, incidentally, the thermistor acts as an oscillator, albeit a very-low-frequency oscillator.

The second curve Fig. 12-2(B), shows that our thermistor does not respond instantaneously but has a time lag. This time lag must be accounted for in your instrument design, or trouble will result. A thermistor intended to respond to a one-hertz stimulus should not have a time constant of several seconds—or even one second for that matter. It should have a time constant of less than the period of the intended stimulus. Repeatable and reliable time constants ranging from milliseconds to seconds are usually available, although one must expect to pay much more for the short-time-constant thermistors.

Thermistors behave differently in two, different, operating modes. In one mode they are biased to a very low current so as to prevent self-heating of the thermistor by its own current. In the other mode just the opposite situation is sought by the designer. A current is caused that just begins to generate self-heating. This mode allows relatively large changes of electrical resistance in response to an external, heat-absorbing stimulus. The manufacturer's-specification manual should be consulted for this application because the self-heating current for different types of thermistors is different. Instruments in this second mode are usually quite sensitive to *change*, but the former is preferred if thermometry is the goal.

Thermistor instruments are almost universally of the Wheatstone-bridge type. At least one arm of the bridge, and usually at least two arms, will be a thermistor. The other resistors are low-temperature-coefficient, fixed or variable types. The voltage applied to the bridge is set to cause a small amount of self-heating for measurements in which dT/dt is important or no self-heating if a measurement of T itself is critical.

Two different forms of bridge are normally encountered. In one form, one of the bridge arms is made variable so the bridge may be offset to either zero or full-scale with the bridge under quiescent, or at rest, conditions. In the other mode, the variable resistor in one arm is calibrated so that it will null the bridge at the measured values of input temperature.

Most thermistor instruments will use at least two thermistors. It is usual to expose one thermistor to the stimulus and the other to a quiescent condition.

There are many examples of thermal transducers that use the two-thermistor technique. For instance, an anemometer design might use one thermistor in the wind and the other in the ambient condition such as a dead-space box that is nonturbulent. Another example is a gas- or vapor-flow meter in which one thermistor is in the pipe and a reference thermistor is in a dead-ended container in a nonturbulent, off-section of the pipe. A practical example of such a thermistor flow meter was given in *Op-Amp Circuit Design & Applications* (TAB book No. 787).

Most thermistor manufacturers are very applications-minded; consequently they offer much good literature on their products. I have found that most of them are very generous with applications notes, catalogues, and technical help.

THERMOCOUPLES

A thermocouple is shown graphically in Fig. 12-3. It consists, in the most elementary form, of two strips of *dissimilar* metals joined to form a junction. If this junction is heated and a millivoltmeter is connected across the free ends away from the junction, there will be found a voltage present. This voltage is caused by the different *work function* of the two metals forming the junction and is dependent upon both the temperature and the types of metal used.

Since the voltage across the thermocouple junction is proportional to temperature we may use it in thermometry. The voltage is not, however, a linear function of temperature, and that tends to limit the application somewhat. It is important to know over what range the thermocouple is linear so that errors will not be excessive. An alternate discussion is the linearization of the thermocouple, a subject discussed in more general terms in the chapter on instrumentation techniques.

Electronic instrumentation using thermocouples tends to be of the two-sensor design, as was also true of thermistor instruments. In this case, however, the sensors are voltage sources in their own right; so they are connected differentially to an operational amplifier. This improves our ability to see small, differential voltages and generates a usable output potential. Pseudo-Wheatstone-bridge configurations are seen in many cases.

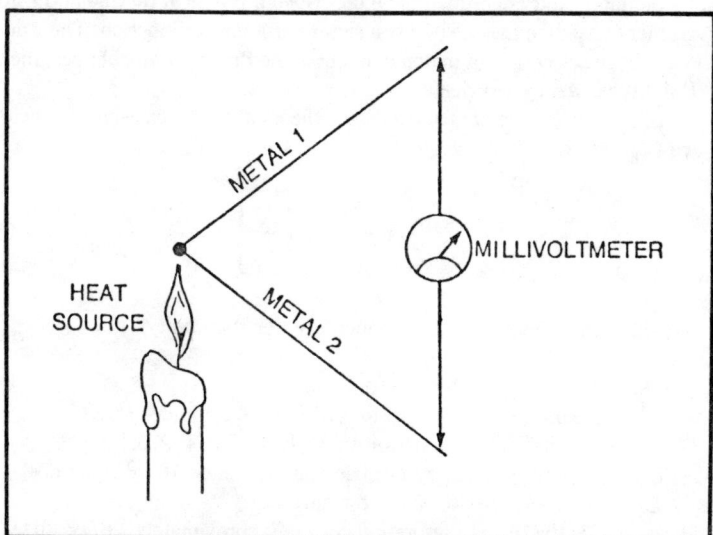

Fig. 12-3. The thermocouple.

Although the classical-bimetallic-strip thermocouple has been known for a century or longer and been available commercially for several decades, it is now taking a backseat to certain solid-state thermocouples more-recently introduced. Many of these are available at relatively low cost yet will linearly cover a wide range of below 0° C to well over 100° C.

TRANSISTOR THERMAL SENSORS

There is another type of solid-state thermosensor, but it is not to be confused with the type discussed above. It is actually a transistor and depends for its operation on the normal temperature sensitivity of the transistor's base-emitter voltage. This potential turns out to be very linear, even in low-cost sensors.

Thermistors and thermocouples will produce a linear temperature curve, but it takes a great deal of effort to make them linear over a wide range of temperatures. In fact, there are a number of thermocouple instruments that require the operator to either change the thermocouple itself or operate a linearization switch when going from one range to another. While this may be satisfactory in many cases it is a definite disadvantage when working with systems that cover a wide range.

The use of a pair of well-matched transistors in a simple electronic circuit results in a stable and highly-linear, thermometry probe that remains competent over a wide dynamic range. This technique is also free of the calibration problems associated with the use of thermistors, thermocouples, and simple single-PN-junction thermosensors.

Simmons and Soderquist have described a practical thermometry instrument based on the use of a transistor-base-emitter junction. Their device uses the closeness-of-match inherent in the Precision Monolithics, Inc., (PMI) type MAT-01-series of dual transistors.

I know from elementary transistor theory that the base-emitter voltage (V_{be}) is given by

$$V_{be} = \frac{KT}{q} \ln \left[\frac{I_c}{I_{rs}} \right] \qquad (12.2)$$

(provided that the ratio I_c/I_{rs} is much greater than unity)

where
$\quad$ K $\qquad$ is Boltzmann's constant (1.38×10^{-23} joules /° K)
$\quad$ T $\qquad$ is the Kelvin temperature (° K = ° C + 273.16)
$\quad$ q $\qquad$ is the elementary electric charge (1.6×10^{-19} coulombs)
$\quad$ I_c $\qquad$ is the transistor-collector current
$\quad$ I_{rs} $\qquad$ is the reverse saturation current (approximately 1.87×10^{-14} amperes in most cases)

A transistor pair that share the same thermal environment (i.e., are on the same monolithic substrate, as in the MAT-01) allow me to rewrite Eq. 12.2 as the difference in two base-emitter potentials, or

$$\Delta V_{be} = \frac{KT}{q} \ln \left[\frac{I_{C1}}{I_{rs1}} \right] - \frac{KT}{q} \ln \left[\frac{I_{C2}}{I_{rs2}} \right] \qquad (12.3)$$

Because of the common term, KT/q, I may rewrite Eq. 12.3 in the form

$$\Delta V_{be} = \frac{KT}{q} \left[\ln \left[\frac{I_{C1}}{I_{C2}} \right] - \ln \left[\frac{I_{rs1}}{I_{rs2}} \right] \right] \qquad (12.4)$$

But the term I_{rs1}/I_{rs2} is very nearly unity in well-matched transistors sharing the same thermal environment. Since I also know that the natural logarithm of unity is zero (ln 1 = 0), the term

$$\ln \left[\frac{I_{rs1}}{I_{rs2}} \right]$$

vanishes, reducing Eq. 12.4 to a simpler form

$$\Delta V_{be} = \frac{KT}{q} \ln \left[\frac{I_{C1}}{I_{C2}} \right] \qquad (12.5)$$

Equation 12.5 is my justification for claiming that thermometry is possible when using cothermal, matched transistors.

Examine Eq. 12.5 for the case where currents I_{C1} and I_{C2} are held constant, but at different values. I must make them different to keep the ratio I_{C1}/I_{C2} from becoming unity, in which case its natural logarithm will vanish to zero. If the ratio is made, say, 2:1, the log of the ratio is a constant equal to ln 2. Equation 12.5 becomes

$$\Delta V_{be} = \frac{cKT}{q} \qquad (12.6)$$

where

c   is ln 2 = 0.6931
K   is Boltzmann's constant
T   is the Kelvin temperature (° K)

This equation is further reduced to

$$\Delta V_{be} = \left[\frac{cKT}{q} \right] T \qquad (12.7)$$

where all terms inside the parentheses are constants. This leaves the base-emitter potential, V_{be}, dependent *only* upon the Kelvin temperature. The constant terms evaluate to

$$\frac{cK}{q} = \frac{(0.6931)(1.38 \times 10^{-23}) \text{ joules/}^\circ \text{K}}{1.67 \times 10^{-19} \text{ coulombs}} \qquad (12.8)$$

$$= 57.28 \ \mu V/^\circ K$$

This allows me to write

$$\frac{V_{be}}{T} = \frac{57.28 \ \mu V}{^\circ K} \qquad (12.9)$$

This relationship allows me to design a complete temperature-measuring instrument using a matched pair of monolithic transistors, a pair of constant-current sources set at I and 2I, and an operational amplifier. An example provided by PMI is given in Fig. 12-4. This instrument is capable of measuring temperatures over the range 218° K to 398° K (−55° C to +125° C). The sensor pair of transistors may be located up to 100 feet remote from the electronics provided that adequate shielded cable is supplied. The extremely close match of the Precision Monolithics MAT-01H made it highly desirable for the sensor. Similarly, a MAT-01GH was selected for the constant-current-source pair.

Current I was selected to be 5 microamperes; so 2I must be 10 microamperes. These currents are regulated by matched pair Q2. The bases of Q2A and Q2B are cobiased by the same resistor-voltage-divider network; so the respective collector currents will be set by the individual emitter resistances. Ideally, the 2:1 ratio of currents dictates a resistance ratio of 1:2, but reality forces us to trim one side to ensure the accuracy of the ratio. Because of this factor, the 200-kilohm side of the dual constant-current source is made variable over the range 180 to 230 kilohms. The sensor pair, Q1, has the two collectors and two bases connected to ground. The respective emitters are connected to the operational-amplifier inputs and the constant-current sources.

Calibration

1. Turn the unit on and wait five minutes.
2. Short the input to the operational amplifier.

3. Adjust the offset for 0 Vdc ± 10 mV output.
4. Allow the amplifier to stabilize for another 10 minutes.
5. Check the offset adjustment; readjust if necessary.
6. Remove the input short from the operational amplifier; place the sensor in a known thermal environment. This could be room air or in a temperature-stabilized water bath.
7. Wait 2 minutes.
8. Adjust R7 (ratio adjust) for the proper output voltage dictated by the temperature being measured. Room temperature is a good thermal environment for this adjustment because it is approximately midrange and is easily measured by other means to good accuracy.

Note that my output scale factor is 10 mV/° K. This makes it very easy to display absolute (Kelvin) temperature; a simple digital voltmeter will do the job.

Display of degrees Celsius is accomplished by circuitry that accounts for the conversion factor

$$° K = ° C + 273 \qquad (12.10)$$

or

$$° C = ° K - 273 \qquad (12.11)$$

A simple operational amplifier connected as a unity-gain summer can be used if your digital-electronics acumen is less than fully developed. Connect the noninverting follower to the output of the thermometer. The inverting input is connected to a stable reference voltage computed from

$$E_{ref} = (273 ° K) \frac{10 \text{ mV}}{° K} \qquad (12.12)$$

$$= 2730 \text{ mV} = 2.73 \text{ volts}$$

COMMERCIAL IC TEMPERATURE-MEASUREMENT DEVICES

Several semiconductor-device manufacturers offer temperature-measurement/control integrated circuits (TMIC). These devices are almost all based on the PN-junction properties discussed earlier in this chapter, although at least one by Analog Devices, Inc. uses an external thermocouple. In this chapter I will discuss the semiconductor TMIC devices offered by National Semiconductor and Analog Devices, Incorporated. In addition, I will discuss a method for converting temperature measurements to frequencies that can be transmitted along a communications link or recorded on a tape recorder.

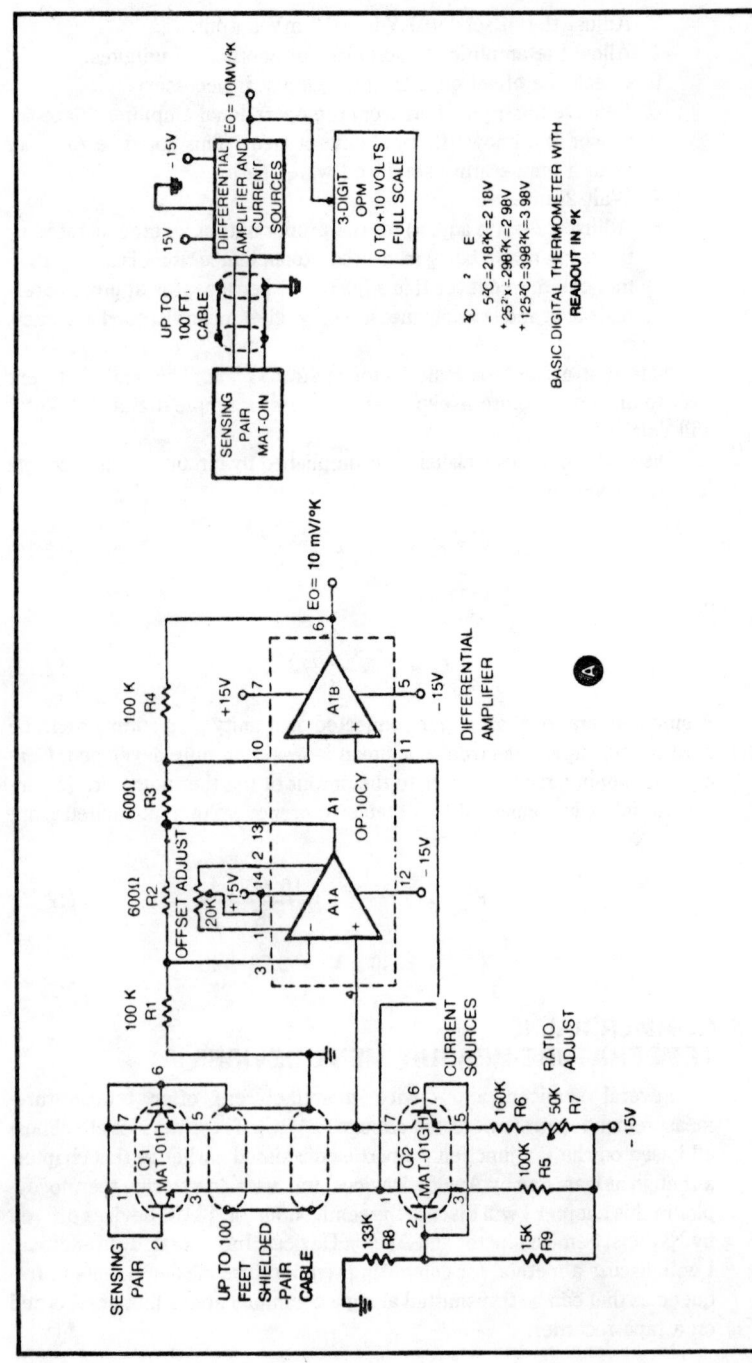

BASIC DIGITAL THERMOMETER WITH **READOUT IN °K**

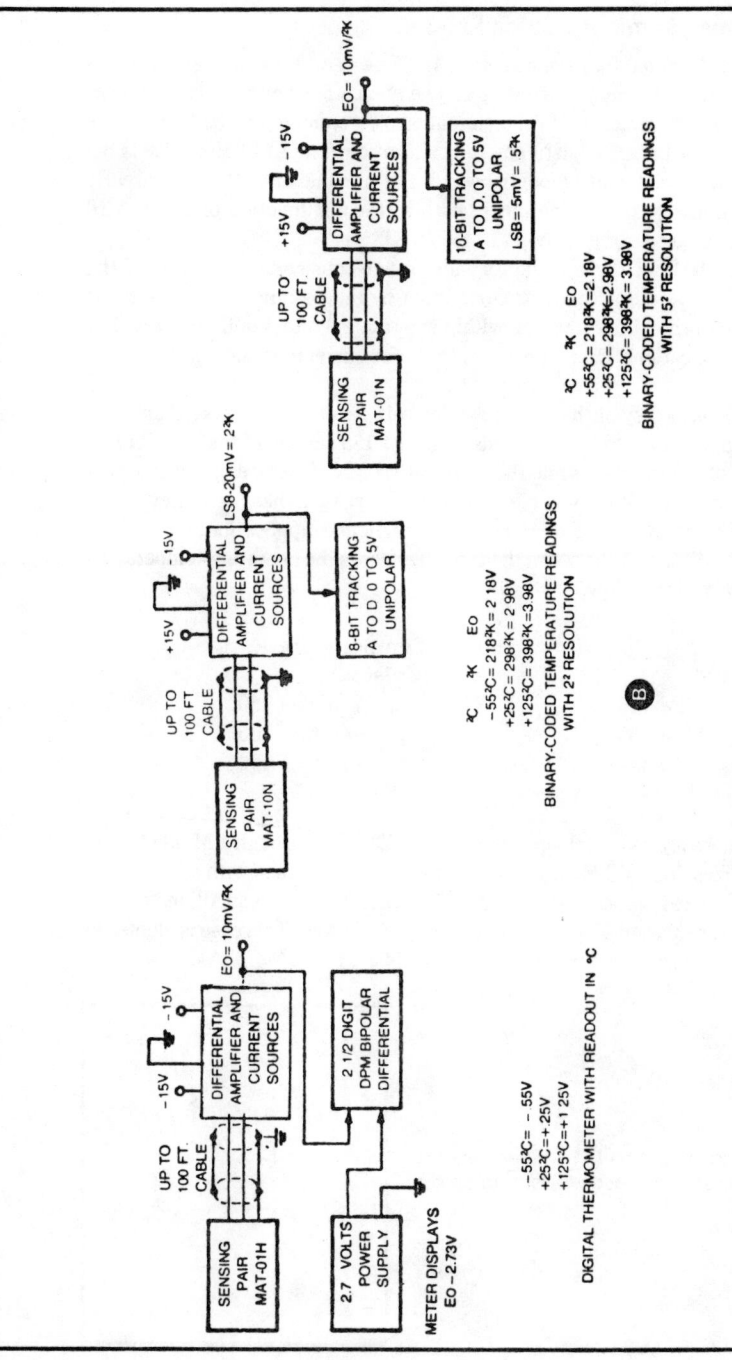

Fig. 12-4. (A), use of a dual transistor as a temperature probe; and (B), different temperature measurement systems. (Courtesy Precision Monolithics, Inc.)

227

National Semiconductor LM-335

The National Semiconductor LM-335 device shown in Fig. 12-5 is a three-terminal temperature sensor. The two main terminals are for power (and output), while the third terminal, shown coming out the body of the diode symbol is for adjustment and calibration. The LM-335 device is basically a special zener diode in which the breakdown voltage is directly proportional to the temperature, with a transfer function of close to 10 millivolts per degree Kelvin (10 mV/° K).

The LM-335 device, and its wider-range cousins the LM-135 and the LM-235, operate with a bias current set by the designer. This current is not supercritical but must be within the range 0.4 to 5 milliamperes. For most applications, designers seem to prefer currents in the 1-milliampere range.

The accuracy of the device is relatively decent and is more than sufficient for most control applications. The LM-135 version offers uncalibrated errors of 0.5 to 1 ° C, while the less-costly LM-335 device offers errors of less than 3 ° C. Of course, clever design can reduce these errors even farther if they are out of tolerance for a particular application.

One difference between the three devices is the operating-temperature ranges, which follow:

Device Type No.	Temperature Range (Celsius)
LM-135	− 55 to + 150
LM-235	− 40 to + 125
LM-335	− 10 to + 100

There are two packages used for the LM-135-through-LM-335 family of devices. The TO-92 is a small, plastic, transistor case with a "Z" suffix to its device number; e.g., LM-335Z, and the TO-46 is a small, metal-can, transistor case (smaller than the familiar TO-5 case). This case is identified with the suffix "H" or "AH"; e.g., LM-335H or LM-335AH.

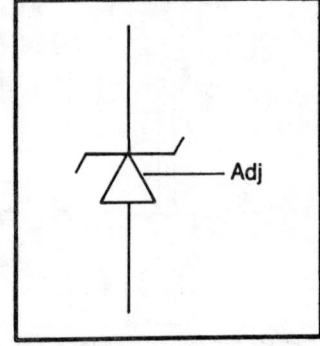

Fig. 12-5. The LM-335 temperature sensor by National Semiconductor.

The simplest, although least-accurate, method of using the LM-335 device is shown in Fig. 12-6(A). The LM-335 is essentially a zener diode, and here it is connected as a zener diode. The series current-limiting resistor limits the current to around 1 milliampere. This value of R1; i.e., 4700 ohms, is appropriate for + 5-volt power supplies as might be found in digital electronic instruments. The resistor value can be scaled upwards for higher values of dc potential according to Ohm's law, keeping I = 1 milliampere

$$R(ohms) = (V+) \times 1000$$

For example, when the power-supply voltage is + 12 volts dc, the value of the resistor in series with the LM-335 is

$$
\begin{aligned}
R(ohms) &= (V+) \times 1000 \\
&= 12 \times 1000 \\
&= 12,000 \text{ ohms}
\end{aligned}
$$

The output of the circuit in Fig. 12-6(A) is taken across the LM-335 device. The voltage has an approximate rate of 10 mV/° K. Recall from earlier that "degrees Kelvin" is the same as "degrees Celsius" except that the zero point is at absolute zero (close to – 273 ° C) rather than the freezing point of water. Using ordinary units—conversion arithmetic will show you how much voltage to expect at any given temperature. For example, suppose you want to know the output voltage at 78° C. The first thing you must do is convert the temperature to degrees Kelvin. This is done by adding 273 to the Celsius temperature per Eq. 12.10

$$
\begin{aligned}
° K &= ° C + 273 \\
&= 78° C + 273 \\
&= 351
\end{aligned}
$$

Next, you convert the temperature to the equivalent voltage per Eq. 12.12

$$V = \frac{10 \text{ mv}}{K} \times 351 \text{ K}$$

$$= (10 \text{ mV}) (351)$$

$$= 3510 \text{ mV} = 3.51 \text{ volts}$$

One problem with the circuit of Fig. 12-6(A) is that it is not calibrated. That circuit works well for many applications, especially those where precision is not needed, but for other cases you might want to consider the circuit of Fig. 12-6(B). This circuit allows single-point calibration of the temperature. The calibration control is obtained from the 10-kilohm poten-tiometer in parallel with the zener diode. The wiper of the potentiometer is applied to the adjustment input of the LM-335 device.

Calibration of the device is relatively simple. One only needs to know the output voltage (a dc voltmeter will suffice) and the environmental temperature in which the LM-335 exists. In some less-than-critical cases, you might take a regular, glass, mercury thermometer and measure the air temperature. Wait long enough after turning on the equipment for both the mercury thermometer and the LM-335 device to come to equilibrium. After that, adjust potentiometer R2 for the correct output voltage. For ex-ample, if the room temperature is 25° C (i.e., 298° K), then the output voltage will be 2.98 volts. Adjust the potentiometer for 2.98 volts under these conditions.

Another tactic is to use an ice-water bath as the calibrating source. The temperature 0° C is defined as the point where water freezes and is recognized by the fact that ice and water coexist in the same spot (the ice neither melts nor freezes, it is in equilibrium). A mercury thermometer will show the actual temperature of the bath. The potentiometer is adjusted until the output voltage is 2.73 volts (note: 0° C = 273° K).

Still another tactic is to use a warmed-oil bath for the calibration. The oil is heated to somewhat higher than room temperature (maybe 40° C) and stirred. Again, a mercury thermometer is used to read the actual temperature and the potentiometer is adjusted to read the correct value. The advantage of this method is that the oil bath can be a constant-temperature situation. There are numerous laboratory pots on the market that will keep water or oil at a constant, preset temperature, a factor that avoids some problems inherent in the other methods.

Another connection scheme for the LM-335 is shown in Fig. 12-6(C). In this variation I am using a National Semiconductor LM-334 three-terminal, adjustable-current source for the bias of the LM-335 device. Again, the output voltage will be 10 mV/° K.

All applications where the sensor is operated directly into its load suf-fers a potential problem or two, especially if the load impedance changes or if it is lower than some limit. As a result, we can sometimes justify us-ing the buffered circuit of Fig. 12-7.

A *buffer* amplifier is one that is used for one or both of two purposes: 1) impedance transformation or 2) isolation. The impedance-transformation factor is used when the source impedance is high (not true of the LM-335). The isolation factor is of somewhat more concern to us here. The opera-tional amplifier in Fig. 12-7 places an amplifier between the sensor and its load. The gain of the amplifier in this case is unity (i.e., 1), but a higher gain could be used if desired. In that case, simply substitute one of the gain-amplifier circuits shown later in this book.

The operational amplifier shown here is an RCA CA-3140 device. The reason for this is simply the freedom from bias currents exhibited by the BiMOS RCA operational amplifiers. The bias currents found on cheaper

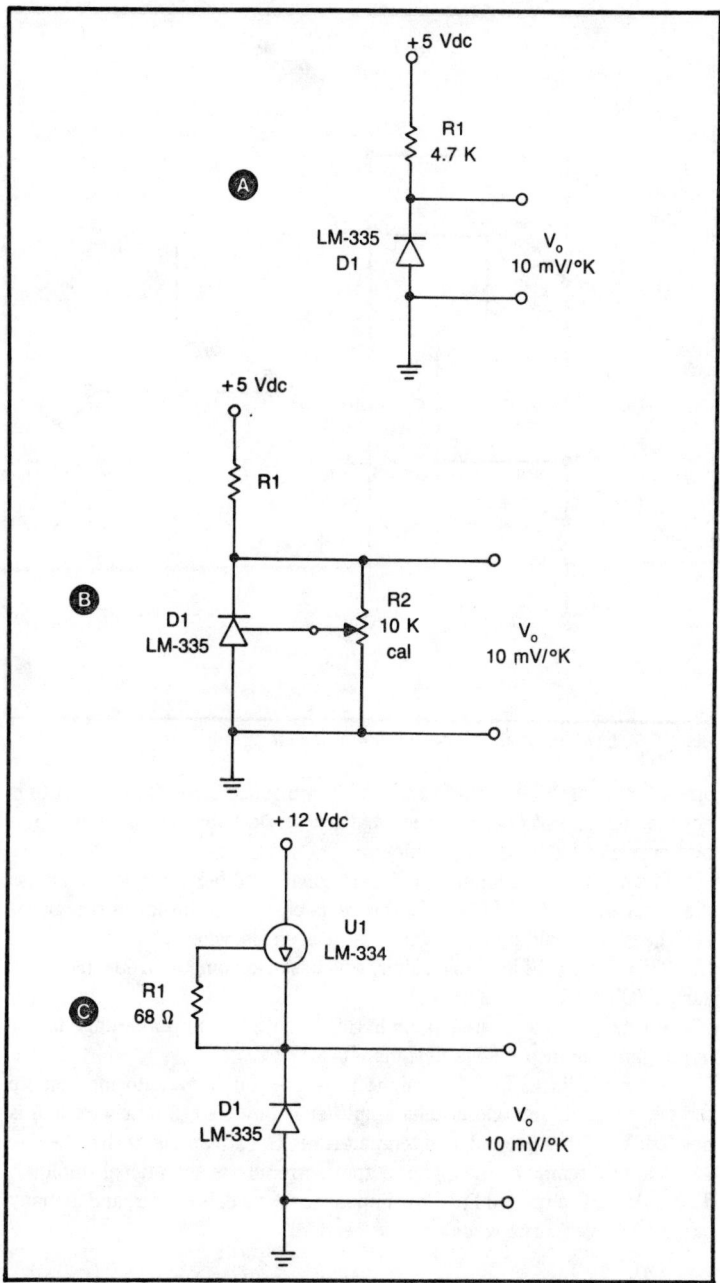

Fig. 12-6. LM-335 circuits (A) fixed calibration, (B) variable calibration, (C) using LM-334 current source.

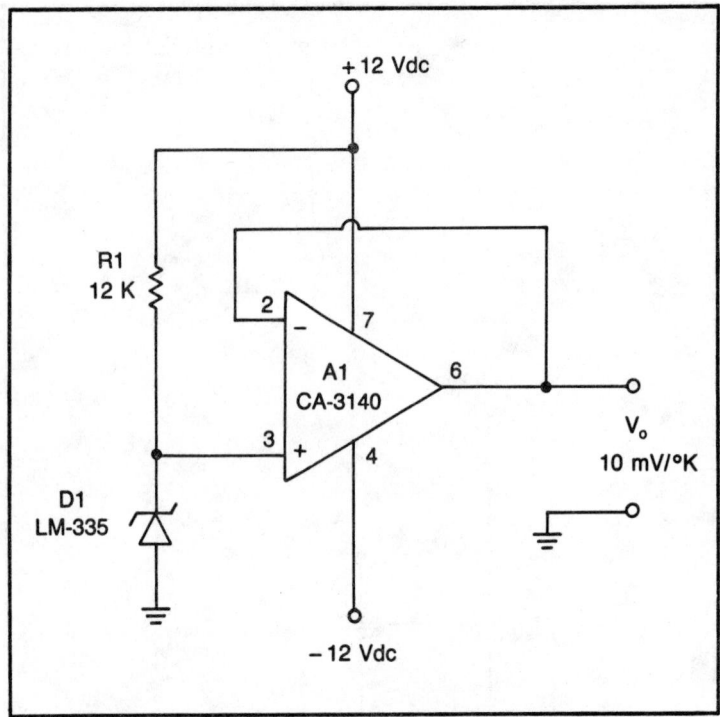

Fig. 12-7. LM-335 used in operational amplifier circuit.

operational amplifiers could conceivably introduce error. The CA-3140 is not the only operational amplifier that will work, however; any low-input-bias-current model will work nicely.

The noninverting input of the operational amplifier is connected across the zener-diode-like LM-335. In this respect, this circuit looks somewhat like the typical voltage-reference circuits seen elsewhere. The bias for the LM-335 is from a 12-kilohm resistor, which is in keeping with our rule given earlier (Ohm's law, remember?).

Since there is no voltage gain in this circuit, the output-voltage factor is the same as in previous designs, 10 mV/° K.

A circuit like Fig. 12-7 might prove useful in monitoring remote temperatures. If the operational amplifier is powered, a four-wire line is needed: V – , V + , ground, and temperature. The advantage is that the line losses are overcome by the higher output power of the operational amplifier. The LM-335 is a rugged little low-impedance device, however, and in many cases such measures would not be needed.

Temperature-Scale Conversions

The Kelvin scale is used extensively in scientific calculations but is

232

not always the most popular in practical measurement situations. In fact, I suspect that most readers of this book will want to make their temperature measurements in either degrees Celsius or degrees Fahrenheit. In this section I will discuss the circuit methods used for both.

If the sensor is being input into a microcomputer, it might be prudent to use the simplest circuit available, which measures in degrees Kelvin, and then let the computer convert the units. The formulae below are useful for this purpose:

$$C = K - 273 \tag{12.12}$$

and

$$F = 1.8\ C + 32 \tag{12.13}$$

Of course, the first job will be to make the computer think it is seeing the correct kind of data. The analog-to-digital (A/D) converter will input (more than likely) a binary number between 00000000 and 11111111. This number must be scaled to the proper value that represents a temperature value. Let's assume that you have an eight-bit, A/D converter, and a temperature range of 0 to 100 degrees centigrade. The input voltage to the A/D will be 2.73 to 3.73 volts. If the A/D converter is able to offer off-set measurements, you can set the maximum range for 1 volt and then off-set it to 2.73 volts. In that unlikely case, 00000000 would represent 0° C, and 11111111 would represent 100° C. More likely, you will use a 5-volt, unipolar-input, A/D converter to measure the narrow range of 2.73 to 3.73 volts and suffer a resolution loss. Of course, this loss is not what it may seem because in many cases it will still be less than the nonlinearity of the transducer/sensor. In such a scheme, the voltage represented by a 1-LSB change in the A/D-output-data word represents approximately 20 millivolts, so would represent 2° K. If all you need to measure is within two degrees, you can use this system. Otherwise, some form of offset measurement is needed.

Figure 12-8 shows a scheme for converting the "degrees Kelvin" output of the LM-335 sensor (D1) into degrees Celsius. Since Celsius degrees are the same size as Kelvin degrees, no change of slope in the output factor is needed: the output is 10 mV/° C, and the circuit gain is unity.

The basic circuit of Fig. 12-8 is a dc differential amplifier based on a common operational amplifier (741-family devices work fine). The gain is set by R4—R2, assuming R2 = R3 and R4 = R5. The noninverting input of the dc differential amplifier (A2) receives the temperature signal, while the inverting input receives a dc offset bias. This circuit is adjusted by using potentiometer R6 to set the voltage at point "A" to +2.73 volts (use a 3-1/2-digit-or-more, digital voltmeter). The result is that the output will be 2.73 volts less than it would have been were the offset not placed in the circuit; thus the output potential is scaled in degrees Celsius.

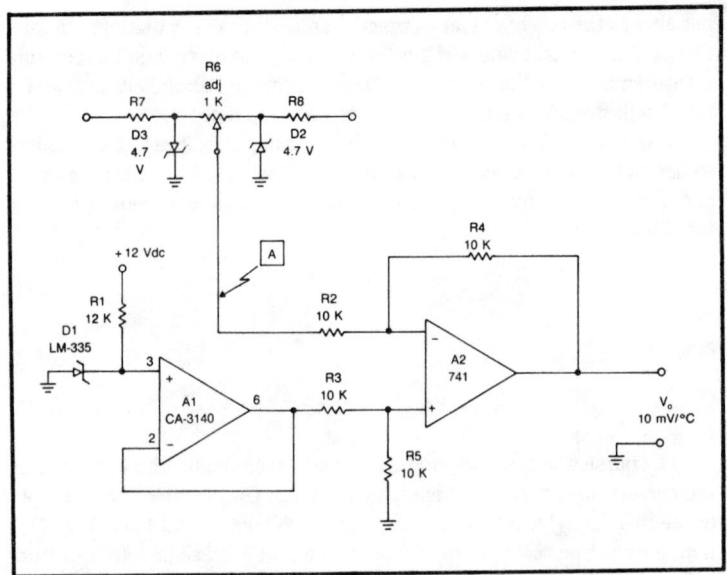

Fig. 12-8. LM-335 circuit used with operational amplifier offset.

Figure 12-9 shows a circuit for converting degrees Celsius to degrees Fahrenheit. The problem here is the two types of degrees are: 1) offset from each other (like Kelvin and Celsius, they have different zero references) and 2) different sizes. Thus, the conversion amplifier must offer both an offset and a change of slope. In Fig. 12-9 you see both. The offset is provided by the potentiometer R5, which is used to set the ice-point (zero degrees Celsius) output level. The feedback potentiometer R2 is used to set a calibration point at some higher temperature; for example, 25° C, room temperature (i.e., 77° F).

Calibration of the two points is performed in a manner similar to above. The zero point is set using an ice bath (adjust R5) in a manner like above; the higher point is probably best set at room temperature. In both cases, the actual temperature could be measured with an ordinary mercury thermometer. Of course, the best accuracy is obtained with a laboratory-grade, mercury thermometer.

Analog Devices AD-590

The Analog Devices, Inc. AD-590 is another form of solid-state temperature sensor. This particular device is a two-electrode sensor that operates as a current source with a one-microampere-per-degree-Kelvin (1 μA/° K) characteristic. The AD-590 will operate over the temperature range -55 to $+150$° C. It is capable of a wide range of power-supply voltages, accepting anything in the range +4 to +30 volts dc. (This range is more than sufficient for most solid-state applications.) Selected versions are available with linearity of ±0.3° C and a calibration accuracy of ±0.5° C.

The AD-590 comes in two different packages. There is a metal can (TO-52) that is recognized as the small size transistor package (smaller than TO-5). There is also a plastic flat-pack available.

Being a two-terminal current source, the AD-590 is simplicity itself in operation. Figure 12-10 shows the most elementary, calibratable circuit for the AD-590. Since it is a current source that produces a current proportional to temperature, you can convert the output to a voltage by passing it through a resistor. In Fig. 12-10, the resistance is approximately 1,000 ohms and consists of the resistance of R2 (950 ohms) and R1 (a 100-ohm potentiometer). From Ohm's law you know that 1 μA/K converts to 1 mV/K when passed through a 1000-ohm resistor. You can calculate the voltage output at any given temperature from the simple relationship below

$$V_o = \frac{1 \text{ mV}}{K} \times \text{TEMP} \tag{12.14}$$

Thus, if you have a temperature of 37° C, which is (37 + 273) or 310° K, the output voltage will be

$$V_o = \frac{1 \text{ mV}}{K} \times \text{TEMP}$$

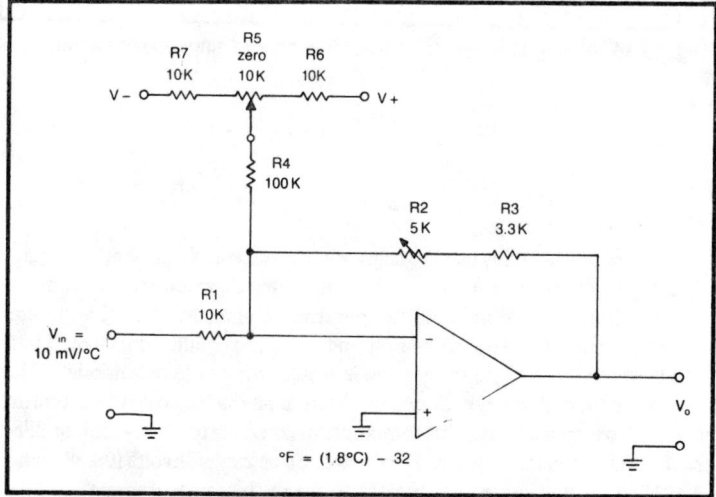

Fig. 12-9. C-to-F converter circuit.

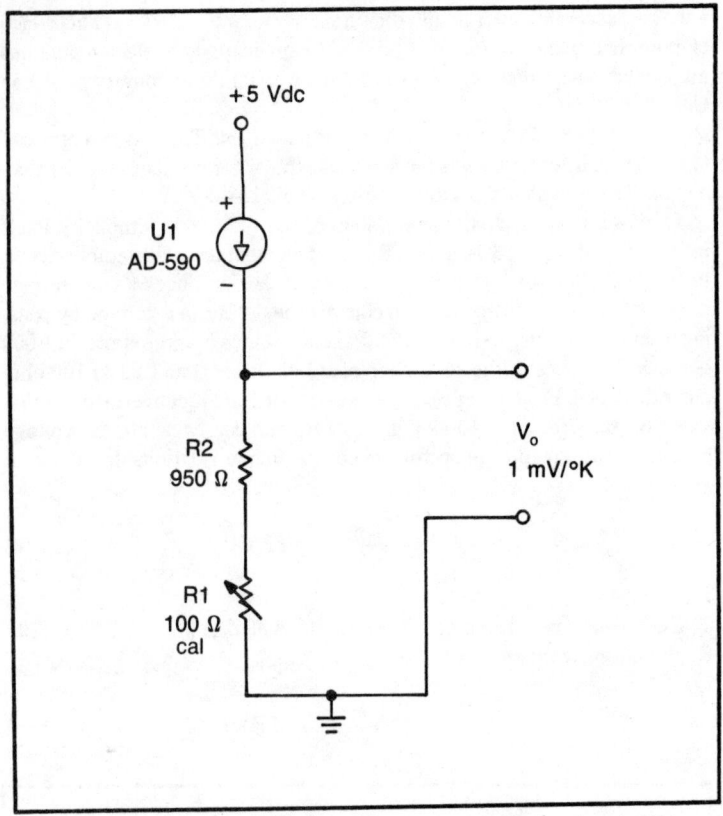

Fig. 12-10. Analog Devices, Inc. AD-590 sensor in single-point circuit.

$$= \frac{1\ mV}{K} \times 310K$$

$$= 310\ mV$$

Potentiometer R1 is used to calibrate this system. You can make a quick and dirty calibration with an accurate mercury thermometer (laboratory-grade recommended) at room temperature. Connect a digital voltmeter across the output and allow the system to come to equilibrium (should take about 10 minutes). Once the system is stable, adjust the potentiometer for the correct output voltage. For example, assume that the room temperature is 25° C, which is 77° F. This temperature converts to (273 + 25), or 298° K. The output voltage will be (1 mV × 298), or 298 millivolts (0.298 volts). A 3-1/2-digit voltmeter is sufficient to make this measurement.

In some cases it might be wise to delete the potentiometer and use a single 1000-ohm resistor in place of the network shown. There might be several reasons for this action. First, the calibration accuracy is not critical for the application at hand. Second, potentiometers are points of weakness in any circuit; being mechanical devices they are subject to stress under vibration conditions and may fail prematurely. If the temperature accuracy is not crucial and reliability is, consider the use of a single, fixed, one-percent-tolerance resistor in place of the network shown in Fig. 12-10.

The circuit of Fig. 12-10 is sometimes used to make a temperature alarm. By using a voltage comparator to follow the network and biasing the comparator to the voltage that corresponds to the alarm temperature, you can create a TTL level that indicates when the temperature is over the limit. A "window comparator" will allow you to have an alarm of either undertemperature or overtemperature conditions. Some electronic equipment designers use this tactic to provide an overtemperature alarm. In one application, a commercial minicomputer generated a large amount of heat. (It used a 35-ampere, +5-volt dc power supply.) The specification called for an air-conditioned room for housing the computer. An AD-590 device was placed inside at a critical point. If the temperature reached a certain level (45° C), the comparator output snapped LOW and created an interrupt request to the computer. The computer would then sound an alarm and display an "overtemperature warning" message on the CRT screen.

The circuit of Fig. 12-10 suffers from one little problem: it allows calibration at only one temperature. Unfortunately, this situation does not allow for optimization of the circuit. You can, however, improve the situation using the two-point-calibration circuit of Fig. 12-11. In this case, you see an operational amplifier in the inverting-follower configuration. The summing junction (inverting input) receives two different currents. One current is the output of the AD-590 (i.e., 1 μA/K), while the other current is derived from the reference voltage, V_{ref}, 10,000 volts. Adjustment of this current provides the zero-reference adjustment, while the overall gain of the amplifier provides the full-scale adjustment.

The operational amplifier selected is the LM-301 device, although almost any premium operational amplifier will suffice. The RCA CA-3140 BiMOS device, or some of those by either Analog Devices or National Semiconductor will also work nicely. If the LM-301 or similar device is used, be sure to use the 30-pF, frequency-compensation capacitor.

The V− and V+ power-supply lines are bypassed with 0.1-μF and 4.7-μF capacitors. The 0.1-μF capacitors are used for high-frequency decoupling and must be mounted as close as possible to the body of the operational amplifier. The value of these capacitors is approximate and may be anything from 0.1 to 1 μF.

Calibration of the device is simple, although two different temperature environments are required. The zero-degrees-Celsius adjustment, R1, can be made with the sensor in an ice-water bath (as described above). The

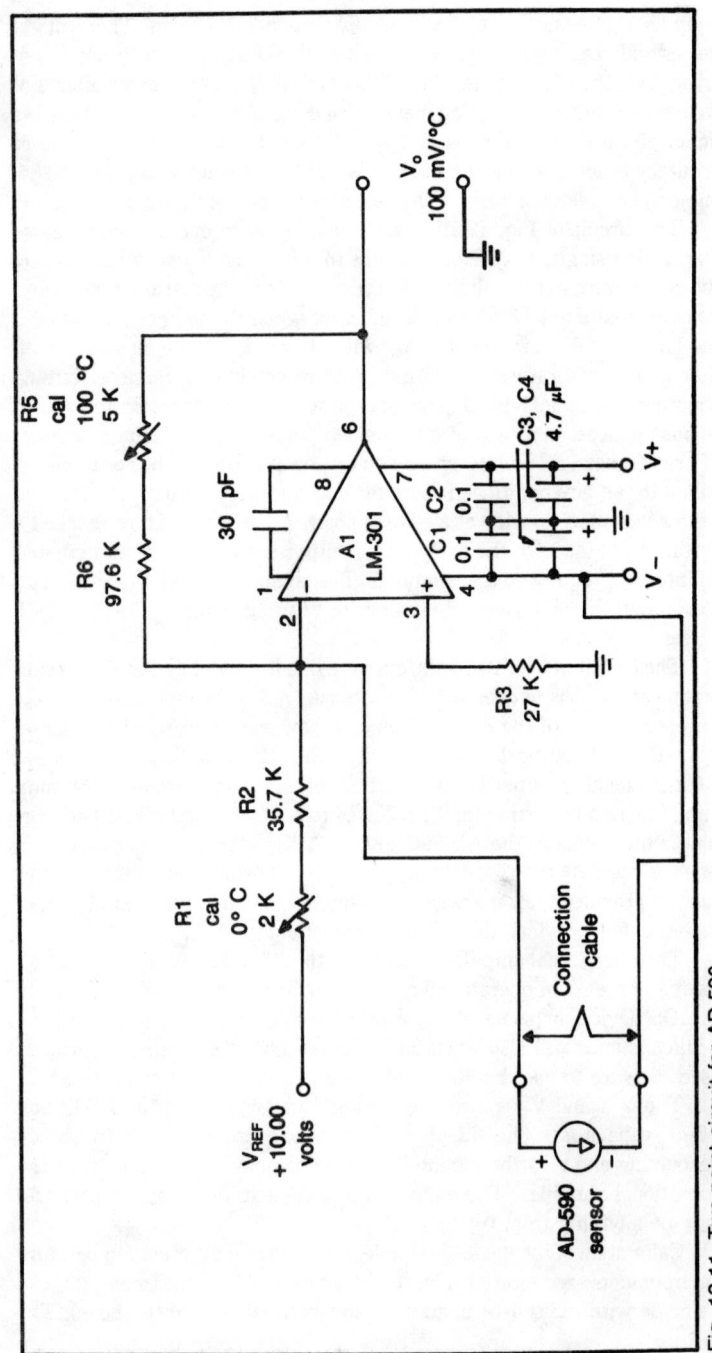

Fig. 12-11. Two-point circuit for AD-590.

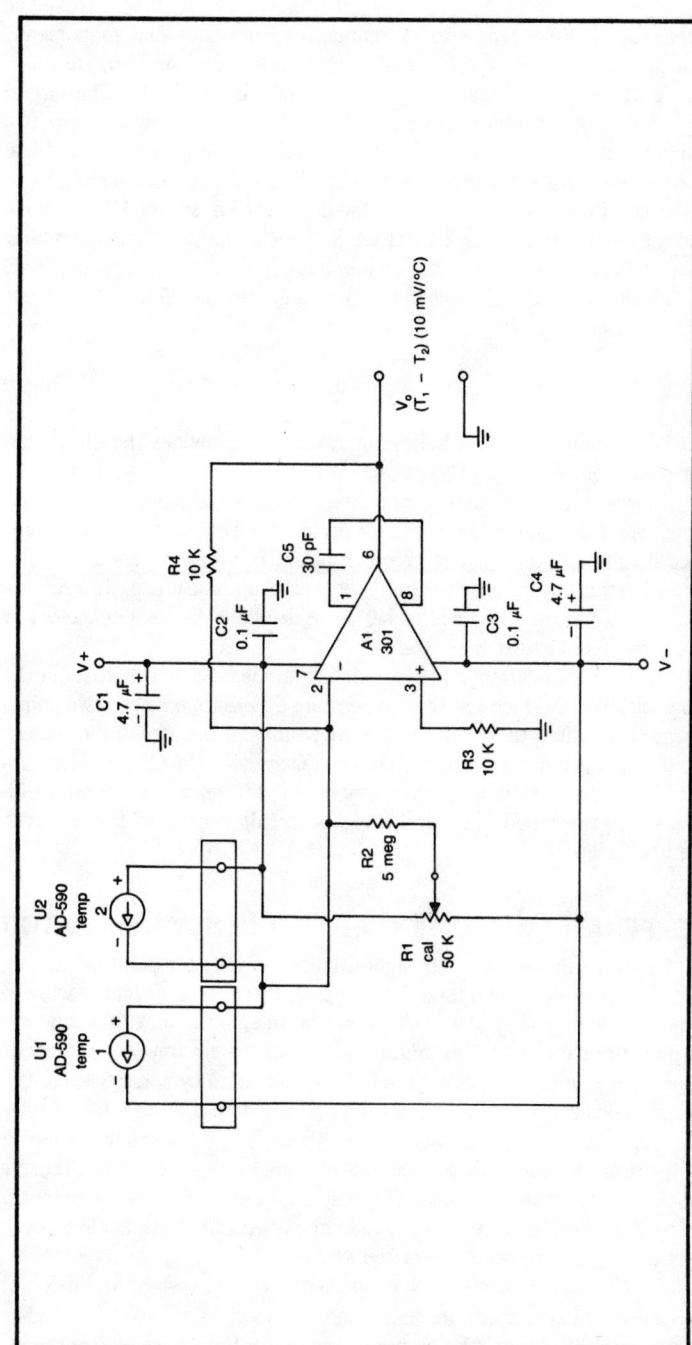

Fig. 12-12. Differential thermometer uses two AD-590 sensors.

239

upper temperature can be room temperature, provided that some means is available to measure the actual room temperature for comparison.

A differential thermometer circuit is shown in Fig. 12-12. This circuit uses a pair of AD-590 sensors to measure the difference between two temperatures. Rather than using a dc differential amplifier in this application we use instead inverting followers. The signal to the inverting input is derived from potentiometer R1, the current from sensor U1 (1 μA/K), and the current from sensor U2 (1 μA/K also). In the operational amplifier chapters you learned that the output voltage of this circuit is proportional to the summation (or difference) of the input currents. Thus, the output voltage is given by

$$V_o = (T1 - T2) \times (10 \text{ mV/}^\circ \text{C}) \tag{12.15}$$

Potentiometer R1 is a balancing circuit that provides the offset that converts from Kelvin to Celsius scales.

So what are differential thermometers used for? In some cases, we find home, environmental-control systems performing heating or cooling chores according to the difference between inside and outside temperatures. Solar water-heating and space-heating devices sometimes use differential-temperature sensors to decide whether to use the solar heat collectors or use fossil fuel to heat the water.

Another application is in biofeedback studies. Some scientists in the biofeedback world measure the temperature difference between two points on the body (often the forehead and the hand) and then teach the subject to balance the temperatures with biofeedback methods. One reported application is in tension headaches, where the differential temperature indicates relative blood flow—and, supposedly, altering blood flow relieves the headache.

TEMPERATURE-CONTROLLED FREQUENCY GENERATOR

There are times when you might want to convert a temperature reading into a proportional audio tone. Examples of this need are transmission of temperature-telemetry data over radio or telephone lines and tape recording the data. You also might want to use audio tones to represent temperature in cases where the sensor is connected by a long wire to the recording instrument, such that the voltage drop in the wire forms a significant error term. Such an application is seen in oceanography where a temperature sensor is let into the sea on a long tether (often hundreds or thousands of feet long). If the sensor output is converted into an audio tone, it can be sent to the surface over a wire, amplified and displayed; line losses play no part in the accuracy of the system.

Figure 12-13 shows a block diagram of a typical system. In this case, we use one of the temperature-sensor and/or amplifier systems shown earlier in this section. The voltage output of the sensor will be applied to the in-

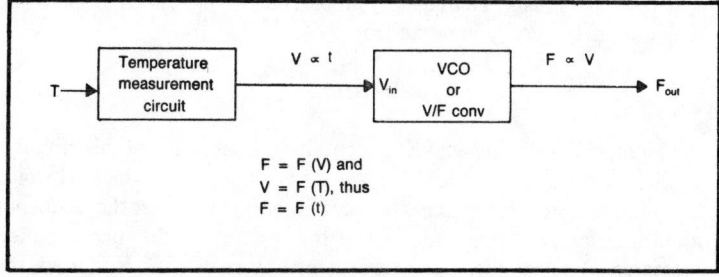

Fig. 12-13. Temperature-to-frequency converter.

put of a voltage-controlled oscillator (VCO) or a voltage-to-frequency converter (VFC). The result is a frequency output (F_{out}) that is proportional to the applied input voltage, which in turn is proportional to temperature.

Figure 12-14 shows a slightly different tactic. The Analog Devices, Incorporated AD-537 is an integrated circuit, voltage-to-frequency converter that has a secondary output that is linearly proportional to temperature. If this output is fed back into the "voltage" input, you have a signal

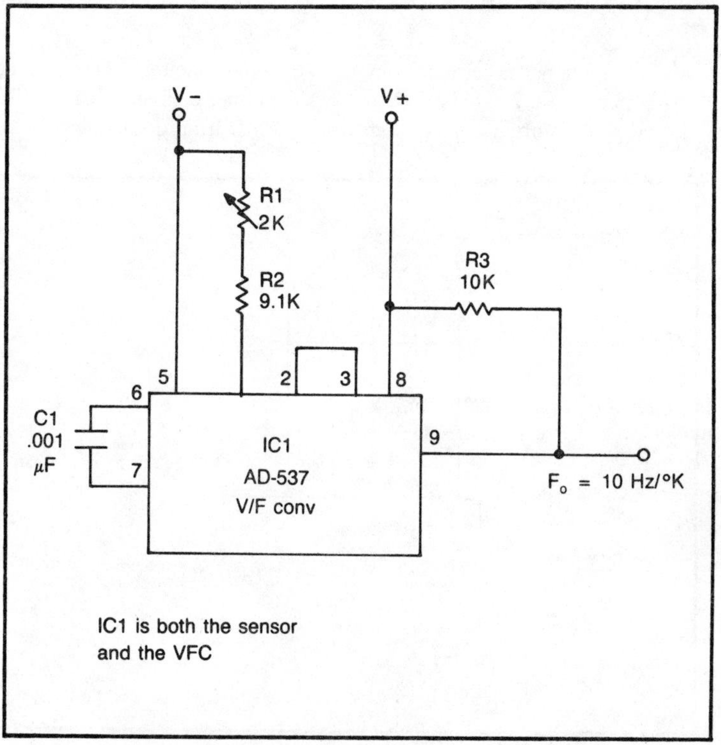

Fig. 12-14. Single-IC temperature-to-frequency converter.

241

generator that produces a frequency proportional to the applied temperature. The output frequency is found from

$$f_o = 10 \text{ Hz/}^\circ\text{K} \tag{12.16}$$

with a maximum frequency of about 10 kHz. Calibration of this circuit is single point, using potentiometer R1. Perhaps the easiest method would be to use air temperature and a mercury thermometer. Set the frequency according to the formula above and the reading on the mercury thermometer. For example, if the room temperature of 298° K is used, the output frequency is

$$f = \frac{10 \text{ Hz}}{\text{K}} \times \text{TEMP}$$

$$= \frac{10 \text{ Hz}}{\text{K}} \times 298 \text{ K}$$

$$= 2980 \text{ Hz}$$

The frequency is set in part by the 0.001-μF capacitor (C1) connected across pins 6 and 7. This capacitor must be either a silver mica or NPO disc ceramic in order to keep temperature drift from occurring.

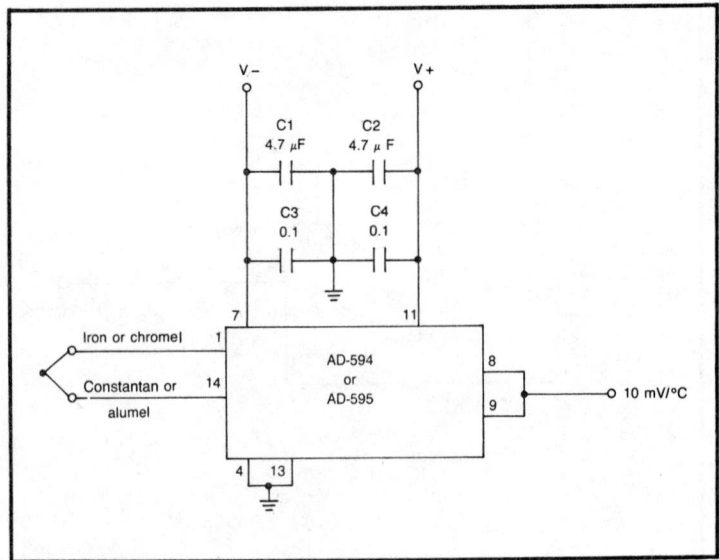

Fig. 12-15. AD-594/AD-595 thermocouple processor.

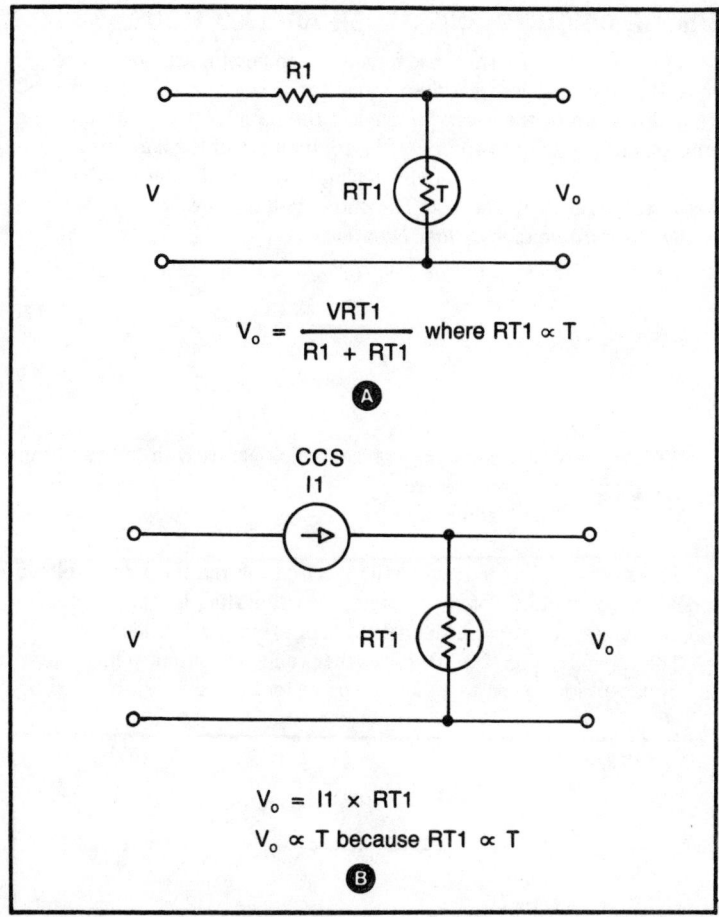

$$V_o = \frac{VRT1}{R1 + RT1} \quad \text{where } RT1 \propto T$$

A

$$V_o = I1 \times RT1$$
$$V_o \propto T \text{ because } RT1 \propto T$$

B

Fig. 12-16. Thermistor circuits (A) standard half-bridge, (B) with current source.

THERMOCOUPLE SIGNAL PROCESSOR

Earlier in this chapter I discussed the thermocouple device. This temperature sensor is a junction of two dissimilar metals. The different "work functions" of the two metals is responsible for generating an output voltage that is proportional to the applied temperature. Figure 12-15 shows a simple integrated-circuit device that produces a linear output voltage of 10 mV/° C from a thermocouple input voltage. There are two devices available: AD-594 and AD-595. The AD-594 is used with type-J thermocouples, while the AD-595 works with type-K thermocouples.

The AD-594 and AD-595 devices operate with supply voltages ranging from a unipolar, +5 volts dc to a bipolar, ±15 volts dc. The power dissipation is typically 1 milliwatt.

BRIDGE CIRCUITS AND OTHER APPLICATIONS

There are several ways to use resistance temperature sensors. Although there are various types (platinum wires, temperature sensitive ceramics, etc.), almost all of them can be used in the circuits below. Two of the simplest circuits are shown in Fig. 12-16; both are of the *half-bridge* variety. The circuit in Fig. 12-16(A) is a simple voltage divider in which the thermistor is across the output. The output voltage of this circuit is given by the standard voltage-divider equation

$$V_o = \frac{V \times RT1}{R1 + RT1} \qquad (12.17)$$

where

R1   is the fixed resistance
RT1   is the thermistor resistance, which varies with temperature
V_o is the output voltage
V   is the excitation voltage

There are several problems with this circuit. First, the output voltage is never zero—with the offset voltage present, scaling is a problem. Second, the output voltage can change with changes in the excitation voltage, V.

The version of Fig.12-16(B) solves the problem of output voltage variation by substituting a constant-current source for the fixed resistance. How-

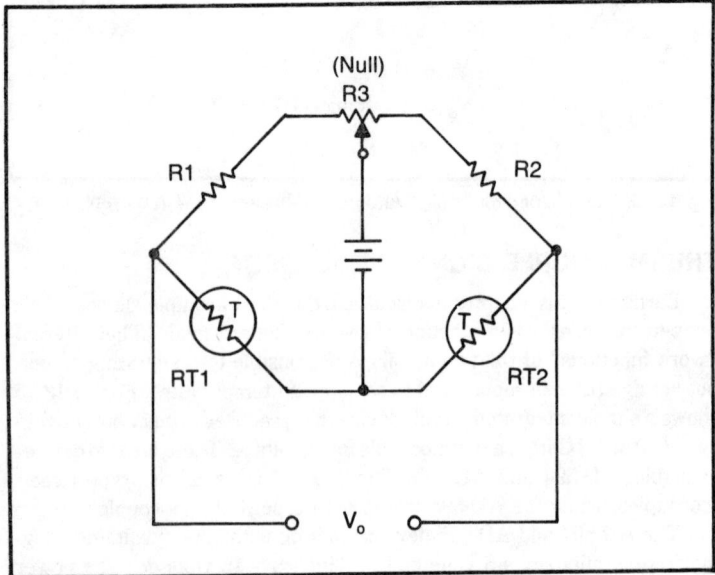

Fig 12-17. Wheatstone bridge.

244

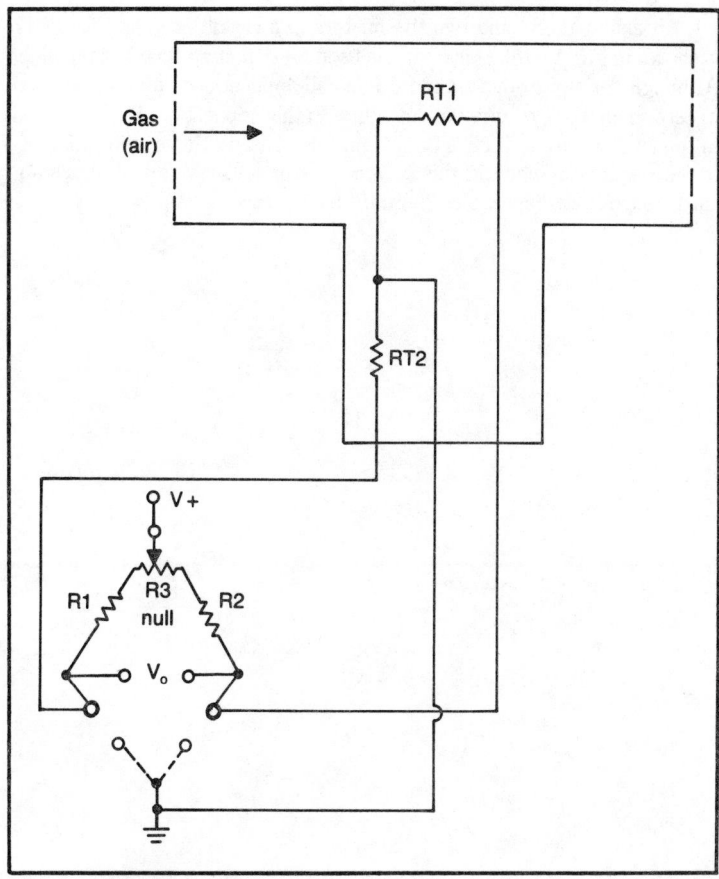

Fig. 12-18. Thermistor flow detector.

ever, this circuit is only a little better than the first; so still another circuit should be examined.

The Wheatstone bridge, our old friend from the 19th century mentioned in other chapters, is the solution (Fig. 12-17). The output voltage is insensitive to drift in the excitation voltage because it is a differential voltage from two, separate, half-bridge, voltage dividers (R1—RT1 and R1—RT2). This configuration shows two thermistors, although some circuits will substitute a fixed resistor for one of the thermistors.

Potentiometer R3 is used for either of two jobs: a) provide an offset when required, and b) allow the user to null the output voltage to zero when the two thermistors are at the same temperature.

In most applications, one of the two thermistors will be used to measure the temperature under examination while the other is "dead-headed" into either an ice-point bath or some other, standard-temperature environment.

245

An example of using two thermistors in a circuit such as Fig. 12-17 is shown in Fig. 12-18. This circuit is used for detecting flow in a pipeline. Although derived from a standard Fenwall-applications note, it was used in certain medical, respiration detectors. In this application, RT1 is placed in the air flow, while RT2 is in the non-turbulent portion of the chamber. In the respiration monitor, this two-thermistor sensor was built in a standard, respirator, disposable, 25-mm, "tee" piece.

Chapter 13

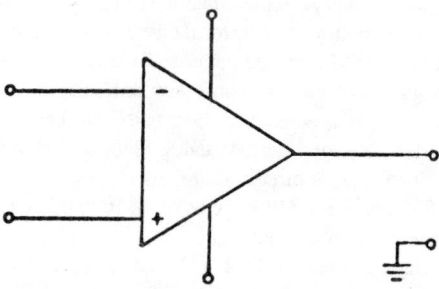

IC Waveform Generators

MOST OF OUR DISCUSSIONS IN THE PREVIOUS CHAPTERS DEALT WITH the processing of signals, but now let's turn our attention to the *generation* of signals—specifically: sine waves, triangles, square waves, and assorted pulses.

MONOSTABLE (ONE-SHOT) MULTIVIBRATORS

A *monostable multivibrator*, also called a one-shot, is an electronic circuit that has but *one* stable state. When it is triggered it will change from its stable state to its unstable state. But it cannot remain in the unstable state indefinitely. After a predetermined amount of time has passed, it reverts back to its original, dormant state.

The states referred to are voltage levels at the output, or at least that is how they manifest themselves to the user. In most such circuits the stable state will produce an output of 0 volts, while the unstable state will produce a positive output voltage. There is no physical reason why these are the levels, except that convention demands it. In the days before such conventions were established by the nature of TTL/DTL chips, designers could, and often did, assign different levels to these states. We, though, are bound by practice to use 0 volts and +5 volts per industry standard.

The 555 and XR-2240 IC timers (see Chapter 3) are often used in the monostable mode. In fact, for most slow-pulse applications they are the devices of choice, especially the low-cost 555.

The principal difference between astable and monostable modes of the 555 chip is that a feedback connection is made so that it becomes self-retriggering. The 555 will produce single, output pulses in the monostable mode that have periods ranging from microseconds to seconds, longer periods being somewhat more difficult because it is a capacitor-charging circuit.

The XR-2240 is a similar circuit and uses the same sort of clock method utilizing a capacitor-charge technique. It is able to handle much longer periods (up to many hours, even several days) because the output is really the wired-OR outputs of a binary counter. In this case the unstable state is 0 volts because the output will stay low until the maximum count (all binary-counter outputs 1) is reached. Because of the counter, the XR-2240 can achieve extremely long periods using modest values for the resistor and capacitor in the clock-circuit stage.

Both the 555 and XR-2240 were covered in another chapter, so they will not be covered again here.

The TTL chips numbered 74121, 74122, and 74123 all are various types of monostable-multivibrator IC devices. These can be a little tricky to use; so the general advice usually offered is to use the 555 unless very-short-pulse durations are required. This means less than 1 μs.

Figure 13-1 shows a typical one-shot circuit using the 74121, and by extension, the other two. This device will trigger when pin 3 is brought low from + 5 volts. It has two different outputs, labeled Q and $\overline{Q}$. These outputs are complementary; so one will be low when the other is high. The Q output will go high for a time set by R and C following each trigger pulse, except those that arrive while the output is still high from a previous triggering.

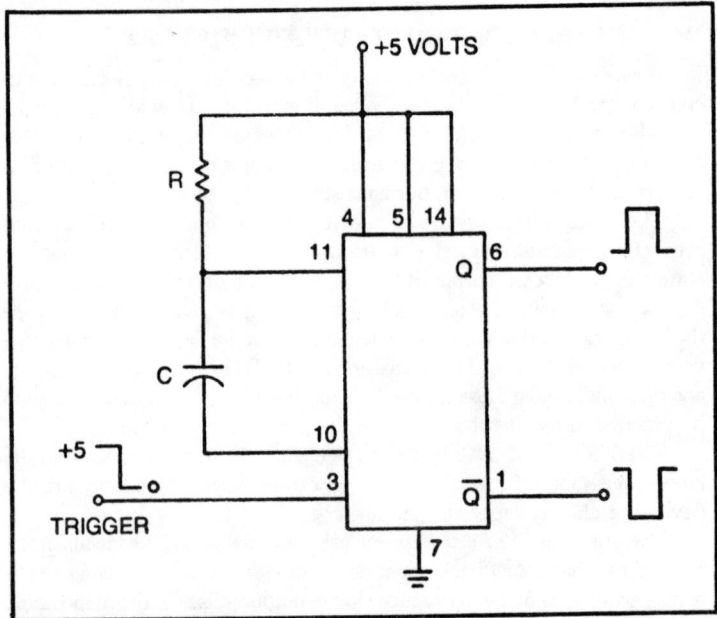

Fig. 13-1. Circuit using the 74121 one-shot IC.

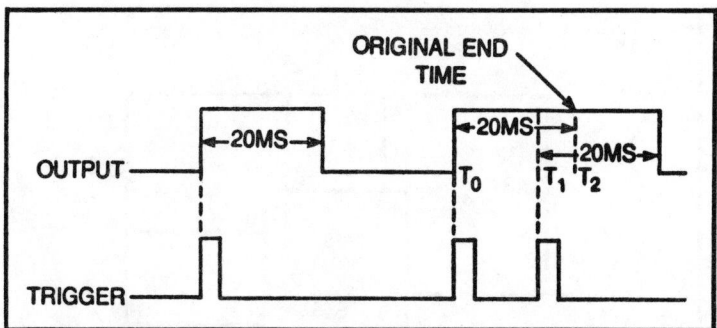
Fig. 13-2. Waveforms showing retriggering.

The 74121 is *not* retriggerable, but the 73122 and 74123 are able to be retriggered before the unstable (high) period is completed. The 74121 will ignore subsequent trigger pulses until after the cycle is finished. Once the Q output is low again, however, it *will* retrigger on the next trigger pulse.

The 74122 and 74123 devices, though, will retrigger if subsequent trigger pulses are applied prior to the completion of the cycle. If the Q output is high when the retrigger pulse arrives, it will remain high for one complete period from that instant. This action is shown in Fig. 13-2. Notice that I am not saying that the total period will be twice the normal period (20 ms), but it is extended by one complete period from the time when the retrigger pulse arrived. The total high time of the output will be 20 ms plus the expired time of the previous period, shown here as t1 – t0.

Figure 13-3 shows a circuit that will allow the 555 to be retriggered. This circuit was designed by engineer Charles McCullough of the George Washington University Medical Center. The circuit operates quite like any other 555 monostable, except that transistor Q2 is present.

In normal operation, transistor Q1 is used as an electronic switch to bring the trigger input terminal to ground whenever a positive-going pulse is applied to the trigger terminal, point A. This grounds the 555 trigger input, initiating the cycle.

Transistor Q2 is connected across the timing capacitor. When the positive pulse is applied to point A, it too turns on and dumps the charge across the timing capacitor. Ordinarily, the charge across capacitor C1 will continue to rise in the normal manner until the RC period set by R3C1 has elapsed. But with transistor Q2 in the circuit, you can dump the charge in C1 and thereby reinitiate the cycle anytime we apply another pulse to point A. Ordinarily, the 555 would ignore the pulse unless it occurred *after* the cycle had been completed.

ASTABLE MULTIVIBRATORS

An astable multivibrator is an electronic circuit that has *no* stable states. An example of such a circuit would be a monostable in which the output

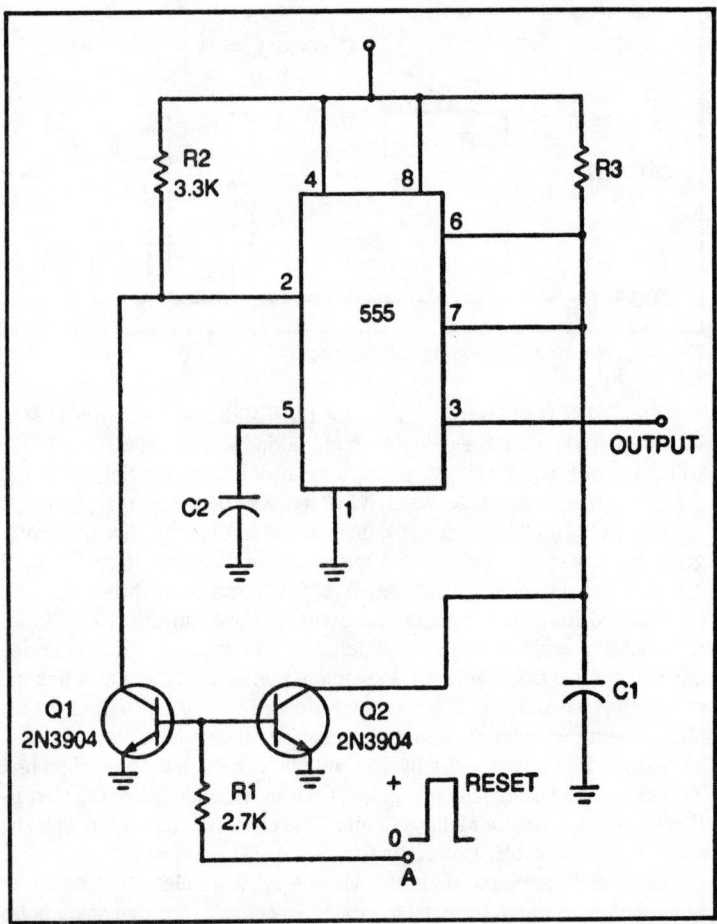

Fig. 13-3. Retrigger circuit for the 555. (Courtesy of C.E. McCullough, the George Washington University Medical Center.)

is fed back to trigger the input. The astable multivibrator is a free-running oscillator that (usually) generates a train of squarewaves.

Figure 13-4 shows an astable circuit made with a single 7400, quad-NAND gate. *The frequency of the oscillators is set by the time constant of R and C.* Generally, R is given a fixed value in the neighborhood of 150 to 270 ohms, while C is given a value dependent upon the required operating frequency. The main effect of the resistor, incidentally, seems to be biasing the input of the first gate into a linear range.

VOLTAGE-CONTROLLED OSCILLATORS

The Signetics 566 circuit shown in Fig. 13-5 is a voltage-controlled oscillator (VCO) integrated circuit that simultaneously produces triangle and square waves.

The frequency of oscillation is a function of four factors: Vcc, R1, C1, and the voltage at pin 5. This frequency is given approximately by

$$F_o \approx \frac{2 (Vcc - V_1)}{R1\ C1\ Vcc} \tag{13.1}$$

provided that

$$2k \leqslant R1 \leqslant 20k$$
$$0.75Vcc \leqslant V_1 \leqslant Vcc$$

Not all applications will require input capacitor C2. If you want it to oscillate at a constant frequency, for example, leave C2 out of the circuit. This capacitor is used to couple ac signals to pin 5. These signals will cause the frequency to shift, so will frequency-modulate (FM) the output frequency.

The 566 is often used to make an audio tape recorder into an FM-instrumentation tape recorder. Low frequency (under about 40 Hz) ac signals so often found in scientific instrumentation will not pass through most audio-

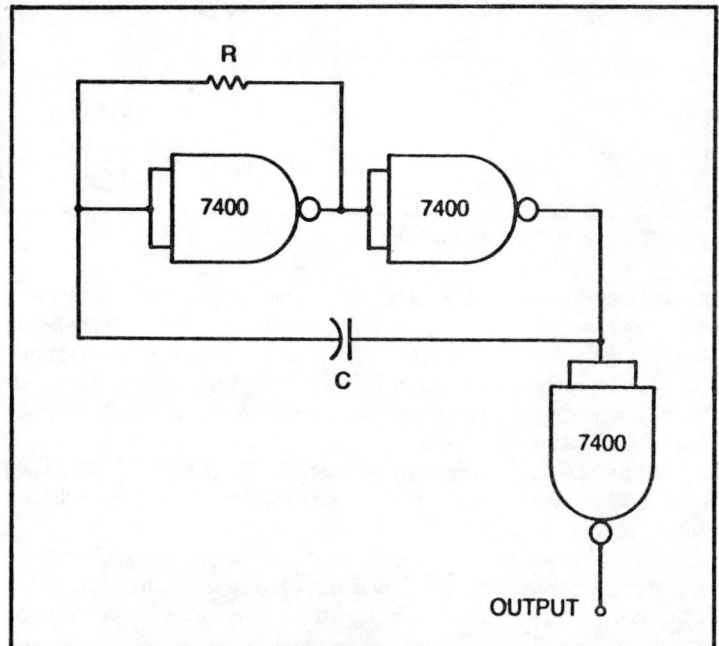

Fig. 13-4. TTL astable multivibrator for use as system clock.

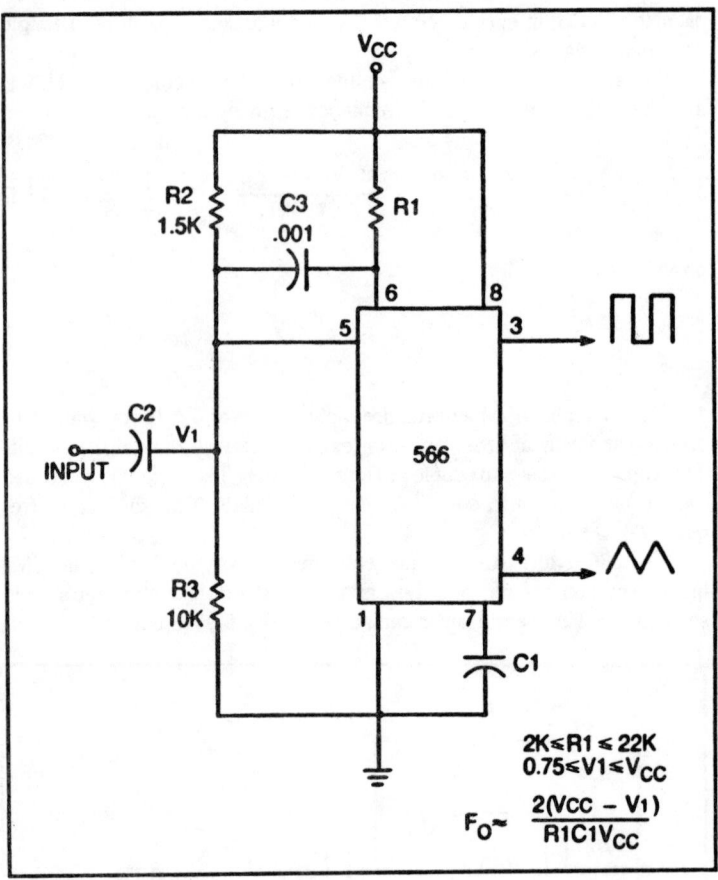

Fig. 13-5. 566 VCO as an oscillator.

grade tape recorders, and even if they did, most audio magnetic tape will not record well at those frequencies. The answer to the problem lies in translating the frequency to some higher point in the audio spectrum that is acceptable to the tape recorder. This is done by having the low-frequency signal frequency-modulate a higher (i.e., carrier) frequency that the recorder can accommodate.

Set the 566 output frequency to something like 1 kHz to 5 kHz, when unmodulated. Apply a substantial signal to C2, and record the resultant output changes on tape.

Set V_1 to approximately 0.88 Vcc, and apply a signal with an amplitude of ± 0.12 Vcc. This will allow V_1 to swing to the limits of Eq. 13.1. The output signal will be too great for most audio recorders; so an attenuator (see Chapter 22) will be required.

Another VCO chip that finds frequent use is the Motorola MC4024.

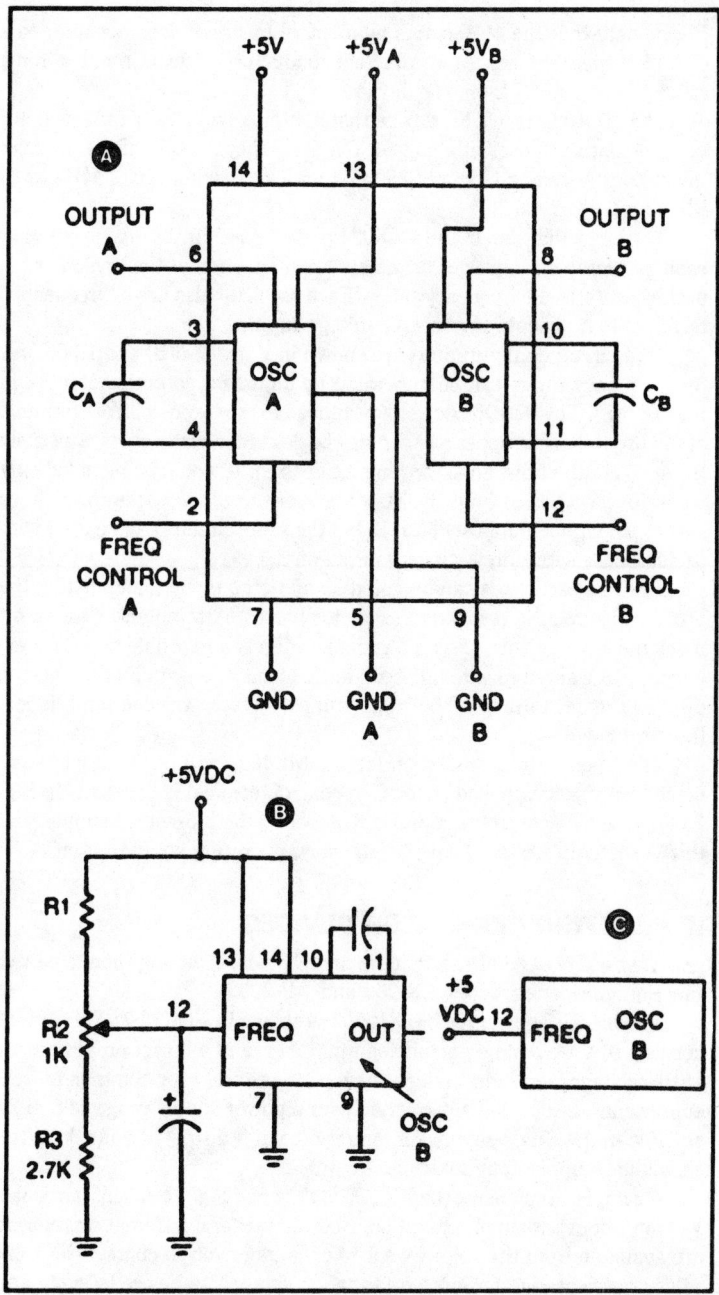

Fig. 13-6. At (A), the internal structure of the MC4024 VCO chip; (B) variable-frequency operation; and (C) fixed-frequency operation.

253

Please note that the 4000-series number, *in this case,* does not refer to a CMOS device but is a Motorola-Semiconductor-Products number for a special-purpose, TTL-bipolar device.

The MC4024 is a TTL device that contains two totally-independent, voltage-controlled-oscillator stages. The frequency of oscillation can be set to anything between 1 Hz and 25 MHz with a capacitor and 2 MHz to 25 MHz with a crystal.

The block diagram of the MC4024 is shown in Fig. 13-6A. Note that each oscillator has a pair of terminals for connection of the crystal or capacitor that sets the frequency range. Each oscillator also has a "frequency" terminal that accepts the control-voltage input.

Two circuit configurations are shown in Figs. 13-6(B) and 13-6(C). A 3:1 frequency range can be generated by adjusting potentiometer R2 in Fig. 13-6(B). This sets the dc voltage applied to the "freq-control" terminal of the MC4024. The upper resistor may be deleted in many cases, especially those where the full frequency range is desired. It would be included only where one wishes to limit the upper end of the range. Note that I have shown pin numbers for oscillator B, but the same circuits can be used with oscillator. A with only a change in pin-outs.

The two oscillators can be gated on and off *only* by manipulating the ground pins for the respective oscillator halves. Attempts to gate on/off using the independent + 5-volt terminals will result in problems. If, for example, you wanted to gate-on oscillator A, connect a switch or the output of a TTL gate, to pin 5. The stage will oscillate only when pin 5 is low (i.e., grounded).

The frequency of oscillation for the MC4024 will be fixed when the dc control voltage applied to the freq-control terminal is constant. In Fig. 13-6(C) the freq-control terminal is strapped to the + 5-volt dc supply, and this is the chip's normal, fixed-frequency, operating configuration.

IC FUNCTION-GENERATOR DEVICES

Exar offers several special-purpose IC devices, among them a pair of function generators: types XR-205 and XR-2206.

The block diagram for the XR-205 is shown in Fig. 13-7. This device consists of a VCO stage, an analog multiplier, and a buffer amplifier. All of these stages are independent of each other except for common power-supply connections and will operate over a power-supply range of 8 to 26 volts dc in the single-supply configuration and ± 5 to ± 13 volts dc in the split (dual)-supply configuration.

A sample circuit using the XR-205 is shown in Fig. 13-8. This particular version is for split-supply operation. Note that several different waveforms are available from the chip. I want to encourage you to contact the Exar sales representative for more information should you desire to apply any of their products. Exar writes some of the best applications literature in the business.

254

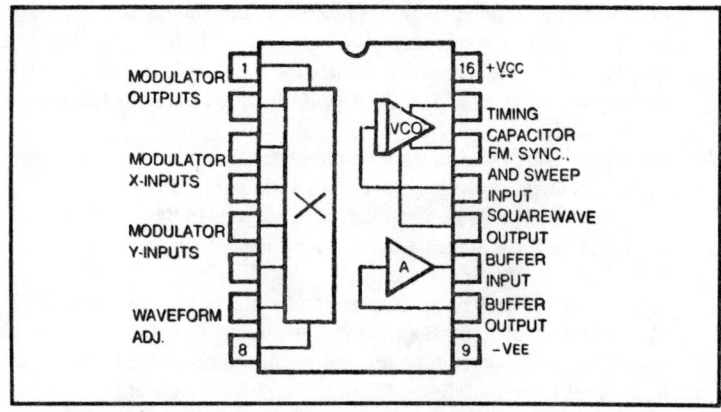

Fig. 13-7. Block diagram of the Exar XR-205. (Courtesy of Exar Corporation.)

The XR-205 oscillation frequency is controlled by capacitor C_o connected between pins 14 and 15. If the sweep input is left open, the oscillation frequency is given approximately by

$$f = 400/C_o \qquad (13.2)$$

where

f	is the operating frequency in hertz
C_o	is the capacitance in farads

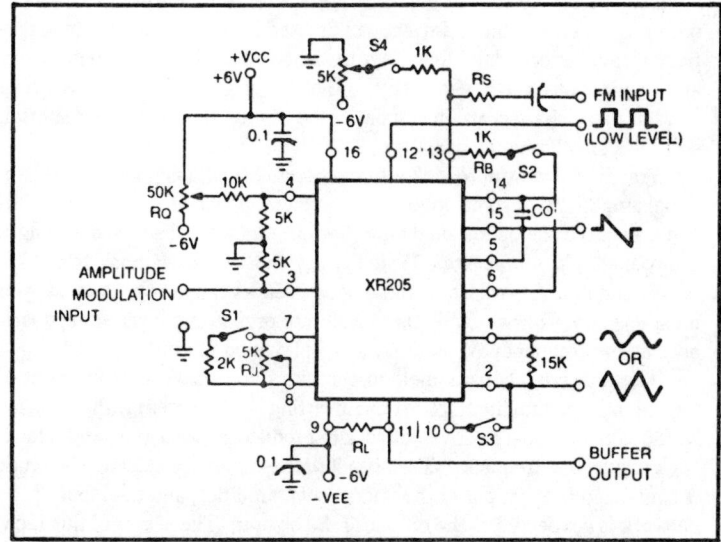

Fig. 13-8. Basic operating circuit of the XR-205.

The analog-multiplier section is labeled as a modulator and comes out on pins 1 through 8. Two of these (pins 7 and 8) are used to control waveform symmetry and will also be used to reduce total-harmonic-distortion (THD) in the sine-wave-output mode. This is accomplished with potentiometer R_J in Fig. 13-8.

Pins 1 and 2 are complementary (differential) modulator outputs. These are usually connected together through a 15-kilohm load resistor but may be used independently as ground-referenced outputs. In this circuit the main output is through the internal buffer amplifier, which has its input connected to pin 2 through a switch.

There are two sets of modulator input connections, two each for V_x and V_y. The X inputs (pins 3 and 4) set the modulator output level, which is controlled by the dc voltage applied to these pins. The X inputs can, then, be used to amplitude-modulate the multiplier output.

Also, according to Exar applications literature, the phase of the output signal is determined by the polarity of the voltage applied to the X inputs. This implies that the XR-205 may be used to provide phase-shift keying (psk) of a carrier tone in binary-control or communications applications.

The Y inputs (pins 5 and 6) of the multiplier can be used to control the type of waveform produced. They are normally connected to the oscillator timing outputs. If the connection is direct, as in Fig. 13-8, the output will be a sine wave or triangle, but if capacitor-coupled, the output will be a pulse or a square wave.

The integrated-circuit devices discussed in this chapter are representative of many devices that can be used for waveform generation. You are advised to keep abreast of manufacturers' literature so that new devices or improvements in these devices can be used if better suited to some particular application. Many manufacturers now charge for their data books and applications notes, especially to those customers who want a whole set, but these charges are just about at the cost of producing the material; so it is a small amount (usually).

You may also want to look into the possibility of using low-cost operational amplifiers, or some other active element, in waveform-generator circuits. Some of the operational-amplifier circuits were discussed at length in my book *Op-Amp Circuit Design & Application* (TAB book No. 787).

In general, regardless of the device used as the active element, you must use either of two methods to generate oscillation: positive feedback only at the frequency of interest, or a relaxation circuit.

The positive-feedback method combines any phase shift inherent in the operational amplifier (180 ° if the inverting input is used) with the phase shift of a frequency-selective network to produce in-phase (i.e., 360 °) feedback at only one frequency. The output of the frequency-selective network is connected to the input of the operational amplifier, and the input of the network is connected to the output of the op amp. The secret of this technique is to provide a total of 360 ° phase shift *only* at the frequency of oscillation, and a total loop gain *at that frequency* greater than unity.

The relaxation oscillator usually depends upon charging a capacitor through a current source such as a resistor connected to a voltage and then discharging the capacitor at a critical time. You may view this as a current through a resistor charging a capacitor. A switch is connected across the capacitor and is designed to close at the appointed time. When this switch closes, the capacitor charge is dumped. The waveform produced by this action will be the normal capacitor-charging waveform on the leading edge and a rapid drop to ground on the trailing edge.

Examples of relaxation oscillators include the unijunction transistor and the neon glow lamp. These work because of the emitter-junction breakover phenomenon in the former case and the ionization potential of the gas in the latter.

There is also a class of relaxation oscillators that depend upon electronic comparators to determine when the firing voltage has been reached. These are the 555 and XR-2240 studied in an earlier chapter. The utility of this method is shown best by the popularity of these devices, especially the 555.

Chapter 14

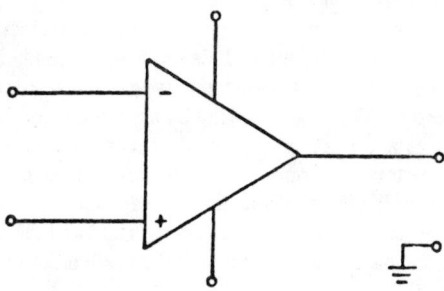

Data Converters

T HE DIGITAL COMPUTER CAN MAKE ALMOST ANY INSTRUMENTATION job easier, more accurate, and sometimes even cheaper. Until recently, though, access to digital computers was limited by economic considerations and by the sheer size of the machines. Prior to now, computers cost a great deal of money and required an air conditioned room of their own—neither of which is possible in most labs. The other alternative is to tape the data from experiments and then hand carry the tape to a computer center where it would often be lost by an inept operator. Even with modern modem data communications and the time-shared terminal we find that the answer to many problems is still the dedicated computer.

Microcomputers are now available that cost very little and can be installed on a desk top or even included inside another instrument. These computers can be used for almost any data-handling job except those of the massive numbers-crunching variety. Before scientific and other electronic equipment can be connected to any digital computer, though, its output must be configured in a form that can be digested by the computer's binary intestines. Similarly, the binary output of a computer input/output (I/O) port must often be converted to an analog format for reading by a human observer. Both types of converters are the subject of this chapter.

METHODS OF ATTACK

There are several approaches to converting data from electronic and scientific instruments into the binary numbers required by a digital computer, but some of them are so tedious as to be ridiculous. For example, one could tabulate values from a display readout or strip-chart recording and then go to a keypunch machine, paper-tape cutter, or CRT terminal and manually enter the data on the keyboard. This, however, soon proves to be insane if more than a very few data points are involved. It is also prone to massive errors if the operator is inattentive or fatigued, creating problems that nobody needs!

The trick is to take the data from the instrument and enter it automatically into the computer. This requires a circuit called an analog-to-digital (A/D) converter. The opposite type of converter is the digital-to-analog converter (DAC) that takes the binary word from the computer and represents it as a voltage or current. In this chapter I will discuss both types of converters, beginning with the DAC. This may seem to be a backwards order, but it actually makes good sense because digital-to-analog converters are simpler than most A/D converters and the DAC is required as a component part of almost all of the so-called feedback A/Ds.

DIGITAL-TO-ANALOG (DAC) CONVERTERS

There are several different approaches to making a DAC, but in most cases they will be of a variety that uses a weighted, current or voltage system that is turned on by the digital word applied to certain switch inputs.

An example is the popular R-2R resistance-ladder system of Fig. 14-1. The active element, A1, is an operational amplifier in a unity-gain, inverting-follower configuration. Although you can get away with using a device from the low-cost 741-family, it is not good practice in a precision DAC. It would pay dividends to specify one of the premium-grade operational amplifiers for this application. I prefer the RCA CA3160 or CA3140 series where both high performance and low cost are desired.

In the circuit of Fig. 14-1 I have illustrated the digital inputs as switches, but in a real world converter the switches would be replaced by a binary counter, computer I/O port, or some other n-bit parallel-data line.

A precision reference voltage is required for accurate conversion, and for most designs this will be at a level of + 10 volts dc. The accuracy of the converter is dependent upon the precision of the reference voltage. To be sure, there are other sources of error, but if the reference is poor, there is no hope for any other factors to be effective in improving the performance of the circuit. Although almost any precision voltage regulator can be pressed into service as the reference, it is a simple matter to use the Precision Monolithics, Inc., (PMI) type REF-01CJ or, in more severe cases, REF-01HJ. These IC devices are especially designed for this type of application and are easily trimmed (see Fig. 4-12) to the reference potential.

Returning to Fig. 14-1, let us consider the circuit action under circumstances where various bits are either high or low. If all bits are low, the output voltage will be zero. The value of the output voltage is given by $I \times R$, and when all bits are low this current is zero. In practical circuits, though, there might be some output voltage under these circumstances due to offsets in the operational amplifier, of which something was said

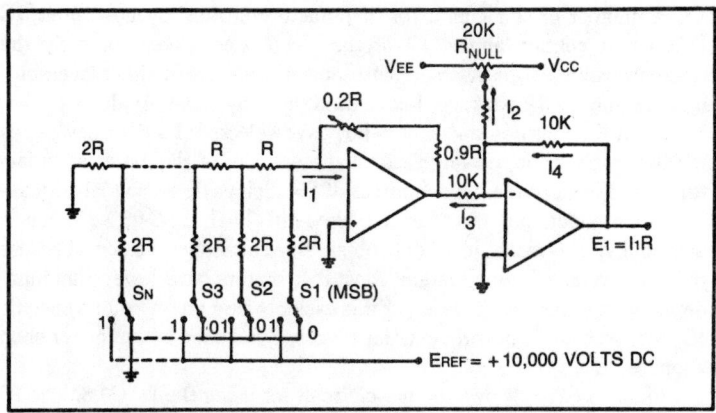

Fig. 14-1. Simple digital-to-analog converter (DAC).

in another chapter. These can be nulled to zero output when all input bits are intentionally set to zero.

If the most significant bit (MSB) is made 1, the output voltage will be approximately $1/2\ E_{ref}$. Similarly, if the next most significant bit is turned on (set to 1) and all others are 0, the output voltage will be $1/4\ E_{ref}$. The least significant bit (LSB) would contribute $(2_{-n})\ E_{ref}$ to the total output voltage.

Let us assume that we have an 8-bit DAC of the type shown in Fig. 14-1. The word applied to the input terminals is 11001011, and the reference potential is precisely +10 volts. What is the output voltage?

The total output voltage of a DAC is the summation of the contributions of each individual bit, or

$$\sum_{i\ =\ 1}^{n}\ \frac{E_{ref}}{2^n}.$$

;so, for our present case,

$$\frac{10}{2^8} + \frac{10}{2^7} + 0 + \frac{10}{2^5} + 0 + 0 + \frac{10}{2^2} + \frac{10}{2^1}$$

$$= (10/256) + (10/128) + (10/32) + (10/4) + (10/2)$$

$$= 0.039 + 0.0781 + 0.3125 + 2.5 + 5$$

$$= 7.93\ \text{volts}$$

261

A number of semiconductor manufacturers offer low-cost, 8-bit, IC DACs that contain almost all of the electronics, except possibly the reference-voltage supply and perhaps some operational-amplifier level shifting, in a single DIP package. I have used those by Datel, Analog Devices, Precision Monolithics, and others but have selected the PMI devices as my illustrations for certain practical reasons. All of the major manufacturers and most of the minor sources of DAC chips do a good job of making the product, but the reason I liked the PMI DAC-08 so much is availability. PMI makes it easier to obtain small quantities (i.e., one) because they have a distribution system. Most distributors have lower minimum-dollar orders than the factories, or it is easier to nest an order for a specific IC, such as a DAC, in with an order for more mundane laboratory or shop supplies.

Figure 14-2(A) shows the basic circuit for using the DAC-08. The IC contains the electronic switches, the current output, the resistance ladder, and a reference amplifier. The current output is complementary, so has both I_o and $\overline{I}_o$ terminals.

Two types of input are required to make this DAC work. One is the reference current, I_{ref}, applied through pin 14. This current may be generated by a precision reference voltage such as the REF-01 and a precision, low-temperature-coefficient resistor. For TTL compatibility and a reference voltage 10 volts, this resistor should be precisely 5,000 ohms.

The other type of input is the 8-bit digital word, which is applied to the IC at pins 5 through 12. Bit 1 (MSB) is applied to pin 5, the next MSB to pin 6, and so on to pin 12 which is the least significant bit, bit 8. The logic levels required are the usually-found TTL levels, namely ground for logical 0 and +5 volts for logical 1.

Figure 14-2(B) shows the connection of the DAC-08 to provide unipolar operation over the range of zero to approximately 10 volts. When the input word is 00000000, the DAC output is in zero state, 0 volts. A half-scale voltage (−5 volts) is given when the input word is 10000000. This occurs when the MSB is high and all others are low. The full-scale output will exist only when the input word is 11111111. The output under full-scale conditions will be approximately −9.96 volts, rather than −10 volts as might be expected.

Figure 14-2(C) shows a circuit that is similar to the unipolar case, but includes an operational-amplifier level shifter to provide bipolar operation. Again, the component values shown are selected for TTL connection. The table in Fig. 14-2(C) gives the binary codes and their associated output levels. Notice that the output levels run from −9.96 volts to +9.96 volts. Also note that there is no true zero but a positive zero and a negative zero. This is because the DAC has only 256 states and these are arranged symmetrically about zero, with 128 for each polarity. You rarely gain something for nothing, and the use of the DAC in a bipolar configuration is no exception. With only a finite number of possible states to represent a voltage range, it only stands to reason that doubling the voltage range will halve

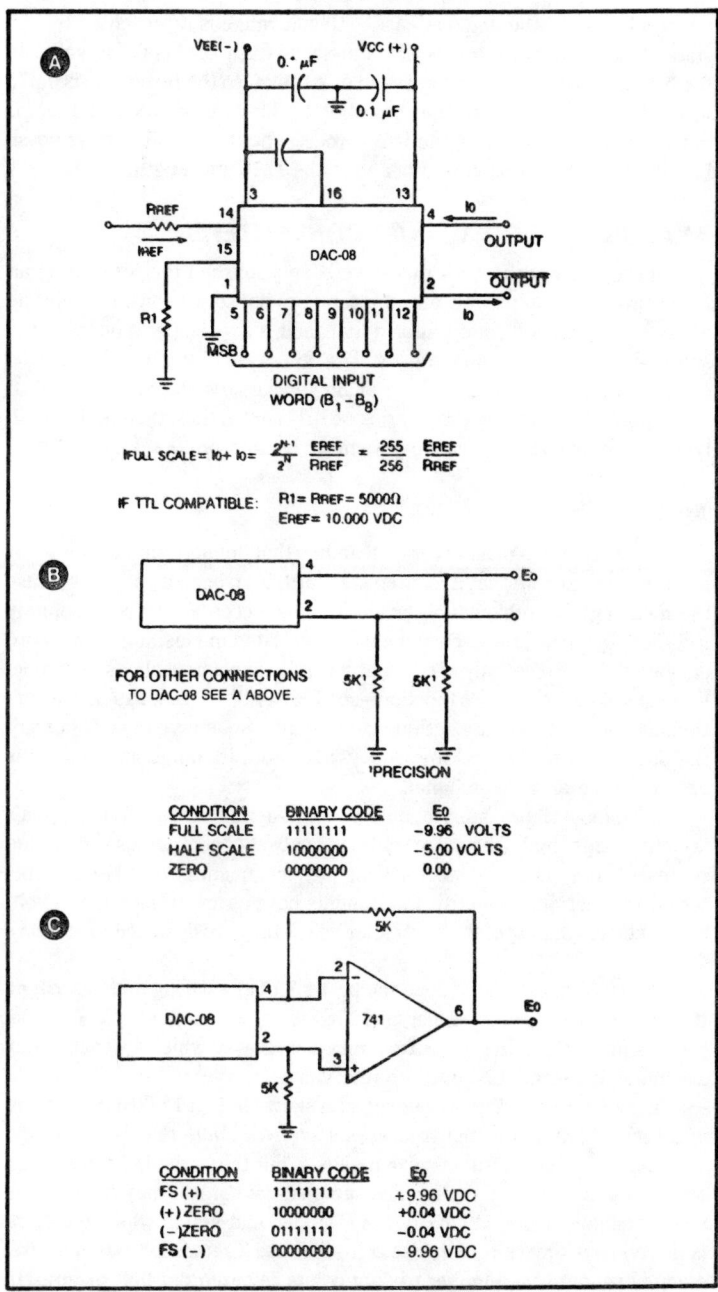

Fig. 14-2. At (A), the circuit for using a PMI DAC-08; (B), unpolar voltage output; and (C), bipolar voltage output.

263

the resolution. In the unipolar case a 10-volt range is represented by 256 states (excluding zero, this is 255 states); so each step is 0.039 volts. In the bipolar case, we have a total span of 20 volts; so the resolution is 0.078 volts. The only answer to the resolution problem, if 39 mV or 78 mV is insufficient, is more bits in the DAC process, but that is not always possible, there being constraints other than the DAC bit length.

ANALOG-TO-DIGITAL (A/D) CONVERTERS

Of the many techniques that have been published for performing an A/D conversion, only a few are of interest to us; so we consider only the voltage-to-frequency, integration (dual- and single-slope), counter, and successive-approximation methods. The others are either not as good, too costly, redundant, or confer little or no special advantage.

Before unraveling the mysteries of A/D conversion, though, it would be profitable to consider just what we mean by the words *analog* and *digital*.

Analog

The word analog has several meanings that connote similar things. It is interesting to note that it is often standard, if erroneous, practice to use the meanings interchangeably when it is not necessary to be absolutely rigorous. In the strictest sense the word is related in meaning to the word analogous, but in the daily vernacular it is used to connote almost any time-varying voltage or current function—see Fig. 14-3(A)—that has both a continuous domain (t) and a continuous range (ft). So any voltage or current that is allowed to take on any value within both its range and domain is usually called an analog signal.

In the more rigorous sense, though, the generic term for all such signals is continuous signal. An analog signal is a subset of continuous signals and represents (i.e., is analogous to) some physical quantity or other parameter. Most electronic scientific instruments have as an output either a voltage or current analog of the parameter which they are designed to measure or detect.

Stanley[1] has, perhaps, done the best job of classifying and describing the differences between analog and digital signals. In his book on signal processing, he describes *quantization* as a process in which a variable may assume only certain distinct, discrete values.

An example of a *discrete-time* signal is shown in Fig. 14-3(B). Here there is a type of waveform that represents *sampled data*—that is to say the variable t may only assume certain values, but the variable f(t) may take on any value within its range. The sampled data signal may be obtained using a sample-and-hold circuit (see Figs. 9-8 and 9-9) in which the clock is driven by a square-wave-pulse train that has a repetition rate (i.e., frequency) equal to the number of data points required per unit of time. In this way, we may call our clocked sample-and-hold circuit an analog/sampled data converter.

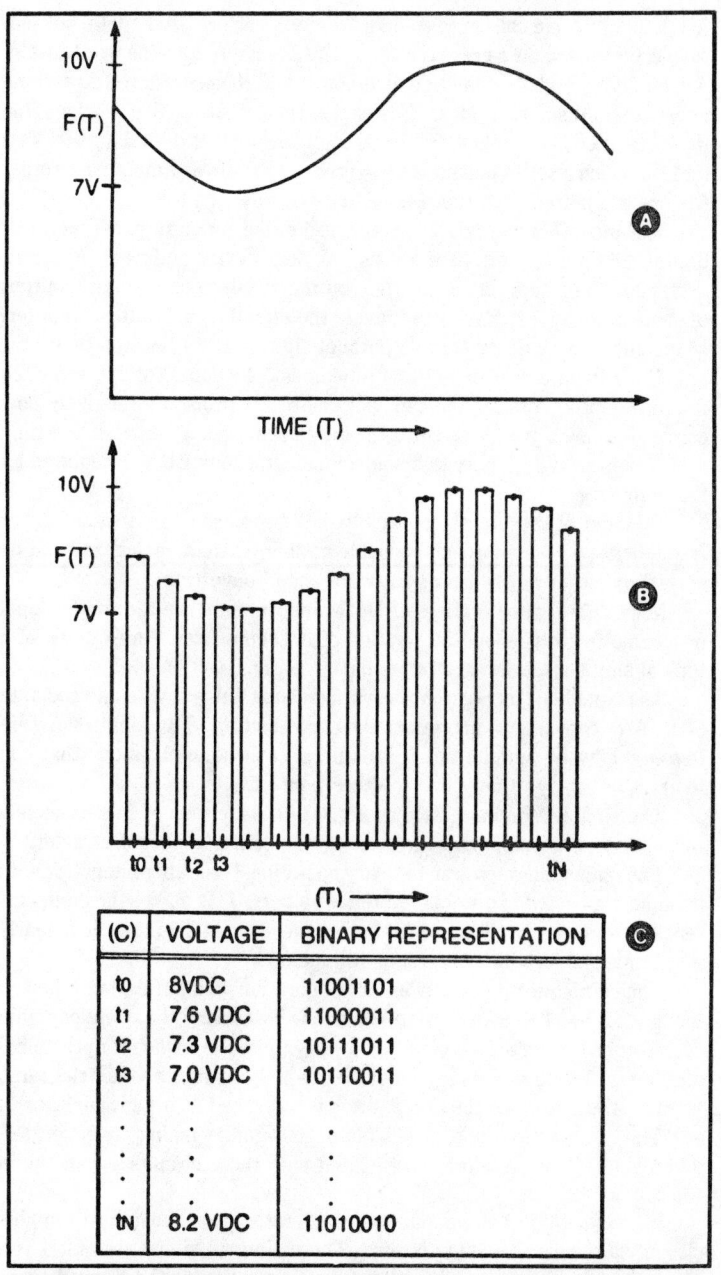

Fig. 14-3. At (A), an analog signal (continuous range, continuous domain); (B), the digitized version of the same signal (discrete domain, continuous range); and at (C), an illustration of a discrete domain, discrete range.

A lot of people call sampled data "digital" or "digitized" data, but that is an error somewhat more serious than the ambiguity over the word analog. A true digital signal is one in which both t and f(t) are allowed to assume only certain discrete values. This is the kind of signal that is digestible in a digital computer. On paper one might represent digital signals as a table in which a series of data words corresponding to an amplitude or other feature are paired with the time signal—as in Fig. 14-3(C).

In some cases you might want to use either a digital panel meter or digital multimeter, both as the visual-readout device and as an A/D converter for the computer. Similarly, you may wish to use a digital output of some existing instrument to provide the digital signal to the computer. Many such instruments provide parallel digital output lines on their rear panels along with control and/or strobe lines. The appeal of this approach falls down rather abruptly, though, because of two problems: relatively slow conversion speed and the fact that the instruments usually have BCD coding rather than straight binary or hexadecimal/octal coding that is required by the computer.

The first of these problems really only becomes a nuisance at higher frequencies; so the approach is usable at near-dc frequencies, and can be solved by either hardware or software code conversion.

Most digital panel meters or DMM instruments use the single-slope, or more often, dual-slope integration of A/D converter. An example of a typical single-slope integrator is shown in Fig. 14-4(A).

As pointed out by Barnes[2] the simultaneous terms simple and accurate (viz., A/D converters) are generally contradictory. The single-slope integrator may be simple, but it is limited to those applications that can tolerate accuracy of only one to three percent.

The single-slope integrator of Fig. 14-4(A) consists of five basic sections: ramp generator, comparator, logic, clock, and output encoder.

The ramp generator is an ordinary operational-amplifier integrator with its input connected to a stable reference voltage. This makes the input current, I_{ref}, essentially constant; so the voltage at point 2 will rise in a nearly linear manner, creating the ramp voltage.

The comparator is merely another operational amplifier, but it has no feedback loop. The gain in that instance is essentially the open-loop gain of the device selected—typically very high even in low-cost operational amplifiers. When the analog input voltage, E_x, is greater than the ramp voltage, the output of the comparator is saturated at a logic high, or 1.

The logic section consists of a main AND gate, a main-gate generator, and a clock. The waveforms associated with these circuits are shown in Fig. 14-4(B).

When the output of the main-gate generator is low, switch S1 remains closed, so the ramp voltage is zero. The main-gate signal at point 1 is a low-frequency square wave with a frequency equal to the desired time-sampling rate. When point 1 is high, S1 is open, so the ramp will begin to rise linearly. When the ramp voltage is equal to the unknown input volt-

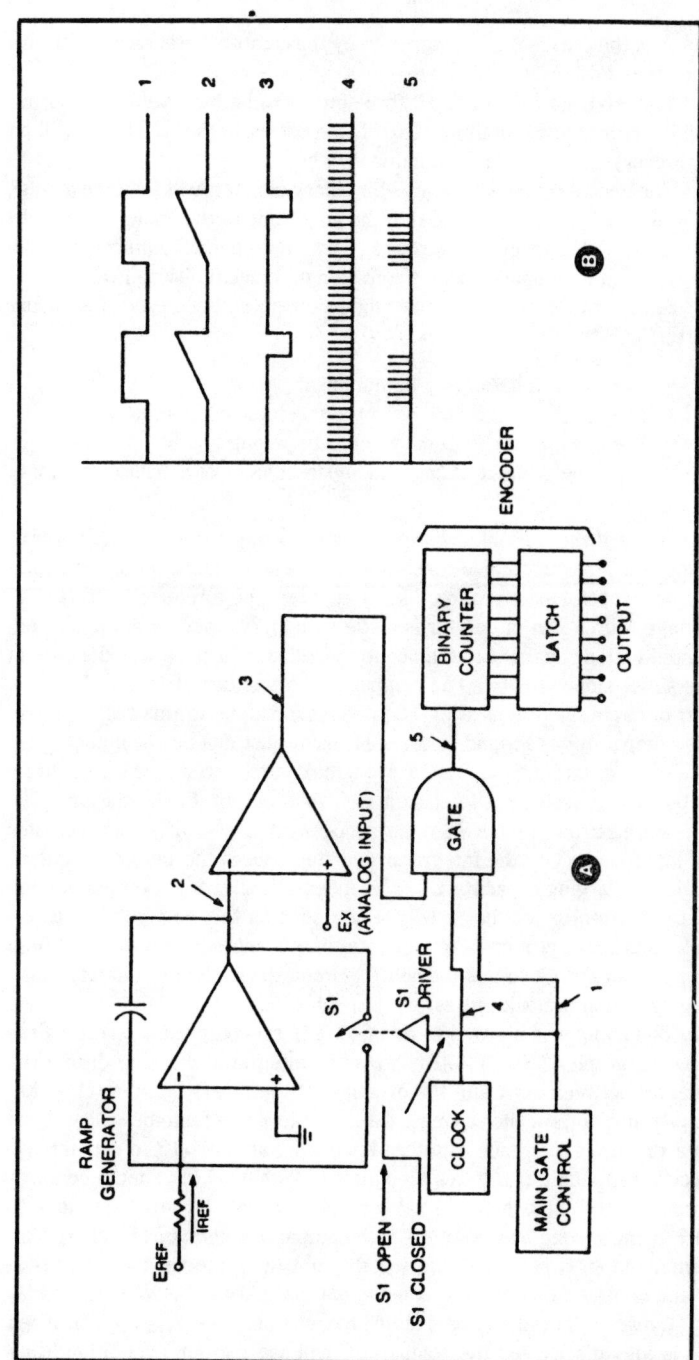

Fig. 14-4. At (A), a single slope integrator; and (B), the timing diagram.

age (E_x), the differential voltage seen by the comparator is zero; so its output drops low.

The AND gate requires all three inputs to be high before its output can be high also, From times t0 to t1, the output of the AND gate will go high every time the clock signal is also high.

The encoder, in this case an 8-bit binary counter, will then see a pulse train with a length proportional to the amplitude of the analog input voltage. If the A/D converter is designed correctly the maximum count of the encoder will be proportional to the maximum range (full-scale) value of E_x.

Several problems plague the single-slope integration type of A/D converter:

- The ramp voltage may be nonlinear.
- The ramp voltage may have too steep or too shallow a slope.
- The clock pulse frequency could be wrong.
- It may be prone to changes in apparent value of E_x caused by noise.

Many of these problems are corrected by the dual-slope integrator of Fig. 14-5. This circuit also consists of five basic sections: integrator, comparator, control-logic section, binary counter, and a reference current or voltage source. An integrator is made with an operational amplifier connected with a capacitor in the negative-feedback loop, as was the case in the single-slope version. The comparator in this circuit is also the same sort of circuit as was used in the previous example. In this case, though, the comparator is ground-referenced, using just one active input.

When a start command is received, the control circuit resets the counter to 00000000, resets the integrator to 0-volts output (by discharging C1), and sets electronic switch S1 to the analog input. The analog voltage creates an input current to the integrator and that causes the integrator output to begin charging capacitor C1. This means, then, that the output voltage of the integrator will begin to rise. As soon as this voltage rises a few millivolts above ground the comparator output snaps high positive. A high comparator output causes the control circuit to enable the counter, which begins to count clock pulses.

The counter is allowed to overflow and this output bit resets switch S1. The graph of Fig. 14-5(B) shows the integrator charging during the interval between start and the overflow of the binary counter (t1 − t0).

At time t1 the switch changes the integrator input from the analog signal to a precision reference source. Meanwhile, at time t1 the counter has overflowed, and again it has an output of 00000000 (maximum count + 1 count is the same as the initial condition). It will, however, continue to increment so long as we have a high comparator output. The charge accumulated on capacitor C1 during the first time interval is proportional to the *average* value of the analog signal that existed between t0 and t1.

Capacitor C1 is discharged during the next time interval (t2 − t1). When C1 is fully discharged the comparator will see a ground condition at its

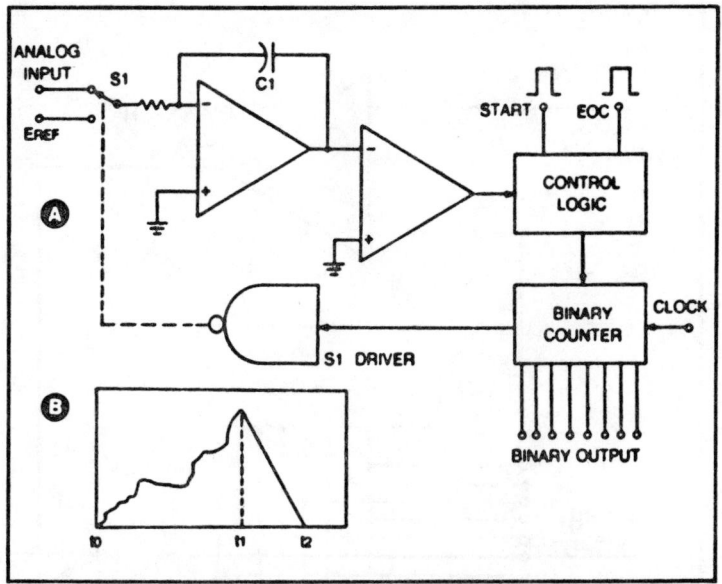

Fig. 14-5. At (A), a dual slope integrator; and (B), the related waveform.

active input, so will change state and make its output low. Even though this causes the control logic to stop the binary counter, it does not reset the counter. The binary word at the counter output at the instant it is stopped is proportional to the average value of the analog waveform over the interval (t1 − t0). An *end-of-conversion* (EOC) signal is generated to strobe the microprocessor or other instrument so that it knows the output data is stable, valid, and ready for use.

VOLTAGE-TO-FREQUENCY CONVERTERS

These are not really genuine A/D converters in the strictest sense, but are very good for representing analog data in a form that can be tape recorded on a machine that is less expensive than next year's salary! Basically, it consists of only a voltage-controlled oscillator (VCO) such as the 566 integrated circuit.

COUNTER TYPE A/D CONVERTERS

A counter-type A/D converter is shown in Fig. 14-6. It consists of a comparator, voltage-output DAC binary counter, and the necessary control logic.

When the start command is received, the control logic resets the binary counter to 00000000, enables the clock, and begins counting. The counter outputs control the DAC inputs; so the DAC output voltage will begin to rise when the counter begins to increment. As long as analog input volt-

269

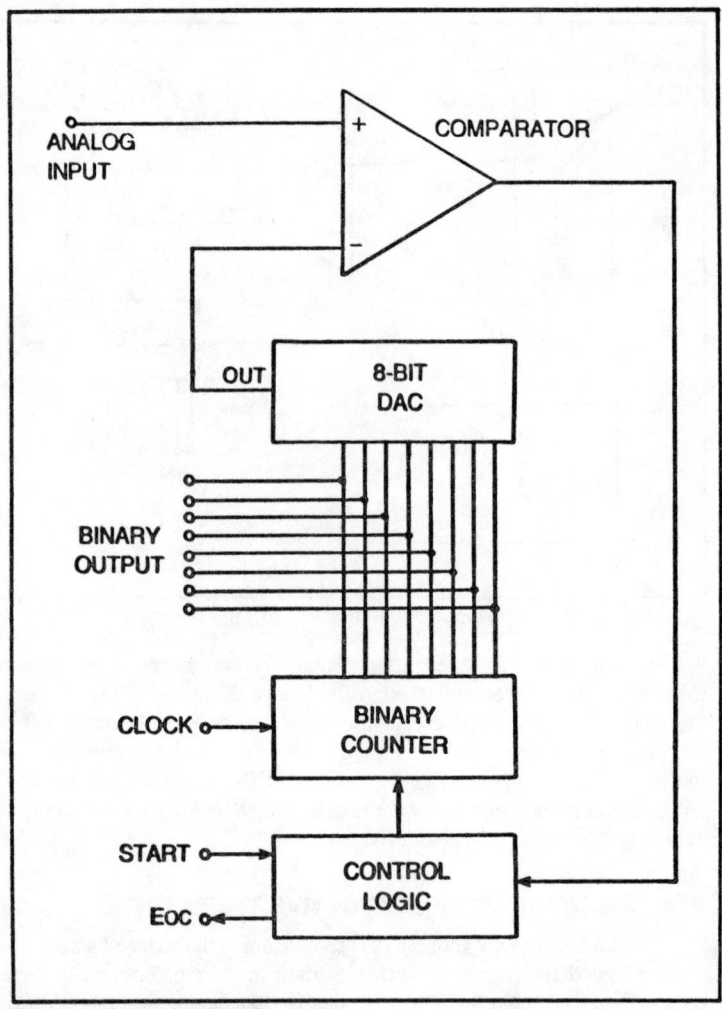

Fig. 14-6. Counter type A/D converter.

age E_{in} is less than E_{ref}, the DAC output and then the comparator output is high. When E_{in} equals E_{ref}, however, the comparator output goes low, and this turns off the clock and stops the counter. The digital word appearing on the counter output at this time represents the value of E_{in}.

Both slope and counter A/D converters take too long for many applications, on the order of 2^n clock cycles (where n = number of bits). Conversion times become critical if the high-frequency component of the input waveform is to be faithfully reproduced. *Nyquist's criteria* require that the sampling rate (i.e., conversions per second) be at least twice the highest frequency to be recognized.

SUCCESSIVE-APPROXIMATION A/D METHODS

Successive approximation has been found to be best suited for many applications, especially those where speed is important. This type of A/D converter requires only n + 1 clock cycles to make the conversion, and some designs allow truncation of the conversion process in fewer cycles if the final value is found prior to n + 1 cycles.

The successive-approximation converter operates by making several successive trials at comparing the analog input voltage with a reference generated by a DAC. An example is shown in Fig. 14-7. This circuit con-

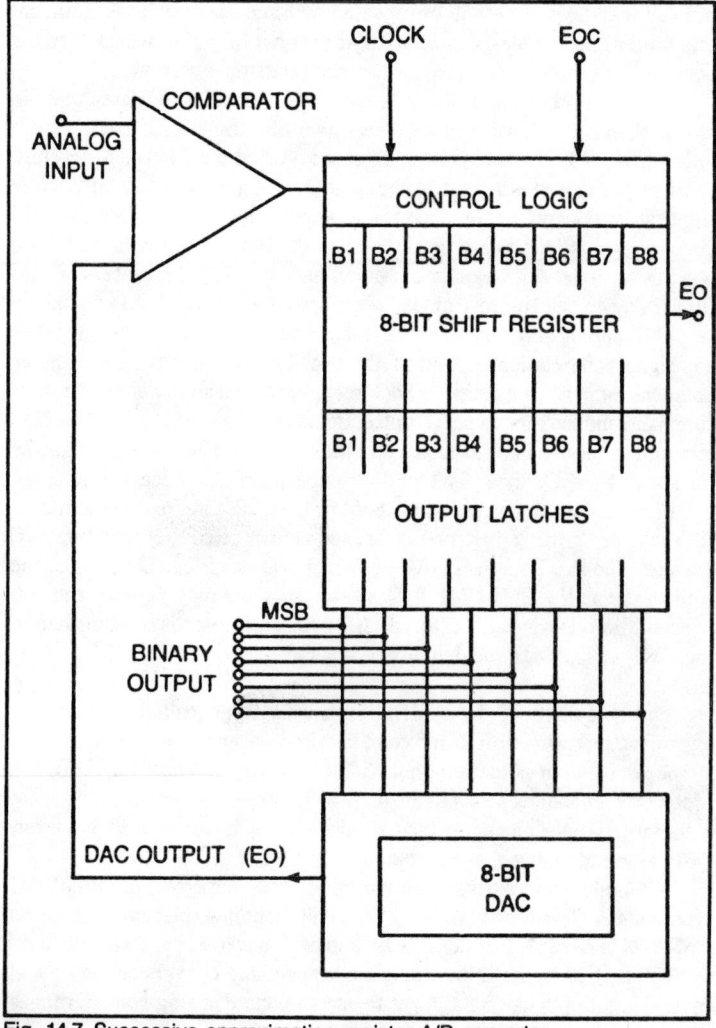

Fig. 14-7. Successive approximation register A/D converter.

271

sists of a comparator, control logic section, a shift register, output latches, and a voltage-output DAC.

When a start command is received, a logic 1 is loaded into the MSB of the shift register, and this sets the output of the MSB latch high. A 1 in the MSB of a DAC will set the output voltage (E_{ref}) to half scale. If the input voltage, E_{in}, is greater than E_{ref}, the comparator output stays high. On the next clock pulse we find that the MSB latch remains high and the 1 in the shift-register MSB position moves to the right and occupies the next most significant but, bit 2. Again the comparator compares E_{in} with E_{ref}. If the reference voltage from the DAC is still less than the analog input voltage, the process will be repeated with successively less-significant bits until either a voltage is found that is equal to E_{in} (in which case the comparator output drops low) or the shift register overflows.

If, on the other hand, the first trial with the MSB indicates that E_{in} is less than the half-scale value of E_{ref}, we find the circuit making trials below half scale. The MSB latch is reset to 0, and the 1 in the MSB shift-register position shifts right to the next most significant bit, bit 2. Here the trial is repeated again. This process will continue as before until the correct level is found or overflow occurs. At the end of the last trial (bit 8 in this case) the shift register overflows and becomes an end-of-conversion (EOC) flag to tell the rest of the world that the conversion is complete.

This, and most other types of A/D converters, requires a starting pulse and signals completion with an EOC pulse. This requires the computer or other instrument to engage in bookkeeping to repeatedly send the start command and look for the EOC pulse. If one ties the start input to the EOC output, conversion is continuous, and the computer need only look for the raising of the EOC flag. This will also speed up the process a little.

Figure 14-8 shows a practical, but low-cost, successive-approximation A/D converter using just one reference source (REF-01) and three IC devices. The DAC is a PMI DAC-08, which was described earlier, and the comparator is the PMI CMP-01C. Faster versions may result if the Advanced Micro Devices (AMD) type AM686 comparator is used instead of the CMP-01C. AMD, incidentally, is a second source for the DAC-08, as well.

The AM-2502 IC by AMD is a special type called a *successive-approximation register* (SAR) and contains almost everything needed to make a successive-approximation-type A/D converter except the DAC-08. Motorola also makes a successive-approximation-register IC, but it is not a pin-for-pin replacement for the one shown here. It can be used, however, if suitable pin changes are made.

A complete successive-approximation, A/D converter, IC is the PMI type AD-02, shown in Fig. 14-9. This chip contains the registers; so no AM-2502 is needed. It is admittedly a little expensive, but its cost can often be justified by its utility. It has a respectable conversion speed and several input options that allow it to accommodate analog-voltage ranges

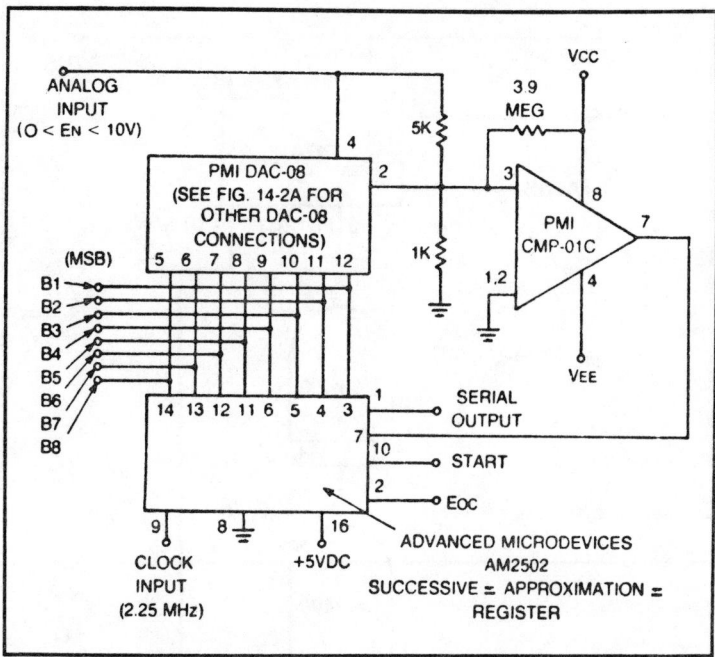

Fig. 14-8. Circuit for an SAR A/D converter using the PMI DAC-08, a PMI CMP-01C (or 311) comparator, and an AMD AM-2502 SAR IC.

of 0 to 5 volts, 0 to 10 volts, -2.5 to $+2.5$ volts, -5 to $+5$ volts, and -10 and $+10$ volts.

A software solution to the design of a successive-approximation A/D converter is shown in Fig. 14-10. This idea is appealing to many, especially those computerists who are more software-oriented. It is also appealing to the hardware person because it reduces the amount of connections and component count—remember one basic design principle: kiss—keep it simple stupid!

The basic circuitry is shown in Fig. 14-10(A), while the software flow chart is shown in Fig. 14-10(B). As long as the comparator is high, indicating that E_{in} does not equal E_{ref}, the program will continue to increment the DAC. In this case the successive-approximation resister and controls are replaced by programming in the computer.

DAC APPLICATIONS

The IC DAC is somewhat interesting as a component class because it can be used for a variety of applications that are seemingly unrelated to their stated purpose. A clever designer should be able to find even more applications for the DAC.

Perhaps the first DAC application to spring to mind is the ramp, or

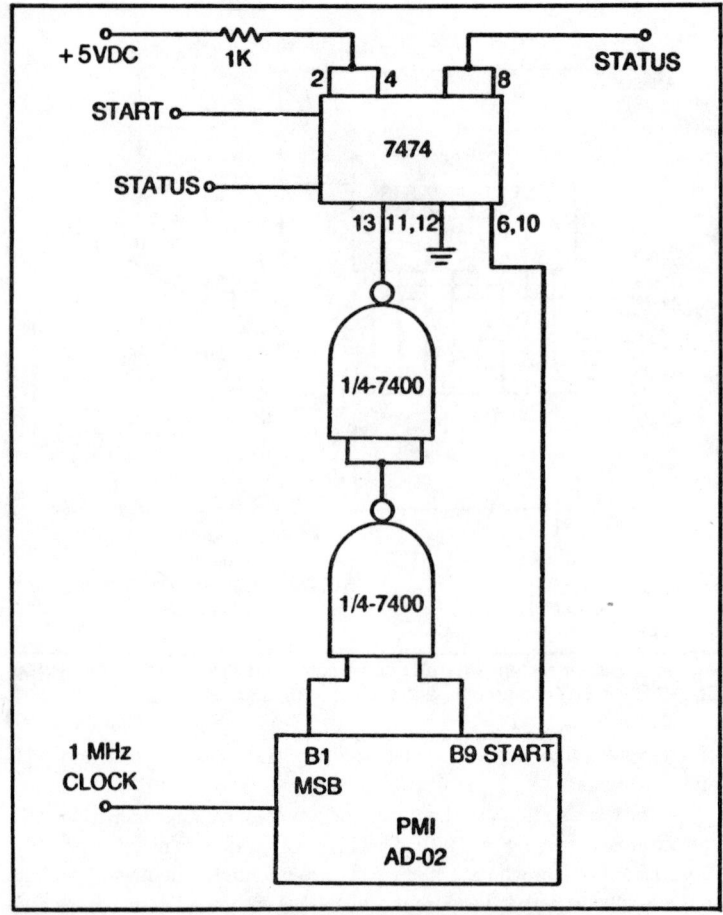

Fig. 14-9. Circuit using the PMI AD-02 SAR A/D converter.

staircase generator, an example of which is shown in Fig. 14-11. Here we have a DAC with its digital inputs connected to the outputs of a binary counter. The input to the binary counter is connected, in turn, to an astable multivibrator, which is usually called a clock. When the counter is reset its output will be 00000000 so the DAC will increase a certain small amount for each count. Each step in output voltage will be equal to

$$\frac{\text{volt}}{\text{step}} = \frac{\text{Full-range output}}{2^{(n-1)}}$$

where n is the number of bits at the DAC input (i.e., 8 for a DAC-08). If n is large enough, each step increase will be negligible; so to the real world it may appear that the step function generated is a continuously rising ramp.

274

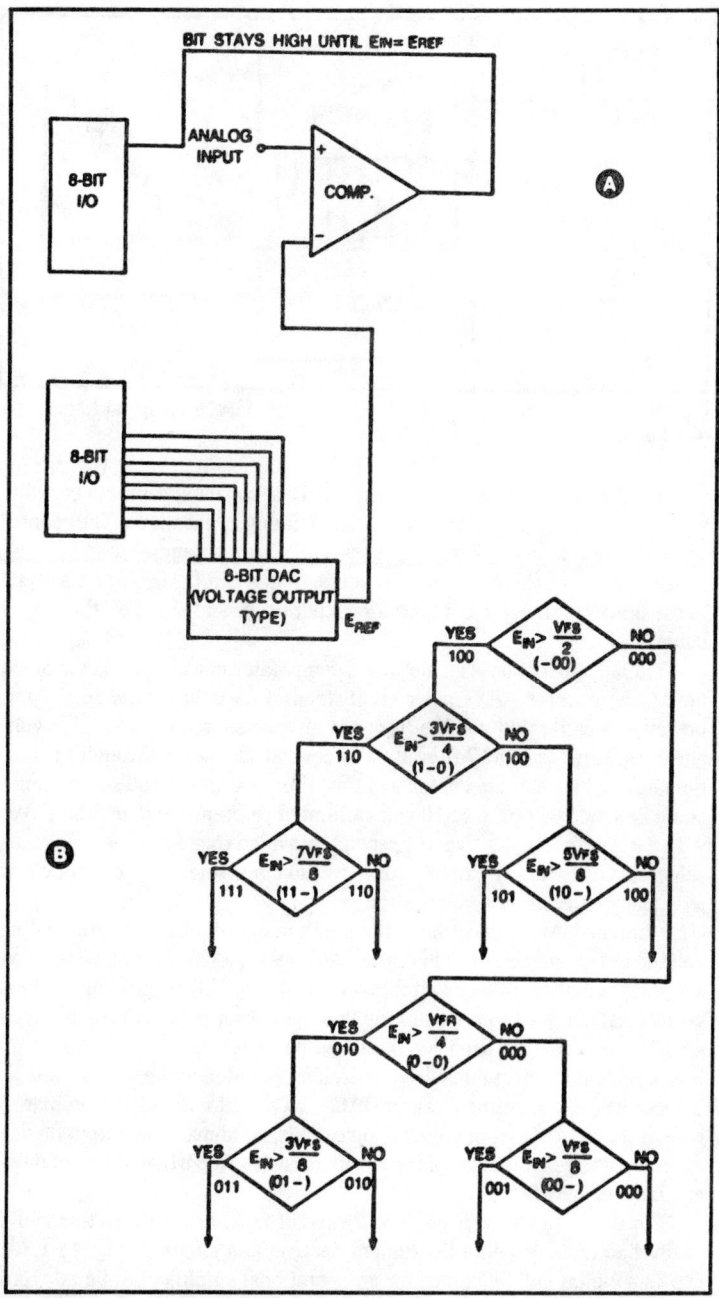

Fig. 14-10. At (A), the SAR A/D converter using a microprocessor and a DAC-08; and (B), the SAR algorithm for using a microprocessor in A/D conversion.

275

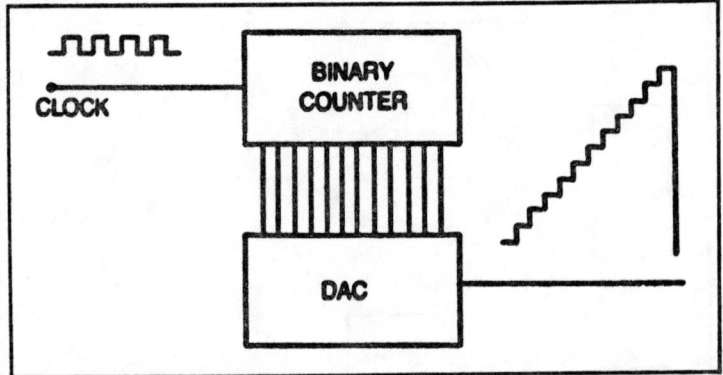

Fig. 14-11. Binary counter and a DAC used to make a ramp and staircase generator.

The DAC ramp generator is used in A/D converter schemes to generate a sawtooth waveform digitally or is used in timing applications. The counter type of A/D converter (Fig. 14-6) uses a DAC ramp generator and a comparator to effect the conversion process. When the DAC output is equal to the input voltage, the comparator output drops and shuts off the binary counter.

Timing applications can also use a comparator in a circuit not too much unlike the counter A/D converter. Instead of an unknown analog input, however, a calibrated voltage from a reference source is used. This voltage could come from a PMI REF-01 or equivalent source, through a potentiometer that is calibrated in units of time. For example, a 100-second timer could be made using a 0-to-10 volt calibrated reference and an 8-bit DAC, with a resolution better than 0.1 second, provided that a suitable clock was selected. Other ranges can be chosen by judicious selection of either clock frequency or reference-voltage range.

Another DAC application is the creation of a digitally-programmable voltage or current source. This can be done using a DAC and, if necessary, a scaling amplifier. Binary switches, or possibly a BCD output thumbwheel switch, can be used to program the voltage output level. Alternatively, a μP I/O port could be used for the same purpose.

A related application is driving oscilloscope-deflection systems, or X-Y recorder pens, by using a pair of DACs (one each for X and Y coordinates) connected to I/O ports or suitable logic. This could make an output device for computer-generated graphics—an increasingly important aspect of modern instrumentation.

Figure 14-12 shows a small collection of DAC applications that could easily find their way into instrument design. The circuit in Fig. 14-12(A) shows a digital null technique for an operational amplifier. A digital word entered into (B0 through B7) is selected to produce a zero output voltage from the operational amplifier. In ordinary circuits this would be an insane waste of money and time because an ordinary 50-cent potentiometer would do the same thing. But in some cases, the DAC approach allows us to automate the null function, for example in an auto-zero, or null-seeking circuit.

Perhaps an illustrative example is in order. In a previous chapter you learned that few Wheatstone-bridge transducers are perfectly balanced under zero-stimulus conditions; there is almost always an offset voltage due to this imbalance, and it appears at the output of the bridge amplifier. As a result, a zero control must be provided to suppress the offset.

Most auto-zero amplifiers use a pushbutton system. The operator will press this button when the transducer is under zero-stimulus conditions.

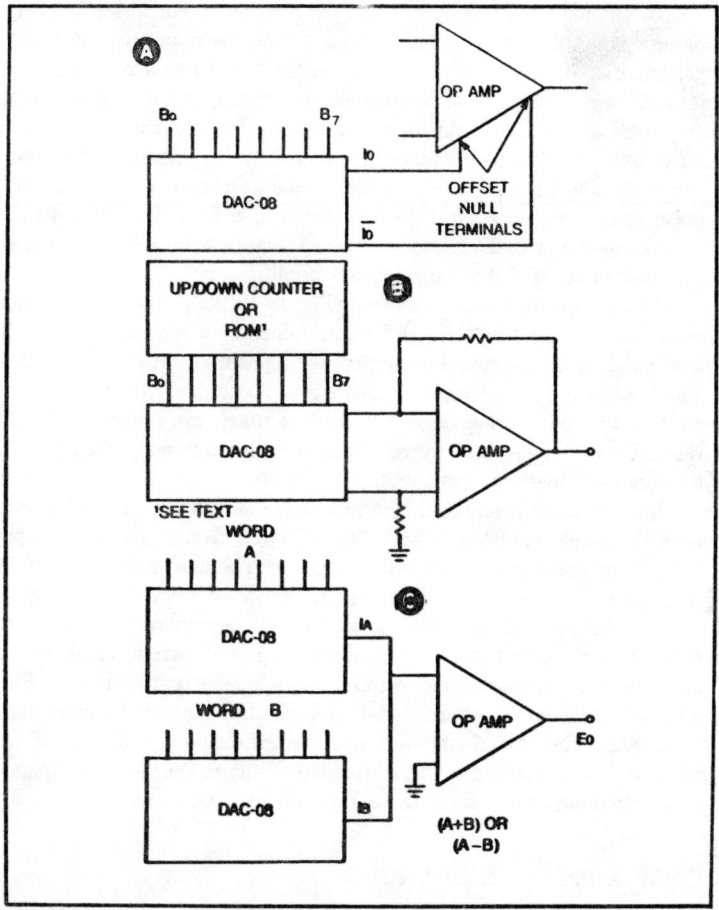

Fig. 14-12. At (A), the use of a DAC-08 for digital offset nulling; (B) waveform generation using DAC-08; and (C) analog output adder or subtractor using DAC-08.

This will charge a capacitor sample-and-hold circuit to a voltage level equal to the offset. The voltage across this capacitor is then fed back to a stage of the amplifier where it bucks the offset, reducing it to zero.

The capacitor approach is fine, and it works, but only for a short time. If the capacitor is required to hold its charge for a long time, droop occurs and the null condition deteriorates. A popular, medical, blood-pressure amplifier uses this approach and nurses find that it must be re-zeroed even if the last measurement was taken as little as a quarter hour previously.

A circuit such as Fig. 14-12(A) can be incorporated into a digital null system, and it is essentially free of droop problems. Such a system could use a binary counter to drive inputs B0 through B7. Pulses into the counter could be gated so that counting could be started or stopped on command. Such a command could be supplied by the output of a ground-seeking comparator which has its free input connected to the output of the bridge preamplifier. Logic circuitry is arranged so that, when a zero button is pressed, the counter is reset and the comparator input is connected to the amplifier output. Also at this time, pulses are gated into the counter. The DAC output rises as the counter increments; so the offset will decrease. When the offset is zero, the comparator snaps low and cuts off the clock pulse gate. The drift from the DAC reference, or from the DAC itself, is small compared with the droop rate of zero circuits that use even very good capacitors and FET input operational amplifiers.

A waveform generator is shown in Fig. 14-12(B). The device or circuit used to drive inputs B0 through B7 determines what type of waveform is generated. A binary counter will produce a positive-going sawtooth if it counts up and a negative-going sawtooth if it counts down. Similarly, if an up/down counter is arranged so that the maximum count (11111111) resets a switch connected to the up/down-mode terminal of the counter, a triangular waveform is generated.

If the counter is replaced by a read-only memory (ROM), almost any periodic waveform up to the limit of the DAC (38 kHz, approximately) can be accommodated. An 8-bit binary counter can be used to step the ROM through its addresses; then the waveform will appear at the DAC output.

A technique used to provide an analog voltage proportional to the sum of two digital words is made by combining a pair of current-output DACs and an operational amplifier. An example is modestly presented in Fig. 14-12(C). Here you see the case where both DAC outputs are connected to the same operational-amplifier input; so the resultant is $(A + B)$. If, on the other hand, each had been connected to a different operational-amplifier input, then either $(A - B)$ or $(B - A)$ would have been produced.

COMPANDING TECHNIQUES

A companding (COMpression/exPANDING) DAC such as the PMI DAC-76 can greatly improve the dynamic range of an A/D-converter scheme. The DAC-76 format is 8-bit, but it is actually a sign bit plus seven

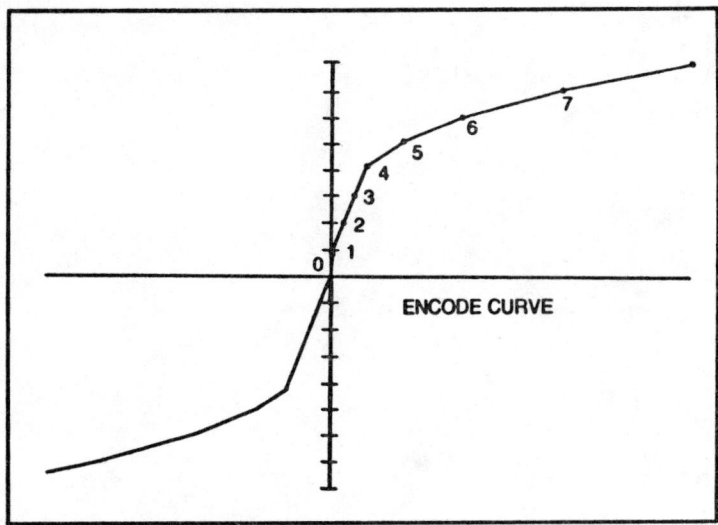

Fig. 14-13. Encode curve for compander DAC.

data bits. Three of these bits are used to select one-of-eight (2^3 = 8) related chords, and the remaining four bits are used to choose sixteen linearly-related steps within each chord (see Fig. 14-13). When used as a successive-approximation A/D converter, the DAC-76 can produce a 72-dB dynamic range, as opposed to around 20 dB for the ordinary A/D converter. The DAC-76 will both encode and decode the signal depending upon the status of certain pins.

References:

1. Stanley, William D; *Digital Signal Processing*, Reston Publishing Co., Reston, Va. 1975.
2. Barnes, J; "A Single-Ramp Analog-to-Digital Converter," Applications Note AN-559, Motorola Semiconductor Products, Inc., Phoenix, 1972.

Chapter 15

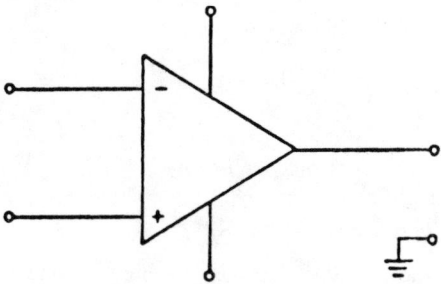

Optoelectronics

O PTOELECTRONICS REFERS TO A CLASS OF ELECTRICAL AND ELEC-
tronic devices that interact with light to produce some effect such
as a flow of current, voltage, or change in resistance. Most authorities also
include those devices which react with infrared and ultraviolet in the same
category as those that react with the visible light wavelengths. In this present study of optoelectronics we will consider photovoltaic cells,
photoresistors, phototransistors, and devices made using these devices such
as optocouplers or optoisolators.

THE PHOTOELECTRIC EFFECT

In 1887 physicist G. Heinrich Hertz observed in an experiment that
certain substances would emit free electrons under the influence of light.
At the turn of the twentieth century, with timing that was ironically pro-
phetic, Max Planck, another physicist, offered the theory in 1900 that light
could exist only in certain discrete bundles of energy which he called *photons*.

Albert Einstein advanced this theory in 1905 when he postulated that
the energy content of a specific photon was proportional to the frequency
(color) of the light. This is expressed by the relationship

$$E = h\nu \tag{15.1}$$

where
- E   is the energy
- h   is Planck's constant (6.63×10^{-34} joule-sec)
- ν is the frequency of light

Materials possess a property called the *work function*, which relates to the energy required to free an electron. If the photon energy (Eq. 15.1) exceeds this level, an electron will be freed. This electron, once freed, will take on a level of kinetic energy equal to the *difference* between photon energy and work function energy, or

$$h\nu = E_o + K.E. \tag{15.2}$$

where

E_o is the work function energy
K.E.   is the kinetic energy of the emitted electron

The photoelectric effect depends, therefore, upon two factors: the frequency of the incident light beam, and the work function of the material. For any given material there will be a light frequency, or color if you prefer, below which no electrons will be emitted because the photon energy will not exceed the work function.

It is interesting to note that Einstein published four papers in 1905, all of which would have earned him a reputation in the world of physics. Of these, he is best known in the popular mind for his theory of relativity, but it was for the paper on the photoelectric effect that he eventually won the Nobel prize in physics. Whether it was excessive conservatism on the part of the committee or just that they failed to appreciate the significance of the other theory seems remote at this time. It does serve to show that sometimes history is a better judge of advances than are individuals.

In a semiconductor material photoconduction occurs if the photon is able to impart enough energy to a valence electron to force it into the conduction band. This phenomenon allows light to affect conduction across a PN junction to a measurable degree. In the forward-biased junction, though, the effect is masked by the normally-large number of charge carriers that flow. In the reverse-bias condition, though, the normal current not due to light is limited to the leakage current across the junction, and this is quite low. When light impinges on the junction there will be a spectacular increase in current. Even an ordinary glass signal diode that is reverse biased will show this effect to some degree.

When the PN junction is part of a transistor that is mechanically constructed to allow light to strike the base, the effect is magnified by the transistor's dc current gain. This action is expressed by

$$I_c = (I \times h_{fe}) + I \tag{15.3}$$

where

I_c is the transistor collector current
I   is the photoelectric base current
h_{fe} is the dc beta of the transistor

PHOTOEMISSION AND LIGHT-EMITTING DIODES (LEDS)

When some form of energy impinges upon an atom there is a possibility that some of it will be absorbed. The atom handles this situation in accordance with the conservation principle by allowing the energy levels of some electrons to increase. In the orbital model of the atom which we use, this corresponds to having the electron move (see Fig. 15-1) to an orbit, or band, that is farther away from the nucleus.

The electron may stay in that excited condition for a length of time, but it must at some time fall back to its *ground state*, or rest condition. This requires, again by the conservation-of-energy principle, that the electron release an amount of energy equal to the difference between the energy levels of the two different bands. This energy can take the form of heat, X-rays, or visible light.

In the forward-biased PN junction, electrons injected by the power source into the N region will take on an excess-energy level to make the crossing at the PN barrier. This energy must be given up when the electron recombines with a hole to become electrically neutral. The energy that is given up can go off in the form of infrared or visible-light radiation.

In silicon and germanium diodes the photons given off by this process are usually in the infrared region, but in the newer gallium-arsenide (GaAs) diodes the energy levels increase to a point where the light is in the red visible region. Light-emitting diodes are currently available in red, yellow, and green colors. Some manufacturers also offer a red/green LED in which a red and a green LED are in the same package and wired back-to-back. The color will depend on the polarity of the applied voltage.

Keep in mind, when selecting light-emitting diodes, that the red types are usually more efficient. The light output of the green and the yellow varieties has often proven disappointing. The red types offer more light at less current in most cases.

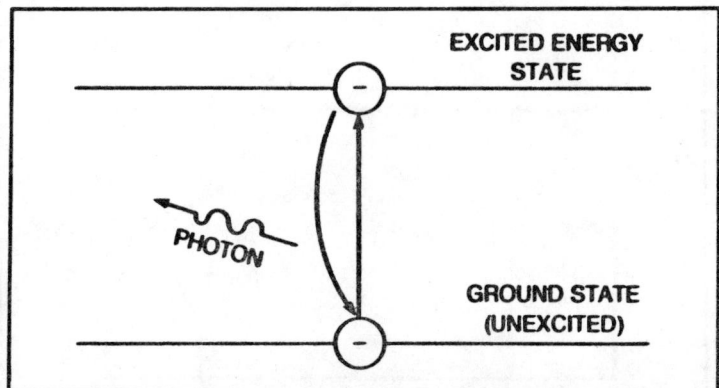

Fig. 15-1. When an electron is excited to a higher energy state, it will be unstable. When it falls back to ground state, an energy photon will be given off.

PHOTODEVICES AND CIRCUITS

The photovoltaic cell, or solar cell, is a device that uses the action of light to create an electric potential, usually something on the order of a fraction of a volt. These devices have been available for many years but have only recently undergone much new development due to pressures of the space program, where they are used to supply power to spacecraft and to help ease the energy crunch.

Of greater importance to my present purpose, electronic instrumentation design, is the application of the photoresistor, also called photoconductive cell, or simply photocell. Be a little wary of that term photocell because many people will use it to describe almost any photosensitive electronic component. In this text, though, it will be used to describe only a photoresistor, also called photoconductive cell, or simply photocell.

A photoresistor is made using materials such as cadmium sulfide, lead sulfide, or cadmium selenide. These materials have the property of changing electrical resistance when exposed to light.

Different photoresistors will have different properties, and these must be considered when selecting a device for a design project. Some, for example, have an extremely high *dark resistance*, and a really tremendous change of resistance as light is applied. One type, for example, has a dark resistance over 500 kilohms, and a resistance under room light of approximately 20 kilohms.

Figure 15-2 shows the basic light meter circuit using a photoresistor device. The resistance of photocell PC1 depends upon the intensity of the light falling on its surface. Battery B1 will set up a current in the series circuit that consists of PC1, R1 and M1. The level of the current registered on M1 will depend upon the resistances of PC1 and R1. When the

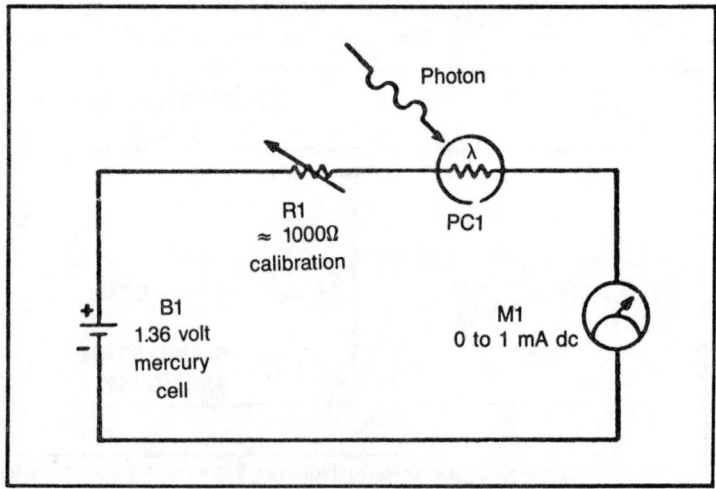

Fig. 15-2. Light meter using a photoresistor.

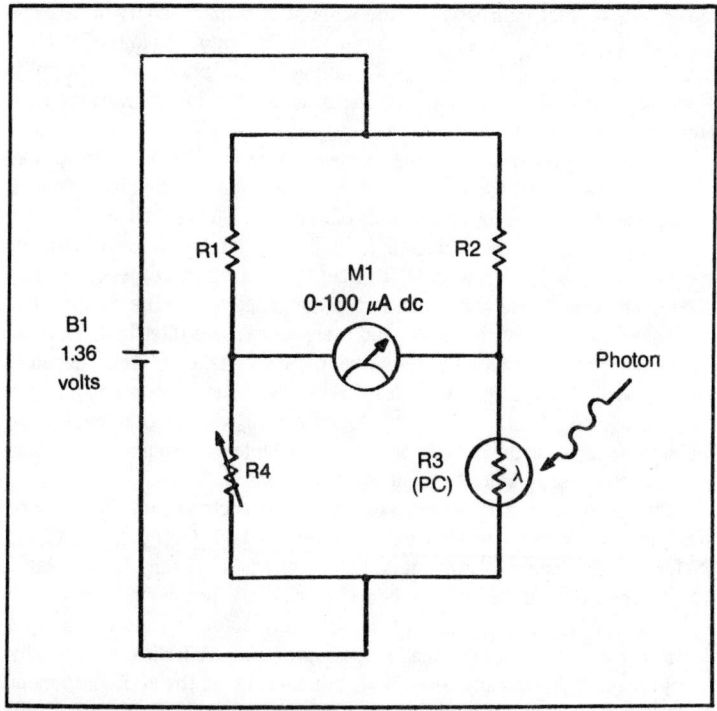

Fig. 15-3. Wheatstone bridge light meter.

photoresistor is dark, the resistance of PC1 will be high; so the current will be low. As the light intensity increases, however, the resistance of PC1 decreases; so the reading on M1 goes up. Resistor R1 may be used to calibrate M1 at some specific light intensity so that the readings will be meaningful to the specific application.

A more sensitive, and potentially more useful, light-meter circuit is shown in Fig. 15-3. Here you see another application of the classical Wheatstone-bridge circuit. Resistors R1 and R2 are usually set equal to each other, although this is merely a convenience and is not essential. These resistors will usually have a value in the range of 500 ohms to 20 kilohms. The precise value selected will depend partially on the value of the photoresistance at the ambient-light intensity where the meter will be operated. Try to keep R1 = R2, and within the range of approximately R3/2 to 2R3.

Again there is a calibration resistor; in this case it is R4, one of the bridge arms. Resistor R4 can be adjusted so the current on the meter will be zero, full scale, or some other particular value under ambient-light conditions.

Quite a large family of electronic instruments and control devices use this type of circuit because there are many physical variables that can be measured, at least indirectly, by the amount of transmissivity or absorptivity of light (either white light or some specific frequency in the infrared to ultraviolet range). This area is sometimes called photometry (literally measurement with light), and the instruments are known variously as photometers or densitometers.

Let me demonstrate this type of instrument configuration by an example. Let's say that some parameter is measurable by the absorption of red light as it passes through a sample of material. The test set-up is shown in Fig. 15-4. A red filter is placed in the light path so as to remove artifacts created by other light colors. Of course, the color and properties of the filter in any given case will depend upon the properties of the material being tested, but I select red. It is good practice to use a filter in any event, even when the material is not sensitive to the frequency of light, because the photocell is. Most photoresistors do not have equal response to all light frequencies; so something must be done (in many cases) to eliminate the problem. In my case the red component of the white-light source is transmitted, and the rest is blocked or at least attenuated.

The photoresistor is connected in a circuit such as Fig. 15-3. There are actually two ways to adjust this circuit. I could, for example, adjust R4 for zero (or near zero) current on the meter when the sample is in place. I could then note the reading when the sample is removed.

A better way is to adjust R4 for precisely 100 μA when the sample is *not* in place. This means that a reading of 100 μA indicates precisely 100 percent light transmission. The full amount of the red component reaches the photocell. When the sample is placed in the light path it will absorb some of the light; so only a fraction of the light falls on the cell. If the circuit is set up correctly, the current reading is the percentage of light transmission.

A variation on this theme is shown in Fig. 15-5 and uses two photocells. This is actually a preferred method and is used in quite a few electronic

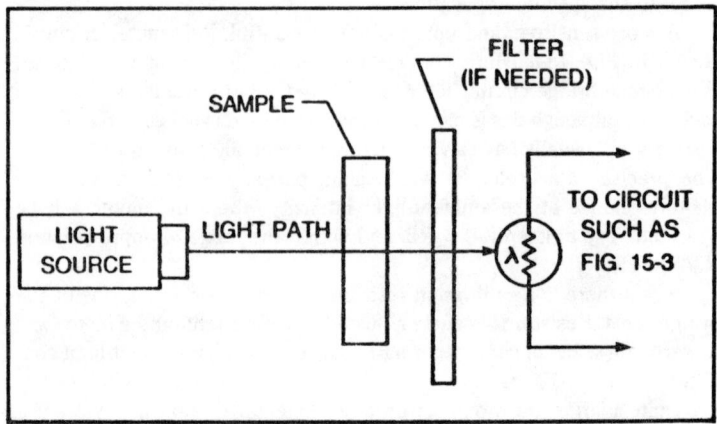

Fig. 15-4. Simple densitometer.

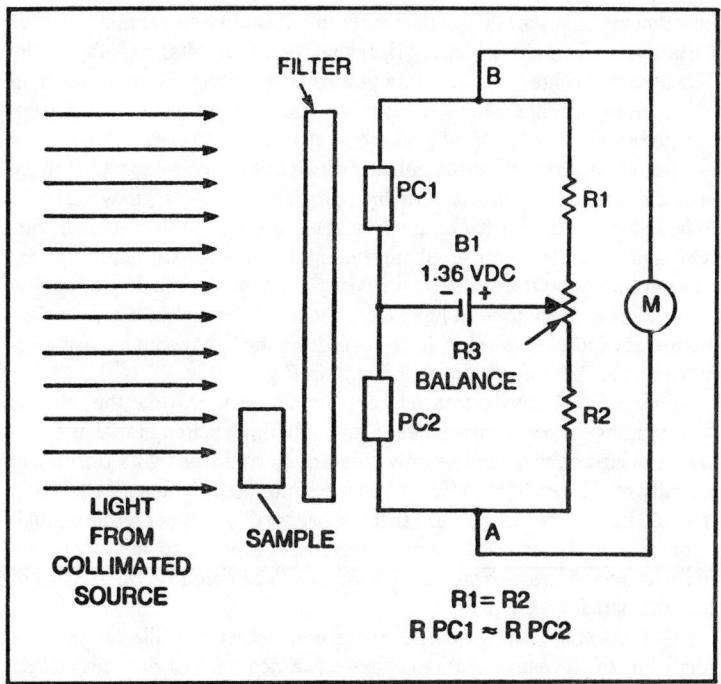

Fig. 15-5. Wheatstone bridge densitometer.

instruments. One cell is used as the reference while the other is the measurement cell. The two photocells are in opposite arms of a Wheatstone bridge. Interestingly enough, meter M1 can be either a microammeter or a millivoltmeter.

Photocells PC1 and PC2 are selected to be a matched pair. If the manufacturer will not match a set for you or wants too much money, a good approximation of a matched pair can be obtained by comparing the dark and light resistances of a collection of photoresistors and then selecting two that have similar readings.

The bridge is balanced with the sample removed by adjusting R3. In this condition both photoresistors see the same amount of light. But when the sample is inserted, part of the light to PC2 will be absorbed so that the PC2 resistance increases to a value higher than the resistance of PC1. This difference upsets the bridge balance; so the point A voltage will be greater than the point B voltage, a fact that will be reflected on the meter.

All of these circuits are examples of optical densitometers because they operate by measuring the optical density of the sample. You may be surprised to find that several classes of very expensive scientific instruments are actually nothing more than precision variations on the circuits shown in this chapter.

One of the biggest design problems associated with simple densitometer instruments is suppression of artifacts in the output signal due to either variations in the light source itself, or the excitation voltage applied to the Wheatstone bridge. In both cases you could use the ratiometric method of instrument design. This was the technique, you will recall, that was developed in my discussion of pressures measurement.

For short-term measurements it may be of little consequence if these artifacts show up, but in long-term (more than a few minutes) they are likely to be bothersome. In any event, strive to use a stable, well regulated, current source for the lamp supplying the light, and a stable, regulated, excitation voltage. If the artifacts still exist you can go to ratiometric circuits.

Ratiometric solution of light-source variations requires a pair of densitometers and a beam-splitting prism to form the light beam into two separate paths. An illustration is provided in Fig. 15-6.

One beam is passed to a reference densitometer, while the other is directed toward the measurement (sample path) densitometer. Since this is a ratio circuit, you need be only moderately concerned with the optical equality of the two light paths. If you were interested in absolute density, it would be necessary to ensure that the beam splitter was precise enough to provide 50 percent of the light to the transmitted beam and 50 percent to the reflected beam. You would also have to be sure that the lengths of the two paths were equal.

One beam is passed to a reference densitometer, while the other is sensed by the measurement densitometer. Since this is a *ratio* circuit you need an analog multiplier/divider circuit connected to provide a transfer function of the form

$$E_o = \frac{k\,E_2}{E_1} \tag{15.4}$$

where

E_o is the analog-divider output voltage
E_1 is the reference-densitometer output voltage
E_2 is the measurement-densitometer output voltage
k is a constant that depends upon the properties of the divider.

The output voltage variations in this circuit are due solely to variations in the optical density of the sample.

A scanning densitometer is a densitometer in which the sample or the light beam is moved in a precision, rectilinear manner, so that variations in sample density can be detected. These instruments are very costly, although electrically they are extremely simple, many of the sort of circuit shown in Fig. 15-2. The reason for the high cost is the very precise mechanics that are required to make the thing accurate. Oddly enough, many popular scanning densitometers cost a great deal, yet make no pro-

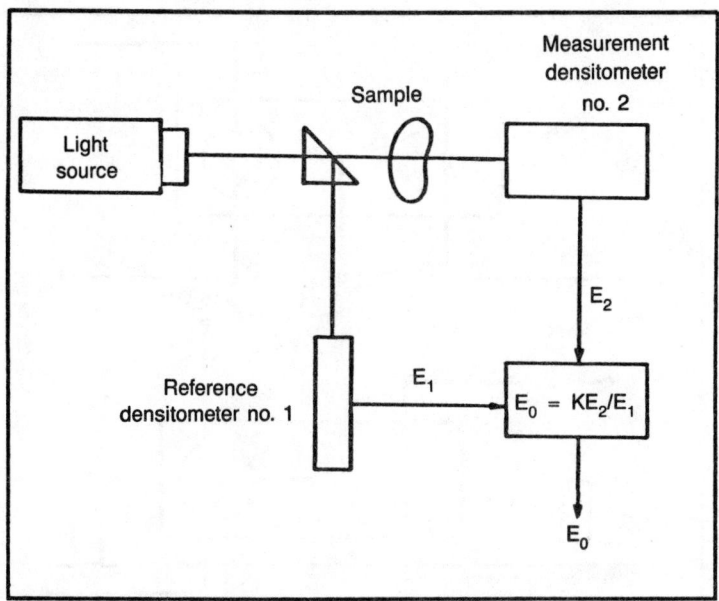

Fig. 15-6. Split-beam densitometer to eliminate artifacts.

vision for the artifacts discussed. These can be improved considerably by the use of ratiometric techniques.

Amplified densitometers can be used to make high-resolution measurements where the output of the Wheatstone bridge is nearly zero. There are two approaches to making an amplified densitometer. You could, for example, merely connect the output of the bridge to the input of an ordinary operational-amplifier circuit that provides the needed voltage gain. Alternatively, you could provide the input of the operational amplifier with a constant-current source and then connect the photoresistor in the negative-feedback loop. The output voltage will be

$$E_o = I_1 \times R_{PC1} \qquad (15.5)$$

PHOTOTRANSISTOR AND LED CIRCUITS

An optoisolator, also called an optocoupler, is a device in which a phototransistor is used as the sensor and an LED is used as the light source. Although mechanical forms differ depending upon application, the basic electro-optical configuration will be as shown in Fig. 15-7. The LED and phototransistor are housed in a common package and are positioned so that the light from the LED lands on the sensitive base region of the transistor.

Optoisolators find uses where direct connection between two circuits is not prudent or possible. Medical instruments, for example, frequently

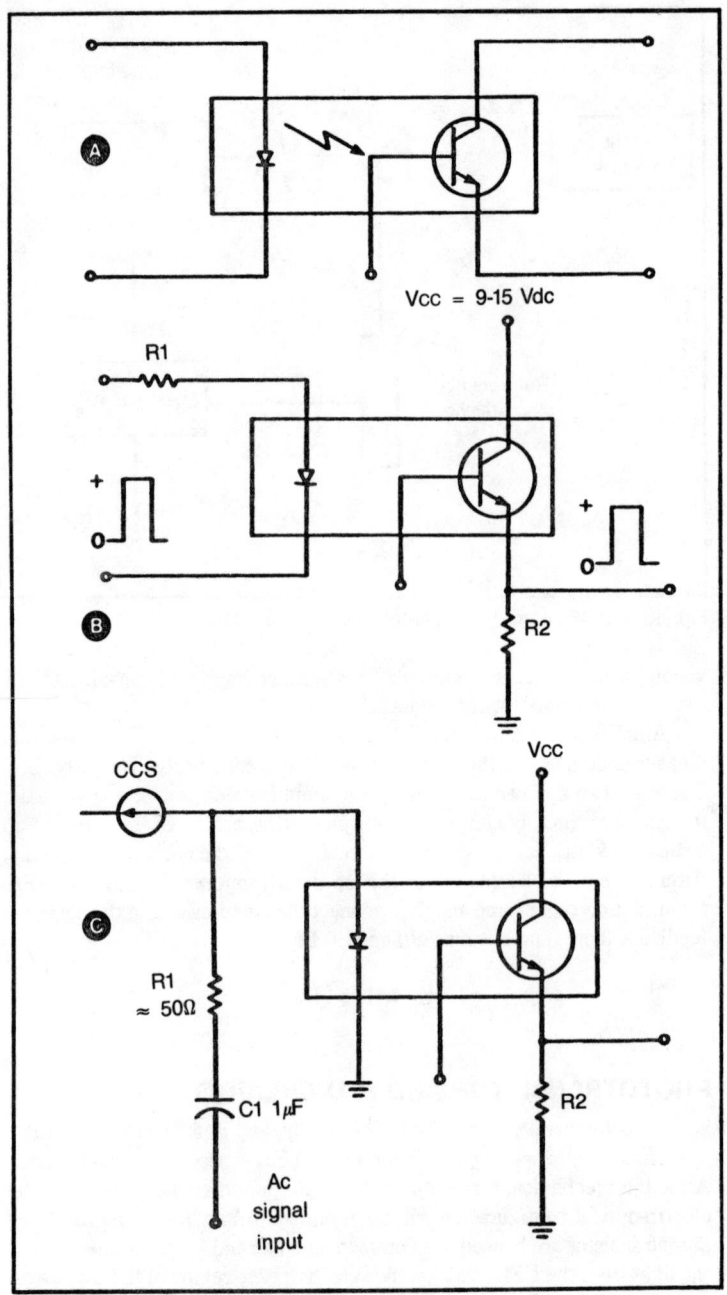

Fig. 15-7. At (A), the basic form of an optocoupler; (B), a pulse coupler; and (C), a linear coupler.

use optoisolators to prevent microshock hazards to electrically suscepti-ble patients (See *Servicing Medical and Bioelectronic Equipment* by Joseph J. Carr, TAB book No. 930).

Figures 15-7(B) and 15-7(C) show two modes for connecting the op-toisolator. The outputs in both the pulse mode, Fig. 15-7(B), and the linear mode, Fig. 15-7(C), will be in phase with the input signal because the tran-sistor is used in the common-collector, or emitter-follower, circuit. If in-version is desired, simply connect a load resistor between the collector of the transistor and the V_{CC} supply voltage. The output is then taken from the collector terminal of the transistor.

One application for the pulse-mode circuit is to perform logic-level translations between systems that are incompatible to avoid a direct inter-connection. You may, for example, want to interface an older logic device that uses ± 12-volt logic levels with a modern TTL system, or vice versa. Or you may want to limit interaction between the two circuits. Another case might be where the environment inside one is dangerous. An exam-ple of this is where the voltage levels are at hundreds of volts above ground but the levels of the measuring instruments, computer, or what have you are ordinary TTL, DTL, or CMOS levels.

The physical form taken by any particular optoisolator or optocoupler will depend almost entirely on its intended application. In some cases, for example, the package will appear to be a 6-pin or 8-pin mini-DIP IC. These are used, as in medical instruments, where the isolation is for purely elec-trical reasons.

Other forms include a reflection model and a vane-blinder model. In the reflection type both LED and sensor are mounted in the same plane, but their light paths are aimed through holes in the package at 45-degree angles so that they will not communicate with each other unless there is an object at a critical point to reflect the light.

The vane-blinder model has a slit, or cut, in the housing between the LED and the sensor so that a blinder can be placed in the light path. These are used in counters, punched-hole readers in computer and control ap-plications, and in alarms (Again, see *Servicing Medical and Bioelectronic Equipment*).

Optoelectronics is one of those electronic areas which I can claim is both infant and elderly. It has blossomed because of modern technology and shows no signs of declining.

Chapter 16

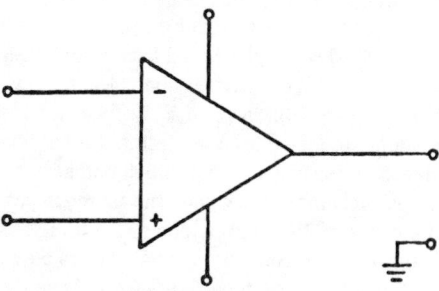

Instrumentation Techniques

A N *ELECTRONIC INSTRUMENT* IS AN APPARATUS CONSISTING OF A collection of circuits and mechanical devices that do some particular job of measurement or control. It might acquire a signal from a transducer of some sort and then process the signal to produce some numerical display or analog waveform at the output. Alternatively, it might take several different inputs, make some computations or comparisons, and then display the result or use them to perform some decision in a control circuit.

TYPES OF INSTRUMENTATION

Instrumentation can be analog, digital, or a synergism of the two separate types. When high precision is mandatory or where extreme complexity might be required, the design is best left to qualified electronic engineers with experience in that area of competence. The amateur or novice (e.g., a student) designer can, though, perform chores in instrumentation that were once regarded as too difficult. So if you are the kind of person who can derive an orgasmic quality from the contriving of contrivances, then start working.

Two main areas of opportunity present themselves as ripe for such people: creation of simple-to-moderately-complex instruments, and interfacing two or more existing, commercially-produced instruments that previously were not compatible.

You will be surprised at the level of complexity that can be achieved by amateurs although only at the expense of some frustration in their early attempts. The frustration quotient, though, can be reduced considerably if the person is willing and able to seek advice from the professionals and the companies (too often overlooked!) who make and sell the products you plan to use.

Even where sophistication is modest there is room for clever application of electronic circuits to save money. A physiologist, for example, might want to use a ×10 preamplifier to acquire cell-action potentials. This instrument costs over $300 when purchased from scientific-instrument suppliers but can be duplicated exactly for about $35 and in function (using other, more modern parts) for around $20. Its chief claim to being a *physiological* preamplifier stems from the fact that it has a high CMRR and an input impedance on the order of 100 billion ohms. But modern BiFET and BiMOS operational amplifiers will do the same job yet offer input impedances on the order of 1500 billion ohms (1.5 teraohms!).

Interfacing two existing commercial instruments may be as simple as constructing an appropriate patch cord. Such will have the connector for one instrument on one end and the connector for the other instrument on the opposite end. Alternatively, interfacing might involve designing electronic circuitry to make the two instruments compatible.

In both industrial and university environments as well as in selected amateur endeavors, one may be asked to make a special system consisting of specially-built instruments or a collection of commercial instruments. You might buy or reallocate a cabinet such as the standard 19-inch relay rack and then go about mounting the instruments and subassemblies. But before much progress can be made, it will be advisable to *study* each instrument to ascertain its capabilities, input/output requirements and specifications, the function and range of each control, and the signal-flow path within the instrument. Any attempt at producing an instrument system without doing this will result in a less-than-optimum device, or at worst, might destroy one or more subassemblies.

SIGNAL ACQUISITION

The *front end* of any instrument consists of those stages that acquire the signal and make it ready for the stages that follow. In the simplest case the front end might consist entirely of a simple operational amplifier-preamplifier. Alternatively, it might consist of a complex analog-signals-acquisition module such as those produced by Analog Devices, Burr-Brown, and others.

If you plan to use the front end or the entire instrument to feed information to a computer, you will need some sort of analog-to-digital (A/D) converter. These are discussed in greater detail in Chapter 14.

Also of use in many applications is a multiplexer circuit. These will allow several inputs to feed a single channel. Multiplexers are available in models that allow any number from 2 to 32 channels to be fed into a single output channel. The cost increases faster than the number of channels in most cases.

Grounding and shielding are two techniques that are little understood even though most workers profess to be knowledgeable. The main cause of trouble in this area seems to be the phenomenon called *ground loops*.

Ground loops are usually caused by currents flowing in the ground conductor. These currents create voltage drops that appear to the system as signals. This is shown in Fig. 16-1(A) where the signal ground, the power-supply return (common) from the amplifier, and the power-supply ground itself all are connected to different points on the ground plane.

The ground plane might be a massive chunk of copper, the earth itself, or a large sheet of metal; yet it still allows these problems to develop. If

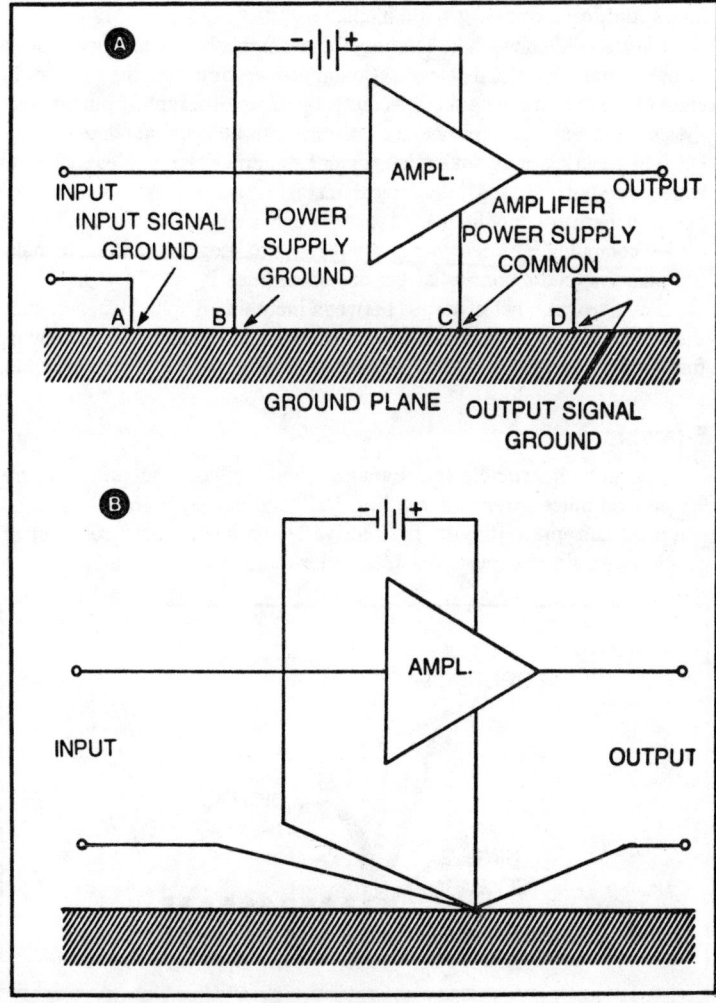

Fig. 16-1. (A), a setting for disaster—ground loops all over the place; (B) shows proper grounding.

the "ground" is a thin piece of wire, ground loop problems are almost certain to develop.

In the case of Fig. 16-1(A) a heavy current (as little as a few milliamperes might be considered heavy) causes a voltage drop between points B and C. This is interpreted as a dc signal by the amplifier input. Although we may think about the input being across point A and the amplifier input line, it is actually referenced to point C as far as the amplifier is concerned.

We often prattle on and on about grounding, but it appears that too much grounding is also possible. The best situation is single-point grounding as shown in Fig. 16-1(B). This prevents, or at the very least inhibits, the formation of ground loops.

Figure 16-2 shows how this may be accomplished when laying out a printed-circuit-board foil pattern. Ground paths return from the various circuits to a point on the card-edge connector. Two different terminals, side by side, are used so that heavier currents can be accommodated.

The function of an analog instrument sometimes resembles that of an analog computer. Indeed, the typical analog measuring instrument (as opposed to mere preamplifiers, etc.) is really only a single-purpose, dedicated, analog computer. The designer is admonished to keep that in mind to make the planning chore somewhat easier—sometimes.

The particular set of stages between input and display depends upon the nature of the job that must be performed. You will have to decide what functions are necessary, and this is done by considering the problem at hand.

Example

Design an instrument (block diagram only, please) that will measure 90-millivolt pulse potentials (see Fig. 16-3) and display both the potential and a signal representing its first derivative on a strip-chart recorder or oscilloscope. Furthermore, the area under the curve of the potential is to

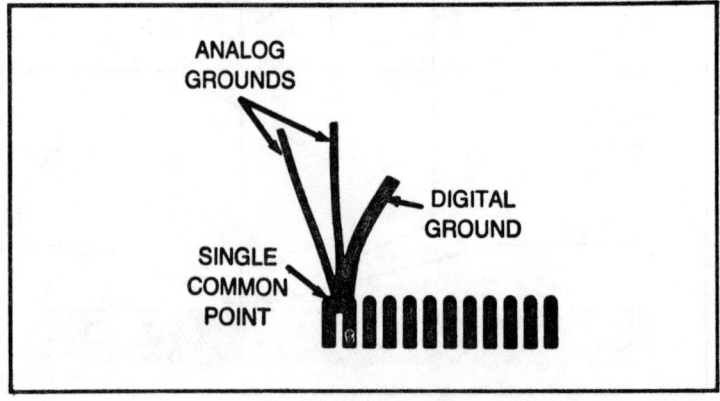

Fig. 16-2. Technique for mixing digital and analog grounds on a printed-circuit board.

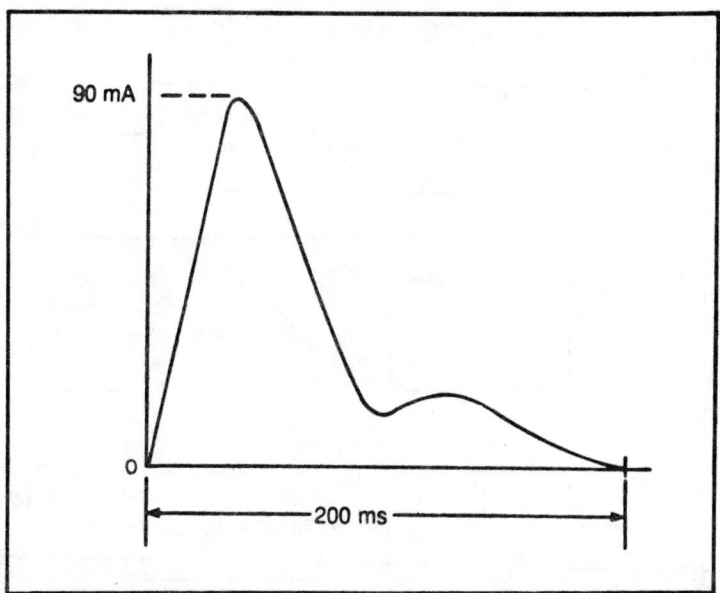

Fig. 16-3. Typical pulse.

be displayed on a digital voltmeter. You should include a method for keeping artifacts above 200 hertz from interfering with the output.

The pulse will be picked up by ohmic electrodes and should be amplified by a factor of 10; therefore you would want to include a gain-of-10 preamplifier. This brings the signal up from a peak of 90 mV to a peak of 900 mV. This signal level is far more comfortable because it is high enough above the level at which drift and noise artifacts are important yet is lower by an order of magnitude than the usual operational-amplifier and linear-IC clipping levels. The circuit is shown in Fig. 16-4.

The job of reducing artifact interference can be performed by a 200-hertz, active, low-pass filter, preferably with unity gain. This keeps the signal at the output of the filter close to the 1-volt range we just defined as desirable.

The filtered pulse signal is passed through a unity-gain buffer amplifier to channel 1 of the chart recorder or oscilloscope. It is simultaneously fed to a differentiator composed of standard op-amp circuits. Similarly, it is also fed to an op-amp integrator.

Example

Create a circuit that will take audio-frequency signals with amplitudes in the 0-to-5 volt range and compute their rms value. Do not use thermal techniques.

First, we get back to basics and consider just what is meant by the

297

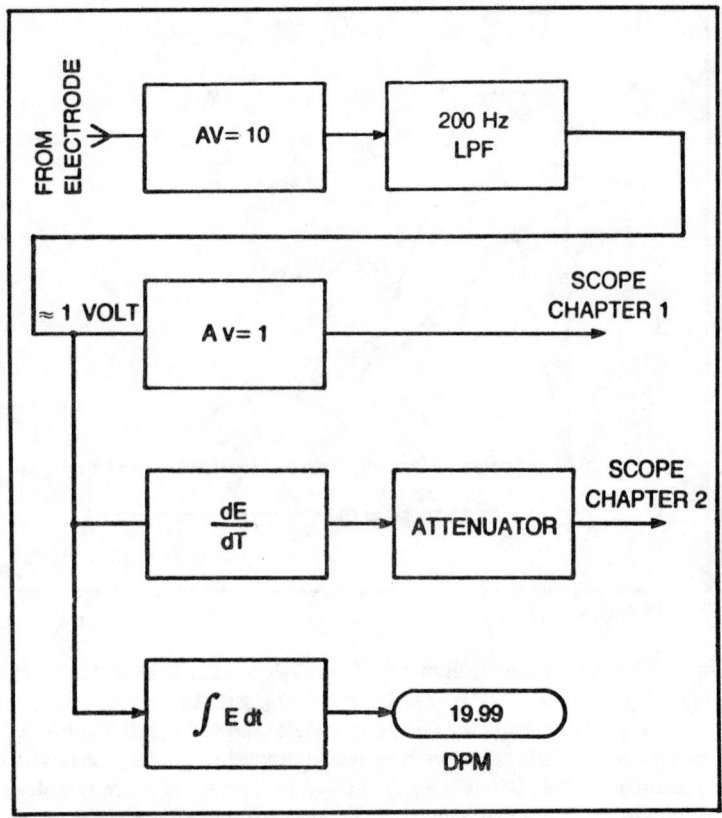

Fig. 16-4. Hypothetical instrument.

rms value of an ac signal. The definition of rms (root mean square) is

$$E_{rms} = \left[\int_0^t (E_{in})^2 \, dt \right]^{1/2} \qquad (16.1)$$

The first step (see Fig. 16-5) is to square the input voltage and then integrate it over time 0 to t. This should be at least one period of the input waveform and defines the lower frequency limit of the circuit. The integrator output is then passed through a square-rooter. In all stages one must consider the input range that must be handled. Because the range of amplitudes will be up to 5 volts, the circuit must be able to process signals in the integrator of 0 to 25 volts! This may require the use of logarithmic compression or some similar technique.

The squarer may be an analog multiplier with the X and Y inputs tied together, while the square-rooter might be a similar circuit connected into the negative-feedback loop of an operational amplifier.

A more expensive, but usually easier-to-implement, approach would be the use of a multiple-function module (i.e., the Burr-Brown 4301/02) that has a transfer function of the form

$$E_o = (X)\left[\frac{Y}{Z}\right]^m \qquad (16.2)$$

By making X a convenient scaling constant and setting Z equal to unity, we may use the Y input for E_{in}. Set m = 0.50 for the square-rooter and m = 2 for the squarer.

Example

Construct a linearization circuit for a pressure-measuring system in the 0- to 2-torr range.

Discussion. Most transducers, regardless of the parameter being measured, meet their linearity specification only over a certain range. In measurement of fluid pressures we find that few transducers are linear in the 0- to 2-torr range. In fact, most have an output-vs.-pressure curve more parabolic than linear.

Figure 16-6 shows both the ideal and the real situations. Ideally, the output of the pressure amplifier will be exactly 1.00 volt at a pressure of 1 torr and 2.00 volts at a pressure of 2 torrs. All values between 0 torr and the 2-torr upper limit are scaled proportionally. This type of function is of the form Y = mX, where m is the scale factor.

But this relationship only holds true in regions where the transducer is linear, and it is *not* linear in the region under 2 torrs. In fact, the region of nonlinearity is probably greater than the region selected for this example. Moreover, most transducers are not trustworthy up to around 10 torrs.

The now-obsolete method of transducer linearization uses a collection of diode-switched breakpoint generators to approximate the slope of the

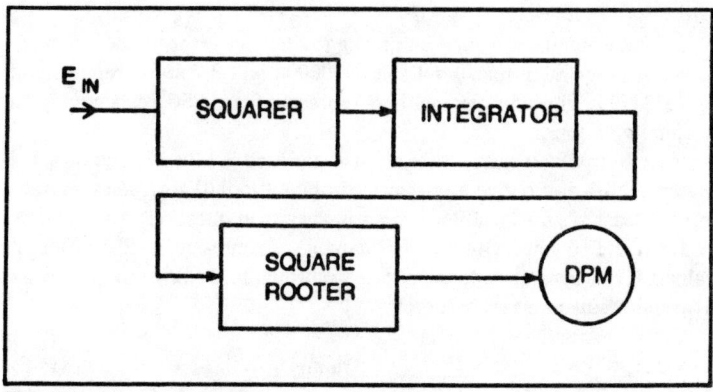

Fig. 16-5. An rms converter.

299

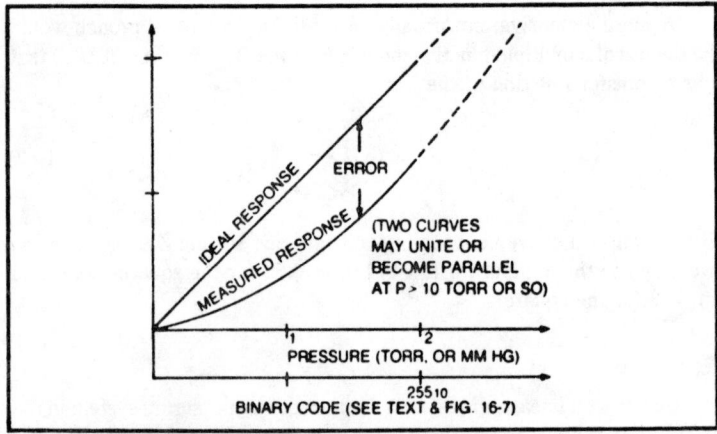

Fig. 16-6. Typical real and ideal transducer curves.

inverse of the real curve. By summing the actual curve with its inverse, we hope to create a linear output.

This technique worked well on paper—but reality occasionally reduced it to a shambles. You may have seen learned discussions on transducer or thermocouple linearization, but few really get into the kinds of problems that you face when actually building the thing.

The biggest limitation resulted from the very nature of the breakpoint method. We must approximate sections on the curve of the inverse by a chord of approximately the correct slope. The problem is that only a few chords can be used without the circuit becoming unbearably large. This requires many switching diodes and operational-amplifier slope generators. Other problems include the necessity of compensating for thermal drift in the diodes and operational amplifiers and of using nonexact values of resistors and other components. Fortunately, modern IC technology brings us a better method at better than competitive cost. This is shown in Fig. 16-7.

This technique uses an 8-bit analog/digital converter to encode the 0- to 2-volt range such that 0 volts is 00000000 and 2 volts is represented by 11111111. These binary numbers represent 0 and 256, respectively, in decimal notation.

The 0- to 2-volt range, if read on a standard 3 1/2-digit digital panel meter (DPM) would give a pressure resolution to 0.01 torr, not considering the least significant digit, which is always in question. An A/D converter with 256 counts (0 to 255) is capable of representing 255 different values. If this type of system is set up according to the protocol just given it would result in a resolution of

$$\frac{2 \text{ torrs}}{255} = 0.008 \text{ torr} \qquad (16.3)$$

We may conclude, then, that our linearization is valid because its potential resolution exceeds the resolution of the readout device by approximately an order of magnitude.

The binary output word of the A/D converter is able to represent discrete values of the input voltage that, in turn, represent pressure values. It is also used to address discrete locations in a read-only-memory (ROM) integrated circuit.

Each ROM location contains an 8-bit binary word that represents the *correct* voltage to represent the pressure applied to the transducer. Following the ROM we have several options. If only digital display is required, a binary-to-BCD IC could be used to convert the ROM output word to a form compatible with most displays. If an analog display is required, use a DAC to generate the linearized voltage from the ROM output.

Control logic is required so that the ROM is disabled during the conversion time of the A/D stage; otherwise incorrect intermediate states result, producing an erroneous output voltage. For the same reason we want an A/D converter that has latched outputs.

The limitation is that the ROM must be programmed for the specific transducer unless you are willing to tolerate *some* error, even though improved over the unlinearized version.

Use a *water* manometer, or one making pressure measurements using an even less dense material, to calibrate the system. Also required is a digital voltmeter with at least 3 1/2-digit accuracy. Adjust the water column for various pressures, keeping in mind the fact that it takes 13.6 mm of water for each torr of applied pressure. This is because 1 torr = 1 mm Hg, and Hg is 13.6 times denser than water.

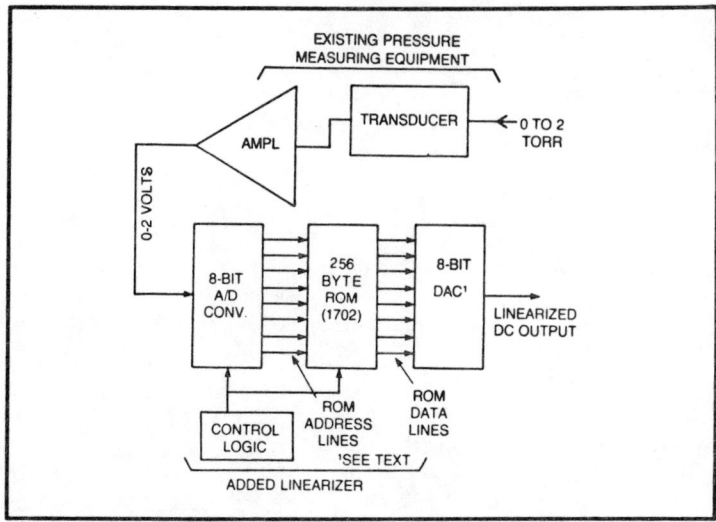

Fig. 16-7. Digital linearizer.

Even water, though, may not be capable of adequately calibrating the system because at 2 torrs the water column will be only 27.2 mm in height, a little over 1 inch. A solution to this problem is to find a liquid that will not corrode the transducer yet is considerably less dense than water.

Another method is to forget about measurement of the pressure against the atmosphere and use a partial vacuum. You can, for example, evacuate the space above the column to a pressure of somewhere near 1/2 atmosphere. This doubles the range of the column.

Chapter 17

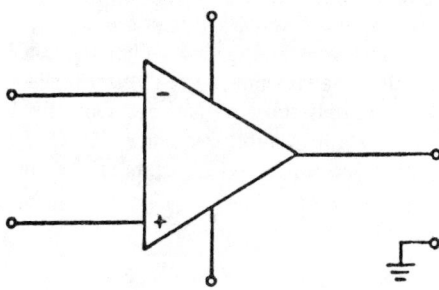

Active Filters

D EVOTING JUST A SINGLE CHAPTER TO A SUBJECT LIKE ACTIVE FIL-
ters may seem presumptuous, or may look just plain ridiculous, once
you know more about active filters than is necessary for this discussion.
This becomes especially apparent when you discover that several thick
engineering books are available on the topic. The idea here is not to make
you into a sophisticated electronic-filter designer but to give you the ability
to put together something that works most of the time, for most applications.

SIMPLE FILTERS

A filter is a circuit that passes some ac signal frequencies but not others.
The basic types are low-pass, high-pass, and bandpass filters, plus the notch
filter.

The low-pass filter passes all ac frequencies between dc and its cutoff
frequency. A 1-kHz low-pass filter, then, will pass all frequencies between
dc and 1 kHz, but above 1 kHz the response drops drastically. In an ideal
filter the response cutoff would be abrupt, like a knife edge placed at the
1 kHz point in the spectrum; but by now you should realize that ideal and
real are not usually mutually compatible concepts. The response of a real
filter will drop off with a given slope that is dependent upon the nature
of the filter circuit and the manner of its construction. Filter rolloff past
the cutoff frequency is usually expressed in decibels per octave (2:1 frequency ratio) or decade per octave (10:1 frequency ratio).

A simple RC low-pass filter consists of a resistor in series with the signal
line and a capacitor across it. If you recognize this simple RC network as
the simple integrator, then congratulations. This passive filter has a slope
of – 6 dB/octave. Several sections may be ganged to increase the rolloff rate.

A high-pass filter is just the opposite of the low-pass filter. It will pass only those signals that have frequencies greater than the cutoff frequency. The simple RC differentiator is an example of the passive RC high-pass filter. If you can appreciate the relationship of the mathematical concepts of integration and differentiation, you may see a glimpse of insight into the beauty of the high-pass and low-pass filter relationships.

A bandpass filter is a combination of the low-pass idea and the high-pass idea. It will pass only those frequencies above the lower cutoff frequency and below the upper cutoff frequency. The Q of a bandpass filter is a property that describes something of its nature and is given by

$$Q = \frac{F_c}{B.W.} \qquad (17.1)$$

where

F_c is the center frequency, assuming the response is symmetrical
B.W. is the bandwidth expressed in the same units of frequency as F_c.

A notch filter is a circuit that passes all frequencies except one specific frequency. It can be likened to a series resonant tank circuit shunt connected across the signal source without torturing reality too much. The main use for notch filters is the elimination of some specific unwanted frequency, such as the interference from the 60-hertz power mains.

Most active filters are built around the operational amplifier. The simplest case for an active low-pass filter is the inverting-follower circuit with a capacitor of suitable value (X_c less than one-tenth R_f at the cutoff frequency) connected in parallel with the feedback resistor. This provides greater negative feedback hence lower gain for frequencies above the cutoff point. This frequency can be approximated, incidentally, by the formula

$$F = \frac{1}{2\pi R_f C} \qquad (17.2)$$

where

F   is the approximate cutoff frequency in hertz
R_f is the feedback resistance in Ohms
C   is the capacitance (in farads) connected **in** parallel with R_f

This type of filter is not, however, particularly effective; so one must look to another circuit if a high cutoff slope is required. Although there are several good filters, I am going to present only one form; so if you require more-detailed information on active filters please see one of the references found in the bibliography, or in the general electrical-engineering literature.

304

SOME MATH BACKGROUND

The transfer function of any electronic circuit will have the form

$$F(t) = \frac{G(t)}{H(t)} \qquad (17.3)$$

where

$G(t)$ is the output-voltage function of time, E_{out}
$H(t)$ is the input-voltage function of time, E_{in}

Considering the definitions of $G(t)$ and $H(t)$ in this context, I may rewrite Eq. 17.3 in the form

$$F(t) = \frac{E_{out}}{E_{in}} \qquad (17.4)$$

In this transfer function, as well as in many functions that describe complex electronic networks, there are several terms represented by each of these general expressions; so one must look more closely to learn the nature of a circuit.

In the transfer function that I will present for active filters in the next section of this chapter, as well as in many other equations used in electronics, there will appear a term designated by s. The term is derived from taking the Laplace transform of a time-dependent function that describes the same thing. Laplace transforms are a method for solving linear differential equations that would be either difficult or less useful if left in the time-domain form. The Laplace moves the expression to the frequency domain.

The definition of the Laplace transform is

$$L(t) = F(s) = \int_{o}^{z} e^{-st} F(t)\, dt \qquad (17.5)$$

The Laplace transforms for several common functions of t are:

$F(t)$	$F(s)$
1	$1/s$
t	$1/(s)^2$
t^n	$n!/(s)^{(n+1)}$
e^{at}	$1/(s-a)$
$\cos at$	$s/(s^2 + a^2)$
$\sin at$	$a/(s^2 - a^2)$

The mysterious s parameter is defined as:

$$s = j\omega = j2\pi f \qquad (17.6)$$

where

  j   is the imaginary operator $(-1)^{1/2}$
  ω   is the quantity of $2\pi f$.

The expressions for a reactance involves Eq. 17.6. The inductive reactance, for example, is $2\pi fL$, while the capacitive reactance is $1/2\pi fC$. In proper notation there would be a $+j$ in the inductive case and a $-j$ in the capacitive. This should tell you that the capacitive and inductive reactance formulae are merely the Laplace transforms of some time-domain equation. The expressions for the reactances in s notation are simply sL and $1/sC$. An expression for an impedance containing only capacitance is $1/sC$, the same expression as for the reactance.

"BETTER" ACTIVE FILTER DESIGNS

The type of active filter that I will now consider is also the most popular of the several types known. It is the two-pole Butterworth filter. The circuit to be considered will have unity gain except in the case of the bandwidth filter in which both unity-gain and greater-than-unity-gain cases are considered. This limitation is practical given the range of filter designs which you will need and greatly simplifies matters for us both. The Butterworth design is chosen because it offers maximally flat response within the passband, although at the expense of a slightly inferior rolloff slope.

Figure 17-1 shows a generalized circuit of the Butterworth using an operational amplifier and some impedances in the feedback loop. These will be resistors and capacitors, although in different designs different positions will be occupied by the specific impedances.

Notice that in Fig. 17-1 I am using admittance notation rather than classical impedance notation. This is a step that makes my analysis somewhat easier. This simplification is almost standard practice in some circuit-analysis textbooks. By definition, admittance (Y) is the reciprocal of impedance (Z), or

$$Y = \frac{1}{Z} \qquad (17.7)$$

The generalized transfer function for the two-pole-Butterworth-filter circuit (for all three of the pass configurations) is of the form

$$\frac{E_{out}}{E_{in}} = \frac{-Y1Y3}{(Y3Y4) + (Y1Y5) + (Y2Y5) + (Y3Y5) + (Y4Y5)} \qquad (17.8)$$

If the admittance is purely resistive, then $Y = 1/R$; and if it is purely capacitive, then $Y = sC$. I can, then, rewrite the general equation in terms

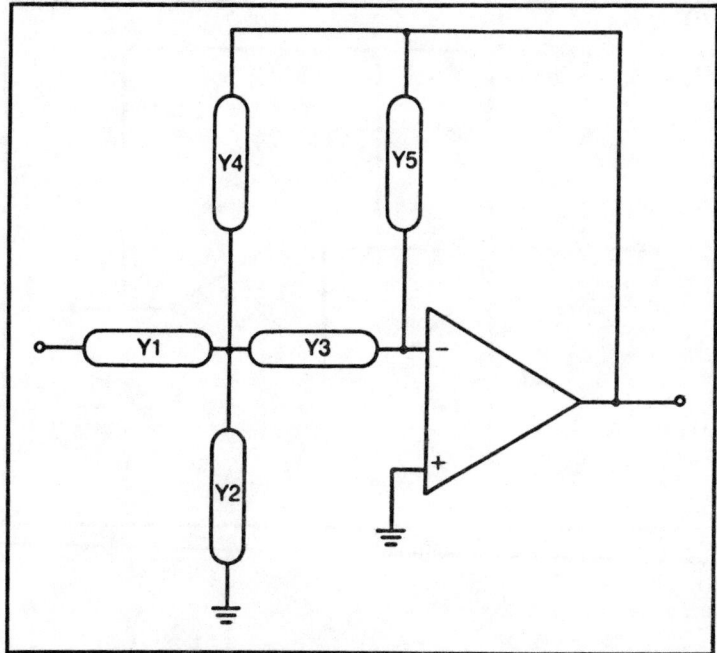

Fig. 17-1. General form of a two-pole Butterworth filter.

of R and C. One such general form stems from the expression (of Eq. 17.8)

$$\frac{-\omega_o^2}{-\omega^2 + j\,(2)^{1/2}\omega(\omega_o) + \omega_o^2} \tag{17.9}$$

ω_o is the frequency where the gain of the circuit is down 3 dB from the unity-gain frequency, ω.

A Low-Pass Filter

An example of a two-pole, low-pass, Butterworth filter is shown in Fig. 17-2. In this case the admittances become:

$$
\begin{array}{ll}
Y1 = 1/R1 & Y4 = 1/R3 \\
Y2 = sC1 & Y5 = sC2 \\
Y3 = 1/R2 &
\end{array}
$$

The general transfer function becomes:

$$f(s) = \frac{-\omega_o^2}{s^2 + (2)^{1/2}s\omega_o + \omega_o^2} \tag{17.10}$$

307

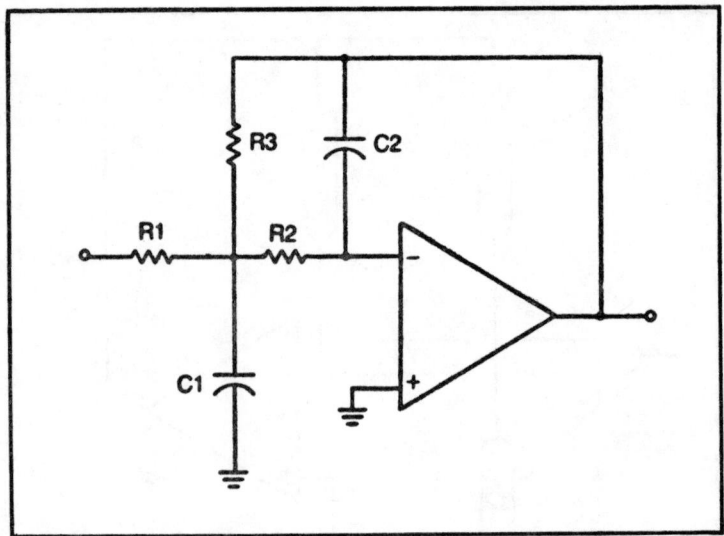

Fig. 17-2. Low-pass Butterworth filter.

or, if we use Eq. 17.8;

$$f(t) = \frac{(-1/R1R2C1C2)}{s^2 + (s/C1)(1/R1 + 1/R2 + 1/R3) + (1/R2R3C1C2)} \quad (17.11)$$

A circuit convention (a rule for which I offer no reason and no apology) is used to simplify this expression. I set certain of the admittances equal to or proportional to certain other admittances, or

$$R1 = 2R2$$
$$R3 = R1$$
$$C1 = 4C2$$

By comparison of Eqs. 17.10 and 17.11, I find that they are of the same form. Notice that (considering my conventions):

$$\left[\frac{s}{C1}\right]\left[\frac{1}{R1} + \frac{1}{R2} + \frac{1}{R3}\right] = \frac{4s}{R1\ C1} \quad (17.12)$$

and that this result is the same as $(2)^{1/2}s\omega_o$ in Eq. 17.10; so

$$\frac{4s}{R1C1} = (2)^{1/2}s\omega_o \quad (17.13)$$

308

I may, then, conclude that

$$\omega_o = \frac{4}{(1.414)(R1)(C1)} \tag{17.14}$$

$$2\pi f_o = 4/(1.414)(R1)(C1) \tag{17.15}$$

$$f_o = 0.45/R1C1 \tag{17.16}$$

where

f_o is the cutoff frequency in hertz
C1 is in farads
R1 is in ohms

Example

Design a two-pole, low-pass, Butterworth with a cutoff frequency of 500 hertz. Use Eq. 17.16 as a rule of thumb and my conventions.

In many design jobs we find it necessary to select some components seemingly out of midair, and then to calculate others. This is not the occult exercise it seems, but merely serves to set a starting point. If it turns out that the selections were not prudent, we can turn the calculator back on and make some new assumptions. It is generally easier to obtain so-called oddball resistor values; so set the capacitances equal to some standard values. In this case let us set $C1 = 0.015$ μF. This assumption will make $C2 = 4 \times 0.015$, or 0.06 μF. From Eq. 17.16

$$500 = 0.45/(R1(1.5 \times 10^{-8}))$$
$$R1 = 0.45/((5 \times 10^2)(1.5 \times 10^{-8}))$$
$$R1 = 6 \times 10^4 = 60 \text{ kilohms}$$

In summary,

$$C1 = 0.015 \text{ } \mu\text{F}$$
$$C2 = 0.06 \text{ } \mu\text{F}$$
$$R1 = 60 \text{ kilohms}$$

and from my conventions,

$$R2 = 30 \text{ kilohms}$$
$$R3 = 60 \text{ kilohms}$$

A High-Pass Filter

To make a high-pass filter of the same form, I need only reverse the roles of the resistances and capacitances found in the low-pass-filter circuit. This is shown in Fig. 17-3. Notice that resistors have become capacitors and capacitors have become resistors. The transfer function is

$$f(s) = \frac{-s^2}{s^2 + (2)^{1/2}\omega_o s + \omega_o^2} \tag{17.17}$$

By an analysis such as we made for the low-pass case, I derive,

$$f_o = \frac{0.225}{R2\ C1} \tag{17.18}$$

By convention

$$R2 = 4R1$$
$$C1 = C2$$
$$C3 = 2C1$$

Example

Design a high-pass, two-pole, Butterworth filter with a cutoff frequency of 500 hertz.

Set $C1 = 0.015\ \mu F$, so $C2 = 0.015\ \mu F$, and $C3 = 0.03\ \mu F$ by Eq. 17.18,

$$\begin{aligned}
R2 &= (0.225)/(5 \times 10^2)\ (1.5 \times 10^{-8}) \\
&= 3 \times 10^4\ \text{ohms} \\
&= 30,000\ \text{ohms}
\end{aligned}$$

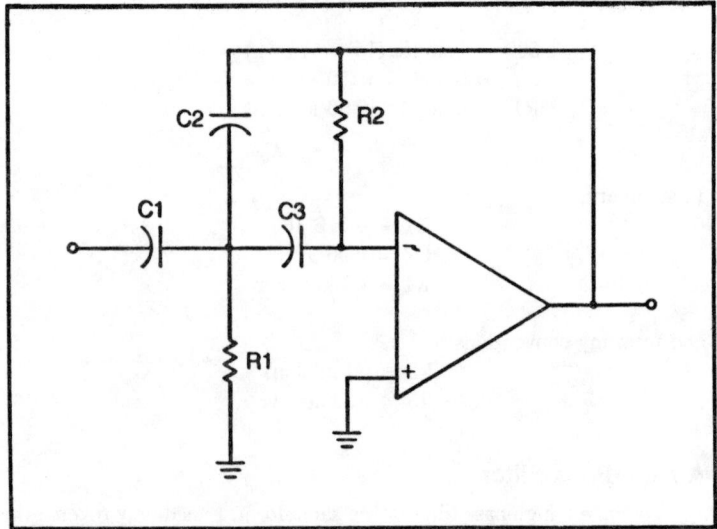

Fig. 17-3. High pass Butterworth filter.

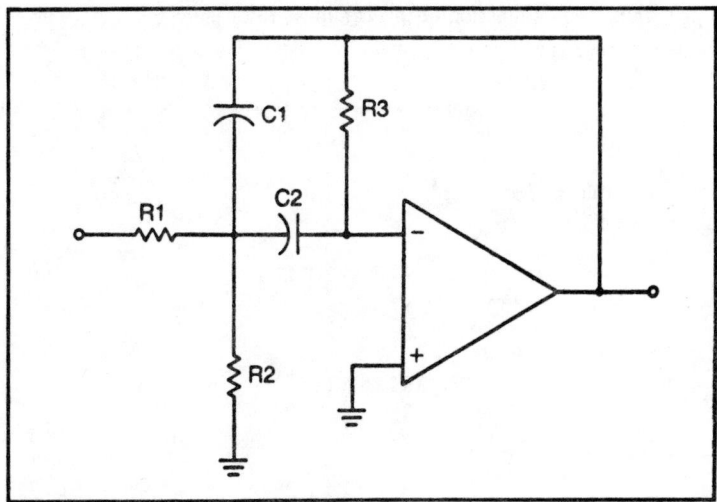

Fig. 17-4. Bandpass Butterworth filter.

By convention,

$$R1 = R2/4$$
$$= 7.5 \text{ kilohms}$$

In summary, then

$$R1 = 7500 \text{ ohms}$$
$$R2 = 30{,}000 \text{ ohms}$$
$$C1 = 0.015 \ \mu F$$
$$C2 = 0.015 \ \mu F$$
$$C3 = 0.03 \ \mu F$$

A Bandpass Filter

One way to make a bandpass filter is to use low-pass and high-pass stages in cascade, provided that both have the same gain. In fact, this approach seems to be the design of choice for many engineers. An example of a single-operational-amplifier bandpass filter is shown in Fig. 17-4.

The so-called "center frequency" is not really a center frequency but is the expression

$$f_o = (f_{ou}f_{ol})^{1/2} \qquad (17.19)$$

where

f_o is the center frequency
f_{ol} is the lower 3-dB-down point
f_{ou} is the upper 3-dB-down point

311

The transfer function of the bandpass filter is

$$f(s) = \frac{(-1/Q)\omega_o s}{s^2 + (\omega_o/Q)s + \omega_o^2} \qquad (17.20)$$

The rule-of-thumb equation is

$$f_o = \frac{0.159Q}{R1\ C1} \qquad (17.21)$$

and by convention,

$$R2 = ((R1R3)/(4Q^2R1) - R3)$$
$$R3 = 2R1$$
$$C1 = C2$$

Example

Design a bandpass filter with an upper 3-dB point of 525 hertz, and a lower 3-dB point of 475 hertz. Set $C2 = 1\ \mu F$ (this was found by trying several different capacitor values and then discarding them because one or more resistor values became ridiculous—it helps to know the stock of your local distributor, as well as the EIA list of standard values). By Eq. 17.19.

$$f_o = (\ (525)\ (475)\ )^{1/2}$$
$$= 500\ \text{hertz (approximately)}$$

By Eq. 17.1,

$$Q = (500)/(525 - 475)$$
$$= 500/50$$
$$= 10$$

By Eq. 17.21,

$$R1 = (0.159)\ (10)/(500)\ (1 \times 10^{-6})$$
$$= 3.18 \times 10^3\ \text{ohms}$$
$$= 3180\ \text{ohms}$$

By convention,

$$R3 = 6360\ \text{ohms}$$
$$C1 = 1\ \mu F$$

$$R2 = (\ (3180)\ (6360)\ )/(\ (4(10^2)\ (3180) - 6360)\ )$$
$$= 16\ \text{ohms}$$

These values are for unity gain, but if gains up to about 20 are desired, use the following equations

$$f_o = (0.159) (Q)/(A_v) (R1) (C2) \qquad (17.22)$$

$$R3 = 2(A_v) (R1) \qquad (17.23)$$

The other equations are the same in the gain filter. Be aware, though, that high-Q figures and high gain are not always totally compatible with ease of construction and good operation. In the case where problems (especially oscillation) occur, reduce gain and make it up in succeeding stages. This is a trade off made because Q is generally more critical in a filter application.

Chapter 18

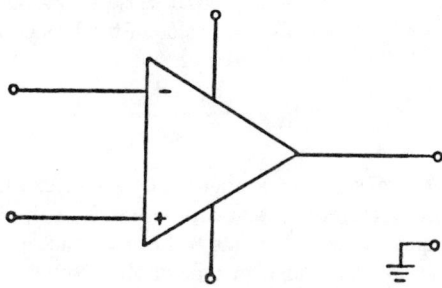

Some Circuit Problems

MANY BOOKS ON TOPICS SUCH AS THIS WILL GLIBLY TELL YOU TO DO this and that, show circuits that are oversimplified, and give great gobs of advice on something called design. But when the reader goes to actually connect these circuits and try them for himself, they just don't seem to perform to expectations. The reason is that most book circuits are based on a false premise: the use of *ideal* components. In this chapter I will discuss some of the more common circuit problems, particularly those of operational amplifiers. This study will probably explain to you the function of some components that are seemingly out of place.

DC OFFSET PROBLEMS

An offset voltage is one that exists at the output of an operational amplifier or other linear IC when the input voltage is zero. You should recall that the output of an operational amplifier should be zero when the differential-input voltage is zero. Two general sources account for most offset-voltage problems: input offset current and input offset voltage.

Input offset voltage creates a spurious output voltage due to an accumulation of voltage imbalances, junction potentials, and so forth. It can be measured by connecting a millivolt power supply across the inputs and then monitoring the output voltage. The input voltage required to force the output to zero is equal to the input offset voltage. The input offset may also be approximated, if the inputs are grounded, by

$$E_{oo} = E_{io} R_f \tag{18.1}$$

The other general source of output offset voltage is input offset current. This comes about as a result of the necessity for biasing the input transistors. Since there are two inputs, there will be two bias currents, I_{b1} and I_{b2}. These two currents flow in the same direction, into or out of the terminals, but since the inputs have opposite sense, they should be mutually cancelled in the output. The average value of the bias current is

$$I_{bias} = \frac{I_{b1} + I_{b2}}{2} \tag{18.2}$$

If these currents are equal, the average value will be exactly equal to either individual bias current. In that case $I_{bias} = I_{b1} = I_{b2}$; so the output voltages generated by these currents will exactly cancel.

Bias current will cause another type of offset voltage at the output. Consider the circuit in Fig. 18-1(A). It is a conventional inverting follower. Assuming E to be zero, you can view the circuit as if the input end of R_{in} were grounded. In this circumstance the output voltage should be zero, but it is not. Current I_{b1} sees the two resistors as a parallel circuit to ground; so a voltage is created at point A equal to

$$E_A = I_{b1} \left[\frac{R_{in} R_f}{R_{in} + R_f} \right] \tag{18.3}$$

This voltage is seen as an input signal and produces an output voltage equal to

$$E_o = I_{b1} R_f \tag{18.4}$$

The cheap cure for this is to apply an equal voltage of the same polarity to the opposite input so that cancellation in the output occurs. This is done by connecting a resistor, (R_c in Fig. 18-1(B), between the other input and ground.

You also have several other alternatives that are especially useful if input impedance is of great importance. You could, for example, use a premium-grade superbeta, JFET or BiMOS/BiFET input amplifier such as the CA3130, CA3140, CA3160, or LM156. Alternatively, you could use another operational amplifier as a current mirror, an example used to provide amplification in a noninverting circuit. This configuration normally has a high impedance, but with the addition of A2 it is even higher. The two operational amplifiers should be part of the same package. Examples are the 1458 (a dual op amp) (See Fig. 18-2) and the 324 (a quad op amp).

The input impedance of any circuit is given by an expression such as $Z = E_{in}/I$. In the normal noninverting follower I is I_b, but in this case it will be $(I_b - I_b)$, which will be close to zero. This, of course, makes the value of Z very high.

In years past one would take a matched pair of JFETs or a dual JFET

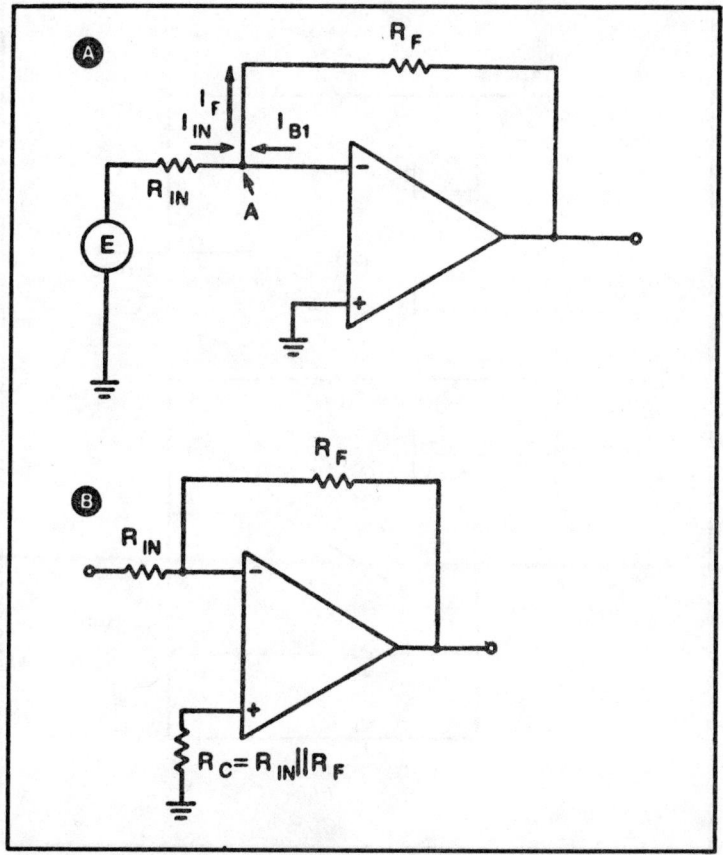

Fig. 18-1. At (A), currents at the summing node in a real op amp; (B), shows a compensation resistor to reduce the problem.

device and connect it into the input circuit of an ordinary operational amplifier. This greatly improves the input impedance, but it is now in disrepute as an effective circuit-design technique because it almost always deteriorates common-mode rejection. This makes the use of BiMOS or BiFET devices very attractive, and because they cost on the order of a dollar or two, they are also cost effective. The performance is better than the hybrid version.

Undesired output offset voltages can be eliminated, and desired offsets can be created, by the same types of circuit. Figure 18-3 shows several versions of the same theme. In all cases, though, I use a dc level shift to supply a counter current that bucks the offset.

In Fig. 18-3(A) I have the complete circuit, including the operational amplifier. In the remaining figures, however, the amplifier has been deleted in the interest of conserving space.

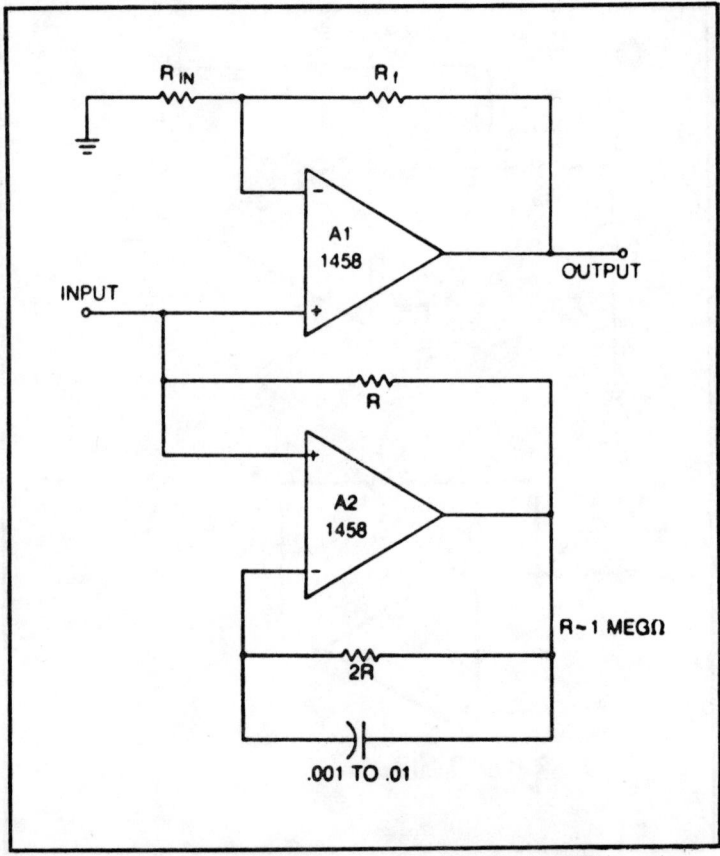

Fig. 18-2. Use of an op-amp current mirror to increase input-Z.

The simplest circuit, shown in Fig. 18-3(A), is also the most frequently used circuit. It consists of a single potentiometer and a fixed resistor. The voltage appearing at the wiper of the potentiometer will be either positive, negative, or zero, depending upon its position, because the opposite ends are connected to Vcc and Vee. This allows you to create a counter current of whichever polarity is required to cancel any offsets or to create a desired offset.

Even if R1 is a 10-turn potentiometer and R2 has a high value, the resolution of the circuit is limited. This is cured with varying degrees of success by the circuits of Fig. 18-3(B) through 18-3(D).

In Fig. 18-3(B) I use the expedient of adding resistors in series with each end of R1. Generally speaking, R1 is a 10-turn potentiometer and has a value equal to approximately 10 percent of R3 and R4. This makes the percentage change of resistance per turn of the pot much smaller than it was in Fig. 18-3(A).

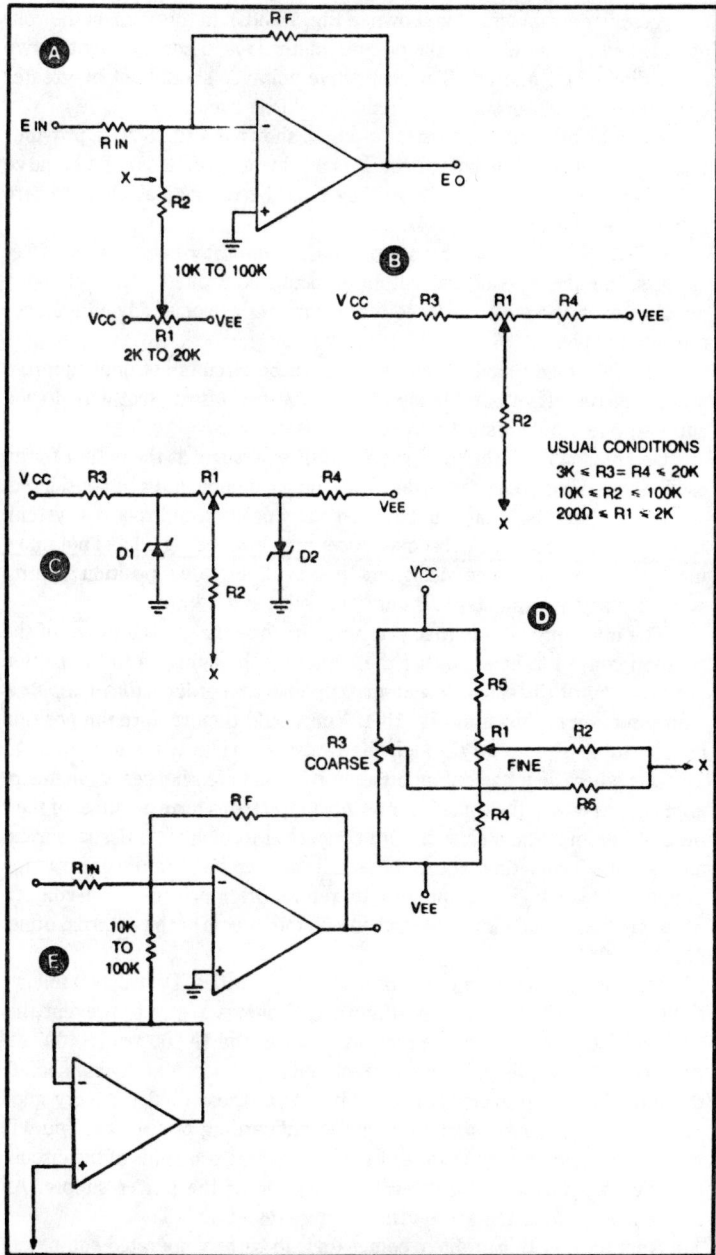

Fig. 18-3. At (A), the regular offset null; (B) . . . with better resolution (C) . . . even better resolution (D) . . . very good resolution; and (E), op-amp isolated offset null.

A similar arrangement is shown in Fig. 18-3(C). In this version the voltage at the opposite ends of the potentiometer is held constant by the two zener diodes, D1 and D2. The respective voltages should not be greater than about Vcc/2, or another circuit should be used.

A combination of two of these circuits, shown in Fig.18-3(D), provides coarse and fine control over offset. It is used where offsets might be large (taken care of by R3), yet where they must be reduced as close to zero as possible, the function of R1.

Figure 18-3(E) shows the use of a buffer amplifier between the offset circuits and the operational amplifier being controlled. It is relatively popular, even though I can't see that it imparts a special advantage over the other types.

Several times I have mentioned that these circuits not only suppress any undesired offset but will also create a *desired* offset. Just why do you suppose one would want to create an offset?

One reason might be to provide a position control at the output for an oscilloscope or strip-chart recorder. Two things, though, must be considered.

One is that it is possible to have too many position controls in a system. If this is true, then some subsequent user may lose the signal and not know how to regain it. Some designers make all but one position control screwdriver adjustments, and only one will have a knob.

The other problem is that you must arrange the overall range of the position control to be such that the signal can be adjusted for a position near the top of the scope screen or strip-chart recorder *without clipping*. This situation is shown in Fig. 18-4. You would then require the position control to be able to go through the range ±3 volts without clipping.

One other use for an output offset is in certain life-sciences or chemistry applications where the transducer or input electrodes form an offset of their own. When metallic electrodes are connected to biological tissue, for example, they form a tiny battery potential that can be a significant fraction of a volt. Physiological amplifiers, therefore, often provide a ±1-volt offset adjustment that allows the user to null the effects of the external offset potential.

Most IC devices will not tolerate reverse polarities. Destruction follows as surely as night follows day. Figure 18-5 shows a way to prevent this problem. Connect a 1-ampere rectifier diode of the 1N4000 to 1N4007 series in each power lead. If an incorrect polarity is accidentally connected, these diodes will prevent damage. The use of these diodes is very good practice during experimentation and breadboarding of circuits. You will most likely be mentally occupied by the electronic problems of the circuit, so may easily make a backwards connection of the power supply. All alligator clips look the same after 30 minutes—Carr's law.

Another way this problem comes up is in battery-operated equipment where it is possible to install the batteries backwards. Avoid battery types which have bidirectional holders. Use those which have, because of their design, little chance for incorrect installation. Also, if fairly large capacitors

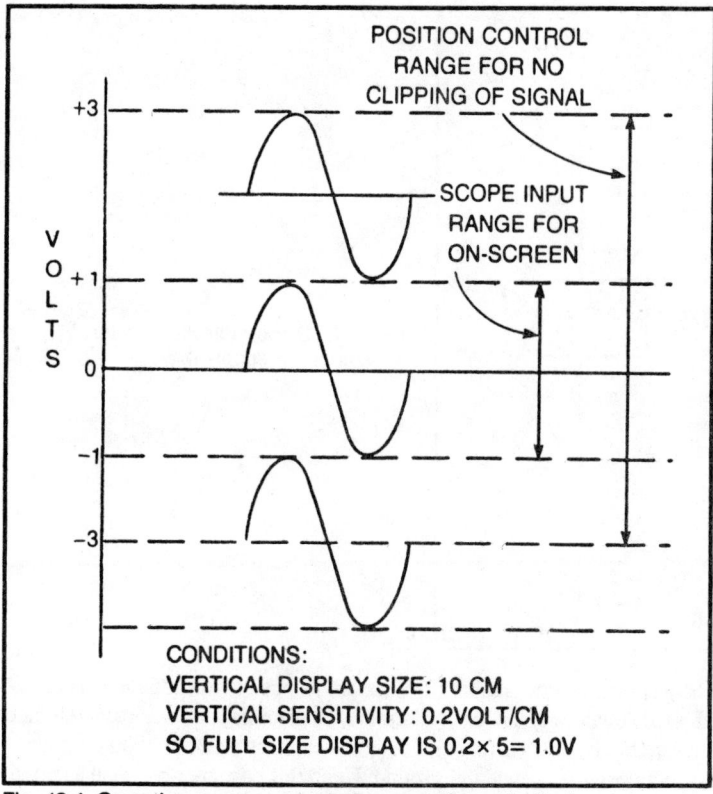

Fig. 18-4. Operating range required of an amplifier with display.

(over a few microfarads) are used at an IC input terminal, use the arrangement of Fig. 18-5. Sometimes, if Vcc and Vee supplies decay at different rates at turn off, the charge on these capacitors may well reverse-bias the IC substrate diode, causing a "mysterious" burn out of the device. In fact, this is a good thing to try if an IC connected to such a capacitor seems to burn out occasionally for no apparent reason.

OSCILLATIONS

Those IC devices with high-frequency response may oscillate if the correct circumstances are met:

1. In-phase (360°) signal feedback
2. Loop gain greater than unity

These criteria can exist because circuit phase shifts can add up to 360° (180° of which may be inherent in the IC design) at a frequency less than

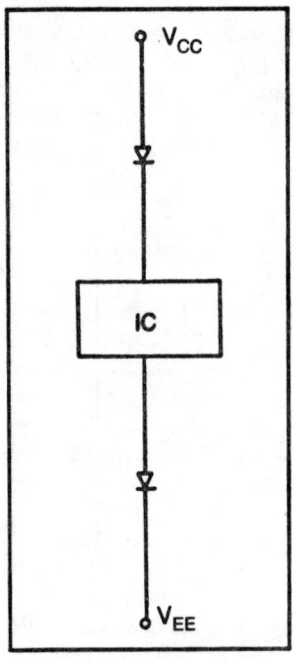

Fig. 18-5. Diode protection against reverse power supply flow.

the gain-bandwidth product (F_t is the frequency at which the gain is unity). In some devices such as comparators, the simple, but unfortunately not-always-totally-successful solution, is to connect a 100-to-1000-pF capacitor between the output and ground. Keep this capacitor as small as possible, though, because many devices object to highly capacitive loads.

The protocol for suppressing oscillations in any given IC device may depend upon the nature of the device. Consult the manufacturer's literature for details.

In operational amplifiers the details are general enough to warrant separate discussion. Figure 18-6 shows two common methods for reducing the gain in the frequency range where oscillations can occur.

In Fig. 18-6(A) I have the use of the frequency-compensation terminals found on many IC operational amplifiers. A small capacitor is connected between these terminals. It will have a value between about 10 pF and 1000 pF, with those in the 50-to-100-pF range being most common.

Another cheap 'n dirty method is shown in Fig. 18-6(B). Here I have connected a small-value (10-to-100-pF) capacitor across the feedback resistor. This is essentially a high-pass filter; so high frequencies are fed back more than low frequencies. The approximate gain of the stage will be dependent upon the impedance of the parallel combination of C1R_f:

$$A_v = \frac{R_f \parallel X_c}{R_{in}} \qquad (18.5)$$

So you can see from Eq. 18.5 how the gain drops with increasing frequency by considering the normal capacitive-reactance behavior; it decreases linearly with increased frequency.

There are a number of other techniques for modifying operational-amplifier frequency response. I refer you to any of the good books on operational amplifiers for details, including my *Op-Amp Circuit Design and Applications,* TAB book No. 787.

THERMAL DRIFT

Almost every type of electronic component exhibits changes in parameter with changes in temperature. The drift in operational amplifiers can be on the order of 1_0 to 10-$\mu V/°$ C, which may not seem like a lot at first glance. It becomes a lot more important, however, when high gain or precision is required.

External components also add measurably to the drift problem. This problem can, though, be minimized considerably by the use of low-temperature-coefficient (Don't you dare call it "tempco.") resistors.

Op-amp drift can be minimized by the use of premium-grade devices, or chopper circuits. Some operational amplifiers have been built with *internal* varactor-diode of JFET choppers inside.

Another technique that seems to work well except in the most critical applications is to take a good grade of micropower, BiFET-, or BiMOS-input operational amplifier (packaged in a metal can, please!) and do three things:

1. Operate the device at its lowest-possible voltages (e.g., ± 5 volts)

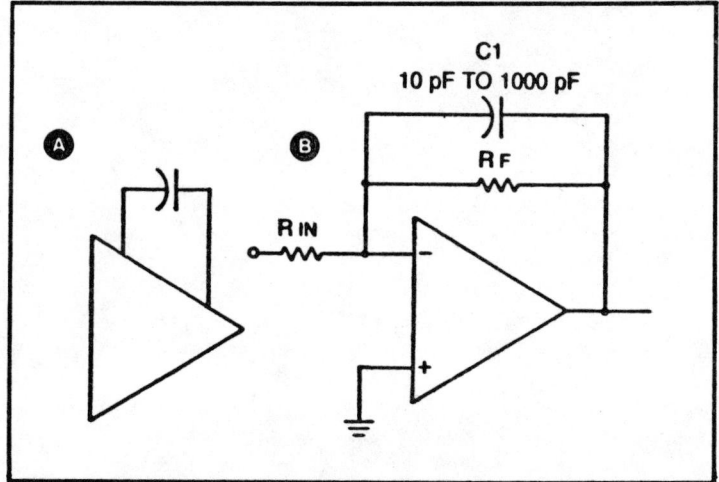

Fig. 18-6. At (A), a compensation capacitor; and (B), a capacitor shunted across feedback resistor.

2. Use a clip-on heat sink such as the type normally used on TO-5 transistor cases.
3. Use minimal gain (i.e., 10) in each of the first two stages.

These "fixes" should apply to the first two or three stages. Drift can also be reduced if the circuit board is thermally isolated. This state can be approximated by building the circuit inside a shielded metal can and then placing the can inside a plastic or Bakelite box. Some instruments such as pH, pO_2, and pCO_2 meters routinely do this for the first $\times 100$ or even $\times 1000$ of amplification. The input signal could be brought in either through coaxial cables or by mounting the box over the input connector.

Chapter 19

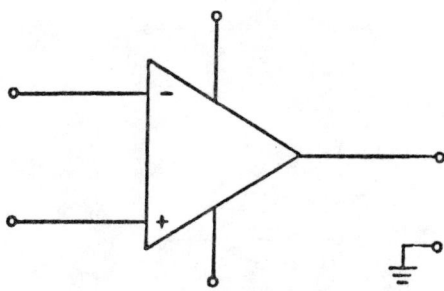

Making Physical
Measurements Electronically

M EASURING PHYSICAL PARAMETERS ELECTRONICALLY IS THE heart and soul of instrumentation in laboratories, factories, medicine, and many other areas. The idea is to acquire some sort of signal, either naturally or through a transducer, process it, and display or use it in some beneficial manner. Some of the material in this chapter is a review of earlier material and is presented in this context so that the discussion is complete. First, let's turn our attention to a review of physical transducers.

There are many different forms of transducer using resistive strain-gauge elements, and most of them are based on the Wheatstone-bridge circuit. Various physical parameters are measured with strain-gauge transducers, including force, displacement, vibration, and both liquid and gas pressure. If you are so unfortunate as to require intensive care hospitalization, the doctor may order continuous blood-pressure monitoring through an indwelling catheter inserted in an artery. The transducer used to measure your blood pressure will probably be a Wheatstone-bridge strain gauge. In this article, I will discuss how strain gauges work and how to make them work in practical cases. Later I will show you how to design the electronic amplifiers used for processing transducer signals.

PIEZORESISTIVITY

All conductors possess electrical resistance, which is opposition to the flow of current; resistance is measured in ohms. The resistance of any specific conductor is directly proportional to the length (see Fig. 19-1), and inversely proportional to the cross-sectional area. Resistance is also directly proportional to a property of the conductor material called "resistivity."

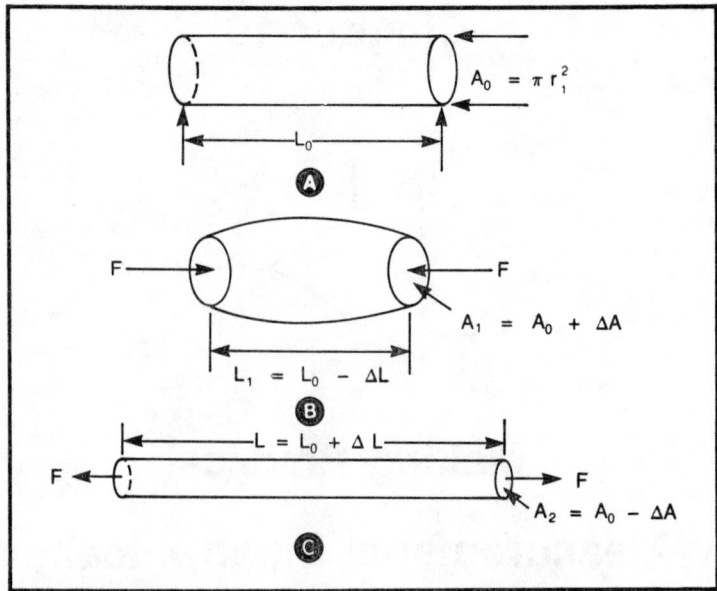

Fig. 19-1. Piezoresistive strain gauge (A) normal, at rest, (B) in compression, (C) in tension.

The relationship between length, area, and resistivity (P) is shown in Fig. 19-1.

The equation in Fig. 19-1 shows clearly that resistance is related to length and cross-sectional area. "Piezoresistivity" is merely a ten-dollar word that means the resistance changes when length and/or area are changed. Figure 19-1(A) shows a cylindrical conductor with a length L_o and a cross-sectional area A_o. When a compression force is applied, as in Fig. 19-1(B), the length reduces and the cross-sectional area increases. This situation results in an increase in the electrical resistance. Similarly, when a tension force is applied (Fig. 19-1(C)) the length increases and the cross-sectional area decreases; so the electrical resistance will increase. Provided that the change is small, the change of electrical resistance is a linear function of the applied force, so can be used to make measurements of that force.

STRAIN GAUGES

A strain gauge is merely a piezoresistive element, either wire, metal foil, or semiconductor, designed to create a resistance change when a force is applied. Strain gauges can be classified as either bonded or unbonded types. Figure 19-2 shows both methods of construction.

The unbonded strain gauge shown in Fig. 19-2(A) consists of a wire resistance element stretched taut between two flexible supports. These supports are configured in such a way as to place a tension or compression force on the taut wire when external forces are applied. In the particular example shown, the supports are mounted on a thin metal diaphragm that flexes when a force is applied. Force F1 will cause the flexible supports to spread apart, placing a tension force on the wire and increasing its resistance. Alternatively, when force F2 is applied, the ends of the supports tend to move closer together, effectively placing a compression force on the wire element and thereby reducing its resistance. In actuality, the wire's resting condition is tautness, which implies a tension force. So F1 increases the tension force from normal, and F2 decreases the normal tension.

The bonded form of strain gauge is shown in Fig. 19-2(B). In this type of device a wire, foil, or semiconductor element is cemented to a thin metal diaphragm. When the diaphragm is flexed, the element deforms to produce a resistance change.

The linearity of both types of strain gauge can be quite good, provided that the elastic limits of the diaphragm and element are not exceeded. It is also necessary to ensure that the change of length is only a small percentage of the resting length.

In the past, the "standard wisdom" was that bonded strain gauges are more rugged but less linear than unbonded models. Although this may have been the situation at one time, recent experience has shown that modern manufacturing techniques can produce linear, reliable units of both types of construction.

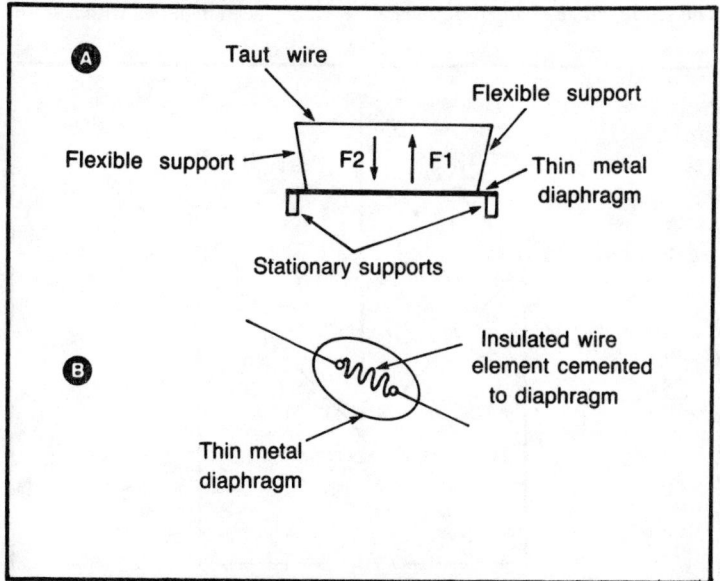

Fig. 19-2. Strain gauges (A) unbonded, (B) bonded.

THE WHEATSTONE BRIDGE

The Wheatstone bridge is a 19th-century holdover that finds a home in many modern electronic circuits. The classic form of bridge is shown in Fig. 19-3. There are four resistive arms to the bridge, labeled R1, R2, R3, and R4. The excitation voltage (V) is applied across two of the nodes, while the signal is taken from the alternate two nodes (labeled A and B). You can consider this circuit as two series voltage dividers in parallel, one consisting of R1 and R4 and the other of R2 and R3.

The output voltage from a Wheatstone bridge is the difference between the voltages at points A and B. When all of the arithmetic is finished, you find that the output voltage will be zero when the ratio R4/R1 is equal to the ratio R3/R2. If these ratios are not kept equal, as is the case when one or more elements is a strain gauge, then an output voltage is produced that is proportional to both the applied voltage and the change of resistance.

STRAIN-GAUGE CIRCUITRY

Before the resistive strain gauge (or other form of resistive transducer) can be useful it must be connected into a circuit that will convert its resistance changes into a current or voltage output. Most applications are voltage-output circuits.

Figure 19-4 shows several popular forms of circuit. The circuit in Fig. 19-4(A) is both the simplest and least useful (although not unuseful!); it is sometimes called the "half-bridge" circuit or "voltage-divider" circuit. The strain-gauge element of resistance, R, is placed in series with a fixed

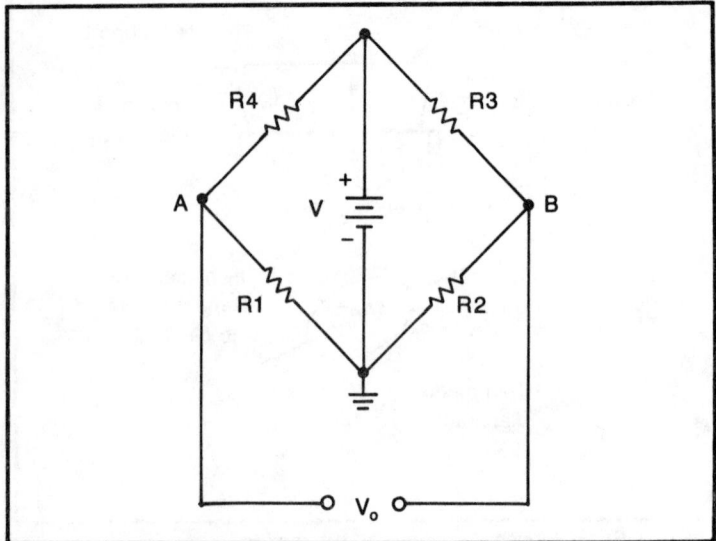

Fig. 19-3. Wheatstone bridge.

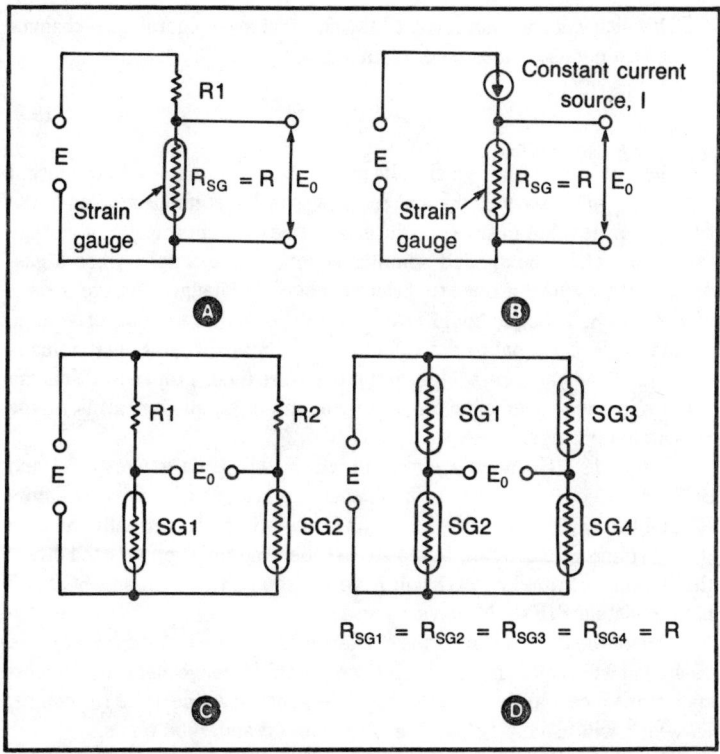

Fig. 19-4. Thermistor circuits: (A) Normal halfbridge, (B) CCS halfbridge, (C) two-device bridge, (D) full bridge.

resistor, R1, across a stable dc voltage, E. The output voltage E_o is found from the simple voltage-divider equation:

$$E_o = \frac{ER}{R + R1} \quad (19.1)$$

Equation 19.1 describes the output voltage, E_o, when the transducer is at rest (i.e., nothing is stimulating the resistive element). When the element is stimulated, however, its resistance changes a small amount, h. The output voltage in that case is:

$$E_o = \frac{E(R + h)}{(R \pm h) + R1} \quad (19.2)$$

Another form of half-bridge circuit is shown in Fig. 19-4(B), but in this case the strain gauge is connected in series with a constant-current source

(CCS), which will maintain current I at a constant level regardless of changes in the strain-gauge resistance. In this case,

$$E_0 = I (R \pm h) \tag{19.3}$$

Both of the half-bridge circuits suffer from one major defect: output voltage E_0 will always be present regardless of the stimulus applied to the transducer. Ideally, in any transducer system, we want the output voltage to be zero when the applied stimulus is zero. For example, when a gas-pressure transducer is open to the atmosphere, the gauge pressure is zero; so the output voltage should also be zero. Secondly, the output voltage should be proportional to the value of the stimulus when the stimulus is not zero. A Wheatstone-bridge circuit can have these properties. You can use strain-gauge elements for one, two, three, or all four arms of the Wheatstone bridge.

Figure 19-4(C) shows a circuit in which two strain gauges (SG1 and SG2) are used in two arms of a Wheatstone bridge, with fixed resistors R1 and R2 forming the alternate arms of the bridge. It is usually the case that SG1 and SG2 are configured so that their motions oppose each other; that is, under stimulus, SG1 will have resistance $(R + h)$, and SG2 will have resistance $(R - h)$ or vice versa.

One of the most linear forms of transducer bridge is the circuit of Fig. 19-4(D) in which all four bridge arms contain strain-gauge elements. In most such transducers all four strain-gauge elements have the same resistance, R, which will usually be a value between 100 and 1000 ohms.

Recall that the output from a Wheatstone bridge is the difference between the voltages across the two half-bridges. You can calculate the output voltage for any of the standard configurations from the equations given in Fig. 19-5. Let's work an example for a bridge with all four arms active (similar to Fig. 19-4(D)).

Example

A force transducer is used to measure the weight of small objects. It has a resting resistance of 200 ohms, and an excitation potential of +5 volts dc is applied. When a 1-gram weight is placed on the transducer diaphragm, the resistance of the arms changes by 4.1 ohms. What is the output voltage?

$$
\begin{aligned}
E_0 &= Eh/R \\
&= (5 \text{ volts}) (4.1 \text{ ohms})/(200 \text{ ohms}) \\
&= 20.5/200 \\
&= 0.103 \text{ volts, or } 103 \text{ millivolts}
\end{aligned}
$$

TRANSDUCER SENSITIVITY

Although there is some practical use for the example worked, most

---

I. One active element

$$E_o = Eh/4R$$

II. Two active elements

$$E_o = Eh/2R$$

III. Four active elements

$$E_o = Eh/R$$

where:
E_o is the output voltage (volts)
E is the excitation voltage (volts)
R is the resting resistance of the bridge ARMS
h is ΔR, i.e. change of R when stimulus is applied

Fig. 19-5. Bridge transfer functions.

readers will probably work with a transducer for which the sensitivity is known. The sensitivity factor (P) relates the output voltage (E) to the applied-stimulus value (Q) and excitation voltage. In most cases, the transducer maker will specify a number of microvolts (or millivolts) output potential per volt of excitation potential per unit of applied stimulus. In other words:

$$P = E_o/E/Q_o \qquad (19.4)$$

or, written another way:

$$P = \frac{E_o}{V \times Q} \qquad (19.5)$$

where
E_o is the output potential.
V   is the excitation potential.
Q   is one unit of applied stimulus.

If I know the sensitivity factor, then I can calculate the output potential as follows:

$$E_o = PEQ \qquad (19.6)$$

The equation above is the one that is most often used for circuit designing. Let's work a practical example. A certain fluid-pressure transducer is often used for measuring human and animal blood pressures through an indwelling catheter. It has a sensitivity (P) of 5 μV/V/T, which means "5 microvolts output potential is generated per volt of excitation potential per torr of pressure (note: 1 T = 1 mmHg). Find the output potential when the excitation potential is +7.5 volts dc and the pressure is 400 torrs (the usual high-end limit for such transducers).

$$E_0 \quad = \quad PEQ$$

$$= \quad (5 \ \mu\text{V}) \times (7.5 \ \text{V}) \times (400 \ \text{T})$$

$$= \quad (5 \times 7.5 \times 400) \ \mu\text{V}$$

$$= \quad 15{,}000 \ \mu\text{V} \ (15 \ \text{millivolts, or } 0.015 \ \text{volts})$$

BALANCING AND CALIBRATING A BRIDGE TRANSDUCER

Few, if any, Wheatstone-bridge transducers meet the ideal condition in which all four bridge arms have exactly equal resistances. In fact, the bridge resistance specified by the manufacturer is only a nominal value, and the actual value may vary quite a bit from the specified value. There will inevitably be an offset voltage (i.e., E_0 is not zero when Q is zero). Figure 19-6 shows a circuit that will balance the bridge when the stimulus is zero.

Potentiometer R1 is usually a precision type with five to fifteen turns to cover the entire range. Alternatively, it is a one-turn potentiometer ganged to a multi-turn vernier dial. The purpose of the potentiometer is to inject a balancing current (I) into the bridge circuit at one of its nodes. R1 is adjusted, with the stimulus at zero, for zero output voltage.

Another application for this type of circuit is injecting an intentional offset potential. For example, on the Heathkit digital scale such a circuit is used to adjust for the "tare weight" of the scale, which is the sum of the platform and all other weights acting on the transducer when nobody is standing on the scale. This is also sometimes called "empty-weight compensation." Being a heavy dude, I always liked the laundry scales in a hospital where I worked. They had a tare weight of −55 lbs to compensate for the weight of empty laundry containers; so I appeared to weigh 55 lbs less than was true! In such a case, a potentiometer is adjusted for zero when the container that would hold the laundry (or whatever) is placed on the scale.

Calibration can be accomplished either the hard way or the easy (and less accurate way). The hard way is to set the transducer up in a system and apply the stimulus. The stimulus is measured and the result is compared with the transducer output. For example, if you are testing a pressure

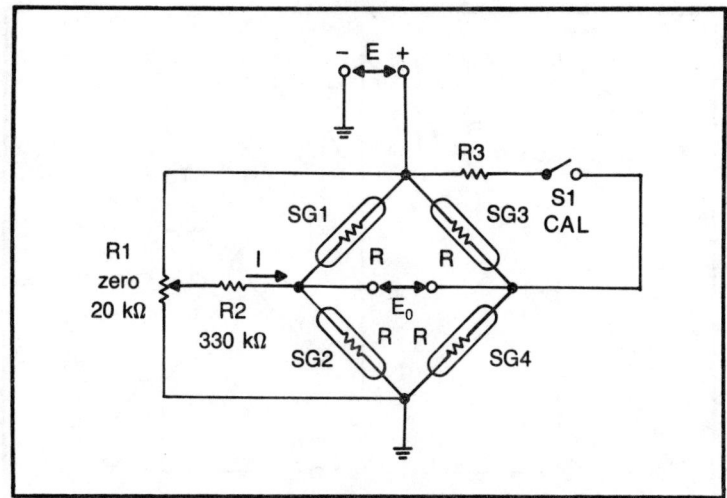

Fig. 19-6. Practical fullbridge circuit with offset balance.

transducer, connect a manometer (pressure-measuring device containing a column of mercury) and measure the pressure directly. The result is compared with the transducer output. All transducers should be tested in this manner when placed in service and periodically thereafter.

In less critical applications, however, you can connect a calibration resistor to synthesize the offset and thereby allow the electronics to be calibrated. The resistor and CAL switch (S1) in Fig. 19-6 is used for this purpose. Resistor R3 should have a value of:

$$R3 = [(R/4QP) - R/2] \qquad (19.7)$$

In the equation above you express the output voltage from the sensitivity factor (P) as volts instead of microvolts.

TRANSDUCER AMPLIFIERS

Figure 19-7 shows the basic Wheatstone bridge with a few extra resistors added for special purposes. The bridge elements "R" are the piezoresistive elements of the transducer. The electrical resistance of these elements changes proportionally to the applied stimulus (force, pressure, displacement, etc). Resistor R1 and thermistor RT1 are used to adjust the sensitivity of the bridge, while R2 and RT2 are used to adjust the dc offset.

Temperature variations will cause changes in the base resistance of all "R" elements, so will cause changes in both sensitivity and offset. Temperature compensation is provided by RT1 and RT2. In most commercial transducers these thermistors are provided by the manufacturer and are "transparent" to the user.

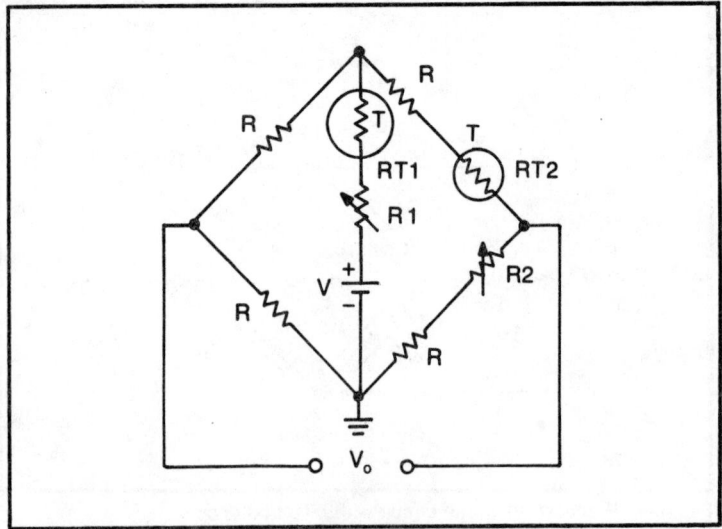

Fig. 19-7. Temperature compensated bridge circuit.

Potentiometers R1 and R2 are used to trim out variations in the normal sensitivity and offset due to differences in the "R" elements. In some cases, these potentiometers will be included by the maker of the transducer, while in others the leads are brought out from the transducer, and you must provide the pots. In still other cases, and this includes most modern transducers, the resistors are actually fixed resistors inside the transducers and are hand-selected by the maker to trim performance. In that case, the resistors are transparent to the user.

Dc Excitation Sources

The Wheatstone-bridge transducer requires a source of ac or dc excitation voltage, with dc being the most common. Most transducers require an excitation voltage of 10 volts dc or less. This voltage is critical, and exceeding it will create a very short life expectancy for the transducer. A typical fluid-pressure transducer requires +7.5 volts dc and operates best (least thermal drift) at +5 volts dc. You must provide a source of dc that is stable, within specifications, and (in some cases) precise.

The simplest form of transducer excitation is the zener-diode circuit in Fig. 19-8(A). A zener diode will regulate the voltage to a close tolerance that is sufficient for many applications. There are two problems with this circuit, however. First is the fact that the zener potential is not a nice even value like 5.0 volts but will have a value such as 4.7, 5.6, 6.2, or 6.8 volts. The second defect is thermal drift. The zener voltage will vary somewhat with temperature in all but certain reference-grade zener diodes. Unless the application is not critical or the diode can be kept at a constant

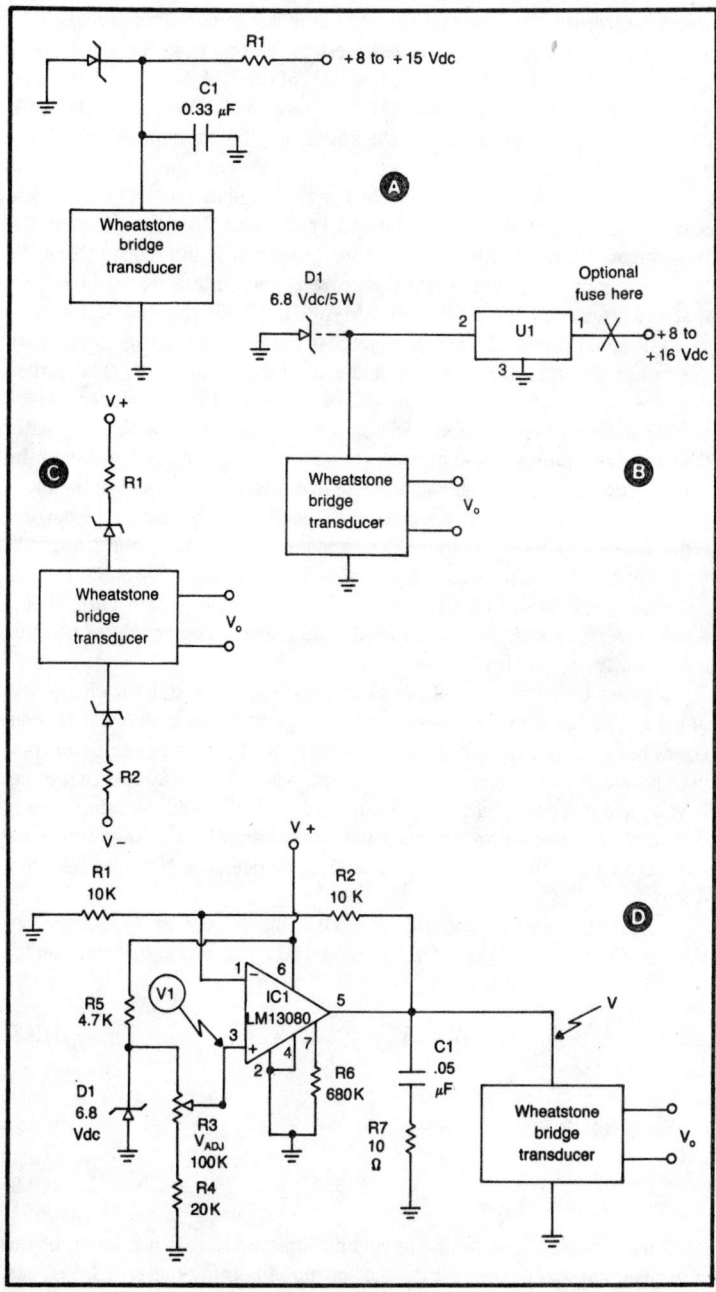

Fig. 19-8. Bridge excitation power supply: (A) zener version, (B) regulator version, (C) improved zener version, (D) best version.

temperature, the method of Fig. 19-8(A) is not generally suitable.

Figure 19-8(B) shows a second method. In this case the regulator is a three-terminal, IC voltage regulator (U1) of the LM-309, LM-340, 78xx, or similar families. In many cases the "H" version of the regulator (100-mA) can be used, although in others the 750-mA, "T" package devices must be specified. The selection depends upon the current normally drawn by the transducer, which is (V/R) where V is the regulator output voltage and R is the resistance of any one transducer element. In a typical case, the transducer will use a +5 volt excitation potential. If the resistance of the "R" elements is 50 ohms, then the current will be less than 100 mA. In that case, you can use a 100-mA LM-309H, LM-340H, and so forth.

The zener diode (D1) in Fig. 19-8(B) is not used for voltage regulation, but rather for protection of the transducer. If the regulator (U1) fails, then +8 to +16 volts from the V+ line will be applied to the transducer—which is fatal! The purpose of D1 is to clamp the voltage to a value that is greater than the excitation voltage but less than the "groan-voltage" rating of the transducer. In some cases, a small fuse is inserted in series with the input (pin no. 1) of U1. The value of this fuse is set to roughly twice the current requirements (V/R) of the transducer and will blow if the zener-diode voltage is exceeded. The fuse will add a certain amount of protection.

Some applications require a dual-polarity power supply. Figure 19-8(C) shows a version in which two zener diodes are used, one each for positive and negative polarities.

Neither the zener circuit nor the three-terminal regulator circuit will deliver precise output voltages. The voltage will be stable (that is, constant) but not precise. A typical three-terminal, IC voltage-regulator, output voltage, for example, may vary several percent from sample to sample. If you need a precise voltage, a circuit such as Fig. 19-8(D) might be used. This circuit is basically a standard operational-amplifier voltage-reference circuit in which the op-amp is a high-current model, the National Semiconductor LM-13080.

The output voltage from Fig. 19-8(D) is determined by R1, R2, the setting of R3, and the value of zener diode D1. The voltage V will be:

$$V = V_1 \times \frac{R2}{R1} + 1 \tag{19.8}$$

or, since R1 = R2 = 10 kohms,

$$V = 2 \times V_1 \tag{19.9}$$

The voltage V_1, at the noninverting input of IC1, is a fraction of the zener voltage that depends upon the setting of potentiometer R3. You can adjust V_1 from 1.13 to 6.8 Vdc; so the transducer voltage, V, can be set at any value from 2.26 to 13.6 volts dc. In most cases, you would probably set V at 5.00 volts, 7.50 volts, or 10.00 volts depending upon the nature of the transducer.

The basic dc differential amplifier is the most commonly used circuit for amplifying transducer signals. Fortunately, such amplifiers are easily constructed from simple operational amplifiers; Fig. 19-9 shows such a circuit. Assuming that R1 = R2 and R3 = R4, the gain of the amplifier will be equal to R4/R2 or R3/R1. The amplifier output voltage will be found from:

$$V_o = V_{in} \times R3/R1 \qquad (19.10)$$

where

V_o is the amplifier output voltage.
V_{in} is the transducer output voltage.
R1
and
R3   are the resistors in the amplifier circuit.

The amount of gain required from the amplifier is determined from a Scale Factor, SF, which is the ratio between the voltage representing full-scale at the output of the amplifier and the voltage representing full-scale at the output of the transducer.

$$SF = \frac{\text{voltage } V_o, \text{ representing full scale}}{\text{transducer-output voltage } V_{in}, \text{ representing full scale}} \qquad (19.11)$$

You may remember that the transducer-output voltage is found from the excitation voltage, the applied stimulus, and the sensitivity factor (P). The sensitivity factor is given in terms of millivolts (or microvolts) output per volt of excitation potential per unit of applied stimulus:

$$P = \frac{V_o}{V \times Q}$$

The output voltage from the transducer is therefore found from

$$V_{in} = P \times V \times Q \qquad (19.12)$$

where

V_{in} is the transducer-output voltage (or amplifier-input voltage!).
V is the excitation voltage.
Q is the applied stimulus (force, pressure, etc.). Let's again work a practical example.

Example

I once worked in a hospital/medical school where one project required a 0-to-100-torrs fluid-pressure transducer/amplifier. The transducer was rated with a sensitivity factor (P) of 50 μV/V/torr and a range of -200 to $+1200$ torrs. The available excitation source was 6.95 volts. At the full scale required (1000 torrs) the output voltage would be:

$$V_{in} = P \times V \times Q$$

$$= (50 \ \mu V) \times (6.95 \ V) \times (1000 \ T)$$

$$= (50 \ \mu V) \times (6.95) \times (1000)$$

$$= 347,500 \ \mu V \ (347.5 \ millivolts)$$

The amplifier-output voltage required will depend upon the desired display method. For example, a strip-chart recorder might have a voltage range of 0.5 volts, 1.0 volt, or some such value. In my case, I needed a digital-panel meter for the output display. Most low-cost DPMs have a 0-to-1999-millivolt range; so you gain a great deal of utility by making the output voltage at full scale numerically the same as the DPM reading; for example, 1000 torrs being represented by 1000 millivolts. In that case, the DPM scale factor would be 1 mV/torr, which is easy for humans (even life-sciences Ph.D's who were my customers!) to read. In that case, the "voltage V_o, representing full scale" in the equation above is 1000 mV. The gain of the amplifier is the scale factor, SF, described earlier:

$$SF = \frac{\text{voltage } V_o, \text{ representing full scale}}{\text{transducer-output voltage } V_{in}, \text{ representing full scale}}$$

$$= \frac{1000 \ mV}{347.5 \ mV}$$

$$= 2.878$$

Thus, a gain (SF) of 2.878 will provide the needed gain; so the ratio R3/R1 in Fig. 19-9 must be 2.878.

A PRACTICAL PROJECT

Now that I have gotten past the theoretical considerations, let's consider a practical project; see Fig. 19-10. This circuit is designed for a dc-excited Wheatstone-bridge transducer and is an outgrowth of the circuit of Fig. 19-9.

The transducer has a connector, P1, and will probably connect to the amplifier through a cable to jack J1. In most cases, the shielded housing

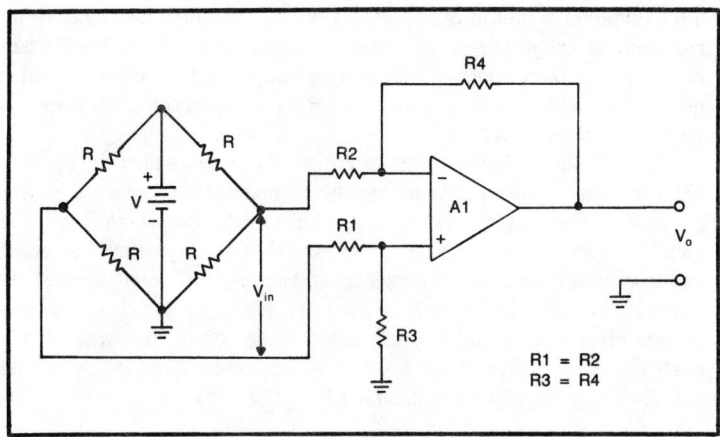

Fig. 19-9. Wheatstone bridge amplifier.

of the transducer will be connected to the amplifier through a separate conductor, so will have to be grounded at the amplifier end. Because I am using a single-polarity power supply, the − EXC voltage terminal (pin D) is also grounded (see J1, pins D & E).

The excitation voltage + EXC is applied through pin A. In this circuit, the current for the transducer was less than 100 mA (the "R" value was 200 ohms); so an LM-309H, three-terminal, IC regulator was selected for a potential of + 5.00 volts.

There are two stages of amplification in this circuit, A1 and A2. Amplifier A2 is simply a post-amplifier, so will have a gain of one. This tactic makes it easier to control gain selection (R6) and zero (R8). The overall gain, when R6 is maximum, is set by R3/R1, assuming that R1 = R2 and R3 = R4 (the same as in Fig. 19-9). The output voltage V_o is given by the product of the sensitivity factor (P), the excitation voltage (5 volts), and the applied stimulus (Q).

$$V_o = 5 \times P \times Q$$

In the example given earlier the output would be

$$V_o = 5 \times 50 \ \mu V \times 1000 \ T$$

or 250,000 μV (250 mV) at full scale (1000 torrs).

Potentiometer R6 is used to set the gain and consists of a 10-kilohm potentiometer and a 2-kilohm fixed resistor. This combination allows you to adjust the output voltage from 0.2 to 1.2 of full-scale. With the potentiometer you are able to account for errors in gain and transducer-output voltage. Similarly, potentiometer R8 is used to adjust the output voltage to zero when the applied stimulus is zero. In the previous example, I wanted

339

a 0-to-1000-torrs pressure amplifier. I would open the transducer to atmosphere (a gauge pressure of 0 torrs) and adjust R8 for 0.00-volts output. I would then apply a 1000-torr pressure (and measure it with a manometer) and adjust potentiometer R6 for an output of 1.000 volts (i.e. 1000 millivolts).

Wheatstone-bridge transducers are widely known and easily applied. With the circuits above you are capable of making your own transducer amplifiers at greatly reduced prices. The scientist for whom I built the amplifier wanted to publish the circuit in a physiology journal. I objected on the grounds that it was so simple and that non-professional, novice hobbyists could easily design such an amplifier. His reply was that hobbyists are not life scientists and that life scientists do not know how to do this neat trick—they would spend $900 from their research budget to buy an amplifier only slightly more complex than Fig. 19-10!

ELECTRONIC INTEGRATORS AND DIFFERENTIATORS

The author of a popular book on operational amplifiers claimed the op-amp made ". . . the contriving of contrivances a game for all," which means that the operational amplifier simplifies design so that almost anyone can do it. The operational-amplifier integrator and differentiator circuits fall easily into that category. I will now explore the active integrator and differentiator circuits and how to design them for most common cases. These circuits are central to many electronic-instrumentation designs.

What are integrators and differentiators? The names come from the time when op-amps were intended almost solely as building blocks in analog computers and refer to the mathematical operations of integration and differentiation. These terms are from the Calculus, but don't let that scare you—their operation is duck-soup simple.

Differentiation is the art (or science?) of finding the derivative of a curve, which is its rate of change. For the simplest case, a straight line as shown in Fig. 19-11(A), the derivative is simply the slope of the line, or (Y2-Y1)/(X2-X1). In this case, we usually write the expression for the slope with the Greek letter delta (i.e. delta-Y over delta-X) to indicate a small change in Y and a small change in X. For the case of a straight line, the derivative is simple to calculate. But in electronics, you very frequently encounter a situation where the line is not so straight, as in Fig. 19-11(B) where you see a voltage that varies with time. Here the curve is really a curve, it is not the simple straight-line case. If you want to know the instantaneous rate of change, i.e. the rate of change at a point, you can take the derivative of a line tangent to that point.

Integration is the inverse of differentiation and is used to find the area under a curve. In electronics, you might want to find the area under a time-varying voltage curve. In Fig. 19-12 you see a voltage that represents a pressure-transducer output, in this particular case the output of a human-blood-pressure transducer. Notice the pressure/voltage varies with time

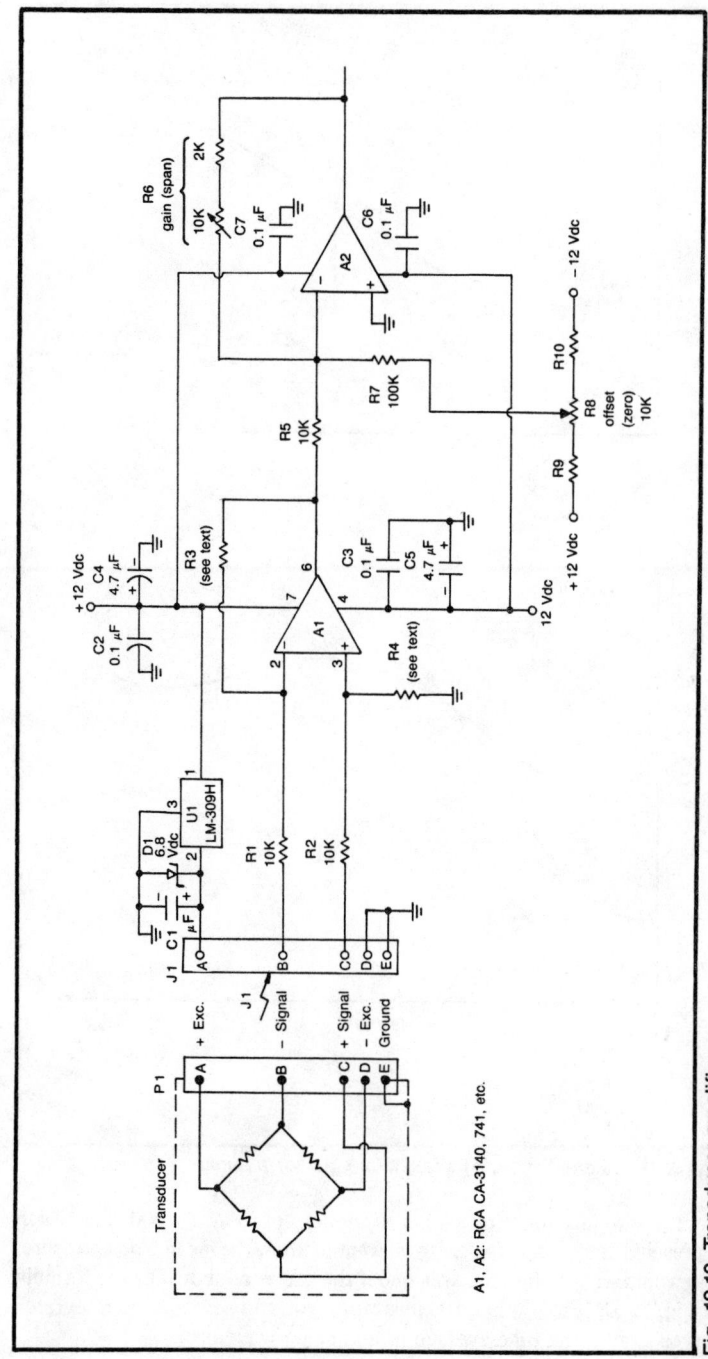

Fig. 19-10. Transducer amplifier.

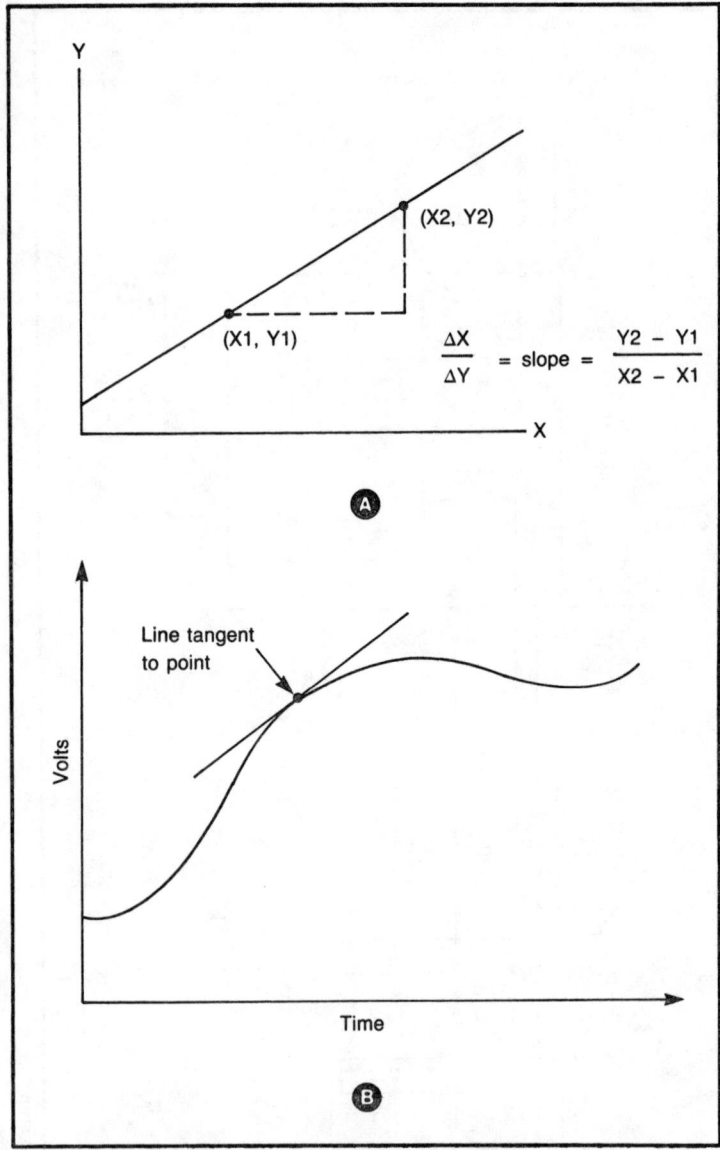

Fig. 19-11. Differentiation (A) straight line, (B) for a curve.

from a low (diastolic) to a high (systolic) between T1 and T2, which represents one cardiac cycle. If you want to know the mean blood pressure, you would want to find the area under the curve, as shown by the formula in Fig. 19-12. From this little illustration you can see that the integrator serves to find the time-average of an analog waveform.

Integrators and differentiators affect signals in different ways. Figure 19-13 shows the example of a square wave (A) applied to the inputs of an integrator (B) and differentiator (C). At time T1 the square wave makes a positive-going transition to maximum amplitude. At this time it has a very high rate of change; so the output of the differentiator is very high (see waveform C at T1). But then the amplitude of the input signal reaches maximum and remains constant until T2, when it drops back to its previous value. Thus, the differentiator will produce a sharp positive-going spike at T1 and a sharp negative-going spike at T2. These spikes are frequently used in circuits such as timers and zero-crossing detectors.

The integrator waveform in Fig. 19-13 shows a constant positive-going slope between T1 and T2. The steepness of the slope is dependent upon the amplitude of the input square wave, but the line is linear. You can see from curve B in Fig. 19-13 that the square wave into the integrator produces a triangle waveform.

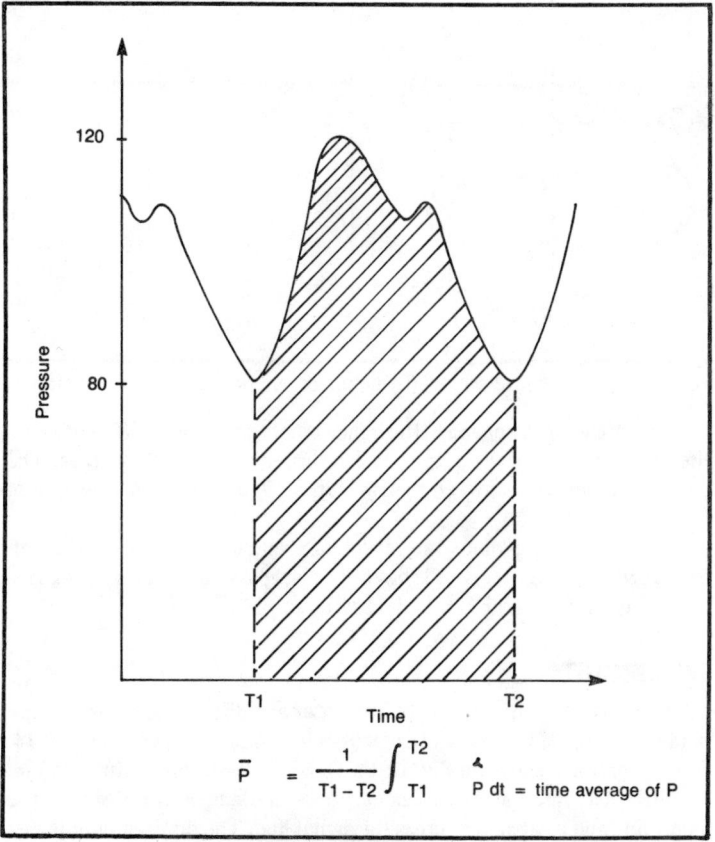

$$\overline{P} = \frac{1}{T1-T2} \int_{T1}^{T2} P \; dt = \text{time average of P}$$

Fig. 19-12. Integration of a complex waveform.

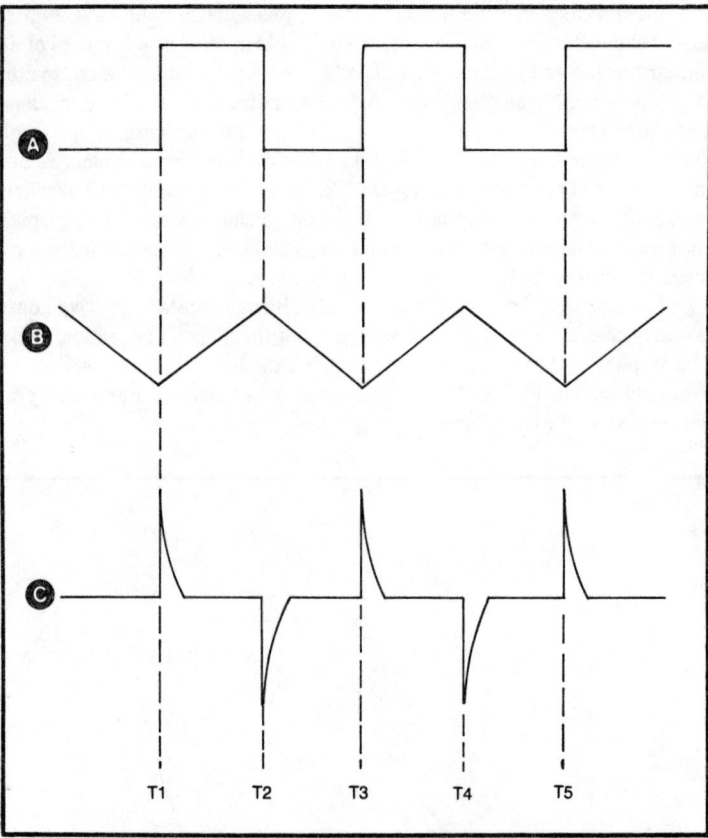

Fig. 19-13. (A) Squarewave input (B) integrator output, (C) differentiator output.

If a sine wave is applied to the inputs of integrators and differentiators, the result is a sine wave output that is shifted in phase 90 degrees. The principal difference between the two types of circuits in the direction of the phase shift.

You can also view integrators and differentiators in terms of their effects as filters. The integrator is basically a low-pass filter for analog signals, while the differentiator is a high-pass filter.

RC CIRCUITS

Perhaps the simplest form of integrator and differentiator circuits are made from simple resistor and capacitor elements, as shown in Fig. 19-14. The integrator is shown in Fig. 19-14(A), while the differentiator is in Fig. 19-14(B). The integrator consists of a resistor element in series with the signal line and a capacitor across the signal line. The differentiator is just the opposite, the capacitor is in series with the signal line while the resistor is in parallel with the line. You may recognize these circuits as low-pass and high-pass filters, respectively. The low-pass case (integrator) has a – 6-dB/octave falling characteristic frequency response, while the high-pass case (differentiator) has a + 6-dB/octave rising characteristic frequency response.

The operation of the integrator and differentiator is dependent upon the time constant of the RC network (i.e. R × C). In most cases, you want the integrator time constant to be long (i.e. ten times) compared with the period of the signal being integrated, while in the differentiator you want the RC time constant to be short (i.e. one-tenth) the period of the signal. You can cascade several integrators in order to enhance the effect and also increase the slope of the frequency-response fall-off.

OP-AMP CIRCUITS

The operational amplifier makes it much easier to build high quality, active integrator and differentiator circuits. Previously, we had to construct a high-gain transistor amplifier for this purpose. Figure 19-16 shows the basic circuit of the operational-amplifier differentiator. Again we have the RC elements, but in a slightly different context. The capacitor is in series with the op-amp's inverting input, while the resistor is the op-amp feed-

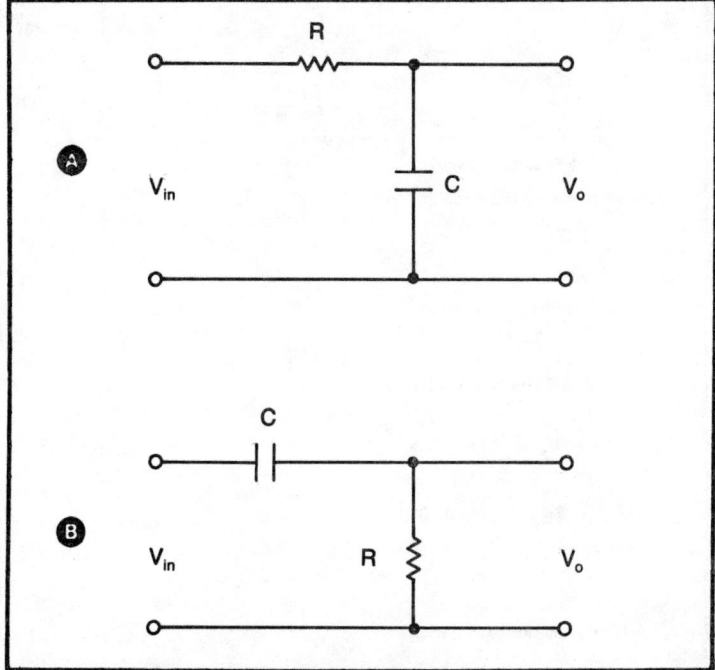

Fig. 19-14. (A) RC integrator, (B) RC differentiator.

345

Derivation of Integrator and Differentiator Equations

FOR INTEGRATORS (refer to Fig. 19-17):

From Kirchhoff's law and op-amp theory we know that:

$$I2 = -I1 \qquad (1)$$

We also know that:

$$I1 = V_{in}/R \qquad (2)$$

and,

$$I2 = C(dV_o/dt) \qquad (3)$$

Substituting (2) and (3) into (1):

$$\frac{CdV_o}{dt} = \frac{-V_{in}}{R} \qquad (4)$$

Integrating both sides:

$$\int \frac{CdV_o}{dt} = \int \frac{-V_{in}}{R} dt \qquad (5)$$

Collecting terms and rearranging:

$$CV_o = \int \frac{-V_{in}}{R} dt \qquad (6)$$

$$V_o = \int \frac{-V_{in}}{RC} dt \qquad (7)$$

which is the final expression.

For differentiators (refer to Fig. 19-16):

$$I2 = -I1 \qquad (8)$$

$$I1 = \frac{C dV_{in}}{dt} \qquad (9)$$

$$I2 = \frac{V_o}{R} \qquad (10)$$

Substituting (10) and (9) into (8):

$$\frac{V_o}{R} = \frac{-CdV_{in}}{dt} \qquad (11)$$

or, with the terms rearranged:

$$V_o = -RC \frac{dV_{in}}{dt} \qquad (12)$$

which is the normal equation.

Fig. 19-15. Derivation of equations.

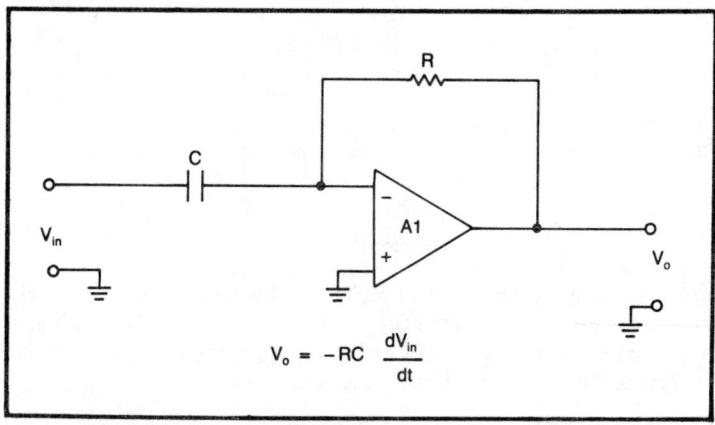

Fig. 19-16. Operational amplifier differentiator.

back resistor. The equation for the output voltage is given below (see Fig. 19-15 for the derivation, if you are interested):

$$V_o = -RC \frac{dV_{in}}{dt} \qquad (19.13)$$

where

V_o and V_{in}	are in the same units (volts, millivolts, etc.).
R	is in ohms.
C	is in farads.
t	is in seconds.

That is a mathematical way of saying that output voltage V_o is equal to the RC time constant times the derivative of input voltage V_{in} with respect to time (the "dV_{in}/dt" part). Because the circuit is essentially a special case of the familiar inverting-follower circuit, the output is inverted, hence the negative sign.

Figure 19-17 shows the classical operational-amplifier version of the Miller integrator circuit. Again, an operational amplifier is the active element, while a resistor is in series with the inverting input and a capacitor is in the feedback loop. Notice that the placement of the capacitor and resistor elements are exactly opposite in both the RC and operational-amplifier versions of integrator and differentiator circuits. In other words, the RC elements reverse roles between Figs. 19-16 and 19-17. That little fact will tell the astute a little bit of truth regarding the nature of integration and differentiation.

The output of the integrator is dependent upon the input-signal amplitude and the RC time constant. The equation is given below, and the derivation for the hardy and interested is in Fig. 19-15.

$$V_o = \frac{-1}{RC} (V_{in} dt) + K \qquad (19.14)$$

where

V_o and V_{in}	are in the same units (volts, millivolts, etc.).
R	is in ohms.
C	is in farads.
t	is in seconds.

This expression is a way of saying that the output voltage is equal to the time-average of the input signal, plus some constant, K, which is the voltage that may have been stored in the capacitor from some previous operation (often zero in electronic applications).

PRACTICAL CIRCUITS

The circuits shown in Figs. 19-16 and 19-17 are classic and appear in numerous textbooks and magazine articles. Unfortunately, they don't work very well (or at all in some cases). The big problem is that these circuits are simplistic because they depend upon ideal operational amplifiers. Unfortunately, the real kind-you-can-go-and-buy op-amps fall far short of the ideal which is in the mind of the textbook writer. In real circuits we find that differentiators ring and oscillate and integrators saturate very shortly after turn-on.

The problem with the op-amp integrators was driven home to me when I worked in a medical school/hospital, bioelectronics lab and had to build an electronic integrator for one of the "customers" of our lab (apologetic quotation marks used because they didn't have to pay us for the work, so were not really customers!). When I used a 741 operational amplifier, the

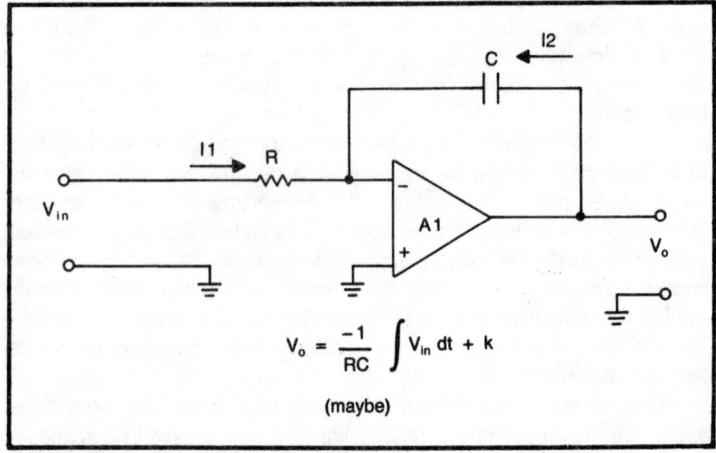

Fig. 19-17. Operational amplifier integrator.

output voltage saturated within milliseconds after turn-on. In fact, it was so fast that I thought the op-amps were bad. The problem was that the input bias currents of the op-amp (which are zero in ideal devices) create a high enough output voltage to fully charge the capacitor in the feedback loop very rapidly.

There is another problem with this kind of circuit, and it magnifies the problem of saturation. Namely, this circuit has a very high gain with certain values of R and C. Let's pick an example and see what this can mean. The gain of this circuit is given by the term $-1/RC$; so what is the gain with a $0.01\text{-}\mu F$ capacitor (certainly not a large capacitor in conventional wisdom) and a 10,000-ohm resistor?

(Note: $0.01~\mu F$ is 10^{-8} farads or 0.00000001 farads.)

$$A_v = -1/RC$$

$$= -1/(10,000~\text{ohms})~(0.00000001~\text{farads})$$

$$= -1/0.0001$$

$$= -10,000$$

In other words, with a gain of $-10,000$, a $+1$ volt applied to the input will want to produce a $-10,000$ volt output. Unfortunately, the operational-amplifier output is limited to -10 to -20 volts, depending upon the device and the applied $V-$ dc power-supply voltage. For this case, the operational amplifier will saturate very rapidly! If you want to keep the output voltage from saturating, you must prevent the input signal from rising too high (not good!). If the maximum output voltage allowable is 10 volts, then the maximum input signal is 10 volts/10,000 or 1 millivolt (sigh)! Obviously, you must keep the RC time constant within bounds.

When I built my first integrator at the university and found that 741 devices were not suitable, I turned to the high-cost premium-grade devices. At that time, a premium 725 device cost $15, and it suffered the same problems as the 741. The only difference between the $15 premium op-amp and the fifty-cent 741 device is that on the $15 op-amp the output saturated slowly enough for me to watch it on an oscilloscope or voltmeter—about 4 seconds—instead of nearly instantaneously. Unfortunately, this was still not acceptable. Applying a waveform to the input of even the premium op-amp integrator allowed me to see the output waveform rise up the screen of the oscilloscope and disappear off the top of the screen!

How to solve the problem? Fortunately, there are some design tactics that will allow you to keep the integration aspects, while getting rid of the problems. A practical integrator is shown in Fig. 19-18. The heart of this circuit is an RCA BiMOS operational amplifier, type CA-3140, or its equivalent BiFET type. The reason why this works so well is that it has

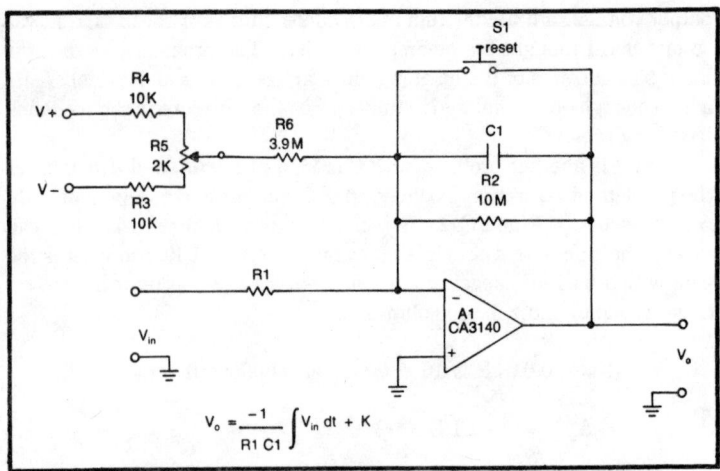

Fig. 19-18. Practical op-amp integrator.

a low input bias current (being MOSFET-input). When I tested close to a dozen different op-amps for the university project, the CA-3140, which cost about two dollars, out-performed devices costing ten times as much.

Capacitor C1 and resistor R1 in Fig. 19-18 form the integration elements and are used in the equation. Resistor R2 is used to discharge C1 to prevent offsets from the input signal and the op-amp itself from saturating the circuit. The RESET switch is used to set the capacitor voltage back to zero (to prevent a "K" factor offset) before the circuit is used. In some measurement applications the circuit initializes by closing S1 (or a relay equivalent) momentarily.

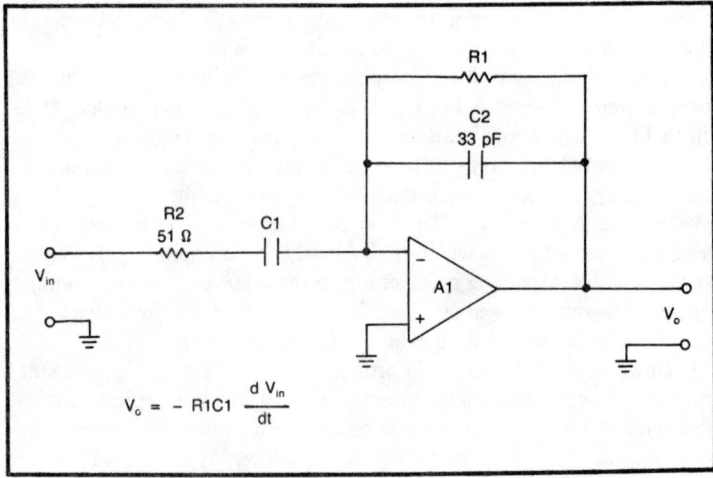

Fig. 19-19. Practical op-amp differentiator.

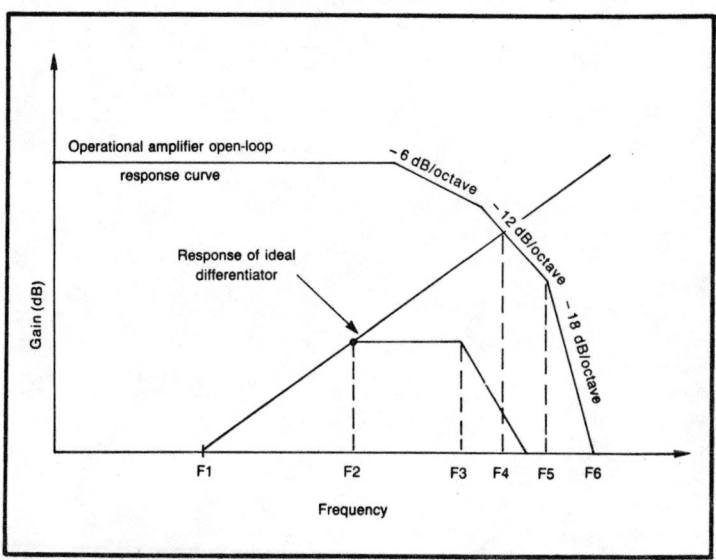

Fig. 19-20. Bode plot.

There is still a minor drift problem; so I added potentiometer R5 to cancel it. This component adds a slight counter-current to the inverting input through resistor R6. To adjust this circuit, set R5 to mid-range initially. The potentiometer is adjusted by shorting the V_{in} input to ground (setting V_{in} = 0) and then measuring the output voltage. Press S1 to discharge C1, and watch the output voltage go to zero. If it does not, turn R5 in the direction that counters the change of V_o after each time S1 is pressed. Keep pressing S1 and then making small changes in R5 until you find that the output voltage stays very nearly zero, and constant, after S1 is pressed (there will be some long-term drift normally).

If you want to build a multiple time-constant version of this circuit, consult my book *Linear IC/Op-Amp Handbook, 2nd Edition* (TAB book 1550, $13.95) for a circuit that I built in the university project mentioned earlier.

Figure 19-19 shows the practical version of the differentiator circuit. The differentiation elements are R1 and C1, and the previous equation for the output is used. Capacitor C2 is a small-value unit (1 to 100 pF) and is used to alter the frequency response of the circuit to prevent oscillation on fast-rise-time inputs. Similarly, a snubber resistor, R2, in the input also limits this problem. The operational amplifier can be almost any type with a fast enough slew rate, and the CA3140 is recommended. R2 and C2 are determined by rule-of-thumb for most people, but their justification is given from the graph in Fig. 19-20.

Chapter 20

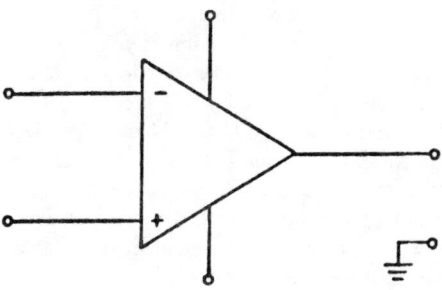

Construction Practices

ELECTRONIC-PROJECT CONSTRUCTION IS A SKILL, AND LIKE MOST skills, it is relatively easy to recognize the difference between the results of the experienced and of the novice. In this chapter I will impart some wisdom, most of which was gained by messing up a lot of my own projects over the past three decades. If you have ever built a kit (e.g., a Heathkit) then you are far ahead, at least in the mastery (?) of certain basic skills. A good writer or a kit-assembly manual will lead you step by step over both the rough spots and the easy. But when you design and build on your own, you are just that—on your own.

SOLDERING

The first thing you should do is learn to solder. Don't laugh. Soldering is a skill that eludes most novices. It looks easy, so receives scant attention until it is too late to help. Please pay attention to the section to follow where I discuss soldering implements, and don't let your own cocky arrogance let you think that you know how to do it, unless you can show me a decade of experience or a certificate from a NASA-grade soldering school.

There are several reasonable ways to learn the art of soldering, and all of them come down to: DO IT! One method favored by many technical-institute instructors is to give the student exactly one foot of 12- or 14-gauge bare bus wire (or insulated wire that has been stripped). This wire is cut up into twelve one-inch lengths, all exactly the same. Why do you suppose that we want exactly twelve pieces, not thirteen or nine? Well, there are twelve edges on a cube, you see. Take these twelve slivers of wire, and construct a cube using solder to hold together the joints. Now, here's the catch—you may use *only* a pair of long nose pliers, soldering iron, and solder. No holders, vices, or any other implement! The idea is to teach you not to move the work while it is cooling. When you finish the cube, let it cool (This should take about fifteen minutes if you have been working diligently), and then crush it in the palm of your closed fist. If any of the joints break, get another twelve inches of wire and do it again, and again, and again, until you do it correctly.

Another alternative is to go to a hobbyist type electronic-parts-and-supplies house and buy a couple fistfuls of terminal strips. The national retail/mail-order chains usually offer a two- or three-dollar assortment of terminal strips that contain a bunch of them that are useless for any other purpose. Practice wrapping the wire around the terminals and then soldering it until you get nice bright, solid joints.

There are only a few rules to ordinary soldering, such as you will do in most projects:

1. Always use quality solder; never reuse old solder. New solder contains a rosin flux that prevents oxidation on the surface of the work from ruining the job. Low-grade solders (off-brands) often have insufficient flux. I prefer Kester and Ersin solders in a 60/40 or 50/50 lead/tin mixture. NEVER USE ACID-CORE, or plumber's solder. If in doubt, do not use it; buy a new roll of rosin or "radio-TV-electronic" solder.
2. Make the *joint* hot enough to melt the solder, and then apply the solder to it. Let the heat of the joint do the melting. A frequent error is to let the iron melt the solder and then hope that it will flow down onto, and adhere to, the joint. This is always a vicious and poor practice but seems to persist no matter what people like me keep saying.
3. Make absolutely sure that the work is clean. Use a pencil eraser, sandpaper, solvent, or whatever it takes to make the joint bright and clean. Even the tip of the screwdriver is better than nothing!
4. Keep the joint still until the molten solder is again solidified. If the joint moves, a cold solder joint will result (not *may*, will!). Such a joint has a dull, brittle, and porous look.
5. Do not overheat the joint. Too much heat will use up the flux, and will often damage the component.
6. Don't use too little solder.
7. Don't use too much solder.

These last two admonitions cover what are probably the most frequent errors made by the new solderer. Use just enough to cover the joint, without overloading it. If two adjacent terminals are overloaded with solder they will short out, and create damage to your hard-built project. I have seen printed-circuit boards in which an unskilled assembler doubled the weight of the work just by improper solder technique! No wonder NASA specified the weight of solder on PC boards. A few milligrams of extra solder on each of over a million joints in a space craft or satellite and the thing simply could not get off the ground!

Now that we have covered all the wisdom of the ages about soldering, let us move on to more weighty matters, like what tools to have.

TOOL SELECTION

"Use the right tool; use the tool right."

This slogan was imparted to me by a wise older technician almost three decades ago, and it is just as true today as then. Many who don't make their living with tools tend to misuse them, often with spectacularly poor results. If you are the kind of person that would pick up a screwdriver and pound nails with it, shame on you. Each tool has a *purpose*, and this purpose should be understood and followed. In electronics there are quite a few different items that would be classified as "the right tool" under the right circumstances.

A collection of small handtools is always a must. While most workers have their own preferences and prejudices, there are some commonalities that are determined by the nature of the work to be done. These items belong in every tool box:

- Long nose pliers, jeweler's size
- Long nose pliers, 4-to-6-inch size
- Diagonal side cutters, jeweler's size
- Diagonal side cutters, 4-to-6-inch size
- Straight blade screwdrivers, 1/16-to-3/8-inch width
- No. 1, No. 2, and (if you work on tanks) No. 3 Phillips screwdrivers
- Complete set of nutdrivers, No. 6 through No. 18
- Crescent wrench, miniature and normal 6-inch sizes
- Gas or slip-joint pliers
- Channel locks or something similar, also good for removing stuck jar lids
- Soldering apparatus

In the screwdriver and nutdriver categories you will soon see the wisdom of having stubby and long-handle versions in addition to the regular sizes. An alternative is to buy something like an Xcelite Master Servicer tool roll which has the blades, points, etc., that fit into common handles. I personally do not like this approach and would rather have the individual tools, but I am usually overruled by associates who *do* like the idea; so use your own judgment.

The jeweler's-size side cutters should be the type with straight outside edges on the cutting nose so that you can get close to the surface on printed-circuit boards. Some side cutters have curved noses, and these are used on what the television ads call "handwired chassis." These are not quite as useful in this type of work.

A good pair of side cutters will have jaws that touch together *only* at the tip when closed. Even if supposedly "off-brand," you can tell a good pair (or at least a good candidate) by performing a little test prior to purchase. Hold the cutters up to the light, hollow side of the nose toward you, and examine the closure of the cutting edges. On a good pair of diagonal side cutters (also called "dikes" by most), you should see some light at the base of the cutting edge, closest to the pivot point, and *none* at the tip. Do not accept any dikes where the edge is irregular, or where the tips don't meet, or where they meet elsewhere but the tip. Shun them, they are junk and should be discarded forthwith!

This brings up another point (I seem to be having a lot of admonitions, but they are the result of my own stupidity over the years. I am giving you the benefit of learning from my mistakes, which helps bolster the view that book learning is condensed experience): Don't buy cheap tools, despite the temptation to save money. Cheap tools are almost never a real bargain and will require replacement often. I paid $15 for a pair of lineman's dikes that lasted twenty years before some #/!* stole them from me. Of course, you can get reasonable quality without buying industrial-grade production tools. A good pair of long-nose pliers or dikes will cost something around or over $10, and really good tools will cost around $15.

A word of wisdom: Wire *cutters* (such as dikes) and wire *strippers* are different types of tool and have different functions. It is often the case that people will attempt to strip the insulation from wire with side cutters, always a bad practice. It is too easy to nick the wire, which forms a break in the (possibly near) future. Also, the cutter tends to stretch and tear the insulation rather than strip it. This results in a dumb looking end that isn't quite what you had in mind.

Oddly enough, it is possible to buy reasonably-high-quality Miller or Xcelite wire strippers for a low cost. These are of simple design and therefore cost little to make. These strippers are remarkably alike and are not at all complex. Fancy, complex wire strippers that have all of the nice whistles and bells are fine, if you want them, but I like simple tools that work all of the time.

Beware of cheapie imitations, though. Those four- and six-dollar types are adequate, but much under that figure the tools available are little more than junk. If you cannot find the brands listed (which is almost impossible to believe), then look for a pair that has nice sharp, clean edges and a nice feel. Poor strippers look dull and have a mushy feel to them. I realize that "feel" is an unscientific method of analysis, but it is quite valid, even though subjective. Once you become familiar with the feel of good tools, you will be hard to please with junk. One thing that almost always gives away a good set of strippers is the cutting edge. It may have a look similar to that of a fine knife and may look blued (barely perceptible) or tempered.

As I mentioned earlier, soldering and *desoldering* are skills required for electronic construction and are frequently neglected until too late by the novice builder. Many tend to dismiss this as trivial or trivially easy to master—until substantial and possibly uncorrectable problems show up due to their poor technique.

One mistake made too often is the incorrect or insufficient selection of apparatus. At least two different types of soldering apparatus are required, even for supposedly light assembly chores. Most PC-board or terminal-strip wiring will be soldered with a pencil iron in the 25-to-50-watt class. For heavier work, use a soldering gun in the 150-to-250-watt class.

Pencil-type soldering irons are so numerous that selection is often very difficult. The only general advice in this respect is to select a type that has a larger number of different interchangeable tips available. My personal preferences are the Weller SP-28 for portable work and the Weller W-TCP soldering "station" (?) on the permanent workbench. In either case, well selected tips will make the job easier. Pick several, thin conical and blade tips, as well as a single broad-blade tip. Make them at least 700° F types.

A good selection for the soldering gun is the Weller D-440. When buying the gun, you may want to also get the kit that comes with it. These are made up to include several widgets and thingwatchees that supposedly make the job easier. At any rate, they are nice to have, even if they don't get used (they can be in your snob's tool kit—those things that are to tools what buzzwords are to language).

Although it may seem difficult to believe, *desoldering* can be almost as hard as *soldering*. Almost any fool can desolder a joint, but the neat trick is to desolder it without destroying the components or the PC track. Alternatively, when desoldering incorrectly, one may tend to spread solder all over the PC board, and of course, Murphy's law and Finagle's principle conspire to hide the most dangerous splatter from even the closest scrutiny until permanent harm has been done to the circuit.

Two general classes of desolderer are good: solder suckers and solder wick. Both have their adherents, but I maintain that both have their respective places and should be available.

There are several types of solder suckers available, but they all depend upon a semivacuum to lift molten solder from the terminal or component being worked. One low-cost type uses a rubber squeeze ball, fitted with an appropriate nozzle. The idea is to squeeze the ball, melt the solder joint, and then suck the solder into the tool by suddenly releasing the rubber ball. These devices look much like the crude ear-irrigation syringes you can buy at the local drug store. They are the device of choice for the occasional user because they are low cost. My own experience, though, is that they deteriorate too rapidly in a busy electronics lab.

Another solder sucker is a long cylinder with a piston inside and a nozzle at the open end of the cylinder. Out of the other end is a drive rod that is attached to the non-vacuum side of the piston face. To cock the device and get it ready for "action," you depress this rod until it clicks; you are then loaded for bear. To use, one must melt the solder with an iron and, while maintaining the temperature, bring the sucker's nozzle close to the joint and press a trigger button, or in some brands, lever. This releases the spring-loaded piston. It flies back up the cylinder, away from the nozzle, creating a partial vacuum that just loves to choke up on solder.

There seem to be two different mechanisms, even from the same company. In one type, the cock rod and the piston are not attached. They make contact only when you are in the act of cocking the spring. At all other times, the rod is in its rest position and harmless. In the other type, however, the rod stays down when the thing is cocked and flies up rapidly when the trigger is released. The only problem with this is that you will be bending over the work to concentrate on what you are doing and the #/&%* thing will strike you in the eye or forehead. I can't say that anybody has ever been seriously injured by this trap, but it is a darn nuisance if you are unaware of it! In other words, when you buy a solder sucker, check the drive rod for operation and, if of the offending type, be careful.

Solder wick is a clever type of copper braid wire that is impregnated with rosin flux. It is placed against the joint to be desoldered and heated with the iron. The melted solder will creep up into the braid by capillary action.

Solder wick leaves the joint remarkably clean but is not too good on large terminal-strip joints. On PC boards, though, it represents a very good way to clean the board of solder without damage to either the components or the PC tracks. Those tracks, incidentally, are merely glued onto the board and will disintegrate if too much heat is applied. Of course, quality boards are better in this respect than El Cheapos, but all can be damaged if the right jerk gets hold of them. Nothing is foolproof, only some take bigger fools than others.

Selection of the correct solder is also essential. Solder is made of lead and tin, and this causes it to melt at a temperature less than the melting point of either metal alone. A few premium solders will have two percent silver, but they are not always desirable because this raises the melting point considerably. Some formulations melt as low as 275° F, but most will melt at some temperature between 300° F and 450° F.

A word of Wisdom:

DON'T EVER use acid-core, or plumber's solder, for any purpose in electronics!!! It eats everything it touches and strains the rest. The mass destruction that can be caused by this type of solder is beyond (shudder) comprehension. If you do not follow this advice, you will get what you deserve. Be reasonable; follow my advice!

The proper solder for electronic, radio-TV, and ordinary electrical wiring is 50/50 or 60/40 lead/tin with a *rosin* core. For small work on IC-laden PC boards use 22-, 24-, 26-, or 28-gauge solder. Larger work requires 14-, 16-, 18-, or 22-gauge solder (these are standard AWG thicknesses). If you are cheap or cannot afford several rolls of solder, choose 22 as a good trade-off; it is common to both groups. Be a little tight with your solder, incidentally; a 1-pound roll of solder costs an amount that is not trivial!

You will also find that metal-working tools are required, unless you have the services of a good, low-cost, machine shop. Metalworking when building electronic projects from scratch is necessary for both practical and aesthetic reasons. After all, you will put a lot of work into the project, and pride of creatorship compels us to cut holes with some tool more viable than a beer-can opener (which has been done, incidentally—he got fired).

Among your arsenal of metalworking tools a hand drill is obligatory, and a saber saw is highly desirable. Both can be bought for relatively low prices, especially on sale days, at your local hardware or Harry-and-Harriet-homeowner do-it-poorly-yourself store. Make sure the drill has a 1/4- or 3/8-inch chuck and no larger. A 1/2-inch-chuck drill is always too heavy for light work and may make a cripple of you before your time. All good hardware stores sell reduced-shank drill bits so that holes up to 1/2 inch can be cut using a 1/4-inch-chuck drill.

Holes larger than 1/2 inch can be drilled with a mean looking tool called a hole saw. Make sure the ones that you buy are for metal (i.e., the Black & Decker). Many are designed for wood only and will cut metal only once or twice, poorly.

All hole saws leave a ragged hole at best and can even be dangerous if you handle drills with the finesse of Attila the Hun. A chassis punch is a far better alternative if you can get to both sides of the sheet of metal being cut. These tools have a stationary mandrel, a cutting block, and a bolt to bring the two concentric pieces together forcibly.

To use a hole punch (also called a chassis punch by many), drill a hole in the metal just large enough to accept the pilot bolt. Thread on the cutting edge, and tighten until you feel the *second* snap (both points on the cutter have to go through the surface) or until the thing falls out on the floor. The punch will make a nice clean, professional-looking, burr-free hole. Low-cost chassis punches are available from Greenlee and certain of the national department-store chains (i.e., Sears, Montgomery Ward, etc.). Round punches over 1 1/2 inches are also available from Greenlee, but the cost is significant; so buy them only if you have to punch more than a few holes of that size—borrow one from an electrician if only occasional work is done.

Greenlee also offers square, D-shaped, and triangle punches, but these also are expensive. Files can be used, after a pilot hole is drilled, to make some of these shape cuts.

Files are one tool where the cost of the good is so close to the cost of the junk that there is no excuse for buying cheap, especially since the results offered by the cheapies are so darn poor, even in moderately skilled hands. Of course, the old, gray machinist, who catches bullets in his teeth—or so it seems—will tell you not to blame your tools for bad work, but just look at his files, first quality all the way!

Keep several different types and sizes of files. Those little jeweler's files are nice for working on PC boards and small components, but will be ruined if larger work is attempted. For working sheet-metal chassis, equipment cabinets, and small brackets use a larger file. The wrong size will either cut off too much metal or will work you to death, depending upon whether it is oversize or undersize. Keep some flat bastard files, some rat-tail files, and a couple of three-cornered files, in at least two different sizes each.

Another useful metalworking tool is actually very clever, even though given the ridiculous name "nibbling tool." The name is, though, highly descriptive of its operation. It actually nibbles its way through the metal. In physical appearance, it resembles one of those hand exercisers one sees at sporting stores. It cleanly munches out little nibbles of metal, each time you squeeze the handle. You can cut almost any shape hole over about 3/8 inch only at the expense of a well-exercised hand.

Before drilling a hole, it is necessary to use a center punch to make a pilot dimple to guide the tip of the drill bit. This keeps it from wandering, thereby putting the hole where you don't want it.

Be wary, though, of those low-cost center punches that have to be hit from the rear with a hammer blow. The force of the hammer, despite your tender touch, will tend to bend and distort the surface. An *automatic* center punch is a clever little tool that eliminates this problem. You merely place it on the correct spot and depress the rear a little. A snap-action trigger will set the punch for you. (Clever, huh?) Most larger hardware stores carry these items.

Before moving out of this discussion of tools, let me introduce you to a pair of companies which seem to have a large selection of tools that just don't seem to be available locally. Write to both and get their catalogues.

Jensen Tools & Alloys
1230 South Priest Drive
P.O. Box 22030
Tempe, AZ 85282

Brookstone Tool Company
Peterborough, NH 03458

PROJECT DOCUMENTATION

One of the rudest, most ill-mannered, lousiest things that an electronics-project builder can do to successors is to make a project, however simple, and *not* provide documentation on what it does and how it works.

Also, when you are designing something, document *as you go*. In other words, keep a laboratory notebook! Almost every college bookstore, large stationery store, or engineering-supplies store will stock either laboratory notebooks or engineer's-computation notebooks. These will have about 100 numbered, quadrilled pages. Unless you are a very busy design engineer, one $10 lab notebook will last you several years, hardly a momentous investment.

When a project is completed, go out and buy some quadrilled drawing paper, and draw the final circuit in a more formal manner, unless, of course, you are a nasty, no-good SOB with a mile-wide mean streak. In which case, you will leave absolutely no documentation.

Having been a university electronics-repair technician, I have developed a sense of justice that would condemn such people to a kind of hell that would keep them occupied for eternity—like maybe having a digital watch permanently attached running incorrectly and no reset button!

PROTOTYPING CHASSIS

Electronic-circuit design is not an exact science, despite the fact that we bandy about esoteric and arcane mathematic formulae to convince ourselves that we are knowledgeable, righteous, and disgustingly clever. The problem comes when the circuit is built, and we find that it does not work correctly. Why? Putting Murphy's law aside, the best explanation is that most of our formulae are either mere *approximations* to begin with, or they are based on ideal electronic components. In the real world, however, we must make do with very much imperfect components and live with our approximations.

What to do (sigh)? We breadboard the circuits first, *before* committing ourselves to any particular design! Making this far easier than was true in the relatively recent past is the solderless breadboard, version of which are made by several companies.

A laboratory breadboard is the Heathkit ET-3300. It has built-in positive and negative 12-volt dc power supplies that can each deliver up to 100 milliamperes and a +5-volt, 1.5-amperes supply. One problem with many of those I looked at before making the decision in favor of the ET-3300 was that they often had 12-volt supplies (or in one case an equally useful 15-volt) but would lack the 5-volt supply. Others had the 5-volt, but not the higher-voltage supplies.

Once the design is nailed down and all of the little glitches are corrected, transfer the design to either a PC board or a wire board, and fix a whole new set of glitches that are specific only to the final configuration.

Almost any electronic-construction project can be given a really professional appearance if it is housed properly (next section) and built on a proper board or circuit card.

Professional printed-circuit layout can be very expensive if done commercially. Alternatively, it can be painfully tedious if attempted by you. If you cannot get a decent (cheap) quotation from a commercial PC house, you may want to at least try once (everybody has to get it out of their system) to make a PC board. One hint, though: many professional PC shops have employees who moonlight. The PC material comes in largish sheets, and

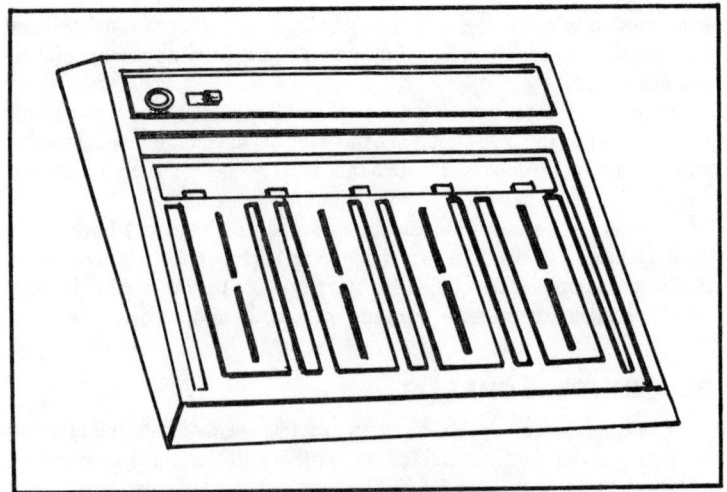

Fig. 20-1. Laboratory breadboard model ET-330. (Courtesy Heath Company.)

if your design is small enough they can sneak it in without cost to their company, even though most do this surreptitiously.

You will find that most local electronic-supply houses that serve the hobbyist market offer PC kits and PC-layout graphics aids that are reasonably professional. If the instructions are followed closely they will produce excellent results. Of those readily available, I have used Vector, Kepro, and G.C.-Calectro with good results.

A PC board is for those projects where several or several hundred are to be made. It is simply too much trouble to make a PC board for a one-of-a-kind project; so you must turn to some other techniques. It is easy to build on perfboard, DIP board, or wire-wrap board.

Perfboard comes in flat sheets of several square inches to dozens of square inches. It is drilled with many rows and columns of 0.042-inch holes on 0.1-inch-by-0.1-inch centers. The only problem with perfboard is that it has no foil pattern to which components can be soldered. At least one company makes stick-on adhesive copper-foil appliques in popular IC patterns, but these are not always available.

A good solution is to use DIP board even though it is a bit expensive. An example is the Vector 3677-2DP pattern. Incidentally, I am not in love with Vector, nor do I sell their products or own stock. I mention them frequently because they have good products, but more importantly, make them available to the general public through local electronic parts distributors. There are other brands, to be sure, some of them with higher quality products, but they seem to shun the hobbyist and small volume commercial or university markets.

The Vector 3677-2DP board, and similar types with slightly different features, is set up to accept DIP IC packages. The pattern of the foil seems to be for digital applications particularly because it has only two power-supply/ground-bus tracks, suggesting a + 5-volt ground configuration. There is no reason why you cannot run extra wires for other buses, reassign these, or use an Xacto knife to cut the foil in critical places to increase utility. Also, the cost of the DIP board can be cut a little bit if you crowd small-

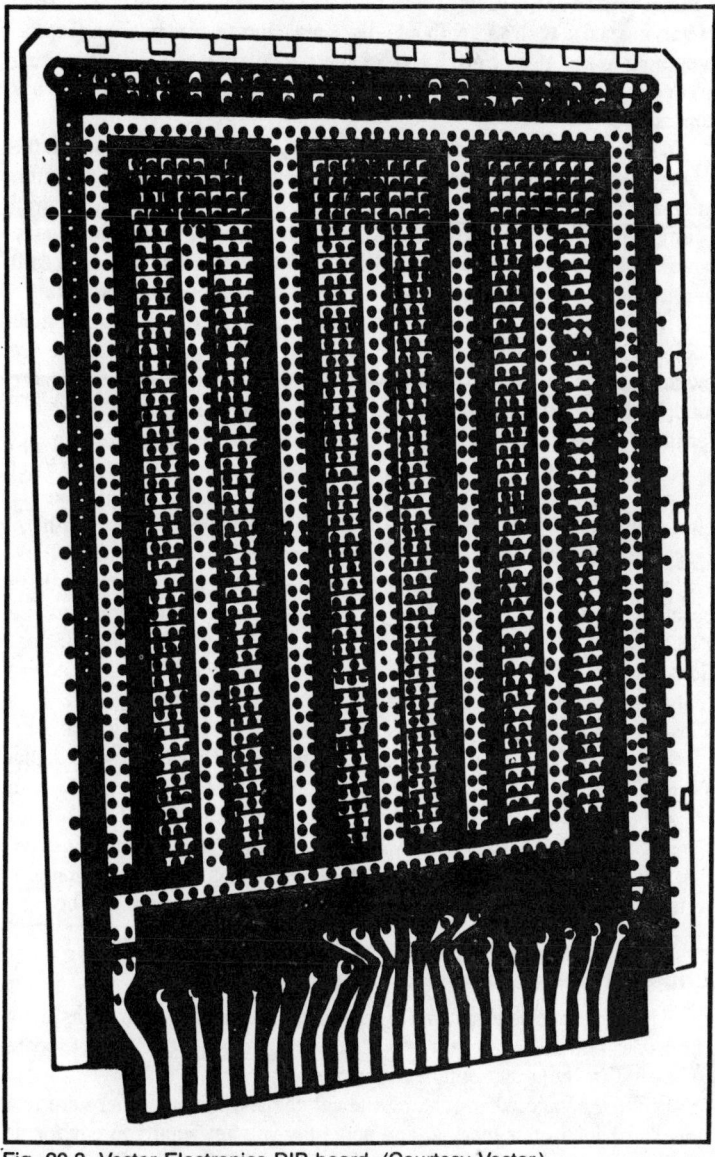

Fig. 20-2. Vector Electronics DIP board. (Courtesy Vector.)

component-count designs to one end and then cut off the unused portion for later projects. This lets you get a lot of mileage per dollar at the expense of a ragged-edge board. The DIP and perfboard can be cut along a row or column of holes using a large pair of sewing scissors—unless it is a real G-10 fiberglass board, in which case you are on your own.

Vector also offers substantially the same sort of pattern on other boards of varying sizes, with and without edge connectors. Large microprocessor-type boards are also offered. One of the two available as of this writing is compatible with the popular S-100 (Altair & Imsai) microcomputer bus connectors.

Wire-wrap is a powerful electronic-construction technique in which the connecting wires are wrapped tightly (without solder) around *square* terminals. The only real drawback to this method is that genuine wire-wrap boards tend to be expensive. This is easily overcome by native cleverness, however. Buy wire-wrap sockets and *solder them* (oh horrors!) to a DIP board or glue them (oh, no!!!) to perfboard. Quarter-watt resistors can be installed in DIP sockets (14- and 16-pin) if the resistor leads are bent flush to their bodies. Other components can be soldered to terminal posts that are mounted to the perfboard or DIP board. Make sure that they are the square kind, however. Vector sells these, as do Radio Shack and others. The inter-terminal connections can then be made using a wire-wrap tool.

Several different types of tool are available. If you really want to get classy, buy an electric wire-wrapper for something over $90. If you are either cheap or broke, buy a $9 type and work yourself to death. A reasonably decent alternative to both approaches is the Vector P180 Slit'n'Wrap tool. This tool is actually more convenient than many others because it does not require you to strip the wire for each connection. The P180 tool has a built-in stripper edge that cuts through the enamel insulation. If you elect to use the Vector wire refill spools, you will have a wire that has insulation that can be easily melted with a soldering iron. This allows you to solder errant joints or those that tend to worry you.

When using any other type of wrapper, buy Kynar insulated wire. Alpha Wire sells this in large spools and in precut lengths. Cambion sells it in blister-pack displays (small quantities) at local electronics-parts houses. This type of wire, by the way, is also useful for regular solder-type construction. I keep a 500 foot spool and cut off small lengths for construction projects. It makes a decent appearance on your solder or wrapped board.

Chassis, Cabinets, and Enclosures

There are many alternatives to instrument packages and almost as many companies that offer different brands. The best advice is to tell you to write to several of them and obtain catalogues.

Be fully awake when reading the catalogues, though. There are few price bargains in this market, and hopefully you get what you pay for. If you buy cheap, expect a nothing cabinet and you shall not be disappointed.

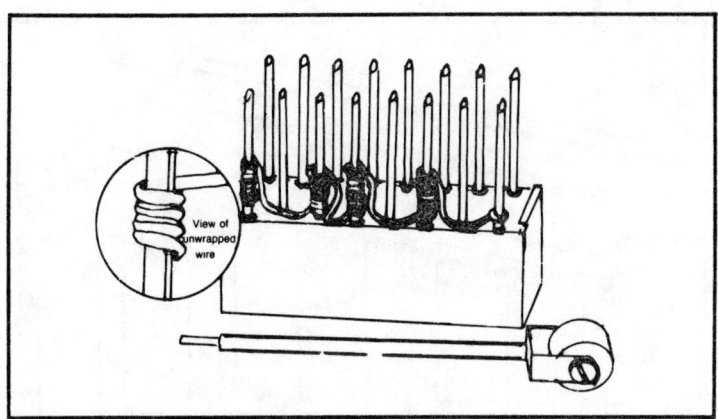

Fig. 20-3. Vector Slit'n ' Wrap tool. (Courtesy Vector.)

If you prefer or demand nice-looking, painted cabinets with top and bottom covers and all the whistles and bells, then expect to lay out a pretty penny. You will pay dearly. If, on the other hand, an ordinary, metal chassis or a gray hammertone, snap-together, sheet-metal box is sufficient, the cost will be modest. These are not particularly rugged; so do not use them where the unit will be abused.

Three basic forms of electronic enclosure are used extensively and seem widely admired. Where pleasant appearance, but not much mechanical strength is required, you might want to buy a Ten-Tec cabinet. These are available in local stores and come in both gray/black and white/walnut color schemes.

A similar product, and just as good in my opinion, are the cabinets by LMB. They offer several styles in several color schemes, at low cost. The type with the black wrinkle finish and white front panel seems especially striking.

All low-cost cabinets can be made even nicer if you take some pains to brace the interiors a little. This is the principal difference between the low-cost cheapies and costlier models. I have used aluminum home-handyman stock and plastic rods for this with good success.

The Pomona blue box is an exceptionally sturdy, yet attractive, diecast aluminum box. There are several very useful sizes, all painted a pleasing blue (or unpainted, if you prefer). They also make a few sizes that to me seem absolutely useless, but I suppose they have their place too. My only complaint with the Pomona line is that they do not make a few sizes which I would personally find exceptionally useful.

Most of the Pomona box models include slotted interior walls to facilitate mounting of PC or wire cards. This makes it much easier to build, because the amount of drilling and mounting of boards is less.

Vector makes a rack-mountable modules cage that will accept either a collection of DIP cards (the sockets fit along a rail at the rear) or Vec-

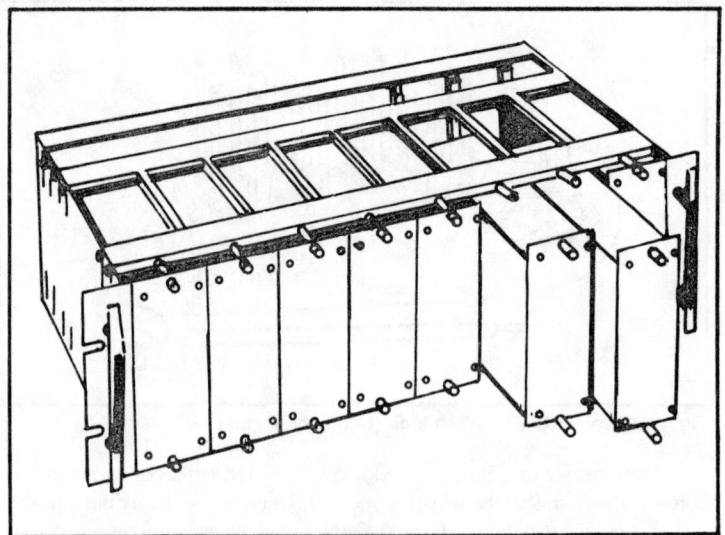

Fig. 20-4. Modules and strut cage. (Courtesy Vector.)

tor's own EFP-series of module cases. If you are building a system of in-struments, use the cage as the mainframe and build the sections modular style in the EFP cases. These may then be used as plug-ins.

Vector actually offers two different styles of module case. One has a closed back, and you must provide appropriate connectors for power and whatever control or signal functions are required. The other type is an open-back and is designed so that a 3677-2DP DIP board will protrude out the open rear panel. The DIP-board's edge connector can then be plugged into a standard 22-pin connector permanently to the cage's rear strut.

The closed-back modules can also be pressed into service as mini-equipment cabinets that are not altogether unattractive. Place the case on its side and install little rubber feet along the edges of the rear sliding cover. I am sure that the person at Vector who designed these would choke if this were made known, but I have found it a useful and low-cost alternative to providing a *sturdy* (key word, that) enclosure for small instruments.

Finishing

The final touch to any professional-appearing project is the finish. At the very least, buy several packages of dry-transfer labels. These can be bought in general sheets from almost all stationery stores and in special electronic-label kits from electronic-supply stores. When the panels are all drilled, punched, and deburred (and painted too, if that route is taken) but before components are installed, label all controls, switches, functions, and so forth. Spray the finished panel with clear plastic lacquer. An especially good type is Datacoat, which is specially formulated for a company that makes some of the more popular dry-transfer packages found in electronic stores. Bake on in an oven.

Spray paint may also be used, but should be baked on for several hours. If you know a friendly artist who will work cheap, have a silk-screen panel made. This will make the project look like you bought it, not made it. Just because a project is made by an amateur or in a small university or commercial model shop is no reason for it to *look* like it was made there! Good luck.

Chapter 21

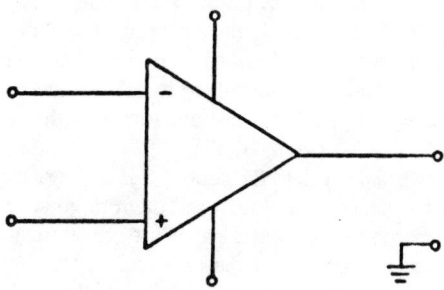

How to Design

THE ONE UNMISTAKABLE SIGN OF THE NOVICE DESIGNER OF ELEC-
tronic instruments is the tendency to jump right into the construction and testing phase of the project without the benefit of the least little
bit of planning. Don't make this time-consuming and expensive mistake.

DESIGN: A PHILOSOPHY

This thing called "design" is a logical process. Indeed, the very word
design connotes activities like planning, thought, intent, and a procedure.
There is little that is really arcane in elementary electronic design, and
anyone with a little knowledge can design *adequate* electronic instruments.

This is not to diminish the first-class design engineer but merely suggests that almost anyone can get something working that will perform the
chore at hand. A good designer exhibits intelligence, insight, knowledge,
and that subjective property known as cleverness. Most of these attributes
are obtained mostly through one process: experience, the art of surviving
repeated attempts by a perverse universe at nailing you to the wall.

You will have to learn good design and laboratory techniques if your
efforts are to be efficient. You will also have to learn some of the more
subjective things such as the upper limits of devices, mostly through the
potentially costly method of converting a lot of silicon to carbon. If you
follow true to form, you will note your mistake only milliseconds before
a puff of smoke indicates a one of a kind sample just evaporated (sigh).
Do not worry if your early efforts seem futile. They are not totally without
merit, if only because you are afforded the opportunity to learn from your
mistakes. This book lets you learn from some of mine and those of others.

Do not be impatient to get started on producing hardware. This is actually one of the last steps in a proper design activity. Unfortunately, other people may not see it that way. If you are designing some electronic widget as part of your employment or to fulfill a requirement in a school course, there may be pressure from above to start producing immediately. There always seems to be some lame-brained supervisor, impatient overlord, or nervous customer who is only too willing to believe that you are not producing anything if you are not spritzing and fussing with wires, capacitors, ICs, and other electronic paraphernalia.

Despite primordial urges to the contrary, if the person who does this has the power to clobber you, do not tell them to go suck a lollipop (or whatever). Be smarter than they are! If they are pressuring you in this manner, you have absolute proof that they are fools. Capitalize on that fact! What is the best defense against a fool? Well, fool them, of course! When you first get the assignment, build something on a breadboard, *anything* at all. If it has blinking LEDs or goes "whirrrrrr," so much the better. Every couple of days you should change it some way that is obvious to the casual inspector. Add an oscilloscope probe (unstable Lissajous figures are most effective!) or a component or two. Always shift its position on the bench a little each day, but not towards the rear (that looks too much like you are avoiding the issue). Above all, place a sign that reads something to the effect "Experiment in Progress: Do Not Disturb." Then return to your desk, a quiet spot in the library, the lavatory, or anyplace else known to produce creative thoughts, and a colorful panorama of ideas should flood forth.

The very first step in any design process involves *knowing the problem that you must solve*. This advice may seem at first glance like a case of runaway cynicism, but in reality it is common sense based on observation. A remarkable number of people will begin something they call designing without really knowing what the device is supposed to do. Studying the problem will involve any or all of several activities including (but not limited to) a literature search, thinking, interviewing the customer, and interviewing people who have had the same problem in the past (especially those who solved it).

Do not underestimate the value of a literature search. Too many alleged designers shun this for some reason. I suspect that they suffer from the N.I.H. syndrome: not invented here. This well known malady afflicts those whose misplaced pride prevents them from using perfectly good solutions from prior art in favor of attempting the new and unknown. They are known to feel ashamed to admit that somebody else once had a good idea. If prior art will solve your problem, use it. Your job is to solve the problem and is not necessarily to prove how clever you are with state-of-the-art designs.

The next step will be to formulate an approach to solving the problem. This will involve trying to figure out several methods or circuits that might do the trick. There is seldom a single "best" way to perform any electronic design job; so be sure to consider several possible alternatives.

370

The word contingencies looms large in this area.

Some wise souls will tell you that your first design in any given project is usually the poorest one that you invent. If this is true even some of the time, it might be wise to collect several approaches before actually beginning to build anything.

One other thing to do prior to building anything is to make a drawing of the entire circuit and every critical mechanical part. I know several people who allege themselves to be electronic-instrument designers, who are often seen in their workshops with an IC-manufacturer's catalogue propped up on a vise or pile of books, copying fragments of their total circuit from first one page and then another. This is extremely poor practice and leads inevitably to burned-out ICs, ragged tempers, unhappy customers, and the well-earned contempt of colleagues. Even if you choose to duplicate a published electronic project, you should copy it *in toto* onto a working sheet, especially if you plan to *change* any part of it.

The drawing and other documentation will play a large role as you begin building and testing the first prototype. Keep detailed records of key voltages, signals, and other parameters that seem important. Change the master drawing to reflect any changes you make in the circuit. Think with a pencil!

Also, please write an alignment and adjustment procedure that can be followed by someone less qualified than yourself. The procedure might be self-evident to you because you originated the concept, but to others it may be mysterious. Do not require your successors to use mental telepathy or the occult sciences to figure out how to adjust, align, or calibrate your creation.

Besides, six months or a year down the road, you may well be the one who is called upon to repair or realign the instrument. Guess who will then be neatly and properly nailed for not knowing how? Serves you right. Good documentation is a fine SYA (save yourself) tactic.

Use the laboratory notebook. Most college bookstores and engineering-supplies companies sell adequate, quadrilled, laboratory notebooks. These, if kept properly, can be your file and may also help you out in a patent question.

Another sign of the inept or novice designer is the tendency to commit even relatively complex or untried designs to the final form without first breadboarding the circuit. Every new design cannot be considered "finished" until it has been tested properly and found not wanting. Every idea that you conceive must be considered merely hypothetical until it has been proven valid. It may appear that certain ideas will work, but when you connect them a big nasty surprise is found. This is why laboratory breadboards are fast-selling items.

The final product will usually be built on a wireboard, DIP-board, PC board, or whatever works best and is cost effective. You will most likely want to package the circuit in as nice a case as can be economically justified. Fancy cabinets add prestige value to your work and are a valid source of pride to the good craftsman. But if you go ahead and build the circuit before testing it, there may appear problems that cannot be solved cheaply. You may, for example, drill an excess number of inappropriate holes. Or you may have selected a 4.5-by-6.5-inch DIP board, only to find that additional circuitry is required and that a 4.5-by-9-inch board would have been more appropriate. The little slivers of DIP board often found hanging onto such poorly-laid-out projects are known rather contemptuously as "kluge boards."

It is a rather long road from concept to finished product; so don't add many curves and detours through poor procedures or inept planning. At the very least you will use up costly resources and generate no small amount of aggravation.

Let's summarize the steps that will most often result in a viable electronic instrument design:

1. Study the problem that must be solved.
2. Formulate several approaches to solve the problem.
3. Make records, drawings, and documents.
4. Prototype on a breadboard or some other medium that is either expendable or reusable.
5. Test the circuit under conditions that are as realistic as possible.
6. Build and test the final product using the best possible craftsmanship.

BE COGNIZANT

Clever designers are often actually merely knowledgeable designers. It is a truism that "the person in the know runs the show." We may conclude from that one bit of wisdom: good designers often make their own luck by knowing integrated circuits and other components that might be available. They will subscribe to the major design magazines and will collect and hoard manufacturer's applications notes, data sheets, and other literature. Both ordinary catalogues and applications notes from manufacturers are an engineering education in their own right, and the clever designer will take advantage of them. Some manufacturers give away their literature free, while others make a nominal charge. In any event, obtain all that you can (be greedy), and keep it on file. Read the stuff, it's well worth it. If you claim that you have no time for reading, then find time. I maintain that you can read most of the tech literature that you need if only your bowels move in the normal manner. By choosing a prudent storage spot for unread material, you can keep up with the industry! The junk mail sent out by electronic component manufacturers often represents some of the best no-nonsense information available anywhere.

Also be aware that most component manufacturers have applications engineers on the staff whose sole job is to help you (well, almost). If a problem develops, call them and ask for advice. Better yet, if time allows, send them a letter describing the problem and include a copy of the circuit and any other pertinent data. Some of these people are actually only sales engineers, but most know their own product well enough to give the troubled designer some good advice.

Read magazines, even amateur radio and hobbyist electronics magazines. Three of the best professional design magazines are

EDN (Electronic Design News)
Cahners Publishing Company
221 Columbus Avenue
Boston, MA 02116

ELECTRONIC DESIGN
Hayden Publishing Company
50 Essex Street
Rochelle Park, NJ 07662

ELECTRONICS
McGraw-Hill Book Company
1221 Avenue of the Americas
New York, NY 10019

The first two of these are free to *qualified* readers. But unless you are a design engineer or are employed in a related capacity, you may not qualify. Even if you are required to pay, though, these magazines are first-class.

Another point to be cognizant of is the matter of free samples from the major companies. This is especially prevalent among the semiconductor manufacturers. Company policies on the matter of samples vary from "Are you nuts!??!#"" to "Sure, take a handful." Some are very open about giving free samples, while others are impossible. I won't tell you who are which, because it might open me to criticism, and it would put you into competition with me for the free goodies! But experience will soon tell you the answer.

Be aware of the reason why a profit-seeking company would want to give something away. If you are an engineer working on a project that could mean a lot of sales in the near future, almost everybody will give away the goodies. They want a shot at the 10,000-piece order that will be made when you finish the design phase and go into production. If your efforts have no earthly hope for profit for them, don't expect them to be excessively generous with free products. Some will, some won't; and it often depends upon the line you give the salesman. A few will arrange for you to buy or to obtain cosmetic rejects that work electrically.

A well-known exception to this rule is the senior in a BSEE or BSET program who must do a senior project, an onerous task. If you level with them and don't try to feed them a line of stuff, it might be in their interest to assist your efforts. They will recognize a potential customer who will be a working engineer, specifying and buying, within about a year. That is good will for them and is hard to beat. Even if the sales manager is doubtful of your future worth to his company, he might take sympathy on you because of memories from his own senior project in engineering school those many years ago.

Chapter 22

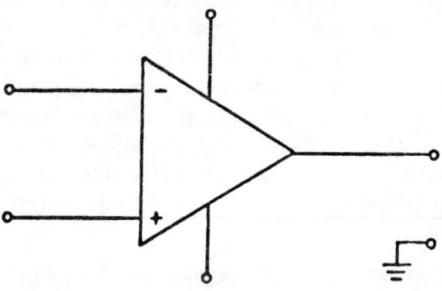

Some Construction Projects

C ERTAIN ELECTRONIC CONSTRUCTION PROJECTS SEEM COMMON-place in electronics facilities that serve researchers, industry, and so forth. This chapter will discuss several of these and give you enough information to allow easy reproduction. It is the firmly held premise of this book that you need neither to obtain an electrical engineering degree nor to hire an electrical engineer in order to design and build certain types of electronic apparatus. Toward the goal of making that premise actually true I have personally built and tested each and every circuit in this chapter.

BEFORE BEGINNING

Some of the circuits are constructed as part of a system of instruments that fit a common rack-mounted main frame. These have been built into Vector EFP204-6.6 modular cabinets. These fit the Vector main frame. Other projects have been built into attractive metal cabinets purchased from local electronic-parts distributors.

Please do not feel that the design of any circuit presented is nailed down just because it is published here as a construction project. Each project is designed to appeal to a large number of readers, but you may or may not have the same problems as the hypothetical average reader. Feel absolutely free to modify any of my circuits to fit your own needs. Copying may well be a form of flattery, and you may direct it toward me if you want. But I would prefer it if you would think a little bit on your own and modify the project to suit your own needs. A better form of flattery would be for you to brag, ". . . just like in Carr's book, except. . ."

To assist you in making modifications to the projects the philosophy of why certain things were done in certain ways is given. Also given, where applicable, is the appropriate design equations.

There is the possibility of problems developing in any electronic construction project. In fact, Finagle's and Murphy's laws are always found in the electronics laboratory. I have discussed the particular problems I found in the article on the project where they were found. Unfortunately, this is just a bookkeeping convention because any one of them is likely to be found in any electronic construction activity. To gain an appreciation of the types of problems to expect, and to gain an insight of just what to do when you do encounter them, please read the entire chapter—in fact the whole book—before attempting to build any project.

A MEDIUM-GAIN SINGLE-ENDED PREAMPLIFIER

The need occasionally arises for an amplifier to boost the level of a signal only a small amount. The amplifier described here will produce a voltage gain between 1 and 100, depending upon how it is adjusted. This situation often arises in facilities that have a wide variety of different instruments of varying ages, manufacturers, and technologies. You might, for example, want to boost the output of an instrument to a level compatible with the full-scale range of a strip-chart recorder, analog input to a digital computer (see A/D converter article in Chapter 14), or the input of an oscillographic display.

There are any number of reasons why such an amplifier might be needed. One cause is that the full-scale output of the instrument or transducer might be only a fraction of the full-scale input of the display device. The amplifier will allow you to increase the span of the display so that better resolution is produced. Another case might be where the instrument or transducer that is used to make a measurement operates only in the lower end of its range during a particular experiment. In that case you may amplify the output, again improving resolution.

The medium-gain, single-ended amplifier is especially suited to interface instruments in situations such as described above. It has two gain ranges, 0 to 10 and 0 to 100, and can handle input voltages up to 1 volt in the times-10 range and 0.1 volt (100 millivolts) in the times-100 range before becoming saturated. Since most oscilloscopes, analog/digital converters, and strip-chart recorders will accommodate signals in the 1-to-10 volt range, you may reasonably expect this amplifier to interface them to almost any type of voltage-output instrument. An untried modification to produce a higher output voltage range might be to use RCA CA3140 BiMOS operational amplifiers and ±22-volt dc power supplies.

This amplifier was designed as part of the author's electronic-instrument package, so was built into a Vector EFP module. You may, of course, select whatever package best suits your own needs or whims.

The layout is not particularly critical since this circuit is low gain and low frequency. You will find, though, that neat wiring in any electronic project will inevitably produce fewer bugs than the so-called rat's-nest approach.

The actual circuit was constructed on a convenient Vector 3677-2DP DIP board, but the full-size board would not allow room inside the module cabinet for front-panel and rear-panel controls, jacks, and switches. This problem was solved by cutting off approximately 1.5 inches of the board at the edge-connector end of the card. Ordinary hookup wire (22 gauge) was used wherever power or ground jumpers had to be made, but a much smaller gauge may be used for the rest of the wiring. Use the heavy type of wire only where mechanical strength, or current-carrying ability, is required.

The electrical circuit for the medium-gain preamplifier is shown in Fig. 22-1. You may notice almost immediately that four operational amplifiers are used, even though some designs are able to get away with fewer. This is not a case of poor, or uneconomical, design practice because of my basic premise that enthusiastic amateurs can reproduce the projects with a good probability of success. Circuits in which several functions are combined in a single stage often require tedious alignment or setup adjustments before they can work; they must be tweaked to specifications. Notice, for example, that I have made the range, sensitivity, and position-control functions in separate operational amplifiers. This is good practice for you, in that setup becomes simple. An industrial designer would probably combine two or more in a single stage at the expense of a more complex procedure. But they are able to save thousands of operational amplifiers at 50 cents apiece. For you, though, building one each of the preamplifiers costs only a buck or two extra.

Power is supplied from an external power pack or from batteries. You may use the circuits from this book to obtain power from the ac mains, a battery pack, or any of the many commercial power supplies on the market. The only requirement is that it be reasonably well-regulated and be free of 60-hertz ripple components.

The power-supply lines are bypassed to ground by a pair of 200-μF, 25-working-volts-dc electrolytic capacitors. You may, however, use anything over 25-μF, and I recommend tantalum types. I used the 200-μF aluminum types because they were in my parts supply, so need not be purchased especially for this project. Scrounging is an old and fine art in electronics; so develop that talent early.

The dc levels are VCC = +15 Vdc and VEE = −15 Vdc. In some instances you will want to use ±12 volts instead, especially if the 12-volt supply is available and a 15-volt type would have to be purchased. The only problem is that the operational amplifier used at the output stage would be unable to deliver the full-scale potential that they would be capable of if 15-volt supplies had been used instead. If this is not a loss, though, don't spend the money for a "proper" power supply.

One almost unique feature of this amplifier is the polarity inversion switch, S1. This switch is used to change amplifier A1 from the inverting

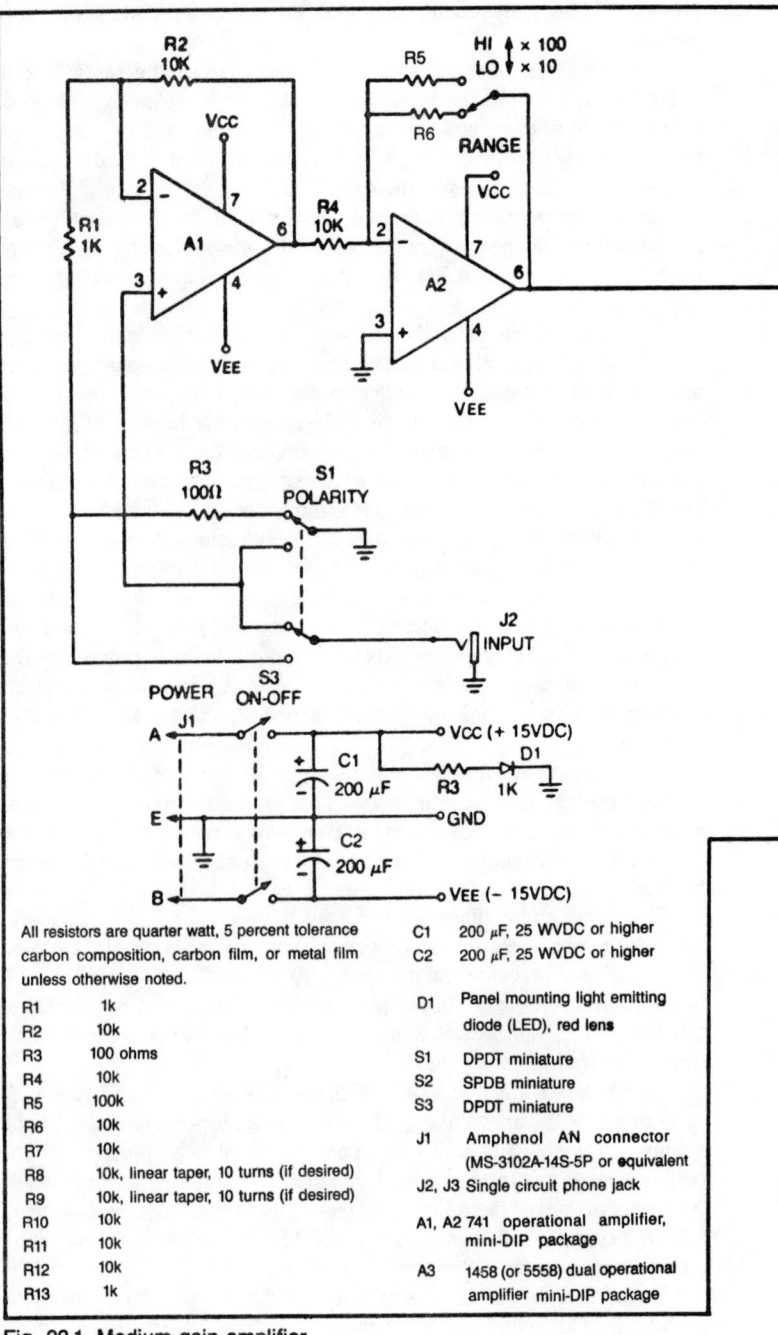

All resistors are quarter watt, 5 percent tolerance carbon composition, carbon film, or metal film unless otherwise noted.

R1	1k
R2	10k
R3	100 ohms
R4	10k
R5	100k
R6	10k
R7	10k
R8	10k, linear taper, 10 turns (if desired)
R9	10k, linear taper, 10 turns (if desired)
R10	10k
R11	10k
R12	10k
R13	1k

C1	200 μF, 25 WVDC or higher
C2	200 μF, 25 WVDC or higher
D1	Panel mounting light emitting diode (LED), red lens
S1	DPDT miniature
S2	SPDB miniature
S3	DPDT miniature
J1	Amphenol AN connector (MS-3102A-14S-5P or equivalent)
J2, J3	Single circuit phone jack
A1, A2	741 operational amplifier, mini-DIP package
A3	1458 (or 5558) dual operational amplifier mini-DIP package

Fig. 22-1. Medium gain amplifier.

378

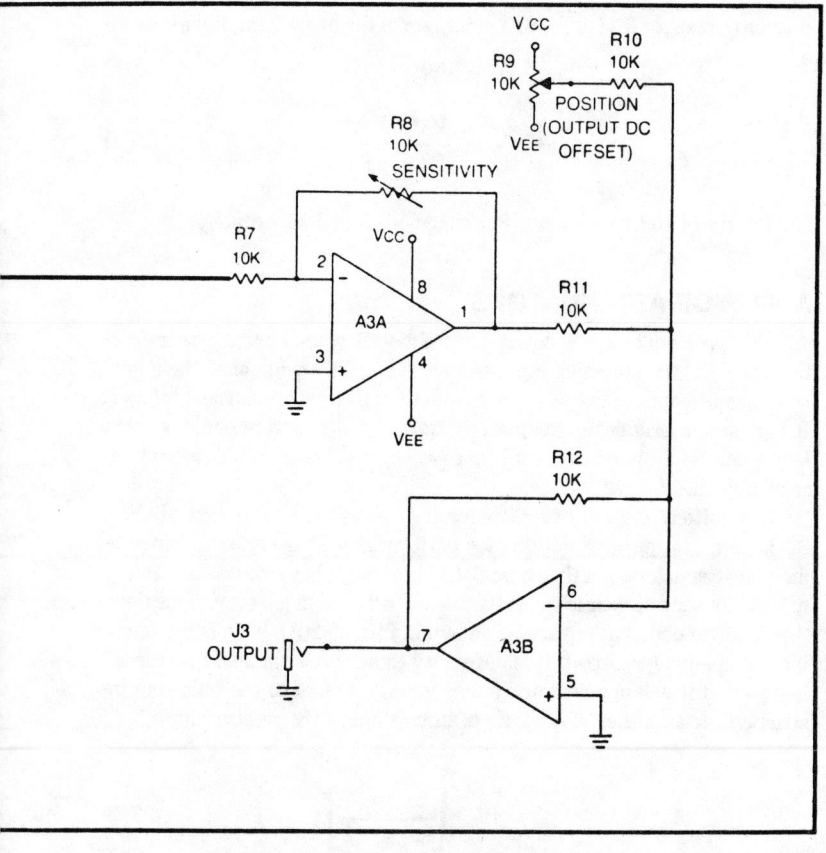

to the noninverting mode at will. Murphy's law implies that any tracing will be upside down some of the time; so the polarity switch will have obvious merit.

When switch S1 is in the upper position, the overall amplifier will produce an inverted output, relative to the input polarity. In that case A1 is operated as a noninverting follower. In the down position of S1, however, the input stage is also an inverter; so an even number of polarity inversions takes place between input and output. This will make the entire amplifier perform as a noninverting stage.

The only significant problem that developed in the design and construction of this project came about as a result of the inclusion of the polarity-switch feature. Recall that the voltage gain of a simple dc follower is given by the relationship (R_f/R_{in}) for the inverting mode and this quantity plus 1 in the noninverting mode. Since the gain of the stage is so low (10), this difference would result in a gain error of 10 percent between the two modes. This error is reduced to approximately 1 percent, which is tolerable, by the use of resistor R3 (100 ohms). The gain of the overall amplifier is given by

$$A_v = \frac{R2}{R1} \times \frac{R5^*}{R4} \times \frac{R8}{R7} \qquad (22.1)$$

*In the times-100 range only. For times-10 range, substitute R6.

A PAIR OF ATTENUATORS

One problem that is encountered frequently when bench testing or interconnecting is excessive input-signal levels. This will cause the signal from one instrument to overdrive the input of the instrument that follows. It has proven prudent to keep a few standard attenuators handy so that levels can be dropped to points that are compatible with the instrument receiving the signal.

Two different types of attenuator are preferred and are shown in Fig. 22-2. Neither is perfect, nor do they perform as well as eighty-dollar commercial attenuators, but then again, they cost only about four dollars apiece to build. In spite of their low cost they do a good enough job most of the time.

A single-ended attenuator is shown in Fig. 22-2(A). This is the familiar voltage-divider circuit. It is necessary to keep the value of resistor R2 low so that it will appear to be a true voltage source to the following instrument or amplifier. This type of circuit obeys the relationship

$$E_{out} = (E_{in}) \left[\frac{R2}{R1 + R2} \right] \qquad (22.2)$$

The term inside the brackets is the *attenuation factor* and will always have a value less than unity. It represents the fraction of the input voltage that will appear at the output.

Determination of the value of R2 is by convention, rather than a hard scientific rule. Make it 100 ohms in most cases, but consider that it must be equal to the input impedance of the stage to follow if power transfer is the desired goal, and less than one-tenth of the input resistance of the following stage if it is to appear as a voltage source.

The other values are determined by plugging in the required attenuation factor and solving for R1. For example, suppose you want to make an attenuator such as this and it is to have an attenuation factor of one-twentieth. Set up the equation:

$$\frac{1}{20} = \frac{R2}{R1 + R2} \qquad (22.3)$$

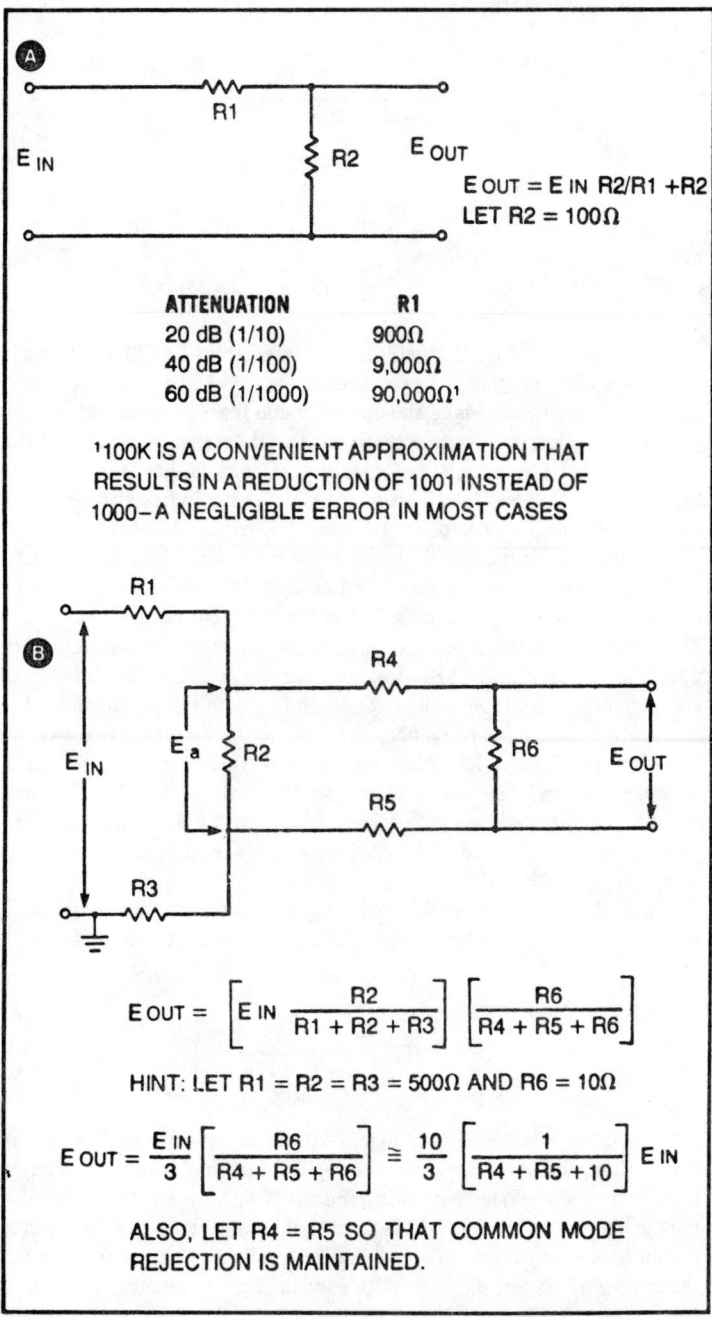

At (A), a single-ended attenuator:

$$E_{OUT} = E_{IN}\ R2/R1 + R2$$
LET $R2 = 100\Omega$

ATTENUATION	R1
20 dB (1/10)	900Ω
40 dB (1/100)	$9{,}000\Omega$
60 dB (1/1000)	$90{,}000\Omega$[1]

[1] 100K IS A CONVENIENT APPROXIMATION THAT RESULTS IN A REDUCTION OF 1001 INSTEAD OF 1000 – A NEGLIGIBLE ERROR IN MOST CASES

At (B), a differential attenuator:

$$E_{OUT} = \left[E_{IN}\ \frac{R2}{R1 + R2 + R3} \right] \left[\frac{R6}{R4 + R5 + R6} \right]$$

HINT: LET $R1 = R2 = R3 = 500\Omega$ AND $R6 = 10\Omega$

$$E_{OUT} = \frac{E_{IN}}{3} \left[\frac{R6}{R4 + R5 + R6} \right] \cong \frac{10}{3} \left[\frac{1}{R4 + R5 + 10} \right] E_{IN}$$

ALSO, LET $R4 = R5$ SO THAT COMMON MODE REJECTION IS MAINTAINED.

Fig. 22-2. At (A), a single-ended attenuator, and (B) a differential attenuator.

Plug in the predetermined value for R2, 100 ohms, and solve for R1

$$\frac{1}{20} = \frac{100}{R1 + 100}$$

$$R1 + 100 = (100)(20)$$

$$R1 = 2000 - 100$$

$$R1 = 1,900 \text{ ohms}$$

Some of the values that will be calculated from Eq. 22.3 will not be found in a table of standard resistance values; so you will not be able to go out and buy them. Make the correct value from either a combination of fixed resistors or from a potentiometer. Of course, you will find that the error caused by using standard values shrinks as the attenuation factor gets very large. A 60-dB attenuator, for example has an attenuation factor of 1/1000. This can be approximated by using a 100-kilohm resistor for R1 and my "standard" 100-ohm resistor for R2. The resultant error will be very small, probably less than 1 percent if the resistor is out of value on the low side of 100 kilohms, but still within tolerance.

A balanced attenuator used to feed instruments with differential inputs is shown in Fig. 22-2(B). It is necessary to keep the circuit balanced and symmetrical, so maintain a condition in which R1 = R3, and R4 = R5. For the sake of convenience, although not an absolute necessity, set R2 equal to R1 and R3. Also, the value of these resistors should be between 200 and 500 ohms. Note that three standard-value, 220-ohm resistors in series will provide a good match for the 600-ohm output impedance ordinarily found on function generators used in most electronics laboratories.

If all three resistors in the series input divider are equal, voltage E_a across resistor R2 will be equal to $1/3 \ E_{in}$. The output voltage from the entire pad will be

$$E_{out} = (E_a)\left[\frac{R6}{R4 + R5 + R6}\right]$$

The balanced attenuator of Fig. 22-2(B) is used mainly for driving high-gain differential amplifiers from the regular function generator. Many of these instruments do not have an attenuator of their own suitable for reducing the output to a low level. In fact, most of them will produce controllable output levels only down to a few hundred millivolts before running into the lower-end stop of the attenuator control. This attenuator, and the one preceding, will allow you to use most of the signal generator's attenuator range, while delivering low-level signals compatible with the input of the instrument you are working on.

Both types of attenuators should be built inside shielded, aluminum boxes. The Pomona blue boxes are especially made for this type of application and are of high enough quality to allow the attenuator to remain in good condition for a long time. You are cautioned against using the sheet-metal type of chassis box for this application because they simply are not durable enough. Chances are about even-up that a critical attenuator box will be stepped on or have a heavy cart rolled over it about five minutes before it is needed. For this reason I always use the die-cast, Pomona type 2417, blue box. It is heavy, die cast, painted a pretty color (if you like blue), and rugged enough to withstand you.

It is also necessary to use appropriate connectors for the attenuator box. In the case of the single-ended version, use BNC connectors or whatever is compatible with the rest of your system. Check the Pomona Electronics catalog for a model the same size as the 2417, but which has the appropriate connectors—several types are made. In the balanced version a BNC will do for the input end, but a stereo (1/4-inch) phone jack is probably the most economical bet for the other end.

You may want to build a variable or switchable attenuator so that you will always have a universal fudge factor (like the specific gravity of peanut butter) at your finger tips. A potentiometer will make a variable attenuator. It may be used at either R1 or R2, but I prefer the former. Similarly, R1 may be replaced by a switch capable of substituting various values. I use one that is switch-selectable between 40 dB (1/100) and 60 dB (1/1000).

"UNIVERSAL" OPERATIONAL AMPLIFIER

There sometimes arises a need for a user programmable operational amplifier. Figure 22-3 shows the circuit for such a device. The power supplies, plus any needed compensation (if premium operational amplifiers are used), are built into the housing so the user need not worry about the actual electronic nitty-gritty; yet the unique advantages of the operational amplifier are still available.

When you have several operational amplifiers of this design sharing a common power supply/main frame, it is often called an *operational amplifier manifold*. The individual operational amplifiers are referred to as *sections*. It is often the case that the op-amp manifold sections will be plug-in units, so that the main frame can be set up as needed for any specified purpose.

Both inputs (– and +), ground, and the output of the operational-amplifier IC used in this project are brought out to five-way banana-plug binding posts on the front panel. These are for the user to program the circuit in any way desired. Also provided are input and output BNC connectors so that the instrument can be connected to other instruments at will.

Switch S2 selects the mode of operation. When S2 is open, the operational amplifier must be *externally programmed* before it can be used. Closing switch S2, on the other hand, makes the stage a noninverting follower which can have a voltage gain between 0 and 100, depending upon the setting of the potentiometer. Should greater gain be desired, you may either cascade several of these amplifiers, or externally program one for voltage gain greater than 100. The dc and low-frequency ac gain of the stage is given by

$$\text{Gain} = \frac{R_f}{R_{in}} \tag{22.4}$$

Operational amplifier manifolds are particularly popular in the fields of chemistry and the life sciences. Users can make them perform any number of jobs normal to the operational amplifier's abilities even though the user may have little in the way of electronic expertise or design acumen. It makes a truly universal (if anything "universal" exists) amplifier that will amplify, isolate, integrate, differentiate, or oscillate as *you* wish.

If you elect to build the op-amp manifold in the form of a larger cabinet, such as the type that accepts the Vector EFP modules, it might be prudent to place a unity-gain, noninverting follower on the output lead—not to the banana post but to the BNC jack—as a buffer. Also, you might want to put such a stage between each of the input banana posts and the inputs of the IC operational amplifier. These are for buffering and isolating. Use of the 741-family of devices or the semi-premium 1456 will prove suitable for all but the most critical applications.

NULL VOLTMETER

Figure 22-4 shows the circuit to a *null voltmeter*. This project is designed to have two ranges and is used primarily for determining the existence of a null point in a Wheatstone bridge external to the instrument. It will also act as an amplifier that will deliver a 1-volt output signal for input signals as low as 10 millivolts. The signal at output jack J5 will be either 10 or 100 times the amplitude of the input signal applied across input jacks J1 and J2.

Also featured are a bridge-excitation output (not mandatory but sometimes convenient to have available), plus both lamp and meter types of null indicator. The sensitivity of the circuit can be varied by potentiometer R10 on the front panel.

Readout meter M1 is a 100-0-100 microampere type converted to a voltmeter by virtue of resistor R15. The particular Simpson meter I used had a coil resistance of about 25 kilohms. DO NOT EVER place a microammeter across the probes of an ohmmeter unless you are absolutely certain that the ohmmeter is a low-current type. Most analog ohmmeters (the kind with a needle pointer) are high-current types and will burn out the

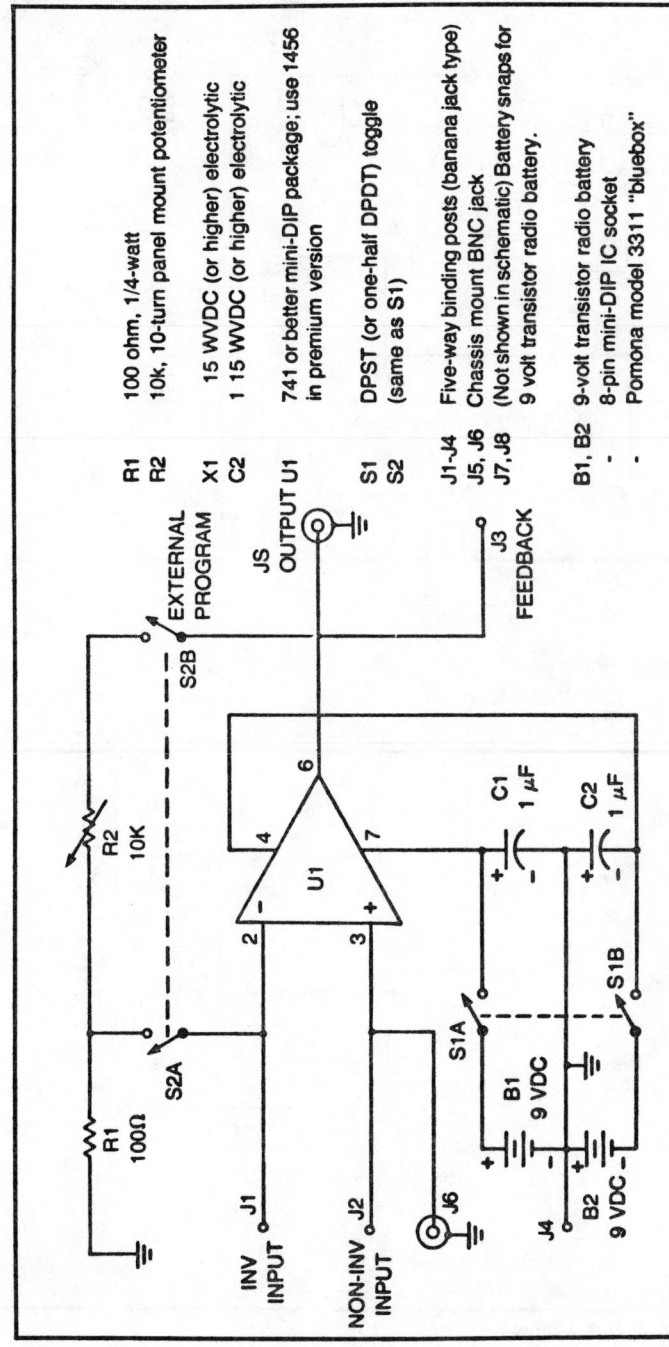

Fig. 22-3. Universal op-amp manifold section.

R1	100 ohm, 1/4-watt
R2	10k, 10-turn panel mount potentiometer
X1	15 WVDC (or higher) electrolytic
C2	1 15 WVDC (or higher) electrolytic
U1	741 or better mini-DIP package; use 1456 in premium version
S1	DPST (or one-half DPDT) toggle
S2	(same as S1)
J1-J4	Five-way binding posts (banana jack type)
J5, J6	Chassis mount BNC jack
J7, J8	(Not shown in schematic) Battery snaps for 9 volt transistor radio battery.
B1, B2	9-volt transistor radio battery
-	8-pin mini-DIP IC socket
-	Pomona model 3311 "bluebox"

385

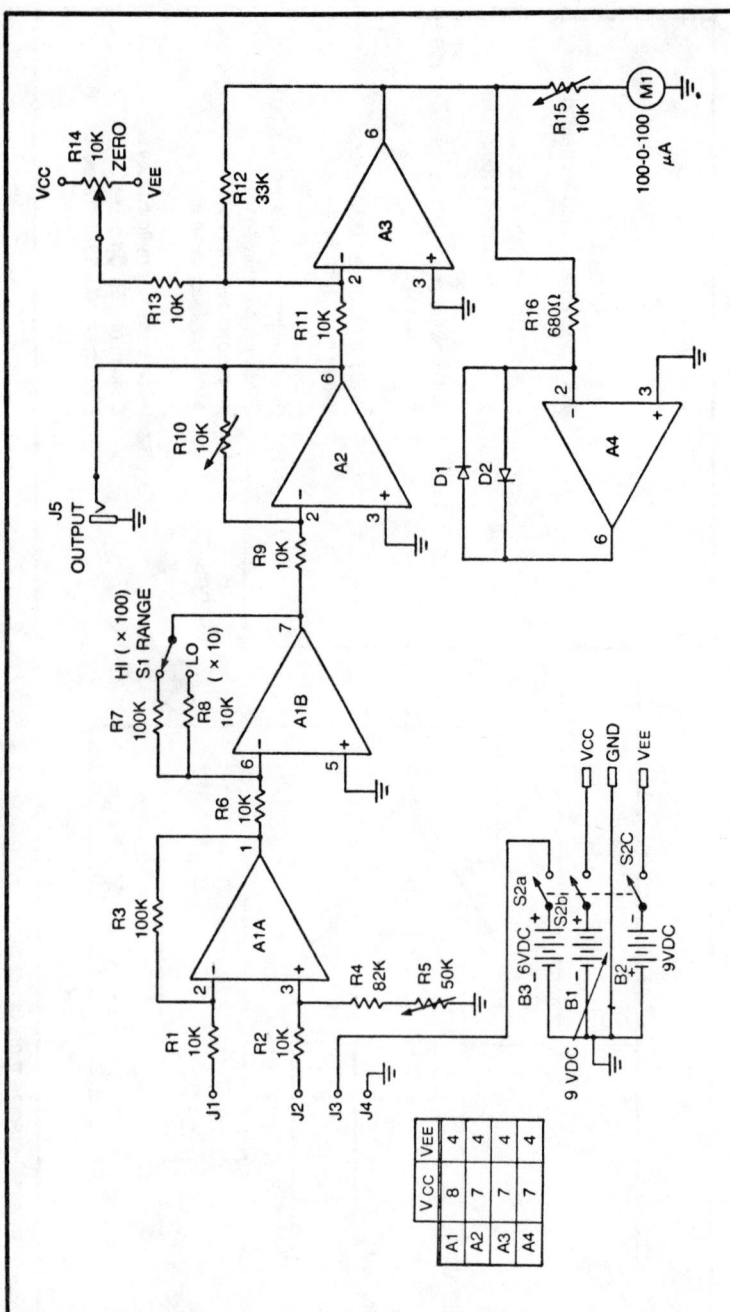

Fig. 22-4. Null voltmeter.

386

All resistors used in this project are 5% or better carbon film, metal film, or precision types rated at 1/4 watt. Best accuracy will result from use of precision resistors.

R1, R2, R6, R8 R9, R11–R13	10k
R3	100k
R4	82k
R5	50k, 10-turn, PC mount trimpot
R7	100k
R10	10k, linear taper, panel mount potentiometer, 10-turn preferred
R14	10k, 10-turn panel mount trimpot
R15	10k, 10-turn PC mount trimpot
R16	680 to 1500 ohm, as desired for proper lamp sensitivity

All ICs are mini-DIP types

A1	1458 dual op amp
A2–A4	741 op amp
D1, D2	Red LED with mounting collar
J1–J4	Large size five-way (banana) binding posts
J5	Phone jack
B1, B2	9-volt transistor radio battery
B3	four, series-connected 1.5 Volt AA penlight cells in a battery holder.
4	mini-DIP IC sockets
1	piece of Vector 3677 perfboard
1	100-0-100 microammeter
3	battery snaps for 9-volt transistor radio batteries
1	4-cell battery holder fitted with the above type snaps
1	SPDT miniature toggle switch
1	3PST miniature toggle switch
1	suitable chassis and cabinet miscellaneous hardware

microammeter as you attempt to find its coil resistance! A *few* modern digital ohmmeters use a very-low-voltage (hence low-current also) design. Any instrument that uses a 1.2 volt-or-higher battery in the ohmmeter section is totally unsuited to this purpose. In that case configure a low-current method for determining resistance by comparison, or look it up in the manufacturer's catalogue or sales literature.

The voltage required to bring M1 to full-scale, by Ohm's law, is

$$E = (1 \times 10^{-4})(2.5 \times 10^{4})$$

$$E = 2.5 \text{ volts}$$

But in the original design concept, I wanted to have a nice, well-behaved set of power-of-10 gains, so that either 10-mV or 100-mV input potentials would produce an output voltage at J5 of 1 volt. Because of this requirement, amplifier A3 had to have a gain of at least 2.5 in order to boost the voltage supplied to the meter to a level high enough to drive M1. It is also desirable to have some extra series resistance to protect M1 from overvoltage conditions. If I used 33 kilohms for the feedback resistor on A3, it allows me to use a standard, off-the-shelf value and raises the voltage applied to M1 to a point where some extra series resistance has to be used. This makes the gain of amplifier A3 equal to 3.3; so the voltage at pin 6 of A3, when the input is seeing a full-scale signal, is also 3.3.

The value for the potentiometer in series with M1 is found by combining Ohm's law with a little common sense. Since the total output voltage will be 3.3 and I want 100 microamperes full-scale, the total resistance of M1 plus the resistance of R15 must be 33 kilohms. Since the resistance of the meter movement was known to be 25 kilohms, the minimum resistance for R15 must be 33, 25, or 8 kilohms. The choice of a 10-kilohm potentiometer allows me to use a standard value and gives me a little over-range to permit minor adjustment for calibrating M1.

Adjustment Procedure

1. Short together J1, J2, and J4.
2. Set R10 to maximum resistance; set S1 to high.
3. Adjust R14 for zero reading on M1.
4. Disconnect short to J4 but leave J1 shorted to J2.
5. Connect a 100-hertz, 1-volt signal source between ground and the junction J1 and J2.
6. Connect the vertical input of an oscilloscope to J5.
7. Adjust R5 to reduce the output signal displayed on the oscilloscope to its lowest value. This might be down in the noise; so use successively lower settings of the scope's vertical input attenuator (range) until no further reduction is possible.

8. Set switch S1 to low.
9. Connect a precise 100-mV potential across J1 and J2 (Remove the short first!).
10. Adjust R15 for precisely 100 μA on M1.

When this procedure is finished, M1 will read 0 to 10 mV in high and 0 to 100 mV in low, provided that potentiometer R10 is at its maximum setting. This makes this project useful as a null voltmeter, a millivoltmeter, or an amplifier for 10-mV-to-100-mV signals.

Resistor R16 sets the sensitivity of the null-indicator lamps. At 680 ohms it was found that the brilliance of the two light-emitting diodes would extinguish entirely within a few millivolts of the actual null. Of course, that value merely reflects the properties of the LEDs in my stock and may be slightly different from the LEDs which you buy; so expect to play with this value.

If you want to use these lamps as a coarse null indicator, set the resistance closer to 1500 ohms. In that case, the lamps may be used when the external bridge is so far out of adjustment that M1 is driven over-range. The meter will be used when the null gets deeper and the bridge is closer to being balanced.

Power is supplied to the internal circuitry by a pair of ordinary, 9-volt, transistor-radio batteries. These supply the VCC and VEE potentials required by the operational amplifier. A set of four series-connected penlight cells, however, can be used to supply the excitation potential for the external-bridge circuitry.

HIGH-GAIN AMPLIFIER

The circuit to a high-gain amplifier is shown in Fig. 22-5. Although it is a genuine instrumentation amplifier, it is a relatively low-performance design due to the type of operational amplifiers and resistors selected. In the parts list are alternatives that will yield an improvement in performance for those who wish to make the necessary investment in parts. The amplifier will perform reasonably well, however, with the 741 line-up specified. You could build it with sockets for the operational amplifiers, specifying the 8-pin, mini-DIP, op-amp package and then try it with 741 devices. They are cheap and ubiquitous. If performance is not up to your needs, substitute a more costly premium op amp that has the same pinouts (so-called industry standard).

Devices A1 through A3 are connected in the classic instrumentation-amplifier configuration. These form the front-end stages of the circuit. Incidentally, it is in these stages that most high-gain amplifiers get into trouble. If excessive noise potentials or drift voltages develop here, they will be magnified by high-gain stages that follow. The common-mode rejection ratio of this circuit is optimized by potentiometer R8. Amplifier A4 is simply an inverting gain block with an amplification factor of either 10 or 27,

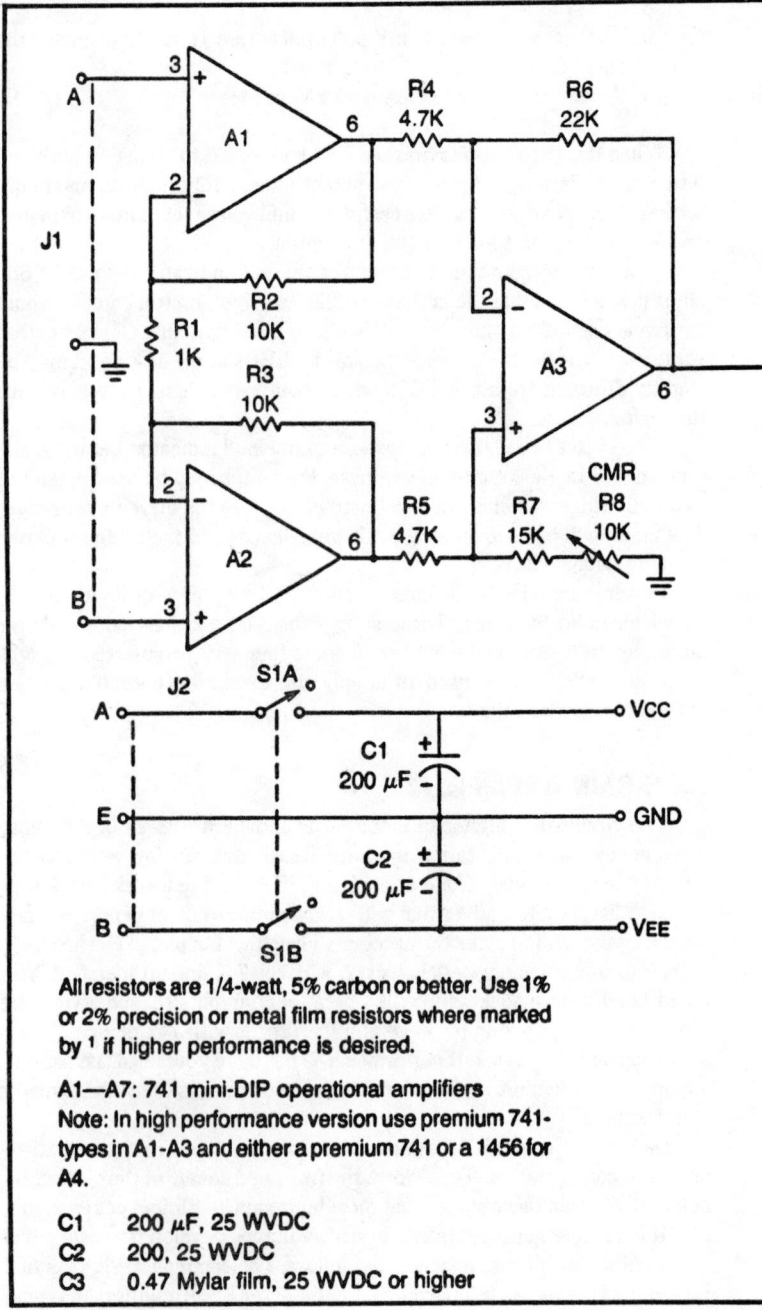

All resistors are 1/4-watt, 5% carbon or better. Use 1% or 2% precision or metal film resistors where marked by [1] if higher performance is desired.

A1—A7: 741 mini-DIP operational amplifiers
Note: In high performance version use premium 741-types in A1-A3 and either a premium 741 or a 1456 for A4.

C1   200 μF, 25 WVDC
C2   200, 25 WVDC
C3   0.47 Mylar film, 25 WVDC or higher

Fig. 22-5. High gain amplifier.

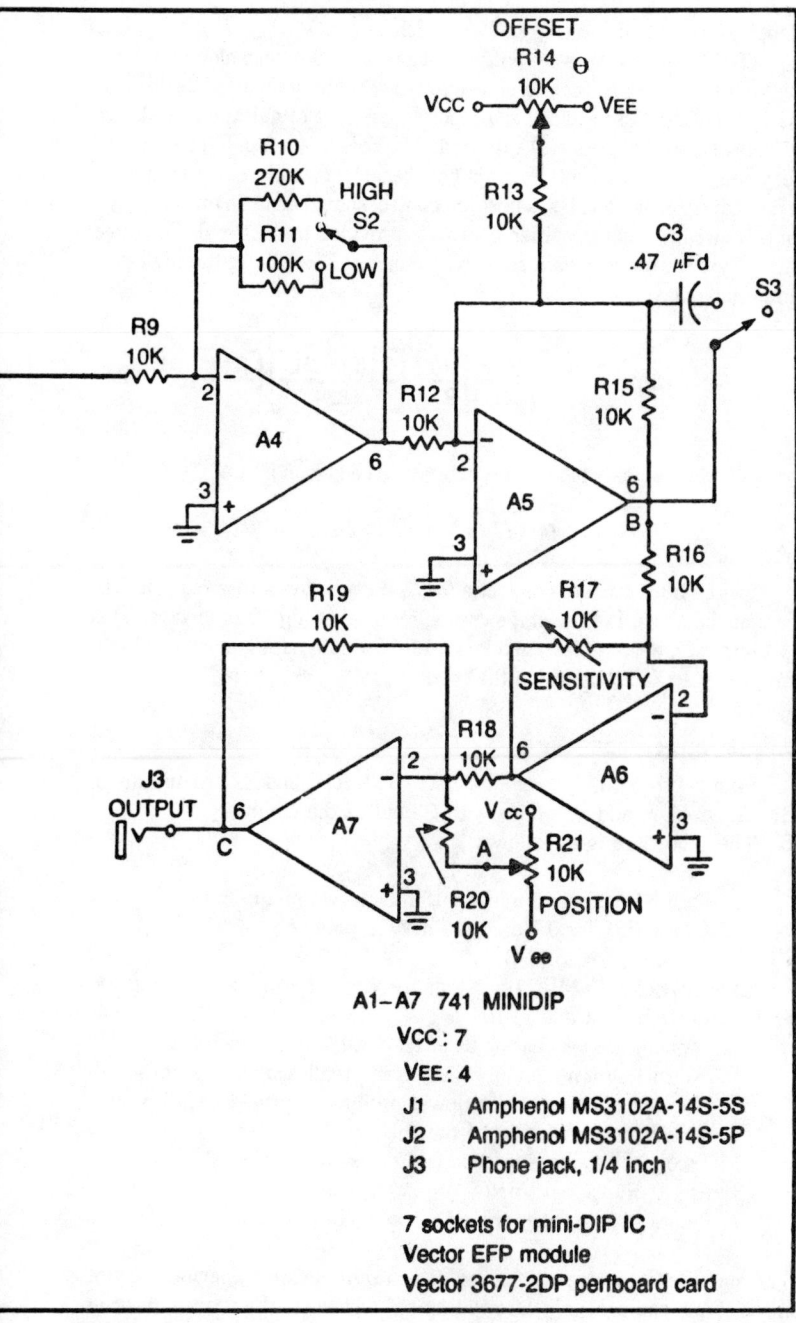

A1– A7 741 MINIDIP

VCC : 7

VEE : 4

J1   Amphenol MS3102A-14S-5S
J2   Amphenol MS3102A-14S-5P
J3   Phone jack, 1/4 inch

7 sockets for mini-DIP IC
Vector EFP module
Vector 3677-2DP perfboard card

depending upon the setting of switch S2.

The last three stages are housekeeping stages designed along the same philosophical lines as similar stages in the medium-gain amplifier of Fig. 22-1. Amplifier A5 provides a trimpot adjustment to null the cumulative effects of the dc offsets of A1 through A5. This may seem to have the effect of a position control, but it is not, because it is placed internally as a null adjustment. Besides, a position control should be in an output stage. In this circuit, output position control is provided by R21 and A7. Stage A6 is the zero-to-unity-gain sensitivity control. The gain of the high-gain amplifier is given by

$$A_v = 1 + \left[\frac{2R2}{R1}\right]\left[\frac{R6}{R4}\right]\left[\frac{R10 \text{ or } R11}{R9}\right]\left[\frac{R17}{R16}\right] \quad (22.5)$$

or, with the values which I selected and R17 at maximum

$$A_v = (21)\ (4)\ (27 \text{ or } 10)\ (1) = 2{,}268, \text{ or } 840$$

Switch S3 is used to connect a 0.47 μF capacitor across resistor R15 so that the ac gain of the stage can be reduced at high frequencies. This is a form of low-pass filter, and it is useful for eliminating noise and high-frequency artifacts riding on the input signal.

Adjustment Procedure

First set the controls as follows: R8, R14, R17, and R21 to midrange; S1 off, S2 high, and S3 open—with C3 out of the circuit.

The procedure is as follows:

1. Turn S1 to on and allow a 15-minute warm-up.
2. Adjust R21 for 0 Vdc (± 10 mV) at point A.
3. Connect J1-A to J1-B.
4. Connect a 100 Hz, 1-volt signal source between ground and the junction J1-A and J1-B.
5. Connect the vertical input of an oscilloscope to point B.
6. Adjust R8 for minimum signal on the oscilloscope. Repeat this adjustment on successively lower oscilloscope ranges until no further improvement can be obtained.
7 Disconnect the oscilloscope and the signal source.
8. Adjust R14 for 0 Vdc (± 10 mV) at point B.
9. Remove the jumper from J1-A and J1-B.

The amplifier is now ready for use. If you were unable to perform any of these steps, you probably screwed up somewhere in the procedure or in construction, and it will be necessary to troubleshoot your work. Don't worry too much about it, if this occurred, because almost everybody fouls up an op-amp project at one time or another.

Whenever high-gain amplifiers are built there is a strong possibility of problems developing. Parameters and those things which would have little importance in lower-gain applications suddenly take on almost critical importance when high gain is attempted.

In all amplifiers it is necessary to use good layout techniques, but in high-gain amplifiers, especially if premium (i.e., high-frequency) operational amplifiers are used, it is critical to having any success at all. Try to keep amplifiers A1 and A2 separated from the high-level stages, A5 through A7. Keep in mind that at gains of 2000 and more, a few millivolts of signal pickup at the A1 and A2 end can result in saturation at A7.

If there appears to be a lot of noise on low-level signals, try either hand-picking from a selection of ordinary 741s or buying a premium 741 device such as the RCA CA6741. It is also possible to use one of the low-noise, high-slew-rate types offered by most manufacturers, but keep in mind that this will probably mean the addition of frequency-compensation capacitors. This problem is discussed in *Op-Amp Circuit Design and Applications* (TAB book No.787). Handpicking, incidentally, may be tedious, there is enough variation in 741 noise levels to make it at least a good possibility.

MULTIPURPOSE DIFFERENTIATOR

A differentiator is a device that delivers an output voltage that is proportional to the derivative, or instantaneous rate of change, of the input signal. In most applications where the differentiator is found, it will follow another instrument that actually acquires the signal from the source. A physiologist, for example, might use a pressure transducer to obtain an arterial waveform from an animal. This signal can be fed through a differentiator to obtain the dP/dt signal. In another situation, we might find an engineer who wants to obtain acceleration and velocity information from an experiment in which a potentiometer is used to generate a position signal. He knows from physics that, for motion along a position axis (designated as x), the velocity is the first derivative of position and that acceleration is the first time derivative of velocity, therefore, the second time derivative of position. That is to say $V = dx/dt$ and $A = dV/dt$. If we pass the position signal generated by the potentiometer through a differentiator will obtain the velocity signal, and if this is passed through a second differentiator, we have the acceleration signal. Figure 22-6 shows the circuit for a differentiator with sufficient time constants to permit differentiation of many differently types of signal.

In this book and most other books that discuss operational amplifier circuitry, you will see the basic operational-amplifier differentiator in which the inverting input of the amplifier is used. This design works fine in fixed-time-constant designs but falls down rather dramatically when multi-time-constant design is attempted.

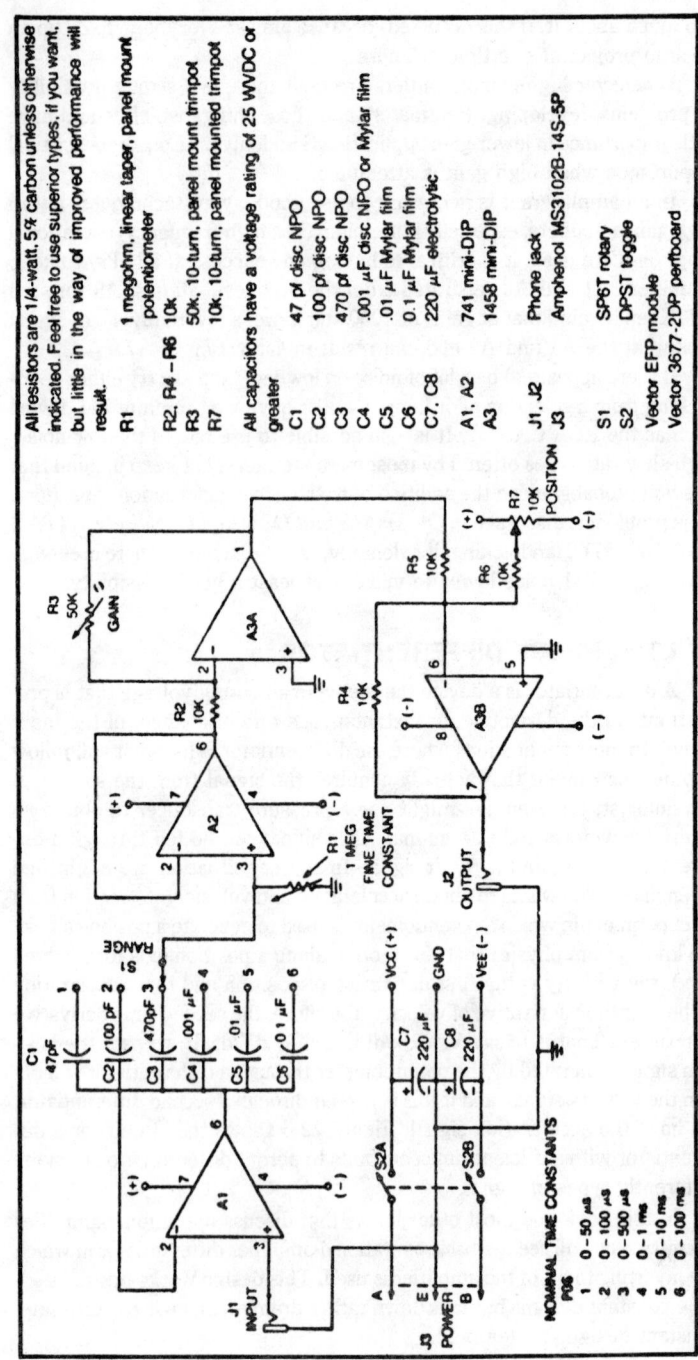

Fig. 22-6. Practical operational amplifier differentiator.

All resistors are 1/4-watt, 5% carbon unless otherwise indicated. Feel free to use superior types, if you want, but little in the way of improved performance will result.

R1	1 megohm, linear taper, panel mount potentiometer
R2, R4 – R6	10k
R3	50k, 10-turn, panel mount trimpot
R7	10k, 10-turn, panel mounted trimpot

All capacitors have a voltage rating of 25 WVDC or greater

C1	47 pf disc NPO
C2	100 pf disc NPO
C3	470 pf disc NPO
C4	.001 μF disc NPO or Mylar film
C5	.01 μF Mylar film
C6	0.1 μF Mylar film
C7, C8	220 μF electrolytic

| A1, A2 | 741 mini-DIP |
| A3 | 1458 mini-DIP |

| J1, J2 | Phone jack |
| J3 | Amphenol MS3102B-14S-5P |

| S1 | SP6T rotary |
| S2 | DPST toggle |

Vector EFP module
Vector 3677-2DP perfboard

394

In the first attempt at this project, which was built for a physiologist, I found that the circuit tended to oscillate somewhat at some settings of R1. This was when the classical op-amp differentiator (along with an inverting input) was used. Apparently, the phase shift of the RC network and the inherent 180° phase shift of the inverting input conspired to produce a total phase shift of 360° at some frequencies within the gain range of the op amp. The classic cure for this is a greater-value resistor in series with the input capacitor, but it was found that this is only good over a narrow range of R1 values. Using other capacitors would bring the oscillations back.

The only effective cure that allowed retention of the many ranges was to use the unity-gain, noninverting-input mode, as shown in the schematic. This eliminated the oscillations for all positions of the range switch and all positions of R1.

The actual differentiator is RC network C1 through C6 and R1. The other stages (A1 and A2) form isolation buffers to keep things well behaved. There is also a sensitivity stage (A3A) and a combination output-buffer and position-control stage (A3B).

All IC devices used in this circuit are 741 operational amplifiers, or at least of the 741-class (i.e., 1458s). These are more than adequate for the time-constant ranges that can be selected by S1. Use of high-slew-rate devices only complicates matters because they may well have a tendency to oscillate as they have a much higher gain-bandwidth product. This allows the mechanism discussed earlier as my objection to the inverting-input type of differentiator.

Notice that the gain of this instrument is rather low, being adjustable only over the range 0 to 5. The reason for this is that input signals tend to be high-level because they are derived from other instruments. In fact, it is very common for most instruments to produce full-scale output signals in the 0-to-1-volt or 0-to-10-volt range. Since the differentiator usually follows such an instrument, it has little need for gain. In fact, if it had too much gain, it may well run into trouble with high-slope input waveforms.

The RC time constants selectable by S1 are:

S1 Position	T = RC
1	50 μs
2	100 μs
3	500 μs
4	1 ms
5	10 ms
6	100 ms

(All positions are actually zero to the given time, which is maximum.)

Keep in mind that these are nominal times due to errors in the actual values of capacitance. Normally, off-the-shelf capacitors are ±10 percent of their marked values, and some are as poor as ±20 percent. Also, the

47- and 470-pF capacitors (C1 and C3) were selected because they are easily-obtainable standard values. To make the ranges actually agree with the calibration, try to obtain 50- and 500-pF capacitors.

In any event, if you have a good capacitance meter or LC Wheatstone bridge available (hint: find a local engineering school), then a far more accurate time-constant range can be realized by hand-picking the capacitors from an assortment of the correct nominal value. Incidentally, it probably won't matter much in most experiments if the actual time constant is off a little bit.

Adjustment Procedure

Set the controls as follows: S1 to position 6, S2 to off, R1 to maximum resistance, and R3 and R7 to midrange.

The procedure is as follows:

1. Connect the power cable to J3.
2. Turn S2 to on, and allow 15 minutes for warm-up.
3. Short J1 to ground.
4. Connect a voltmeter or calibrated dc oscilloscope to J2.
5. Adjust R7 for 0 Vdc ($\pm$10 mV) at J2.
6. Check for differentiation by applying a square wave to the input and looking for the output waveform in Fig. 22-7 on an oscilloscope. The degree of differentiation will depend upon the frequency of the square wave (use less than 400 Hz) and the setting of S1. Try to adjust the RC time constant through manipulation of R1 and S1 so that it is approximately 1/10 of the period of the input square wave.

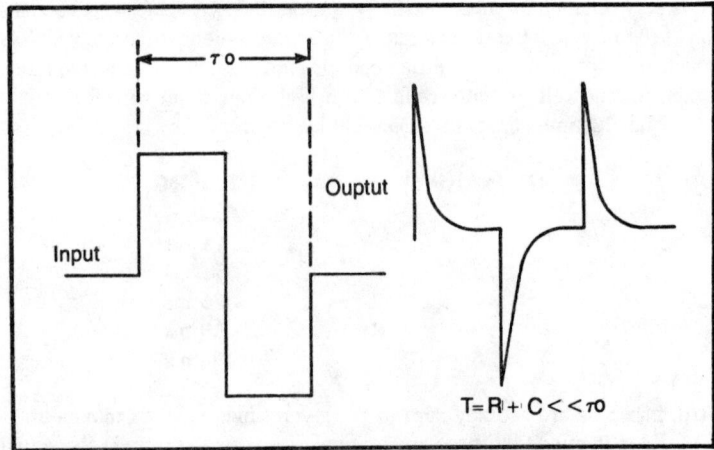

Fig. 22-7. Waveforms from differentiator. It is usual to set the differentiation RC time constant, T, to $\approx$ 1/10 the period of the input waveform, t.

In most cases the output of the differentiator need not be calibrated. Most applications only require the relative differential so that the shape can be examined. If you want numbers, though, some means of calibration is required.

Most users will calibrate the differentiator with a ramp voltage at the input. This can be obtained from a triangle generator or a genuine sawtooth/ramp generator. It is necessary to determine the slope rise of the ramp on an oscilloscope. Adjust the amplitude and frequency of the sawtooth so it fills most of the screen, and find the slope in volts/time. This can be determined from the voltage along the vertical scale and the time along the horizontal.

If a ramp function is applied to a differentiator the output will be a constant voltage. This is because the slope of the ramp is constant. If a ramp with a known slope has been applied, the output voltage from the differentiator will be proportional to a derivative with the same volts/time units. Best precision is obtained if the scale is checked at several points using different slope-ramp functions.

WIDEBAND GAIN BLOCK

The amplifier circuit shown in Fig. 22-8 is a wideband gain block designed to be used between the output of certain electronic instruments and the inputs of recording or display devices such as chart recorders or oscilloscopes. It is assumed that the driving instrument has a relatively low output level that is not compatible with the input requirements of the display device. Without the gain block it is found that the tracing would be too small for practical use.

The heart of this circuit is the RCA type CA3140 operational amplifier. This device is of the BiMOS family which uses diode-protected MOSFET transistors at the input to improve input impedance and a bipolar transistor at the output in order to be compatible with other operational-amplifier requirements, the best of both worlds, as it were. This operational amplifier features relatively low noise when operated at VCC/VEE of ± 5 volts and a frequency response somewhat greater than that of the 741. A lower-noise operational amplifier similar to these is the National Semiconductor BiFET series of devices in the LM-150 family.

The gain block is constructed in a Pomona 2901 blue box. Input and output connectors (J1 and J2) are BNC types in my version, but you can use whatever fits your own instrument system. Chances are pretty good, though, unless your stuff is old enough to have voted in several presidential elections, it will have BNC input and output connectors.

Mount the two BNC connectors on opposite ends of the Pomona box. The Amphenol 126-197 chassis connector is mounted on one of the sides remaining free. A suitable power cable must be made using a mating Amphenol 126-196 connector. Battery power is suggested.

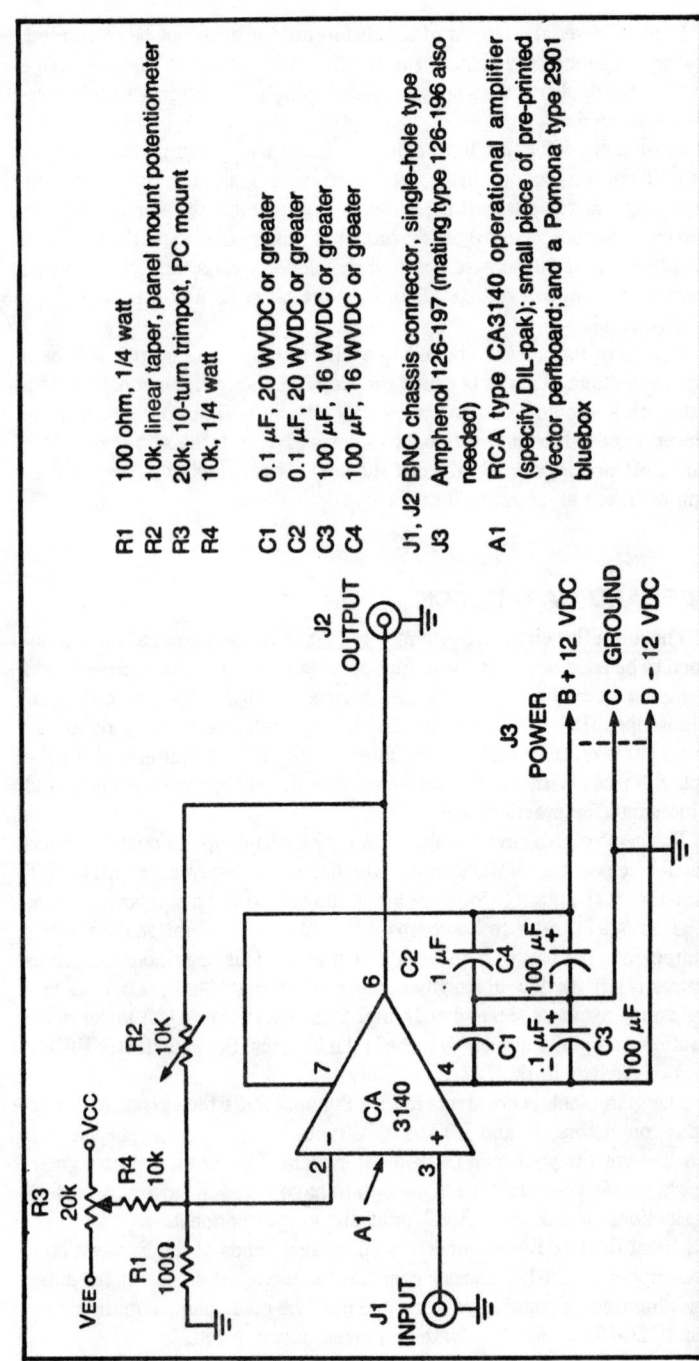

Fig. 22-8. Wideband gain block.

R1 100 ohm, 1/4 watt
R2 10k, linear taper, panel mount potentiometer
R3 20k, 10-turn trimpot, PC mount
R4 10k, 1/4 watt

C1 0.1 μF, 20 WVDC or greater
C2 0.1 μF, 20 WVDC or greater
C3 100 μF, 16 WVDC or greater
C4 100 μF, 16 WVDC or greater

J1, J2 BNC chassis connector, single-hole type
J3 Amphenol 126-197 (mating type 126-196 also needed)

A1 RCA type CA3140 operational amplifier (specify DIL-pak); small piece of pre-printed Vector perfboard; and a Pomona type 2901 bluebox

Adjustment Procedure

1. Connect J1 through a cable to the driving instrument, but make sure that no signal is being delivered. Alternatively, connect a 1000-ohm resistor from J1 to ground.
2. Connect a dc oscilloscope to J2. Calibrate the oscilloscope such that the zero-volt line is known accurately.
3. Connect power to J3; set R2 for maximum resistance.
4. Adjust potentiometer R3 so that the output is 0 Vdc (± 10 mV). It may be necessary to start on a higher range of the oscilloscope and work down to the lowest range, recalibrating for zero on each range.
5. Turn on a driving signal (Disconnect the 1000-ohm resistor, if used)
   . and check for amplification.

The universal gain block is now ready for use.

TRANSDUCER PREAMPLIFIER

It is often the case that experimenters in the life sciences, as well as engineers in certain industrial environments, will use Wheatstone-bridge strain-gauge transducers to generate electrical analog signals of certain physical or biological parameters. It is the nature of transducers to produce very weak differential signals, even in the presence of relatively large input stimuli.

It is the usual practice to connect the output of the strain-gauge transducer to an amplifier such as those discussed in Chapter 11. Such amplifiers, though, tend to be expensive, and this causes many to attempt to operate directly into an oscilloscope or strip-chart recorder without the use of an intermediate amplifier. The transducer amplifier of Fig. 22-9 is a low-cost alternative for those for whom a lack of grant money makes cleverness of even more than usual importance.

This preamplifier was built not for eventual publication, but for a physiologist who was trying to view the 50 μV output of a Grass force-displacement transducer on an elderly Tektronix oscilloscope.

The circuit for this project is nothing particularly interesting. It is merely an RCA CA3140 operational amplifier connected in the simple dc differential-amplifier configuration. This operational amplifier was selected because of its low offset voltage. The object was to avoid the necessity of a CMR-adjust potentiometer. It is, however, quite possible to find a 741 with a low offset, even though you may have to try several in the socket before a good one is found. The gain for this circuit is about 67. This is just about the highest that should be attempted, lest offset problems pop up. If additional gain is required, mate the output of this amplifier with something like the medium-gain amplifier shown in the earlier part of this chapter.

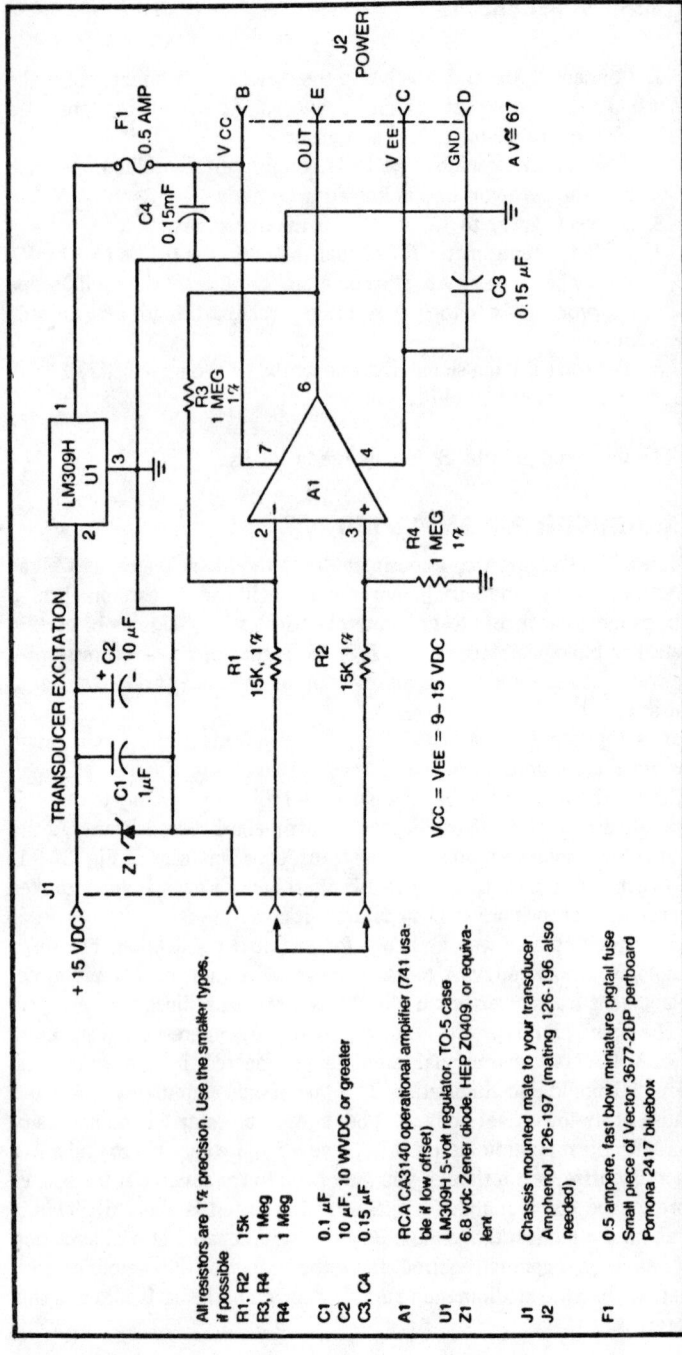

All resistors are 1% precision. Use the smaller types, if possible.

R1, R2   15k
R3, R4   1 Meg

C1   0.1 μF
C2   10 μF, 10 WVDC or greater
C3, C4   0.15 μF

A1   RCA CA3140 operational amplifier (741 usable if low offset)
U1   LM309H 5-volt regulator, TO-5 case
Z1   6.8 vdc Zener diode, HEP Z0409, or equivalent

J1   Chassis mounted mate to your transducer
J2   Amphenol 126-197 (mating 126-196 also needed)

F1   0.5 ampere, fast blow miniature pigtail fuse
            Small piece of Vector 3677-2DP perfboard
            Pomona 2417 bluebox

VCC = VEE = 9 – 15 VDC

Fig. 22-9. Transducer preamplifier.

Transducer excitation could be brought in through power connector J2 if it is desired to use a separate 6-volt battery. In this case, though, an LM309H, 100-mA, 5-volt, three-terminal regulator IC is used to drop the Vcc(+) power supply from the 9-to-15-volt range to something in the 5-volt range that the transducer will accept.

Zener diode Z1 is used for protection of the transducer. Most commercial strain-gauge transducers will only tolerate voltages up to some figure in the 5-to-8-volt range. If the LM309H were to short out or for some reason lose its ground, the voltage applied to the transducer will rise to the full value of Vcc, and that will cause a great deal of damage. The zener voltage of Z1 is selected to be between the normal output level of U1 and the maximum voltage tolerated by the transducer. In this case, a 6.8-volt diode was used. When the voltage rises above the zener point, it will break over and short out the excitation line. This will cause F1 to blow. This technique is another type of crowbar protection circuit because of its crudeness.

One of the best features of this preamplifier is the fact that it is built inside a Pomona, type 2417, blue box that had been fitted directly onto the transducer. See Fig. 22-10. This particular preamplifier was built to be used with a Grass transducer; so J1 was a Cannon WK-21C-6-5/16 chassis-mounting connector that is the mate to the connector of the body of the transducer.

The transducer connector is mounted on the other end. It was found feasible to mount the small piece of perfboard containing the amplifier circuit in a vertical position inside the box.

(ALMOST) UNIVERSAL PREAMPLIFIER

One of the most popular pieces of equipment in any laboratory seems to be a preamplifier that features high gain and input connectors designed for easy connection to wire leads or electrodes. The project in Fig. 22-11 shows such a preamplifier.

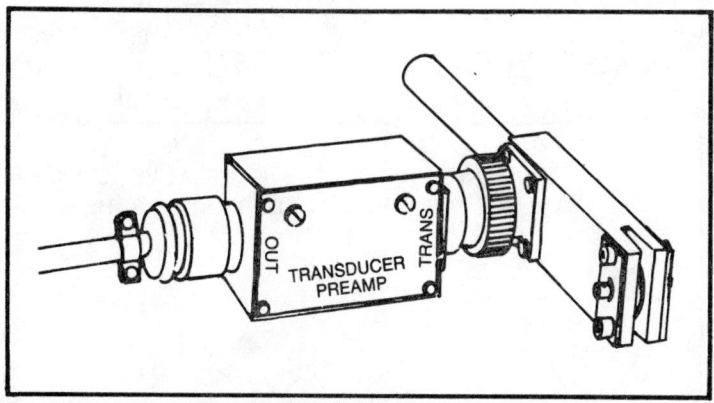

Fig. 22-10. Transducer preamplifier mounting on transducer.

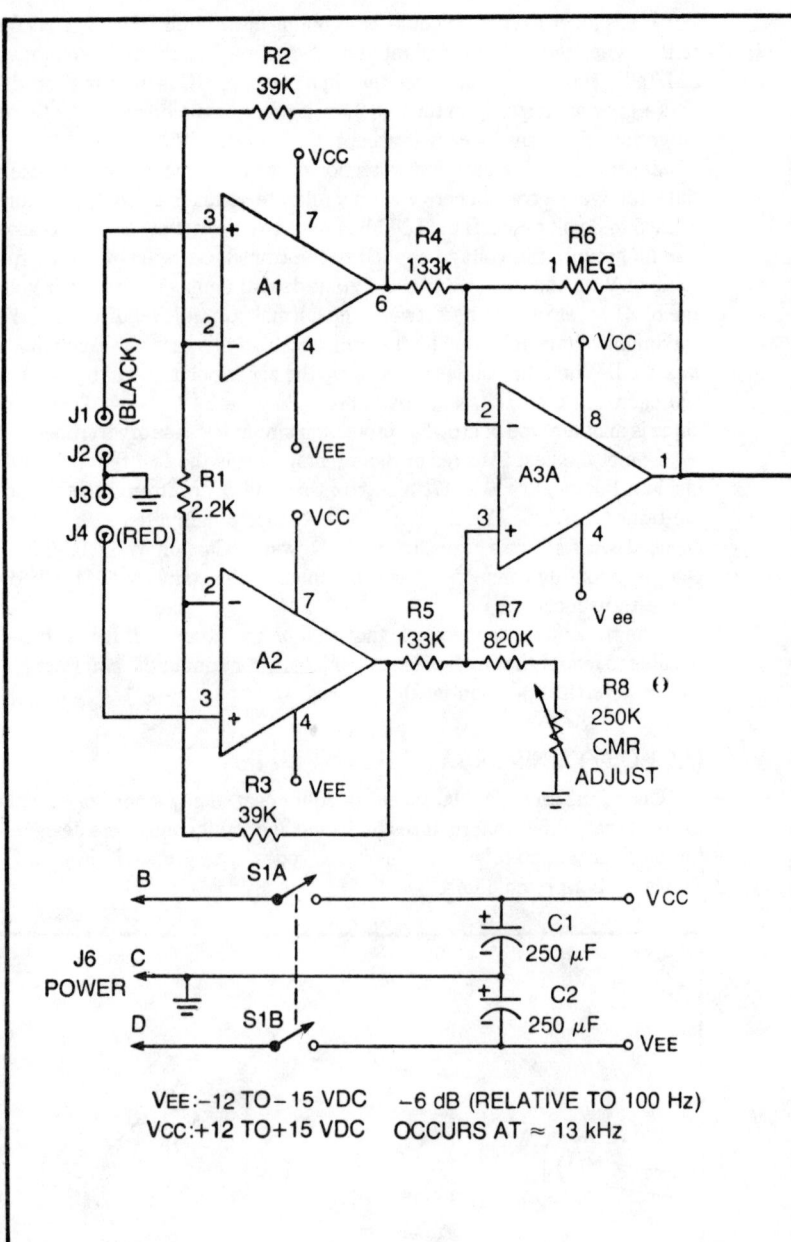

Fig. 22-11. Universal preamplifier.

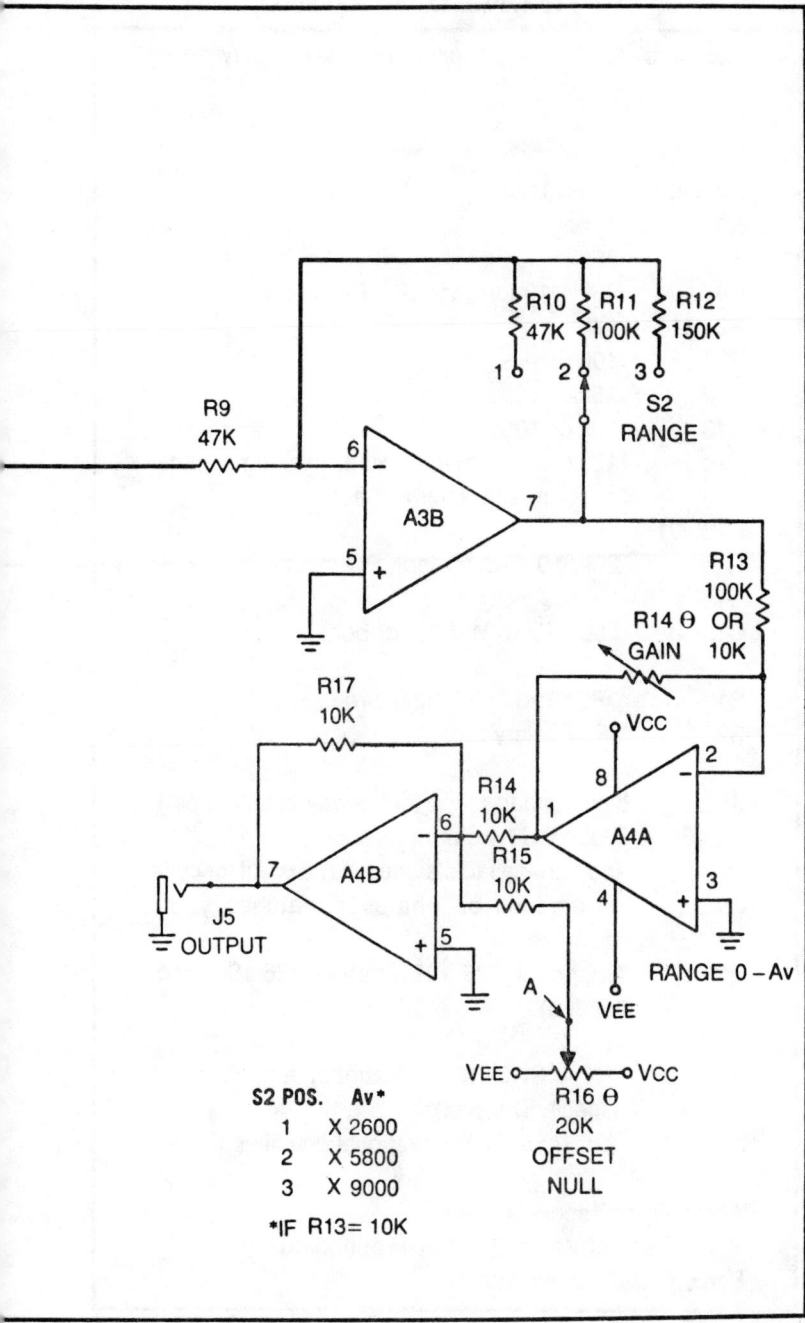

R10 47K R11 100K R12 150K

1 2 3

S2 RANGE

R9 47K

A3B

R13 100K

R14 Θ OR GAIN 10K

Vcc

R17 10K

R14 10K

R15 10K

A4A

A4B

J5 OUTPUT

A

RANGE 0 – Av

VEE

VEE o——————o Vcc

R16 Θ 20K

OFFSET NULL

S2 POS.	Av*
1	X 2600
2	X 5800
3	X 9000

*IF R13= 10K

Table 22-1. Parts List for Fig. 22-11.

Resistors are 1/4 watt, 5 % or better, unless otherwise noted.

R1	2.2k
R2, R3	39k
R4, R5	133k, 1%
R6	1 meg
R7	820k
R8	250k, 10-turn, trimpot, PC mount
R9, R10	47k
R11	100k
R12	150k
R13	10k or 100k
R14	100k, screwdriver adjust, miniature, panel mount potentiometer
R15, R17	10k
R16	20k, 10 turn, trimpot, PC mount
C1, C2	250 μF, 16 WVDC, or better
S1	DPST or DPDT miniature toggle
S2	SP3T rotary
J1	black banana jack (not 5-way binding post)
J2, J3	Grounded type banana jack
J4	Red banana jack, same as J1 except for color
J5	Phone jack or whatever matches your equipment.
J6	Amphenol 126-197 (mating 126-196 also needed)
A1, A2	RCA CA3140 operational amplifiers (specify DIL-pak)
A3, A4	MC1458 dual operational amplifier

4-mini-DIP IC sockets
Small piece of Vector 3677-2DP perfboard
Pomona 2901 bluebox

Biopotentials tend to be tiny voltages and are derived from living things. Signals sought by researchers are therefore low-level, on the order of a few microvolts to about 100 millivolts.

This preamplifier can accommodate the smaller of these because it has gains ranging from zero to as much as 9000, or more, depending upon the setting of switch S2.

It will, though, have excessive gain when recording action potentials, which range in the 100-mV region. In the lowest position of S2 this would result in an output of (0.1 × 2600, or 260 volts), clearly beyond the range of the operational amplifiers used in the circuit. If it is required for recording action potentials or similar signals, you change the values of several resistors in the early part of the circuit. The gain of the preamplifier is given by

$$A_v = \left[1 + \frac{2R2}{R1}\right]\left[\frac{R6}{R4}\right]\left[\frac{R10 \text{ or } R11 \text{ or } R12}{R9}\right]\left[\frac{R14}{R13}\right] \quad (22.6)$$

The high gains presented by this preamplifier exist when R13 is equal to 10 kilohms. This makes the gain of stage A4A equal to 10. Lower gain, by one order of magnitude, occurs when this resistor is raised in value to equal the full range of R14, or 100 kilohms in this case. Alternatively, the same thing may be accomplished if the values of R6 and (R7 + R8) are reduced.

The input connectors for this project are banana jacks, or if you prefer pin-tip jacks, making it an almost universal package. These allow the amplifier to be connected to any number of different electrode or transducer configurations.

The output connector shown in the parts list was chosen for one very good reason: it fit my need at the time and was in the junkbox; so it didn't have to be purchased. Select a connector such as the BNC or PL-259, as you require.

Adjustment Procedure

This project requires pretty much the same procedure as any other classic, operational-amplifier, instrumentation amplifier. Because of the high gains, however, the CMR-adjust control may be treated as if it were a dc balance control.

Set the controls as follows: R8 and R16 to midrange, R14 for maximum resistance, and S2 to position 3.

The procedure is as follows:

1. Short together J1 and J4.
2. Connect an oscilloscope or dc voltmeter to J5.
3. Connect a dc voltmeter to point A.

4. Adjust R16 for 0 Vdc ($\pm$ 10 mV) at point A.
5. Run R14 through its range from the maximum down to minimum and then back to maximum. Note whether any baseline shift showed up on the oscilloscope.
6. Adjust R8 for zero shift of baseline when step 5 is repeated. This step may have to be repeated several times at increasingly more sensitive settings of the oscilloscope so that the best null can be found.
7. Adjust R16 for 0 Vdc ($\pm$ 10 mV) at J5.
8. Remove the short from J1 and J4.
9. Ground J4.
10. Connect a signal between J1 and ground. This signal should have an amplitude under either 10 mV or 1 mV, depending upon which value you selected for R13.
11. Check for distortionless amplification.
12. Disconnect the signal source, and unground J4.
13. Ground J1 and connect the signal source between J4 and ground.
14. Repeat step 11.
15. Use the amplifier—it is ready.

UNIVERSAL OUTPUT STAGE

One goal of this book is to let you design many of your own electronic instruments. The circuit in Fig. 22-12 is a universal output stage that helps attain that goal. It is reasonably universal and can be used to follow almost any instrument design.

Please note the similarity between this circuit and the latter stages of many other projects in this chapter. It is designed along the same lines and for the same reasons.

The voltage gain of the circuit is given by the equation

$$A_v = \frac{R2}{10 \text{ kilohms}} \tag{22.7}$$

If resistor R2 is set at 10 kilohms, the total gain will be unity (1). Practical values for R2 range up to 1 megohm, which would make the total gain of the section equal to 100, but you are admonished to design the rest of the circuits to provide the bulk of the circuit gain. This is really an output-control section and should be kept to low-gain figures.

The real purpose of R2 is to provide the capability to match almost any input requirement and make it into an opportune output voltage. You might, for example, wish to scale an unusual input-voltage function to a linear function with a slope of 1, 10, or 100 so that it may be read directly on something like a digital voltmeter, or as power-of-10 divisions on a strip-chart recorder. It may also be useful to make R2 a potentiometer in order

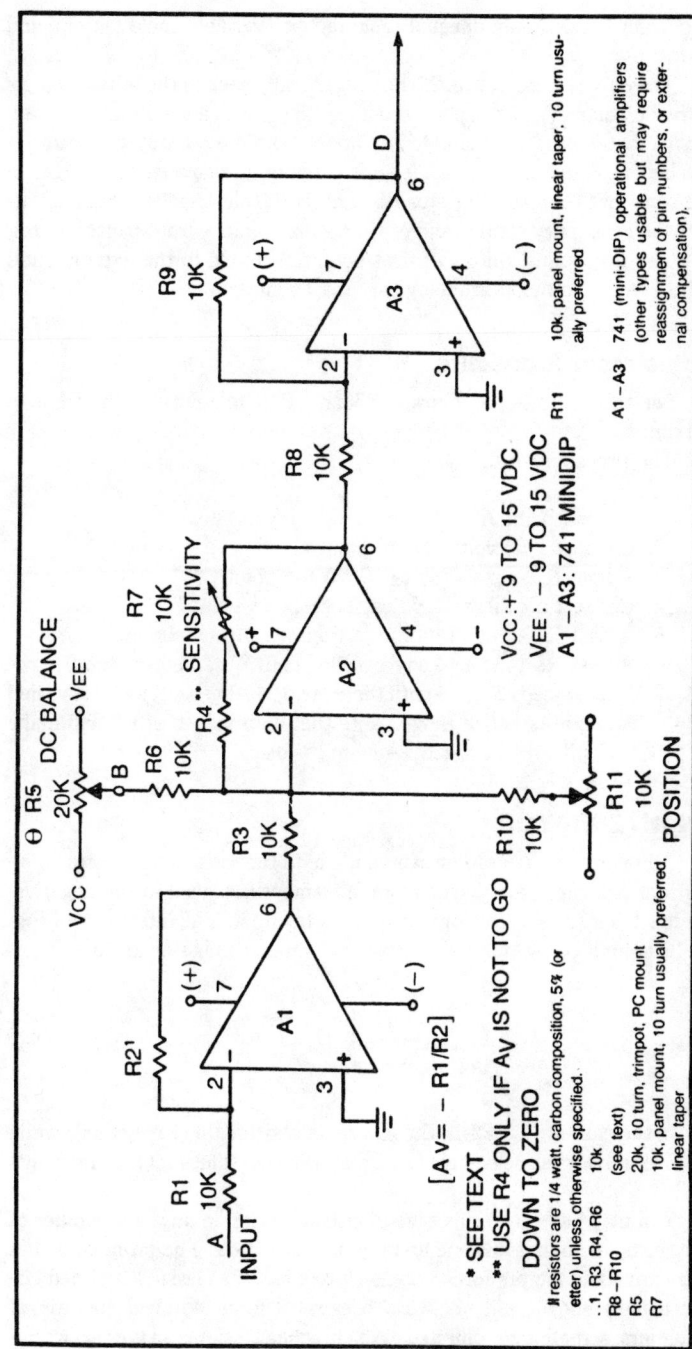

Fig. 22-12. Universal output section.

407

to accommodate really unusual scale factors, variable scale factors, and so forth.

The only feature of Fig. 22-12 not generally seen in the other projects is the dc balance control, potentiometer R5. This is a small-range dc null circuit that is used to eliminate any offset from previous stages before it can be translated into a baseline-shift variation in response to changes in the sensitivity control. Resistor R5 cancels this dc offset so it cannot affect the sensitivity-control stage. If it is made into a front-panel control, the user may adjust out any offset potentials found in the experimental set-up—most common in biological instrumentation.

Adjustment Procedure

Set the controls as follows: R5 and R10 to midrange and R7 to maximum.

The procedure is as follows:

1. Ground point A.
2. Connect a dc voltmeter to point C.
3. Adjust R11 for 0 Vdc (± 10 mV) at point C.
4. Connect a calibrated dc oscilloscope to point D.
5. Adjust R7 through its range from zero to maximum.
6. Adjust R5 to cancel any baseline shift observed in step 5.
7. Repeat steps 5 and 6 until there is no baseline shift as R7 is varied through its entire range. Note that high values of R2 will make this adjustment touchy—or impossible.

AN INTEGRATOR

An operational amplifier can be made to function as an electronic integrator if a capacitor-charge storage element is placed in the negative-feedback loop and a resistor is placed at the input. The integrator of Fig. 22-13 is built using this idea. It has an output voltage equal to

$$E_{out} = \frac{-1}{RC} \int_0^t E_{in}\, dt + C \qquad (22.8)$$

In the case of Fig. 22-13 the R term is resistor R3 (1 megohm), while the C term is switch-selectable by S2 and will have values set by C1 through C8.

You may have seen this type of circuit before, in any of a number of textbooks, including this one and my previous book. The problem is that there are some factors forced on us by reality. The classic simplified circuit usually shown will not work because it assumes ideal operational amplifiers, something prohibited to the practical designer in the real world.

It seems that the usual circuit given in most operational-amplifier textbooks leaves you without sufficient information to allow you to make the circuit work. This circuit should clear up some of the problems caused by the simplified approach.

If the operational-amplifier integrator were connected as is usually the case, you would find the output voltage rising as soon as power was applied or as soon as the reset switch operated. The amplifier would saturate in only a few seconds. The output in that case will be pegged to either the Vcc or Vee supply rails. The cause of this problem is the sum of all offsets normally found in *real* operational amplifiers. These cause a current to charge the capacitor, and it is not in response to an input-voltage condition.

These offsets will cause the output to rise at a rate that is determined by the RC-time constant of the circuit and the magnitude and polarity of the offset potentials. Potentiometer R5 is used to cancel this type of drift. Its function is to supply a current that has a magnitude and polarity that cancels the offset voltage. With the offset nulled to zero, the output should remain stable unless an input signal is present.

Also helping to reduce the severity of this problem is the correct choice of operational amplifier. I tried a number of operational-amplifier devices and found quite a variation between them with respect to offset-caused integrator drift. Not unpredicted was the fact that the 741 caused the worst drift problems, on the order of 4 or 5 volts/second. This rate would, of course, put the integrator in saturation in about three seconds if a ± 15-Vdc power supply were used.

A bit of a surprise, though, was the fact that most of the premium-grade operational amplifiers that I tried were also very poor in this respect, some no better than the lowly 741. Several manufacturers of the μA725 were tried with rather disappointing results. Although the thermal drift was quite good, as advertised, the integrator showed a drift of about 1.5 volts/second, better than the 741 but not good enough for use as an integrator. Oddly enough, the low-cost RCA CA3140AH proved to be the best of the devices in my parts supply. It had a drift of only 0.4 volts/second in the circuit shown. The aspect of this that surprised me the most was the cost, about one order of magnitude less than the average cost of the 725s. Sometime after learning how to make an integrator actually work, I found out that the CA3140 will work even better (drift-wise) at low voltages if a heatsink is provided. Most TO-5-type-transistor finned heatsinks will also fit over the case of the op amp, and this will help prevent drift. The device also works best at Vcc and Vee potentials of ± 5 volts, but that would tend to reduce the total output range.

In really bad cases of drift it might be wise to make potentiometer R5 into a pair of series-connected potentiometers labeled coarse and fine. The coarse control would be a 10-turn potentiometer with a value of 2000 ohms, while the fine control would be a 200-ohm, 10-turn potentiometer.

A second form of output drift, although smaller in magnitude than the

409

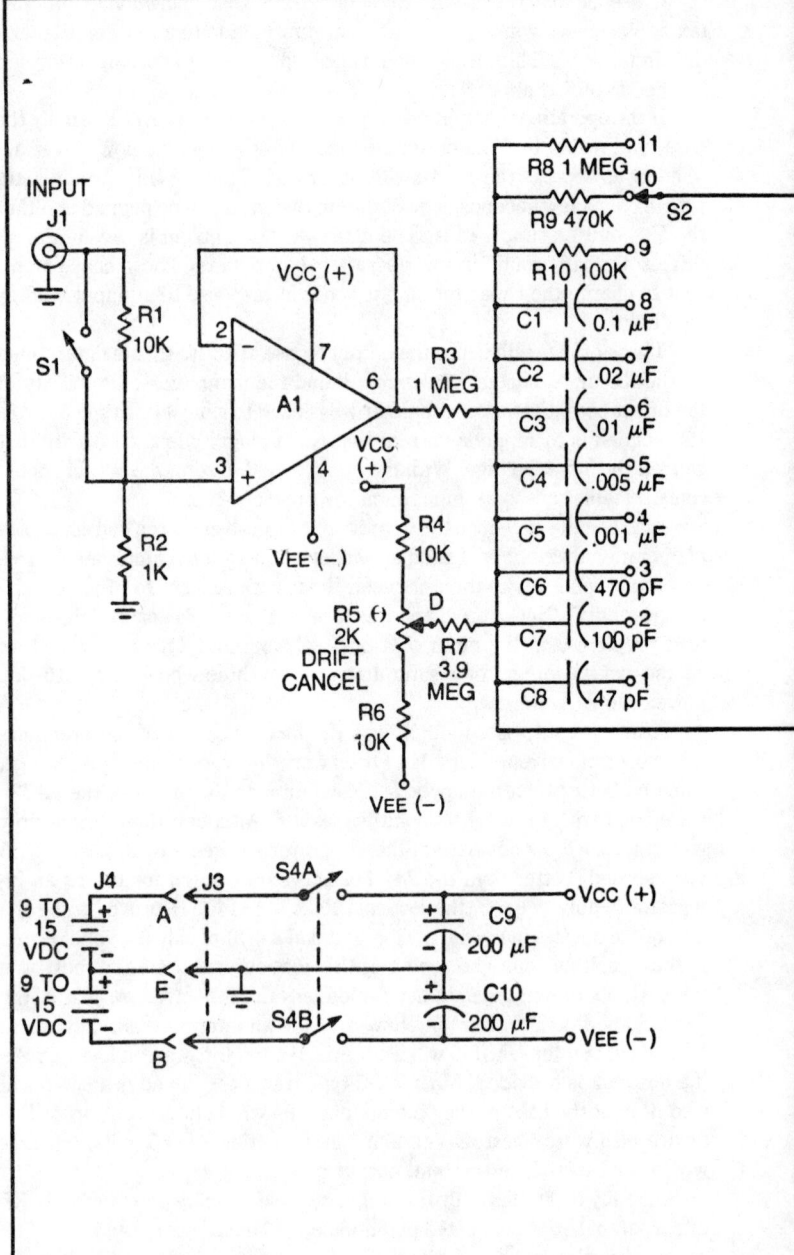

Fig. 22-13. Integrator.

410

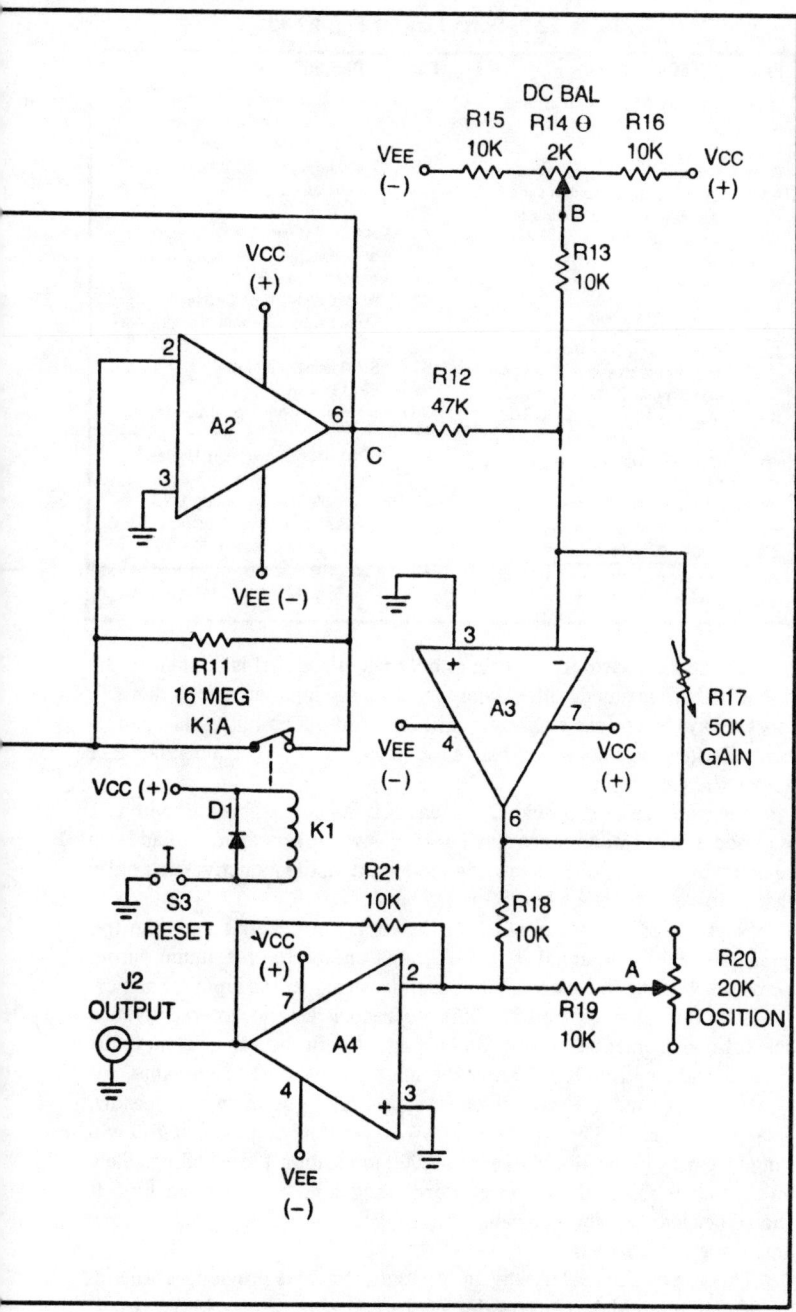

411

Table 22-2. Parts List for Fig. 22-13.

PART	DESCRIPTION	PART	DESCRIPTION
R1, R4, R6, R13, R15, R16, R18, R19,		C6	470 pf, silver mica or better
R21	10k	C7	100 pf, silver mica or better
R2	1k	C8	47 pf, silver mica or better
R3, R8	1 meg	X9, C10	200 μF, 25 WVDC or better
R5	2k, linear taper, 10 turn trimpot	C10	Same as C9.
R7	3.9 meg, 1% (or 5% metal film)		
R9	470k	A1, A3, A4	741 mini-DIP operational
R10	100k		amplifiers, or same as A2 for superior
R11	16 meg		performance.
R12	47k	A2	RCA CA3140T or CA3140AH.
R14	(same as R5)		Do not substitude with impunity.
R17	50k, linear taper, 10 turn panel		
	mount (1 turn usable if lower gain resolu-	S1	SPST miniature toggle
	tion is permissible, as it often is).	S2	SP11T rotary
R20	20k, otherwise sample as R17	S3	Normally open, pushbutton, SPST
C1	0.1 μF, mylar or better	S4	SPST or DPDT miniature toggle.
C2	.02 μF, mylar or better		
C3	.01 μF, mylar or better	J1, J2	single-hole mounting BNC
C4	.005 μF, mylar or better		chassis connectors
C5	.001 μF, mylar or better	J3	Amphenol MS3102B-14S-5P
All resistors are 5% carbon film, 1/4 watt unless otherwise specified		J4	Mate of J3 in suitable or prefered shell (e.g., MS-3106B-14S-5S)

offset drift discussed above, occurs only when a signal is applied to the input. If there is any dc offset component on the input signal, as there almost always is, you see a baseline shift in the output circuit as this component charges the integrating capacitor. It is reduced by connecting R11 across the capacitors.

Integrator reset is provided by relay K1. When pushbutton switch S1 is closed, the relay energizes and discharges whatever capacitor had been selected by S2. Diode D1 is provided to eliminate the inductive-kick pulse that is to be expected when S3 is opened.

Selection of a suitable value for the capacitor depends upon the frequency of the input signal to be integrated and on the maximum output amplitude that can be tolerated. In general, the higher the input frequency, the lower the value required for C. At the frequencies encountered in much scientific instrumentation you can be content with the values from 0.001 μF to 0.1 μF, with 0.01 μF being the most commonly chosen value.

If too low a value is selected for the capacitor, the output will chip badly. Examine Eq. 22.8. The constant term 1/RC multiplying the integral will range from 10 in position 8 to over 22,000 in position 1, depending solely on the values selected for the resistor (1 megohm) and the capacitors. It should be clear that this can cause the amplifier to saturate with even low-amplitude input signals.

This is also the reason why an input attenuator is provided. If the input signal is too high for the value of the capacitor selected, this circuit must reduce it prior to integration.

This circuit has demonstrated integrator behavior at frequencies between dc and 100 kHz. At higher frequencies, though, the natural roll-off of the 741 operational amplifiers used in most of the stages provides severe attenuation, not altogether undesirable in light of the high value of the multiplying constant.

An incidental benefit of the integrator is that it also functions as a low-pass filter. The use of a capacitance around the negative-feedback loop of an operational amplifier causes the circuit to see higher frequencies fed back more than low frequencies. This means that they will be *attenuated* more than low frequencies. It is usually possible to find a trade-off setting of S2 that will attenuate unwanted high-frequency components of the input signal without showing integrator behavior. The key, then, is judicious selection of the RC-time constant of the integrator.

Adjustment Procedure

Set the controls as follows: S1 and S4 open; S2 to position 11; R5, R14, and R20 to midrange; and R17 to maximum resistance.

The procedure is as follows:

1. Close S4 and allow the circuit to warm up for 15 minutes. Do not become alarmed if the output rises to maximum and the circuit seems saturated (it is, of course, but that is unimportant at this time).
2. Adjust R20 for 0 Vdc (± 10 mV) at point A.
3. Adjust R14 for 0 Vdc (± 10 mV) at point B.
4. Adjust R5 for 0 Vdc (± 10 mV) at point D.
5. Apply a 1-volt, 100-Hz (Frequency is not too critical), square wave to J1.
6. Use an oscilloscope to check the output for signal.
7. Disconnect the signal source; turn S2 to position 8 and ground J1.
8. Press reset switch S3.
9. If the oscilloscope trace of the output signal (scope set to 1 volt/cm) shifts when step 8 is performed, adjust R5 to cancel this drift.
10. Repeat steps 8 and 9 until no further improvement is noted. A slight amount of offset is permissible, provided that it forms a stable baseline after S3 is released. This offset can be canceled by the position control, so is unimportant.
11. Reduce R17 to zero; then adjust it back to maximum.
12. If a baseline shift occurs in step 11, adjust R14 to cancel it.
13. Repeat steps 11 and 12 until no further improvement is obtained.
14. Apply a 10-Hz, 1-volt, square wave to J1. Remove the input short.
15. The output waveform should be a triangle. If it is clipped, close S1 and reinspect the waveform. In low settings of S2, the gain of integration will be so high as to re-clip the triangle, making it appear as a square wave. This can be eliminated through proper setting of S2.

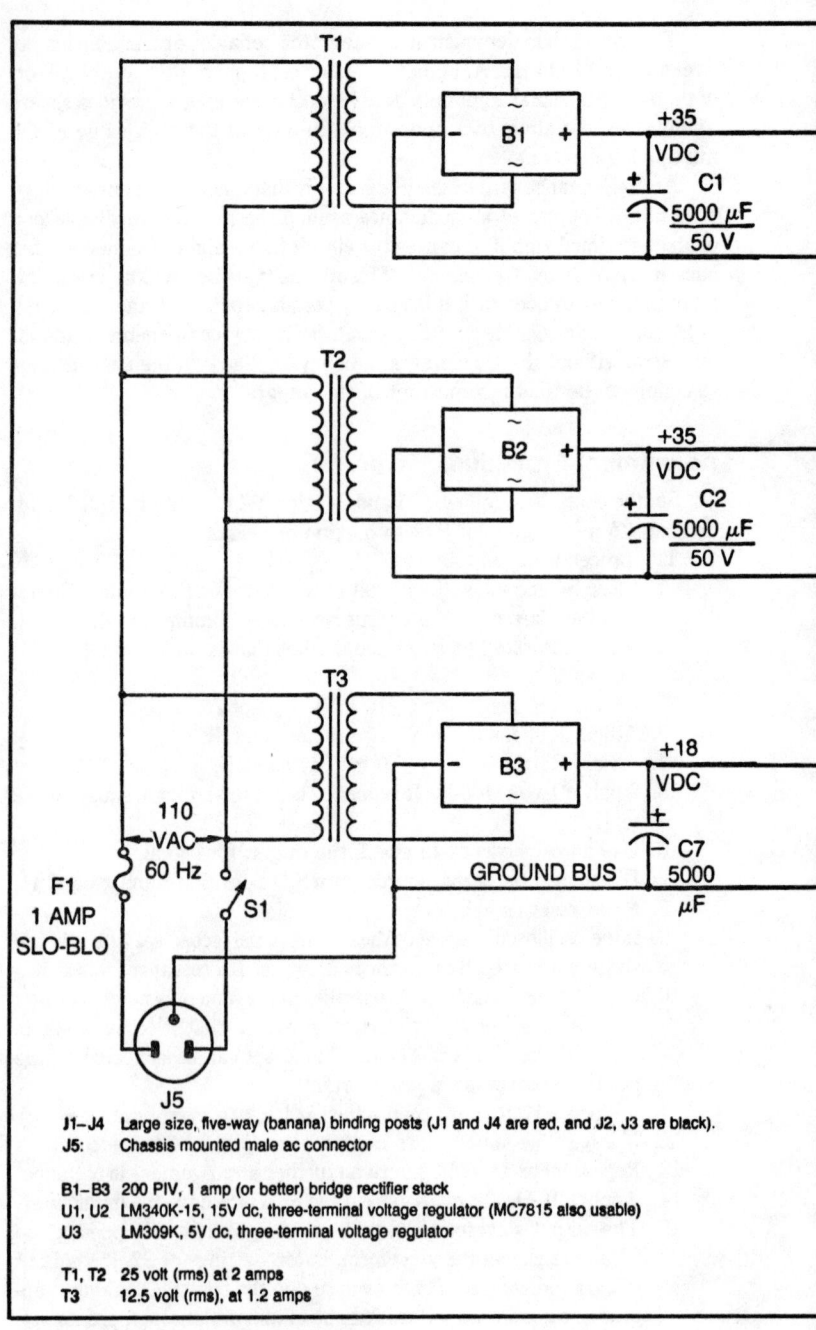

Fig. 22-14. Universal power supply 1.

J1–J4   Large size, five-way (banana) binding posts (J1 and J4 are red, and J2, J3 are black).
J5:   Chassis mounted male ac connector

B1–B3   200 PIV, 1 amp (or better) bridge rectifier stack
U1, U2 LM340K-15, 15V dc, three-terminal voltage regulator (MC7815 also usable)
U3   LM309K, 5V dc, three-terminal voltage regulator

T1, T2 25 volt (rms) at 2 amps
T3   12.5 volt (rms), at 1.2 amps

414

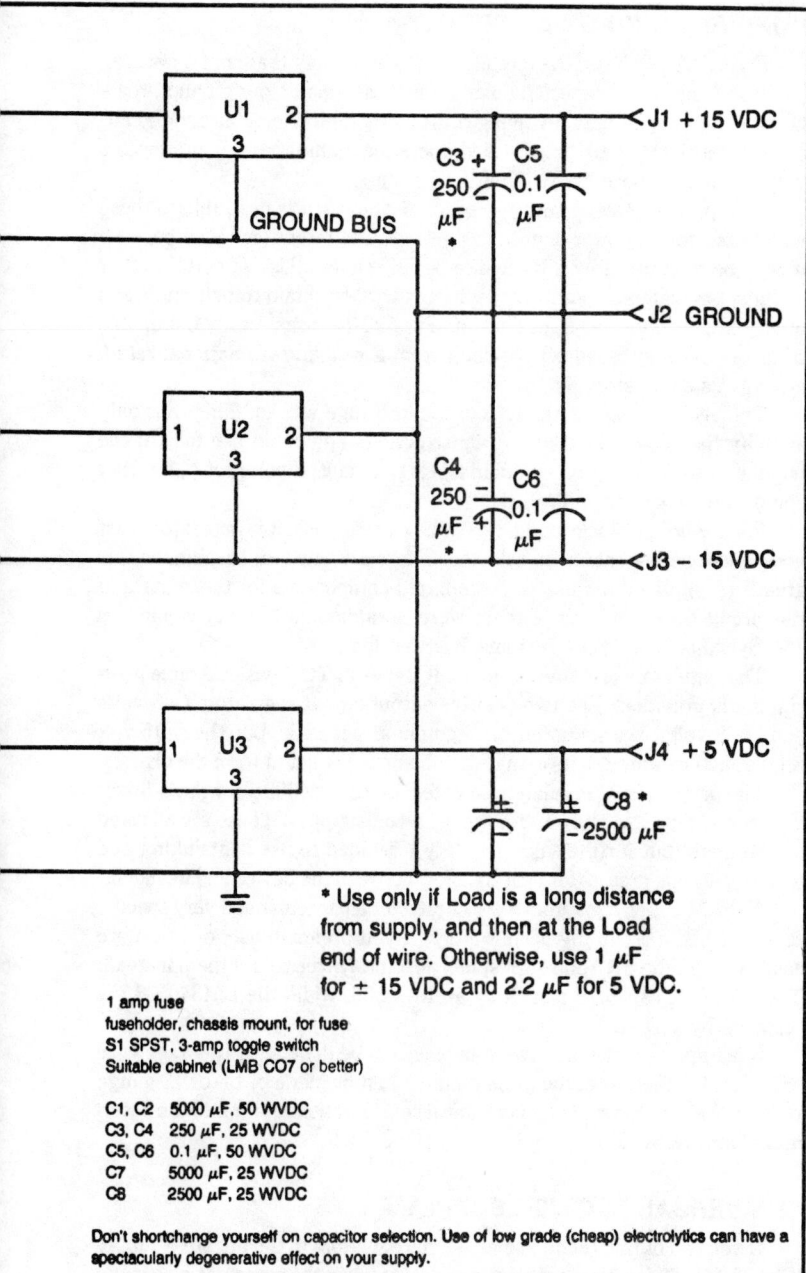

1 amp fuse
fuseholder, chassis mount, for fuse
S1 SPST, 3-amp toggle switch
Suitable cabinet (LMB CO7 or better)

C1, C2   5000 μF, 50 WVDC
C3, C4   250 μF, 25 WVDC
C5, C6   0.1 μF, 50 WVDC
C7   5000 μF, 25 WVDC
C8   2500 μF, 25 WVDC

Don't shortchange yourself on capacitor selection. Use of low grade (cheap) electrolytics can have a spectacularly degenerative effect on your supply.

"UNIVERSAL" POWER SUPPLY 1

Figure 22-14 shows the circuit of a power supply that is of a general-purpose design. It is especially useful for those who do design and bread-board electronic circuits involving transistors, digital logic integrated circuits, linear integrated circuits, and operational amplifiers. It produces 5 volts dc at 1 ampere and ± 15 Vdc at 1 ampere.

This project may appear less sophisticated than is desirable to those who know clever power-supply design, but it is intentionally so because it is to be duplicated by a large number of readers. Use of certain other circuits, for example, might require locally-hard-to-obtain transformers and other components. The circuit of Fig. 22-14 uses easily-obtainable transformers purchased off the shelf from a well-known, national retail, electronics chain store.

The rectifiers are actually four-diode bridge stacks. They not only simplify the design but also the construction. You could use four of the popular 1N4000-series rectifiers in a bridge circuit, if you prefer, but that complicates construction a little.

For the bridge stacks and large-value electrolytics, it is mandatory that good-quality components be selected. There are few real bargains in this area. I recommend the use of name-brand components for these parts of the circuit. In the prototype there were Sprague and Mallory capacitors and Sylvania ECG-160-series and HEP bridges.

The regulators are three-terminal IC types in TO-3 cases. Ample heat-sinking is provided. The two positive-output-circuit regulators (+ 5 volts and + 15 volts) are mounted on a common heatsink, but the – 15-volt regulator is mounted to its own heatsink and is insulated from the chassis.

The particular IC regulators selected are the LM309K for the + 5-volt supply and type LM340K-15 for the ± 15-volt supplies. These are all rated at 1 ampere, but for the sake of safety I decided to use heatsinking and to apply silicone grease (for heat transfer) between the device and its mount.

The LM-series regulators were chosen for a particular and very specific reason: I had a set in my parts supply. If you prefer to use, or can more easily obtain, the MC7800-series, feel absolutely free to use them instead. The LM309K can be replaced by an MC7805, while the LM340K-15 is replaced by a 7815.

It is important that all low-voltage wiring be done with 18-gauge wire or larger. Furthermore, the ground bus should be made of 14- or 12-gauge wire. In this latter case, only mechanical considerations enter into the determination of what is too large.

"UNIVERSAL" POWER SUPPLY 2

If you can obtain them, use negative regulators in the negative supply (Fig. 22-15). These were not readily available when the prototype was made (and they were costly), but now almost any electronics parts store will carry them, at least in the Motorola HEP line. Beware though, the pinouts are

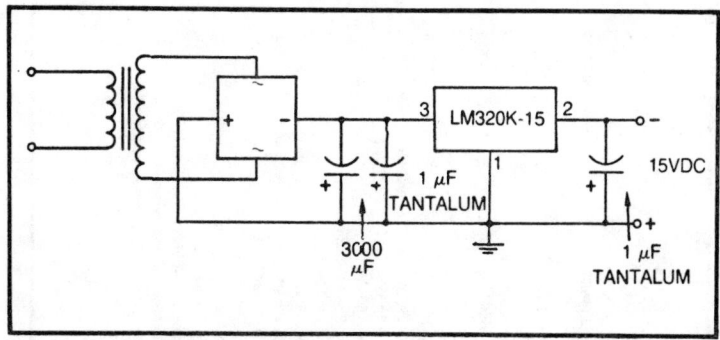

Fig. 22-15. Universal power supply 2.

different on these types. In general, the output is the same, but the input and common-ground terminals are *reversed*. The case or metal heat-sink plate in TO-220 types is NOT GROUNDED in the negative regulators!

A HIGH-CURRENT +5-VDC POWER SUPPLY

Modern microcomputers and other large-scale TTL systems require an awful lot of current at +5 volts dc. Furthermore, most such systems require that the output voltage from the power supply be within ±5 percent of 5 volts, which translates to a range of 4.75 volts to 5.25 volts. Generally, the nearer it is to actually being +5 volts the better the circuit will perform. The circuit of Fig. 22-16 will supply +5 volts at 10 amperes.

Transformer T1 is a 6.3-volt ac filament transformer rated at a secondary current of 10 amperes or more. Before you go out and buy one of these new, check the local surplus electronics market for bargains. There are an awful lot of 6.3-volt filament transformers around, it seems. In one trip I located two identical transformers rated at 13 amperes and one rated at 20 amperes. The cost was only $12 for the lot!

Rectifier D1 is a bridge-rectifier stack rated at 50 PIV and 25 amperes. The rule to follow whenever selecting a rectifier is to buy one rated at substantially more than the expected load, in this case 2 1/2 times.

It is always a good idea to heatsink high-current rectifiers. The heatsink can only extend the life of the device. Most of the high-current bridges are housed in a block that is designed to be bolted directly to the chassis or a heat sink. Be sure to use silicone heat-transfer grease underneath the bridge when it is installed.

At these current levels a large electrolytic filter is required for proper operation. In this case capacitor C1 is rated at 15,000 microfarads or greater. If a smaller type is used the input to the regulator will see too low a voltage when high currents are drawn.

Integrated circuit U1 is the Motorola MC1469R regulator, also available from Motorola under their HEP program as the type C6049R. This device may oscillate under some circumstances; so follow good, tight, layout prac-

417

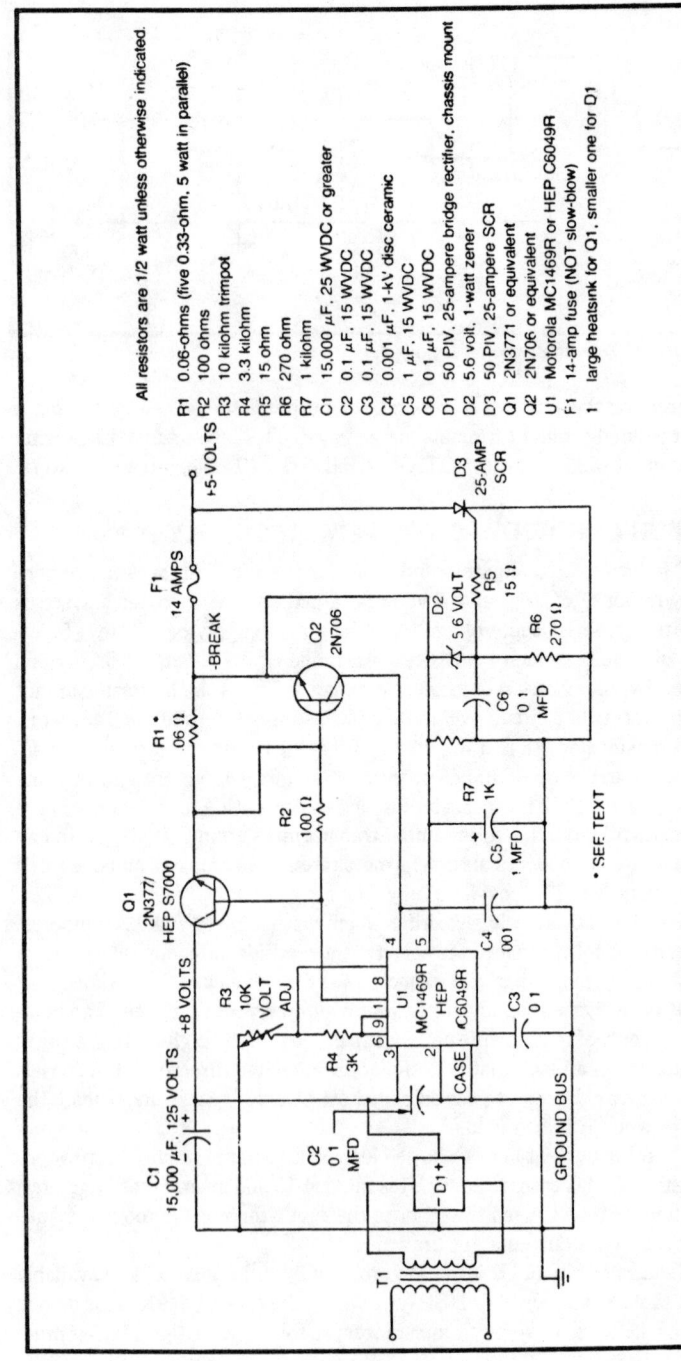

All resistors are 1/2 watt unless otherwise indicated.

R1   0.06-ohms (five 0.33-ohm, 5 watt in parallel)
R2   100 ohms
R3   10 kilohm trimpot
R4   3.3 kilohm
R5   15 ohm
R6   270 ohm
R7   1 kilohm
C1   15,000 μF. 25 WVDC or greater
C2   0.1 μF. 15 WVDC
C3   0.1 μF. 15 WVDC
C4   0.001 μF. 1-kV disc ceramic
C5   1 μF. 15 WVDC
C6   0.1 μF. 15 WVDC
D1   50 PIV. 25-ampere bridge rectifier, chassis mount
D2   5.6 volt. 1-watt zener
D3   50 PIV. 25-ampere SCR
Q1   2N3771 or equivalent
Q2   2N706 or equivalent
U1   Motorola MC1469R or HEP C6049R
F1   14-amp fuse (NOT slow-blow)
1   large heatsink for Q1, smaller one for D1

Fig. 22-16. High current, +5-volt power supply.

418

tice. Keep leads short and mount the critical components (C2, C3, C4, C5, R2, R3, R4, and Q1) close to the IC.

The main current is handled by series-pass transistor Q1. This transistor has a 25-ampere collector-current rating and is a type 2N3771 or HEP S7000. Transistor Q1 MUST be mounted on a substantial heatsink, with silicone grease, and, if possible, in the breeze from a whisper fan.

One feature of this particular circuit, which incidentally was adapted from a Motorola applications data sheet on the MC1469, is current limiting. The current drain creates a voltage drop across resistor R1, which is a 60-*milli*ohm unit. Since you are not likely to go out and buy a 0.06-ohm resistor, I suggest you make one by paralleling five 0.33-ohm auto-radio-fuse resistors. These are available from most electronic-parts distributors who deal with consumer electronics parts. The voltage drop across R1 biases transistor Q2, which is connected to the current-limiting terminal on the regulator IC.

If the load is located more than a few inches from the power supply, excessive voltage drop may occur. If that is the case, you might want to connect this circuit in the remote-sense configuration by breaking the circuit at the emitter of Q2 and rerouting it through a second wire to the remote +5-volt dc point.

Also featured is an SCR crowbar protection circuit. If the voltage at the output goes over +5.6 volts dc, SCR D3 is gated on, blowing the fuse.

Chapter 23

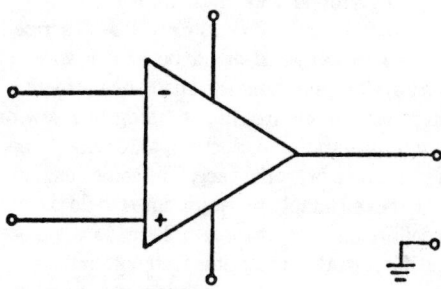

Microprocessors

THE WORD *MICROPROCESSOR* LOOMS LARGE IN THE FIELD OF ELEC-
tronic-instrument design. The introduction of the integrated-circuit
microprocessor was as important as the introduction of the transistor in
the fifties. Its impact has been acutely felt in all areas of electronic
technology.

BACKGROUND

The Intel Corporation is generally credited with the invention of the
microprocessor integrated circuit under an agreement with a Texas small-
computer and computer-peripherals company. Although the device they
produced failed to meet the specifications of the original sponsor, it was
marketed to the general market. Intel apparently had the good sense to
realize that a huge potential existed for their products, which were then
called the 4004 and 8008 microprocessor for the 4- and 8-bit products,
respectively.

WHAT IS A MICROPROCESSOR?

What is a microprocessor? It is often called a computer on a chip, but
that is not quite correct. The microprocessor IC contains almost all of the
logic circuitry necessary to form a central processing unit, but most require external memory and some sort of other circuitry to operate. At least
one microprocessor offers CPU, logic, I/O, and memory on a single chip.

A *microcomputer* is a device, often on a single, printed-circuit board,
containing a microprocessor chip and all of the necessary memory, I/O,
and logic circuitry needed to make a genuine digital computer. In actual

practice, though, many people tend to use the terms microprocessor and microcomputer almost interchangeably. They may hold up a $10 IC and say "this is a microcomputer," or they may point to a pile of electronic-circuit cards costing around $800 and call it their microprocessor.

Any programmable digital computer will be described as a universal logic element or as an *analytical engine* because it will do almost any job depending upon how it is programmed. Theoretically there is little difference in the computers used to control the stitching of a sewing machine, the operation of a multiple-station, industrial, automatic lathe, or the heating and cooling of your home or work space. The same computer could possibly be reconfigured or even simply reprogrammed to do the monthly data processing for accounts payable or receivable or do the laborious statistical work in a scientific setting. In fact, it is well within the realm of possibility to use a microprocessor to control an experiment and then use it under a different program control to perform the mathematical analysis on the data obtained.

MICROPROCESSOR COMPATIBILITY

There are actually several different forms taken by the microprocessor hardware that are available on the open market. Most of them are available ready-built, or you may buy them in kit form and do the tedious part yourself. A popular microcomputer main frame produced by one of the pioneer companies in the hobbyist and small-system market was the Altair 8800B produced by MITS of Albuquerque, NM. Incidentally, the word Altair is used by many hobbyists as a generic term, but it is properly used only as a trademark of MITS. Many of the so-called "Altair" boards on the market are actually merely Altair-compatible. Most are of good quality, but they are not MITS products and may lead to problems. One that has shown itself to be particularly troublesome is the peripheral-interface area, where timing anomalies exist.

The now-obsolete Altair computer was based on a motherboard concept in which a large number of 100-pin printed-circuit connectors, prewired, are in parallel on a board. Each pin on the 100-pin Altair bus (also called S-100 bus by non-MITS sources) has a specific assignment, or no assignment is made so that special designs may be accommodated. The beauty of this system is that all PC cards in the Altair bus system use the same connector and pinouts that depend solely on their function. The motherboard is not the least bit concerned about the order in which the cards are inserted into the machine. Incidentally, one of the difficulties with nonstandard S-100 PC boards is that some manufacturers have become a little maverick in their assignment of the unassigned pins. When two special-purpose cards are inserted, again unless they are original MITS cards, there is a possibility that they are incompatible because of conflicting pin assignments. This situation, of course, defeats the concept of a standard bus.

The use of a standard bus allows you to custom tailor the main frame to the job to be done. A business billing machine, for example, might require an I/O interface to a printer or teletypewriter. Such a machine might require all 64 K bytes of memory that can be addressed by the 16-bit address bus.

In other applications, you may find a small machine dedicated to some instrumentation or control problem where you want less than 10 K bytes of memory and only A/D converters or D/A converters for I/O devices.

Another type of computer is the so-called single-board computer. Examples of this include the KIM-1 and several others by a dozen manufacturers. Most single-board computers will have from 256 to 1024 bytes of *read only memory* (ROM) and up to 4 K bytes (usually 1 or 2 K bytes) of *random access memory* (RAM) (e.g., 2102s).

The KIM-1 is possibly the most popular of the single-board computers because it has a modest price tag. It offers a 1 K byte of RAM, a cassette interface, and a teletype interface. The cassette interface is to an ordinary audio-cassette tape recorder and is used to store programs for later use.

Such computers are ordinarily considered to be training devices, but they are easily applied to applications where the computer is merely a component in a larger system.

The third class of microcomputer is the microcontroller variety which is really a single-board computer. It is a PC board with almost all of the features of the above type except that it is intended specifically for application inside a dedicated instrument. Training-type single-board computers usually have a hexidecimal keyboard attached and possibly some hexidecimal LED-readout displays. The controller lacks these.

In controller or other dedicated applications, the ROM is used to permanently store the program, while the RAM accepts data from the outside world. Many new instrument designs feature microprocessors; so I recommend that you read the appropriate titles given in the bibliography.

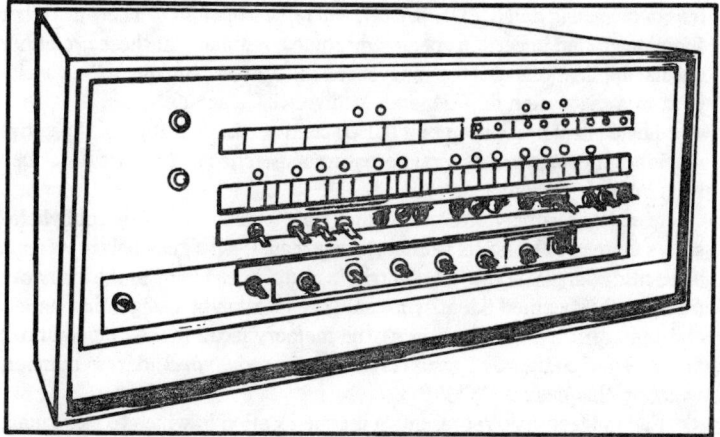

Fig. 23-1. Altair 800B.

423

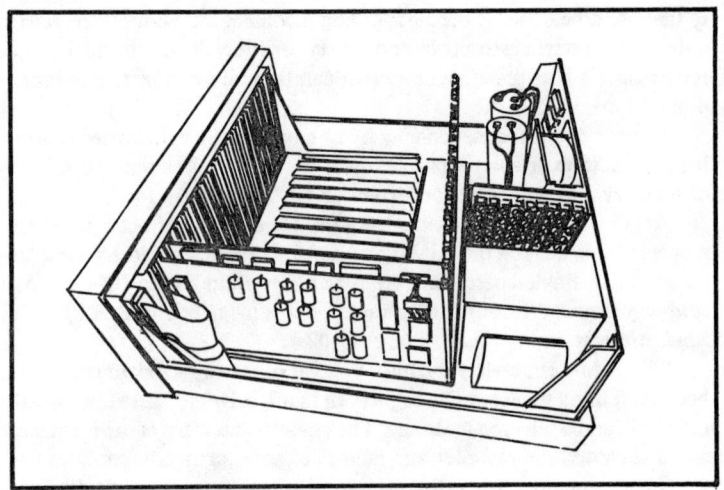

Fig. 23-2. Altair 8800B.

THE TAB AUTOMATIC COMPUTER (TABAC)

Microcomputer books frequently assume a familiarity with computers in general that is simply not there in all readers, even in an era of widespread computer literacy. Thus, I have a need for a section that discusses the programmable digital computer as a generic device. The "machine" of this section is a universal computer, representative of the entire set of programmable digital computers, and not any one manufacturer's offerings. It is a hypothetical machine and not available anywhere. It, like an androgynous being, has elements of all kinds within its body. I call this hypothetical computer the TAB Automatic Computer, or TABAC for short.

Like any programmable digital computer, TABAC has three main parts: *central processing unit* (CPU), *memory*, and *input/output* (I/O). There are other functions found in certain specific machines, but many of these are either special applications of these main sections, or are too unique to be considered in a discussion of a general, "universal" machine.

The central processing unit (CPU) controls the operation of the entire machine. It consists of several necessary subsections which are described in greater detail later on.

Memory can be viewed as an array of "pigeon holes" or cubbyholes such s those used by postal workers to sort mail. Each pigeon hole represents a specific address on the letter carrier's route. An address in the array can be uniquely specified (identifying only *one* location by designating its *row* and **column**. If I want to specify the memory location (i.e., pigeon hole) at row 3 and column 2, I would create a *row-3, column-2* address number, which in this case is "32."

Each pigeon hole represents a unique location in which to store mail. In the computer, the memory location stores not pieces of paper but a single

424

binary "word" of information. In an eight-bit microcomputer, for example, each memory location will store a single eight-bit binary word (e.g., 11010101). The different types of memory device are not very important to us here, except in the most general terms. To the CPU and our present description of how it operates, it doesn't matter much whether memory is *random access read/write memory* (RAM), *read only memory* (ROM), *dynamic*, or *static*.

There are three main lines of communication between the memory and the CPU: *address bus, data bus*, and *control-logic signals*. These avenues of communication control the interaction between memory and I/O on the one hand, and the CPU on the other, regardless of whether the operation is a read or write function. The address bus consists of parallel data lines, one for each bit of the binary word that is used to specify the address location. In most eight-bit microcomputers, for example, the address bus contains sixteen-bits. A 16-bit address bus can uniquely address up to 2^{16}, or 65,536, different, eight-bit, memory locations. This size is specified as "64K" in computerese, not "65K" as one might expect. It seems that lower-case "k" is used almost universally in science and engineering to represent the metric prefix *kilo*, which represents the multiplier 1000. Thus, when someone tells you that their computer has "64K" of memory, you might expect it to contain 64×1000, or 64,000, electronic pigeon holes. But you would be wrong in that assumption because long ago computerists noted that 2^{10} was 1,024; so they determined that their "K" would be 1,024, not 1,000! This means that a 64K computer will contain a total of 64×1024, or 65,536, electronic pigeon holes in which to stuff data. To differentiate "big-K" (1024) from "little-k" (1000), standard computerist shorthand uses upper-case K rather than lower-case k.

Be aware that the size of memory which can be addressed doubles for every bit added to the address bus. Hence, adding one bit to our sixteen-bit address bus creates a seventeen-bit address bus that is capable of addressing up to 128K of memory. Some "8-bit" machines which have 16-bit address buses are made to look larger by tactics which create a pseudo-address bus. In those machines, several 64K memory banks are used to simulate continuously-addressable 128K, 256K, or more machines.

The data bus is the communication channel over which data travels between the main register (called the *accumulator* or *A-register* in the CPU and the memory. The data bus also carries data to and from the various input and/or output ports. If the CPU wants to read the data stored in a particular memory location, that data is passed from the memory location to the accumulator (in the CPU) by way of the data bus. Memory-write operations are exactly the opposite: data from the accumulator is passed over the data bus to a particular memory location.

The last memory signal is the *control-logic* or *timing* signal. These signals tell memory if it is being addressed and whether the CPU is requesting a *read* or *write* operation. The details of the control-logic signals vary considerable from one microprocessor chip to another; so there is little that

I can say at this point. Later in this book you will be introduced to both the 6502 and Z80 standard signals (which are representative of two different architectures), but for others I recommend either the manufacturer's literature or my own book titled *8-bit & 16-bit Microprocessor Cookbook* (TAB book No. 1643).

The input/output (I/O) section of the TABAC computer is the means by which the CPU communicates with the outside world. An input port brings data in from the outside world and pass it over the internal data bus to the CPU where it is stored in the accumulator. An output port is exactly the opposite: it passes accumulator data from the data bus to the outside world. In most cases, the "outside" world consists of either peripherals (e.g., printers, video monitors) or communications devices (e.g., MODEMs).

In some machines, there are separate I/O instructions that are distinct from the memory-oriented instructions. For example, in the Z80 machine there are separate instructions for "write to memory" and "write to output port" operations. In Z80-based computers the lower eight bits (A0 through A7) of the address bus are used during I/O operations to carry the unique address of the I/O port being called (the accumulator data will pass over both the data bus and the upper eight bits (A8 through A15) of the address bus). Since there are eight bits in the unique I/O address, we can use up to 2^8, or 256, different I/O ports that are numbered from 000 through 255. In the 6502-based computer there are no distinct I/O instructions. In those machines (e.g. Apple II), the I/O components are treated as memory locations. This technique is called *memory-mapped I/O*. Input and output operations then become memory-write and memory-read operations, respectively.

Central Processing Unit (CPU). The CPU is literally the heart and soul of TABAC. Although there are some differences between specific machines, all will have at least the features shown of my TABAC computer. The principal subsections of the TABAC computer are: *accumulator* or *Arithmetic Logic Unit* (ALU), *Program Counter* (PC), *Instruction Register* (IR), *Status Register* (SR) or *Processor Status Register* (PSR), and *Control-Logic Section*.

The accumulator is the main register in the CPU. With a few exceptions, all of the instructions use the accumulator as either the source/destination of data or the object of some action (e.g., unless otherwise indicated, an ADD instruction always performs a binary addition operation between some specified data and the data in the accumulator).

Although there are other registers in the CPU (the Z80 is loaded with them!), the accumulator is the main register. The main purpose of the accumulator is the temporary storage of data being operated on or transferred within the machine. Note that data transfers to and from the accumulator are non-destructive. In other words, it is a misnomer to tag such operations as *transfers* because in actuality they are *copying* operations. Suppose, for example, the program says to "transfer" the hexadecimal number $8F from the accumulator to memory location $A008. After the proper instruction ("STA $A008" in 6592 assembly language) is executed, the hex number $8F will be found in *both* the accumulator *and* memory location $A008.

It is important to remember that accumulator data changes with every new instruction! If I have some critical datum stored in the accumulator, it is critical that I write it to some location in memory for permanent storage. This function is sometimes performed on the *external stack* or in some portion of memory set aside by the programmer as a *pseudo-stack*.

The arithmetic logic unit (ALU) contains the circuitry needed to perform the arithmetic and logical operations. In most computers, the arithmetic operations consist of addition and possibly subtraction, while the logical operations consist of AND, OR, and Exclusive-OR (XOR). Note that even subtraction is not always found! In some computers, there is no hardware arithmetic function other than addition. The subtraction function is performed in software using two's-complement arithmetic (a method of making the computer think it's actually adding!). Multiplication and division are treated as multiple additions or subtractions unless the designer has thought to provide a hardware multiply/divide capability.

The program counter (PC) contains the address of the next instruction to be executed. The secret to the success of any programmable digital computer such as TABAC is its ability to fetch and execute instructions *sequentially*. Normally, the PC will increment appropriately (1, 2, 3 or 4) while executing each instruction (i.e., 1 for a one-byte instruction, 2 for a two-byte instruction, etc.). For example, the instruction "LDA,N" on the 6502 microprocessor loads the accumulator with the number N. In a program listing, you will find the number N stored in the next sequential memory location from the code for the LDA portion (called the operations code, or op-code for short):

| 0205 | LDA |
| 0206 | N |

At the beginning of this operation, the PC contains 0205, but after execution the PC will contain the number 0206 because LDA,N is a two-byte instruction.

There are several ways to modify the contents of the PC. One way is to let the program execute sequentially: the PC contents will increment for each instruction. You can also activate the *reset* line, which forces the PC to either location $0000 or some other specific location (often at the other end of the memory; for example, $FFFA in the 6502). Another method is to execute either a JUMP or a conditional-JUMP instruction. In the latter case, the PC will contain the address of the "jumped-to" location after the instruction is executed. Finally, some computers have a special instruction that will load the PC with a programmer-selected number. This "direct-entry" method is not available on all microprocessors, however.

The instruction register (IR) is the temporary storage location for the instruction codes that were stored in memory. When the instruction is fetched from memory by the CPU, it will reside in the instruction register until the next instruction is fetched.

The instruction decoder is a logic circuit that reads the instruction register contents and then carries out the intended operation.

The control-logic section is responsible for the "housekeeping" chores in the CPU. It issues and/or responds to control signals between the CPU and the rest of the universe. Examples of typical control signals are memory requests, I/O requests (in non-memory-mapped machines), read/write signalling, and interrupts.

The status register (SR), also sometimes referred to individually as status flags, is used to indicate to the program and sometimes to the outside world the exact status of the CPU at any given instant. Each bit of the status register represents a different function. Different microprocessor chips use different sets of status flags, but all will have at least a *carry flag* (C-flag) and a *zero flag* (Z-flag). The C-flag indicates when an arithmetic or logical operation results in a "carry" from the most significant bit of the accumulator (B7), while the Z-flag indicates when the result of the present operation was zero (00000000). Typically, Z equals 1 when the result is zero, and C equals 1 when a carry occurs.

I have now finished my tour of the CPU of the TABAC. This discussion in general terms describes most typical microprocessors. Although various manufacturers use different names for the different sections and some will add sections, almost all microprocessors are essentially the same inside the CPU.

Operation of TABAC

A programmable digital computer such as TABAC operates by sequentially fetching, decoding, and then executing instructions stored in memory. These instructions are stored in the form of binary numbers. In some early machines, there were two memory banks, one each for program instructions and data. The modern computer, however, uses a single memory bank for both instructions and data.

How, one might legitimately ask, does dumb ol' TABAC know whether any particular binary word fetched from memory is data, an instruction, or a binary representation of an alphanumeric character (e.g. ASCII)? The answer to this instruction is the key to the operation of TABAC; cycles!

TABAC operates in *cycles*. A computer will have at least two discrete cycles: instruction-fetch and execution. In some machines the process is more sophisticated, and cycles are added. While the details differ from one machine to another, the general operation is similar for all of them.

Instructions are stored in memory as binary numbers, op-codes. During the instruction-fetch cycle, an op-code is retrieved from the memory location specified by the program counter, and stuffed into the instruction register inside the CPU. The CPU assumes that the programmer was smart enough to arrange things according to the rules so that the datum fetched from location ABCD during some instruction cycle in the future is, in fact, an instruction op-code. It is the responsibility of the programmer to arrange things in a manner so as not to confuse the poor, dumb TABAC.

During the first cycle, an instruction is fetched and stored in the instruction register (IR). During the second (i.e., next) cycle, the instruction-decoder circuit inside the CPU will read the IR and then carry out the indicated operation; this is called the execution cycle, while the first cycle was the instruction-fetch/decode cycle. The CPU then enters the next instruction-fetch cycle, and the process is repeated. This process is repeated over and over as long as TABAC is operating. Each step is synchronized by an internal clock that is designed to make things remain rational.

From the above description, you might be able to glimpse a truth concerning what a computer can and cannot do. The CPU can shift data around, perform logical operations (AND, OR, XOR), add two N-but numbers (sometimes—but not always—subtract as well)—all in accordance with a limited repertoire of instructions encoded in the form of binary words. Operations in TABAC (and all other microcomputers) are performed *sequentially* through a series of discrete steps. The secret to whether or not any particular problem is suitable for computer solution depends entirely on whether a plan of action, called an algorithm, can be written that will lead to a solution through sequentially executed steps. Most practical instrumentation, control, measurement, or data processing problems can be so structured—a fact which accounts for the meteoric rise of the microcomputer in those fields. A scholarly field that seeks to define sequential solutions for problems is *numerical methods*.

The art of making instruments with microcomputers at their heart depends a lot on knowledge of interfacing methods. That subject is a bit beyond the scope of this book, but TAB Books, Inc. (Blue Ridge Summit, PA 17214) offers several titles on that subject; they tell me they would be happy to send you a catalog or sign you up for their various, discount, book clubs.

Chapter 24

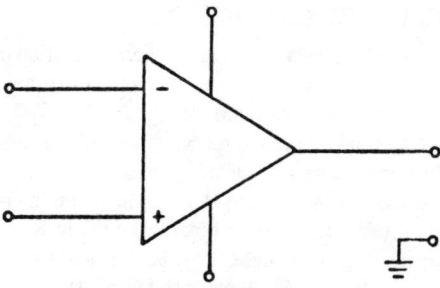

Signal Sources
and Waveform Generators

A LMOST EVERYONE WHO WORKS IN ELECTRONICS EVENTUALLY needs to build an oscillator, multivibrator, or digital-clock circuit. These circuits are used to produce a variety of electronic waveforms that are needed in many different circuits. A simpler oscillator, for example, might be used in a radio transmitter or receiver. Such circuits are generally considered sine wave generators even though the word *oscillator* is correctly applied to circuits that produce other waveforms. The multivibrator could produce square waves, triangles, or other waveforms. Similarly, a digital clock is a special case of the multivibrator and/or oscillator that is used in digital-logic and computer circuits.

In general, let's keep our jargon straight by adopting the term oscillator to cover all three cases, including astable multivibrators and digital clocks. The term oscillator can be defined as a circuit that will produce a periodic waveform (i.e., one that repeats itself). The output waveform can be a sine wave, square wave, triangle, sawtooth, pulse, or other waveshape. The important part is that it is *periodic*.

There are two forms of oscillator circuits: relaxation oscillators and feedback oscillators. Some relaxation oscillators use one or another of several *negative-resistance* devices (e.g., tunnel diode). Such devices operate according to Ohm's law at certain potentials and opposite Ohm's law at other potentials. There are other relaxation oscillators that pass little or no current at voltages below some threshold and pass a large current at voltages above the threshold. Examples of these devices are neon glow lamps and unijunction transistors (UJT).

Feedback oscillators use an active device as an amplifier and then provide feedback in a manner that produces regeneration instead of degeneration. These circuits account for a large number of oscillators used in practical electronic circuits. In this chapter, I will discuss both forms of oscillators.

RELAXATION OSCILLATORS

There are two main devices used in practical relaxation oscillators: neon glow lamps and UJTs. There are also some circuits that are based on SCR-like devices, but these are not common.

Figure 24-1(A) shows a simple relaxation oscillator based on the neon glow lamp. The neon glow lamp has a low-pressure inert gas inside a glass envelope that also contains a pair of electrodes. When the potential across the electrodes exceeds the ionization potential (V_t in Fig. 24-1(B)) of the inert gas, the gas will give off light. Popular types of neon glow lamps include the NE-2 and the NE-51. For these lamps, the threshold potential is somewhere between 40 and 80 volts, although the lamp will maintain its ionized state (and continue to glow) at a somewhat lower *holding potential* (V_H).

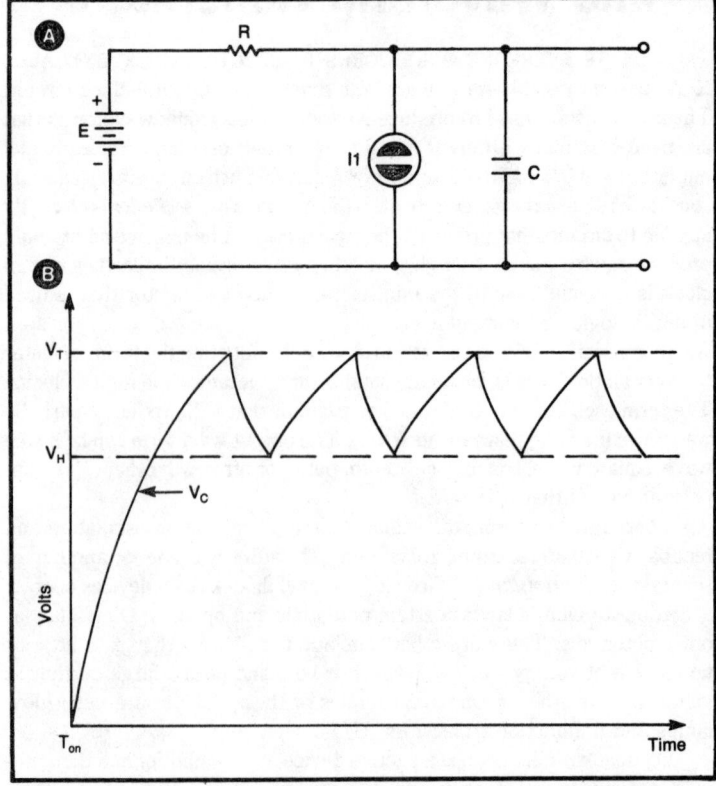

Fig. 24-1. Glow-lamp relaxation oscillator (A) circuit, (B) waveform.

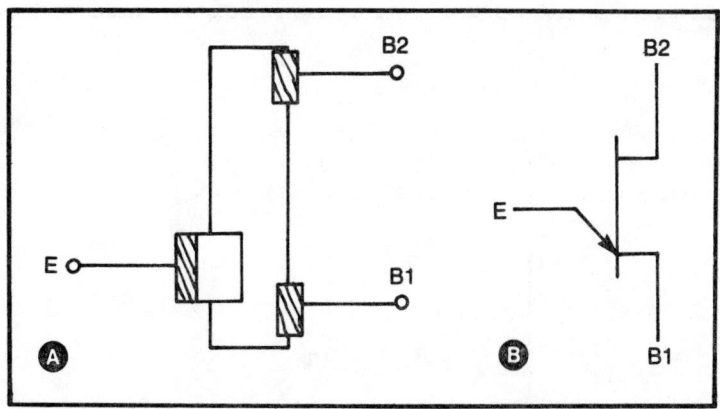

Fig. 24-2. (A) Unijunction transistor (UJT) and (B) UJT symbol.

The circuit for the relaxation oscillator is shown in Fig. 24-1(A), while its operating waveforms are shown in Fig. 24-1(B). The frequency of oscillation is set by the ionization and holding potentials, acting in concert with the RC time constant of the resistor and capacitor. The resistor is connected in series with the lamp, and the capacitor is connected in parallel with the lamp.

You must keep in mind the two states of the neon lamp. When the lamp is not ionized, it conducts no current. This is the situation at all potentials below V_t. At currents above the threshold, the lamp resistance suddenly drops to nearly nothing and, in fact, is so high that it will blow the lamp if the resistor is too low in value. When the circuit is turned on, the capacitor will begin to charge and the capacitor voltage (V_c) will begin to build up. When the voltage across the capacitor reaches the threshold potential, the lamp ionizes and assumes a very low resistance. Since this low resistance is across the capacitor, it will discharge the capacitor. But the discharge only continues until V_c reaches the threshold potential. When this potential is reached, the lamp goes out and reverts to its high-resistance state. The cycle continues, with the capacitor voltage varying between threshold and holding potentials. The frequency of oscillation is determined by the difference between these potentials and the values of R and C.

Unijunction Transistors. The UJT is a special form of transistor that has one emitter, two bases, and no collector (see Fig. 24-2). The structure of the UJT is shown in Fig. 24-2A, and the circuit symbol is shown in Fig. 24-2(B). The UJT is constructed of a single channel of silicon, and the "bases" are merely the electrodes at either end of the channel. The emitter forms a pn junction with the channel. Like most pn junctions, the UJT pn junction will not pass current when reverse- or zero-biased but will pass a current when forward-biased (usually 0.6 to 0.7 volts is required to forward-bias the pn junction).

An example of a UJT relaxation oscillator is shown in Fig. 24-3. Tim-

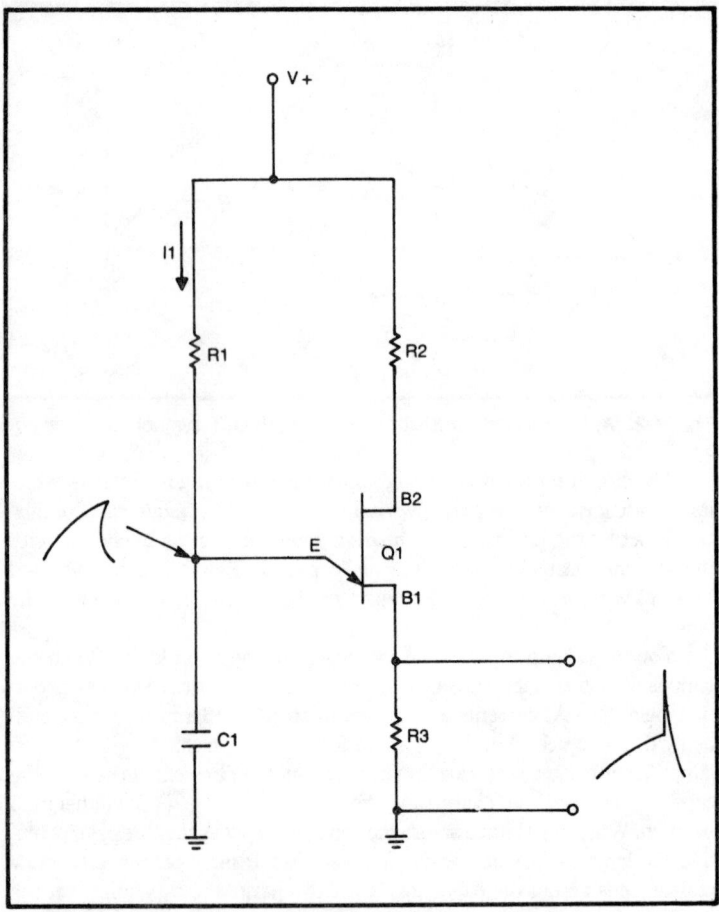

Fig. 24-3. UJT relaxation oscillator.

ing, hence the frequency of oscillation, is set by the combination R1 and C1, acting with the properties of the UJT. When the circuit is turned on, the B1-E pn junction is unbiased; so there will be no current in resistor R3. At power-on the capacitor will begin to charge from V + through timing resistor R1. When the threshold potential is crossed, the UJT will "fire" and a pulse of current flows in R3. This produces (by Ohm's law) a pulse output from across R3; this is the output signal.

Negative-Resistance Devices. There are several semiconductor devices on the market that have a so-called negative-resistance property. What does this mean? Regular, or positive resistance, operates according to Ohm's law. When the voltage across the circuit is increased, the current in the device increases proportionally; when the voltage is decreased, the current also decreases. In a negative-resistance device, however, exactly the opposite occurs. An increase in voltage will produce a decrease in current, and vice versa.

An example of a negative-resistance device is the tunnel diode (also called an Esaki diode after the inventor), the symbols for which are shown in Fig. 24-4(A). The I-vs-V curve for the tunnel diode is shown in Fig. 24-4(B). In the positive-resistance zone (PRZ), the tunnel diode operates much like other Ohm's-law devices. But in the negative-resistance zone (NRZ), it behaves just the opposite from Ohm's law!

An example of a tunnel-diode oscillator is shown in Fig. 24-5. This particular circuit was popular in the mid-60s for small Citizen's-Band beacons and hobbyist oscillators. Supposedly, the circuit could be used clandestinely as a homing device for sneaky people to follow others! I doubt that it was used for such, but it was very popular. With the circuit constants shown, the oscillator will operate in the 27-mHz region using an overtone crystal. Other circuits, similar in concept but different in form, can cause a tunnel diode to operate into the UHF or even microwave region. In those cases, strip-line tuners and cavities are used to establish resonant frequency.

Relaxation oscillators are simple and are good circuits for students and

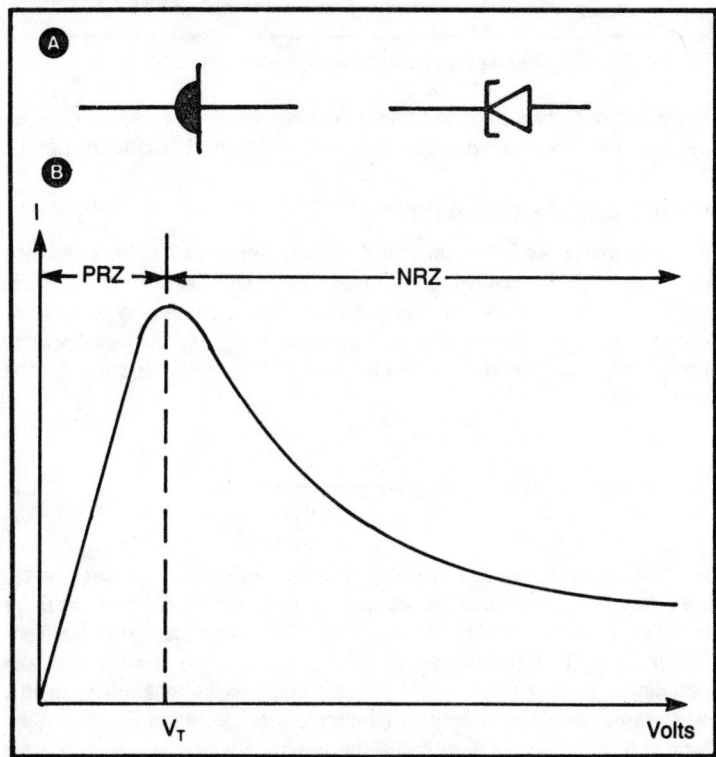

Fig. 24-4. (A) Tunnel diode symbols, (B) I-vs-V curve.

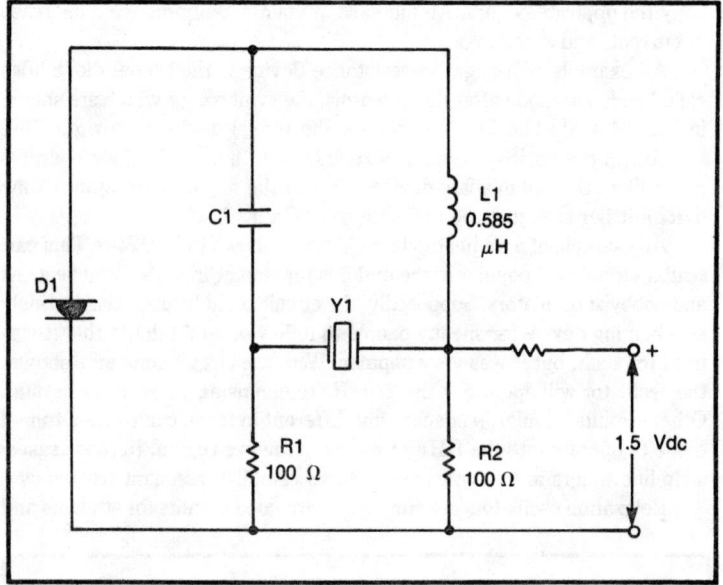

Fig. 24-5. 27-MHz tunnel diode oscillator.

beginning hobbyists to build (if only to learn how they work). Most other applications, however, will use one or another feedback-oscillator circuits.

FEEDBACK OSCILLATORS

A feedback oscillator consists of an amplifier and a feedback network (see Fig. 24-6). It is called a *feedback oscillator* because the output signal of the amplifier is fed back to the amplifier's own input by way of the feedback network. We call the gain of the amplifier A_{vol} and the transfer function of the oscillator B. The overall gain of this circuit is given by the relationship

$$A_v = \frac{A_{vol}}{1 + A_{vol}B} \qquad (24.1)$$

The amplifier can be any of many different devices. For some cases, it will be a common-emitter bipolar transistor (npn or pnp devices). In others it will be a junction field-effect transistor (JFET) or metal-oxide semiconductor field-effect transistor (MOSFET). In older equipment it was a vacuum tube, most often operated in the common-cathode mode. In still other cases, it will be an integrated-circuit operational amplifier or other form of IC amplifier. In most cases, the amplifier will be an inverting type; so the output is out of phase with the input by 180 degrees.

436

The feedback network will provide a sample of the output signal to the input of the amplifier. This network may be LC, RC, or RLC. Rarely is it a simple resistance network because such networks are not frequency-selective. They will produce oscillation but at some frequency that is not controlled. The problem is that there are distributed (stray) circuit capacitances in the circuit, and these act with the resistance to produce an RC time constant at some unspecified frequency; it is rarely a frequency that you want!

This circuit will oscillate if two conditions are met: 1) the loop gain (as above) is greater than unity, and 2) the feedback signal is in phase with the input signal. Because the typical amplifier inverts the signal (180 degrees) and because we need a total phase shift of 360 degrees (in order to have an in-phase feedback signal), the feedback network must provide an additional phase shift of 180 degrees. If the network is designed to produce this required phase shift at only one frequency (as in an LC or RC network), the circuit will oscillate at that frequency.

We have determined that there are two basic forms of oscillator: relaxation oscillators and feedback oscillators. The relaxation oscillator depends upon the properties of certain devices such as the neon glow lamp and the unijunction transistor. Such devices will present a high impedance at applied potentials (usually the voltage across a timing capacitor) less than a certain threshold. Once the threshold is passed, however, the device breaks down in a manner that discharges the capacitor. Feedback oscillators, on the other hand, depend upon applying a sample of the output signal of an inverting amplifier to its own input, thus causing regeneration rather than degeneration.

Figure 24-7 shows the block diagram for the feedback oscillator. The forward block is an amplifier element that has an open-loop gain of $-A_{vol}$ (the minus sign implies 180-degree phase inversion between input and out-

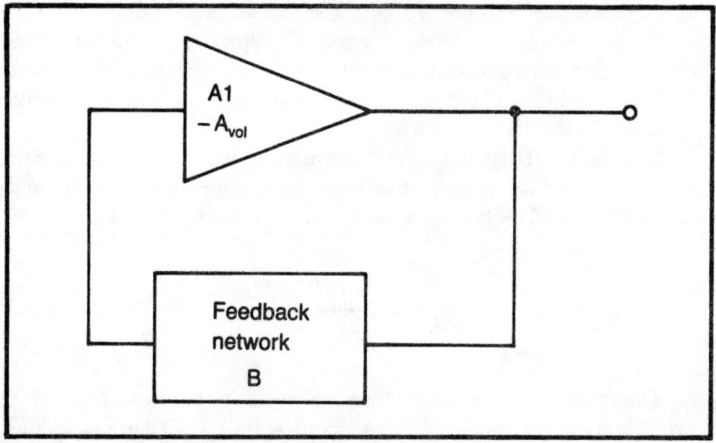

Fig. 24-6. Feedback oscillator.

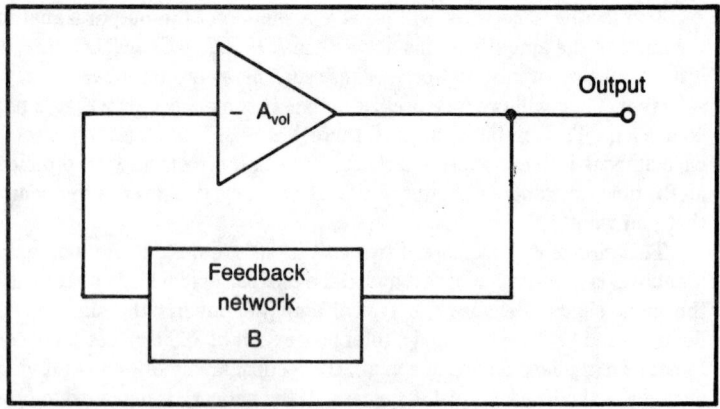

Fig. 24-7. Feedback oscillator.

put). The feedback network has a "gain" designated by "B" and may consist of a simple resistance network or a complex impedance. In most cases, its "gain" is really a loss. The value of the closed-loop gain (which takes into account both A_{vol} and B) is

$$A_v = \frac{A_{vol}}{1 + A_{vol}B}$$

If this circuit is to oscillate, you must make the closed-loop gain greater than unity. It is also necessary to ensure that the fed-back signal is in phase with the input. Since the amplifier has a 180-degree phase shift, the feedback network must provide an additional phase shift of 180 degrees at the frequency of oscillation. The combination of phase shifts produces the 360-degree phase shift needed for the in-phase condition.

Figure 24-8 shows a block diagram of a typical LC oscillator circuit. The amplifying element is an operational amplifier, but it could be any form of inverting voltage amplifier. The use of the op-amp here is for convenience of illustration rather than an endorsement.

In this general case, I designate the reactive elements with the general form Z1, Z2, and Z3, meaning that they can be either capacitive or inductive reactances. The feedback factor, B, is given by

$$B = \frac{Z2}{Z1 + Z2} \qquad (24.2)$$

The usual arrangement is to use the same form of reactance for Z1 and Z2 and the opposite for Z3. For example, if Z1 and Z2 are capacitors, then Z3 is an inductor; if Z1 and Z2 are inductors, then Z3 is a capacitor.

438

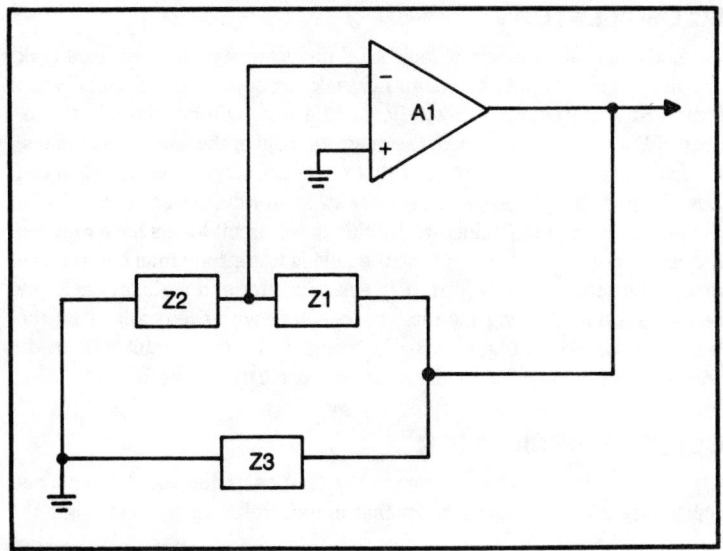

Fig. 24-8. Feedback oscillator model.

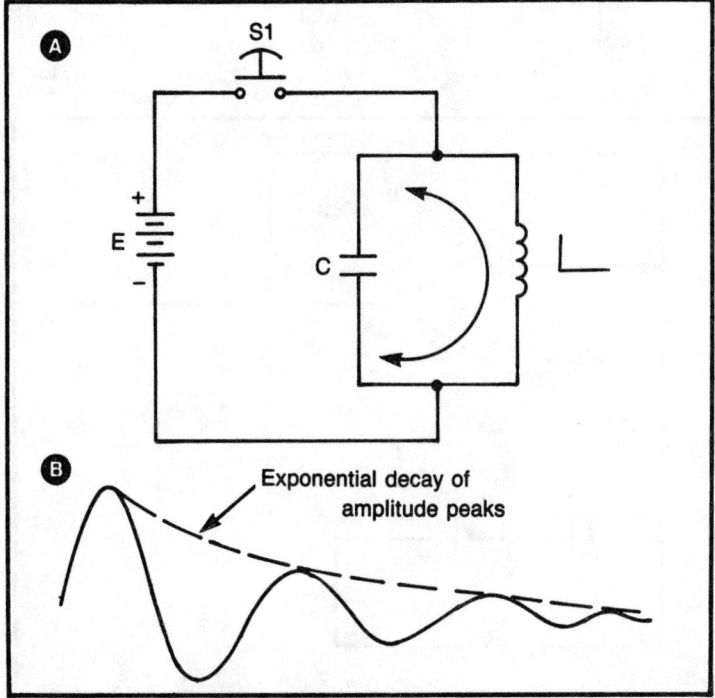

Fig. 24-9. (A) LC tank circuit, (B) ringing waveform.

LC OSCILLATORS

Let's digress a moment to look at LC (inductor/capacitor) resonant tank circuits. Figure 24-9(A) shows an LC tank circuit in a test circuit. When switch S1 is closed, current will flow and energy will be stored in the inductor. But when S1 is opened, the magnetic field of the inductor collapses, inducing a current into the circuit that tends to charge capacitor C. When C is charged, the process reverses; the capacitor discharges and reforms a magnetic field in the inductor. But this time, circuit losses have reduced the amount of available energy; so the field is less strong than before. The fields alternate back and forth between capacitor and resistor, each one being a little less strong than its predecessor—with the result being the sampled waveform of Fig. 24-9(B). Each time the LC tank circuit is "pulsed" with a bit of energy from closing S1, the oscillations will begin.

COLPITTS OSCILLATOR

The *Colpitts oscillator* is shown in Fig. 24-10 with the L and C elements replacing the Z terms. We know that certain relationships are true:

$$Z1 = \omega L1 \tag{24.3A}$$

$$Z2 = -1/\omega C1 \tag{24.3B}$$

$$Z3 = -1/\omega C2 \tag{24.3C}$$

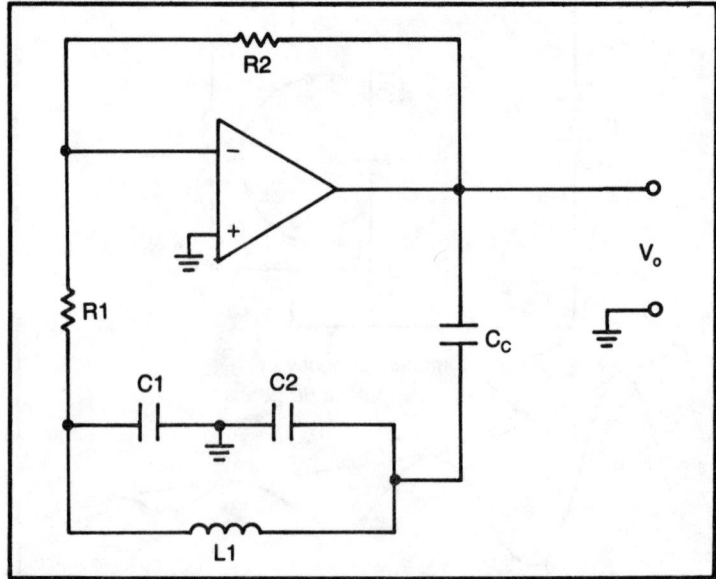

Fig. 24-10. Colpitts oscillator.

440

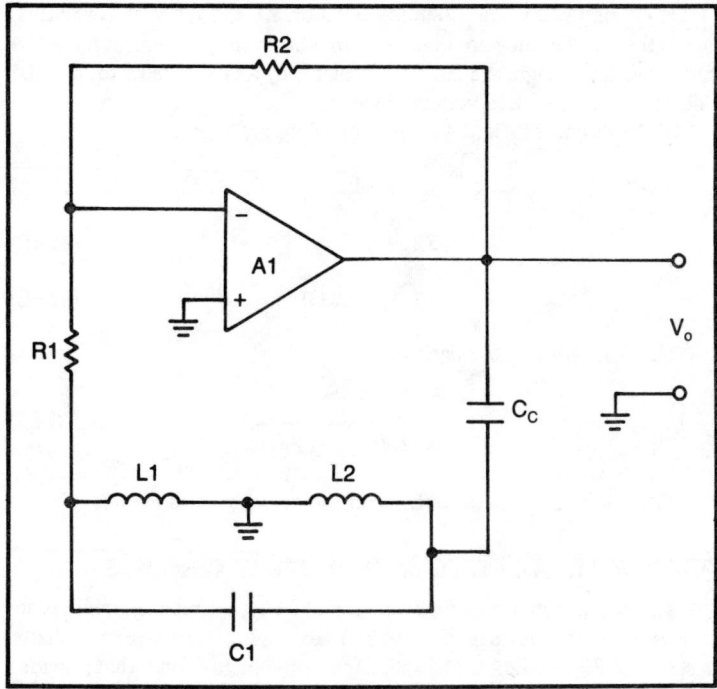

Fig. 24-11. Hartley oscillator.

I can derive an equation for resonance, i.e., the oscillating frequency, by summing these expressions to zero, with the result of

$$f = \frac{1}{2\omega\ L1C} \tag{24.4}$$

The unit C is defined as the series combination of C1 and C2, which is found from the standard expression

$$C = \frac{C1 \times C2}{C1 + C2} \tag{24.5}$$

The minimum gain of the circuit is C1/C2, and I must make C2 greater than C1 if the gain is to be less than unity.

HARTLEY OSCILLATOR

The classical *Hartley oscillator* is shown in Fig. 24-11. The principal difference between the Hartley and Colpitts oscillators is the arrangement of Z1, Z2, and Z3. In the Colpitts oscillator, Z1 and Z2 were capacitive, and in the Hartley they are inductive. In other configurations of these circuits, you can recognize a Hartley oscillator by a tapped inductor and the Colpitts by a capacitive voltage divider.

In the circuit of Fig. 24-11 the following are true:

$$Z1 = \omega L1 \tag{24.6A}$$

$$Z2 = \omega L2 \tag{24.6B}$$

$$Z3 = -1/\omega C1 \tag{24.6C}$$

The frequency of oscillation is

$$f = \frac{1}{2\omega(L1 + L2)C1} \tag{24.7}$$

The minimum gain is L2/L1, and the actual gain is R2/R1.

PRACTICAL COLPITTS AND HARTLEY CIRCUITS

The circuits shown in Figs. 24-10 and 24-11 are models, while practical versions are shown in Fig. 24-12. An example of the Colpitts oscillator is shown in Fig. 24-12(A). The split-capacitor voltage divider that provides feedback consists of C1 and C2; in general, C2 is larger than C1 (typical values are 82 pF for C1 and 0.001 μF for C2). Resistor R3 is used for biasing the transistor, while R1 provides a little stability; the output signal is developed across R1. The frequency of oscillation is provided by a parallel-resonant tank circuit.

A variation on the circuit called the *Clapp oscillator* is the same as the Colpitts, except that the LC tank circuit is series-resonant (see Fig. 24-12(A)).

The Hartley version is shown in Fig. 24-12(B). Again, a bipolar transistor is used for gain purposes. Resistors R1 and R2 are used for biasing the transistor, but play no other role. The output signal can be taken either from the emitter of the transistor or from a separate, link-coupled coil to the main, frequency-setting inductor.

The resonant frequency is set by capacitor C1 and the total inductance of L1A and L1B. The setting of the tap that divides L1A from L1B is a trade-off between stability and output amplitude.

Both Hartley and Colpitts/Clapp versions are common-collector circuits. In both cases, the collector of the transistor is set to ac ground by using a bypass capacitor (C4 in Fig. 24-12(B) and C5 in Fig. 24-12(A)).

OTHER LC OSCILLATORS

There are at least two additional forms of LC oscillator, and both of

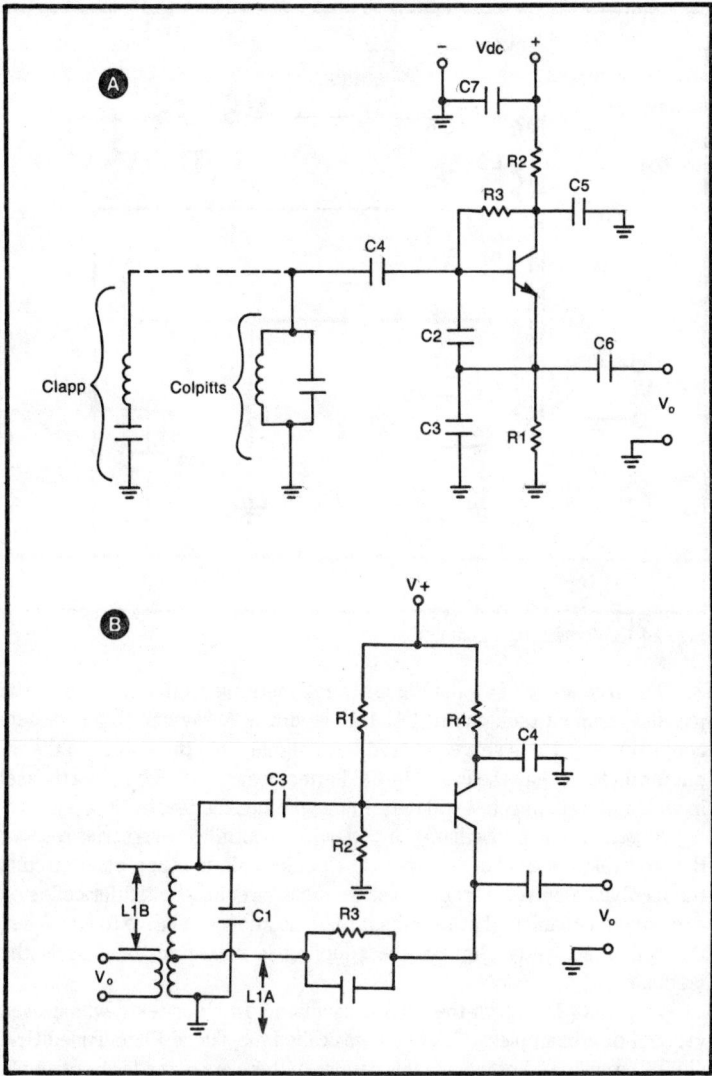

Fig. 24-12. (A) Colpitts and Clapp oscillator circuit, (B) Hartley oscillator.

them are among the oldest forms of electronic oscillator: the Armstrong oscillator is shown in Fig. 24-13, and the Tuned-Base-Tuned-Collector (TBTC) is in Fig. 24-14.

The Armstrong oscillator is named after Major Edwin Armstrong, who invented the regenerative detector, superheterodyne radio, and frequency modulation, among other things. In another form, the Armstrong oscillator becomes the regenerative radio detector.

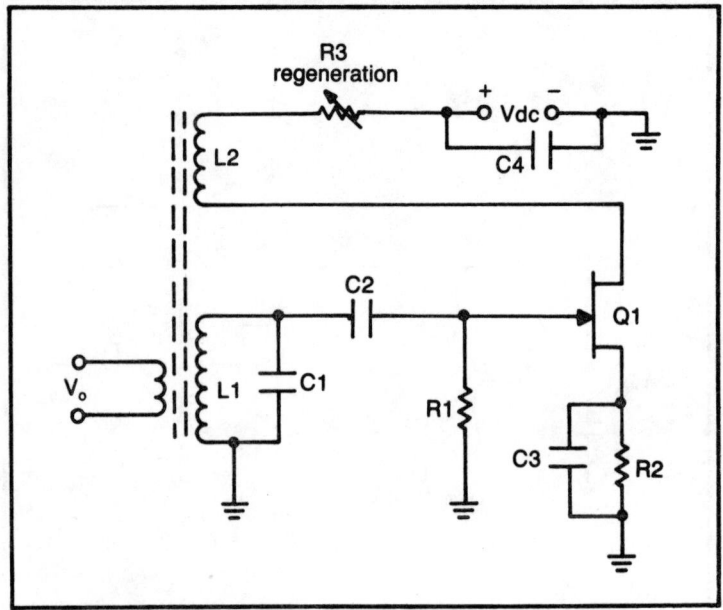

Fig. 24-13. Armstrong oscillator.

The frequency of oscillation for the Armstrong oscillator is set by the parallel-resonant tank circuit C1L1. A feedback *tickler* coil (L2) is closely coupled with L1 and serves to feed back signal from the output to the input (drain to gate, in the case of a field-effect transistor). The *sense* (phase) in which L2 is connected will determine whether the feedback is positive.

Regeneration, or oscillation if you will, is controlled by a series resistor that controls the level of current in the tickler coil. In other, older circuits the feedback control is mechanical. In those circuits, the tickler coil is on a moving mechanism that permits it to be positioned relative to L1. When the coil is perfectly aligned, the coupling is maximum (and so is the feedback).

Figure 24-14 shows the TBTC oscillator. In older texts which used vacuum-tube examples, this circuit was called the Tuned-Plate-Tuned-Grid (TPTG) oscillator. Perhaps, in this day when there are many forms of amplifying devices, this circuit should be renamed the "Tuned-Input-Tuned-Output (TITO) oscillator." The oscillating frequency is set by both L1C1 and L2C2. Most textbooks recommend that these two resonant tank circuits be at slightly different, but almost the same, resonant frequency.

Feedback in the TITO oscillator is determined by the interelectrode capacitances of the transistor. These capacitances are not physically in the circuit, but are inherent in the devices: C_{be} is the capacitance between base and emitter of Q1, and C_{cb} is the capacitance between collector and base.

The TITO oscillator is rarely used these days, even though some VHF and UHF, TV- and FM-receiver local oscillators are implicitly TITO.

RC SINE-WAVE OSCILLATORS

In this section I will look at RC-timed oscillators. Such oscillators produce an output frequency that is determined by one or more RC time constants in the circuits. Some of my examples are based on discrete components, while others on operational amplifiers.

RC PHASE-SHIFT OSCILLATORS

The basic RC phase-shift oscillator uses a three-section RC network to provide 180-degrees of phase shift. When added to the 180-degrees provided by any inverting amplifier, this phase shift will meet the 360-degree criterion for oscillation. Figure 24-15 shows two basic forms of RC phase-shift oscillators. The circuit in Fig. 24-15(A) is based on a junction-field-effect-transistor (JFET) amplifier, and the circuit in Fig. 24-15(B) is based on the operational amplifier. Both of these circuits produce a sine wave output signal.

In both cases, there is a three-section, phase-shift network: R1/C1, R2/C2, and R3/C3. These three networks are identical to each other and are designed to provide 60 degrees of phase shift at the desired oscillating frequency. When the three phase shifts are added together, you find the

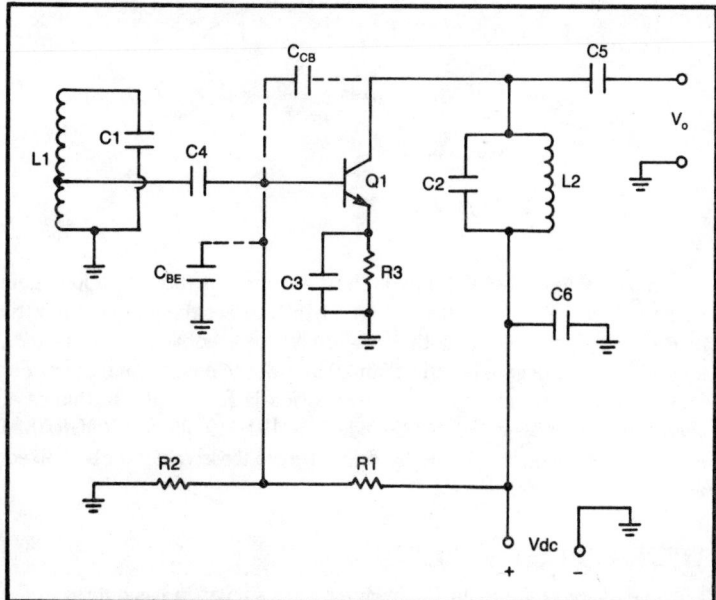

Fig. 24-14. TITO oscillator.

445

required 180-degree phase shift. In this example, it is assumed that R1 = R2 = R3 = R and C1 = C2 = C3 = C (where R and C are the resistance and capacitance values used in the equations). As in most cases, you will use farads for C, ohms for R, and hertz for f.

The oscillating frequency is given by

$$f = \frac{0.408}{2 \, \omega \, R \, C} \qquad (24.8A)$$

Of course, your problem in practical designs will be to find a resistor/capacitor combination that will produce a desired (and known) frequency; so this expression is backwards. Because there are fewer standard, capacitor values, one tends to select a capacitor, plug in the frequency, and then find the value of resistor that will produce that frequency with that capacitor. Because of that I will rearrange the equation to read:

$$R = \frac{0.408}{2 \, \omega \, f \, C} \qquad (24.8B)$$

Example

Find the resistance required to make a 1000-Hz oscillator when a 0.01-μF capacitor is used for C1, C2, and C3 in Fig. 24-15(B).

$$R = 0.408/(2 \, \omega \, f \, C)$$

$$= 0.408/(2 \times 3.14 \times 1000 \times 1 \times 10^{-8})$$

$$= 0.408/6.28 \times 10^{-5}$$

$$= 6497 \text{ ohms}$$

In all feedback oscillators you must ensure that the closed-loop gain is unity or more. In the case of Fig. 24-15(B), I set the gain as the ratio of R_f/R. Analysis will show that the loss in the feedback circuit is 1/29; so I will make the gain greater than 29 in order to ensure oscillation. For most practical circuits, the recommendation is $R_f > 30R$. In this case, therefore, the value of $R_f > (30) (6497) > 194,910$ ohms. PROGRAM I is a BASIC computer program that will calculate these values for both fixed- and variable-frequency cases.

WIEN-BRIDGE OSCILLATOR

The *Wien bridge* is an RC bridge circuit similar in basic form to the Wheatstone bridge (OK purists—sit down!). Two of the arms of the Wien

446

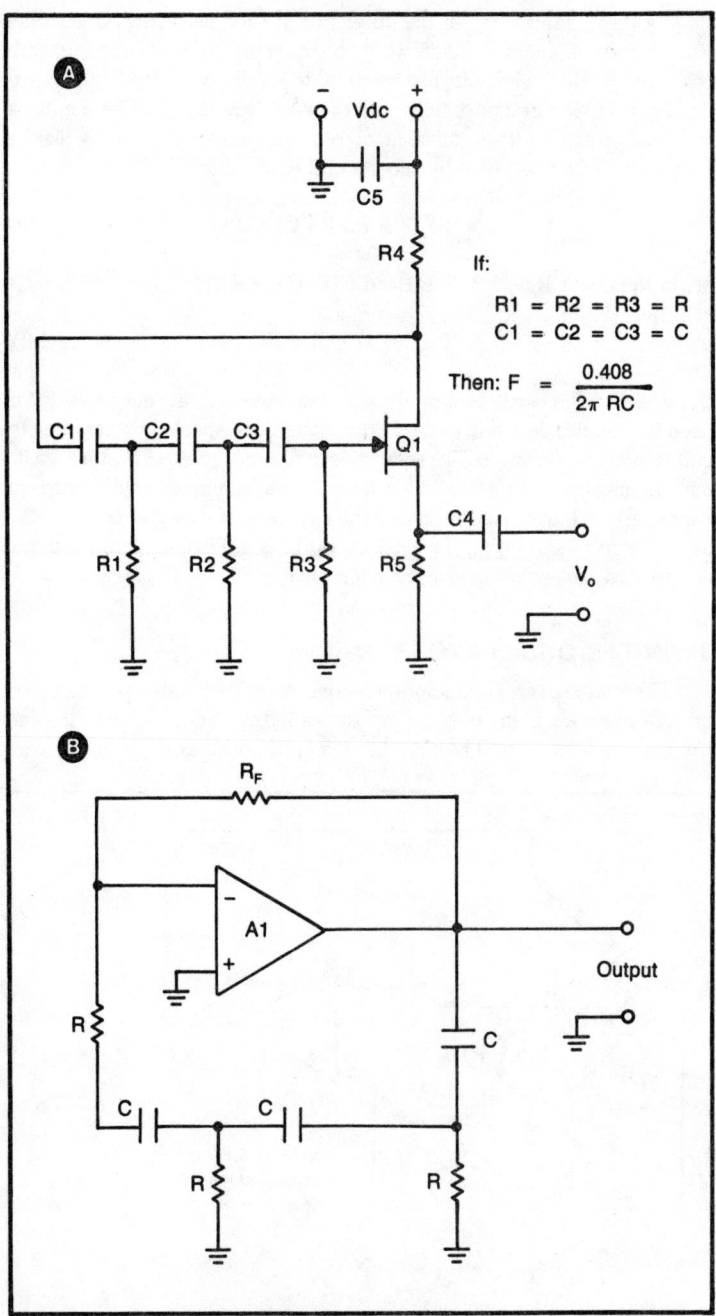

Fig. 24-15. (A) RC phase-shift oscillator based on JFET, (B) on op-amp.

447

ridge are resistances, while the other two are RC networks. One of the RC networks is a series circuit, while the other is a parallel circuit; in both cases the R and C elements are equal between the two networks. An example of a Wien-bridge oscillator is shown in Fig. 24-16. The feedback loop is degenerative (thus stable) at all frequencies other than the oscillating frequency, which is given by the general expression

$$f = 1 / (2 \omega R3 R4 C1 C2) \qquad (24.9A)$$

or, in the event R = R3 = R4 and C = C1 = C2,

$$f = 1 / (2 \omega R C) \qquad (24.9B)$$

Wein-bridge oscillators produce a sine wave output, but when R2 is used the amplitude tends to be a little unstable—especially when the circuit is frequency variable. The solution is to replace resistor R2 with a small-current incandescent lamp. This lamp has a nonlinear voltage-current characteristic that tends to stabilize the output amplitude and prevents the amplifier from saturating. In most cases, I1 is a 0.040-ampere lamp and is operated somewhat below incandescence.

TWIN-TEE OSCILLATORS

There are several other forms of sine wave oscillator that are based on RC networks. Some of these are shown in this section as twin-tee and bridged-tee oscillators. The circuit in Fig. 24-17 is an example of the twin-

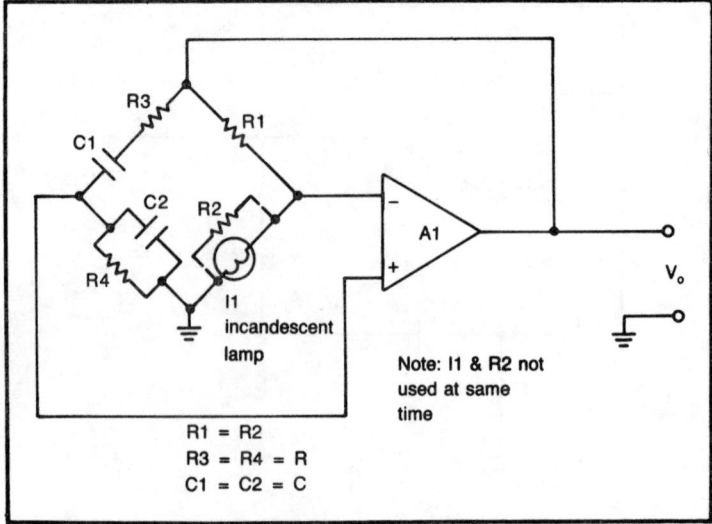

Fig. 24-16. Wein bridge oscillator.

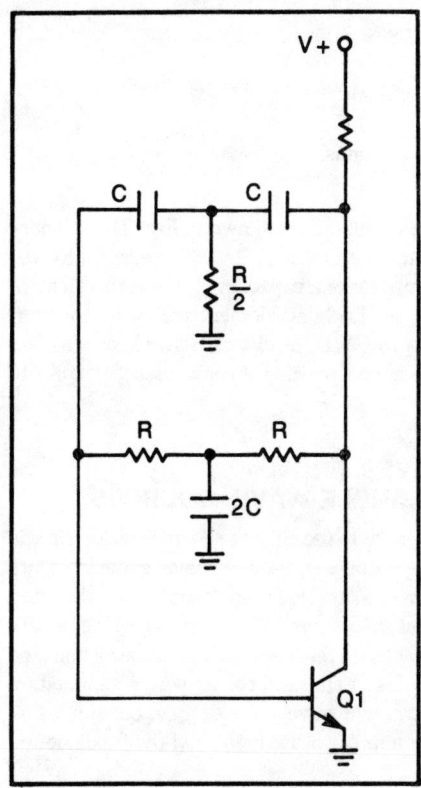

Fig. 24-17. Twin-tee oscillator.

tee oscillator, so-called because its feedback network consists of two tee networks. Note that these networks are of opposite types. One of them uses series resistors and a shunt capacitor, while the other uses series capacitors and a shunt resistor. The frequency is approximated by the equation below

$$f = 1/17.8 \, R \, C \qquad (24.10A)$$

Again, I can rearrange the equation to the more practical, but less familiar, form:

$$R = 1/17.8 \, f \, C \qquad (24.10B)$$

Example

Find the resistance needed to make a 500-Hz twin-tee oscillator when the capacitance C is 0.01 μF.

$$R = 1/(17.8 \times f \times C)$$

$$= 1/(17.8 \times 500 \times 10^{-8})$$

$$= 1/8.9 \times 10^{-5}$$

$$= 11,234 \text{ ohms}$$

Two additional forms of tee oscillator are shown in Fig. 24-18. These are two alternate forms of bridged-tee oscillator. In both cases, an RC tee network is bridged by either a resistor or a capacitor. If the series element of the tee network is a resistor, the bridging element will be a capacitor, and vice-versa. As in the case of the Wien-bridge oscillator shown above, the bridged-tee oscillators use a resistor and an incandescent lamp to stabilize the output amplitude.

GENERATING SINE WAVES FROM SQUARE-WAVE OR TRIANGLE-WAVE SOURCES

Some instrument designers prefer to use either a square wave or triangle wave to generate a sine wave output. Most sine-wave generators are unstable with respect to amplitude, so are less than desirable. On the other hand, many square-wave oscillators operate in a saturating mode, so are inherently amplitude stable. We can do this neat trick because all non-sine waveforms are made up of a series of sine and cosine waves summed together. The square wave and triangle wave, for example, contain a fundamental sine wave and a large number of harmonics of the fundamental waveform.

If we conspire to filter out all of the harmonics, except the fundamental frequency, the result will be a sine wave at the fundamental frequency. The purity of the sine wave can be quite good, especially if a high order of filtering is used. In most cases, either a notch filter (which passes only the fundamental) or a series of low-pass filters is used for the conversion.

Besides amplitude stability, you gain one other advantage by using the filtering method. The other forms of RC sine-wave oscillator are a little hard to make variable over a wide range of frequencies. The square-wave oscillator, on the other hand, is much easier in this respect. You can make a variable square-wave oscillator with only one resistor variable and then use a low-pass filter to remove the harmonics. It is feasible to use just one low-pass filter per octave of frequency change.

WIDE-RANGE OSCILLATORS

It is difficult to design and build wide-range audio oscillators using just RC components operating at normal audio frequencies. You can, however, solve the problem using a method that was popular in the 1950s (see Fig. 24-19). In this circuit, there are two oscillators (often LC). One of them is fixed-frequency (F2), operating at 100 kHz; and the other one is variable-

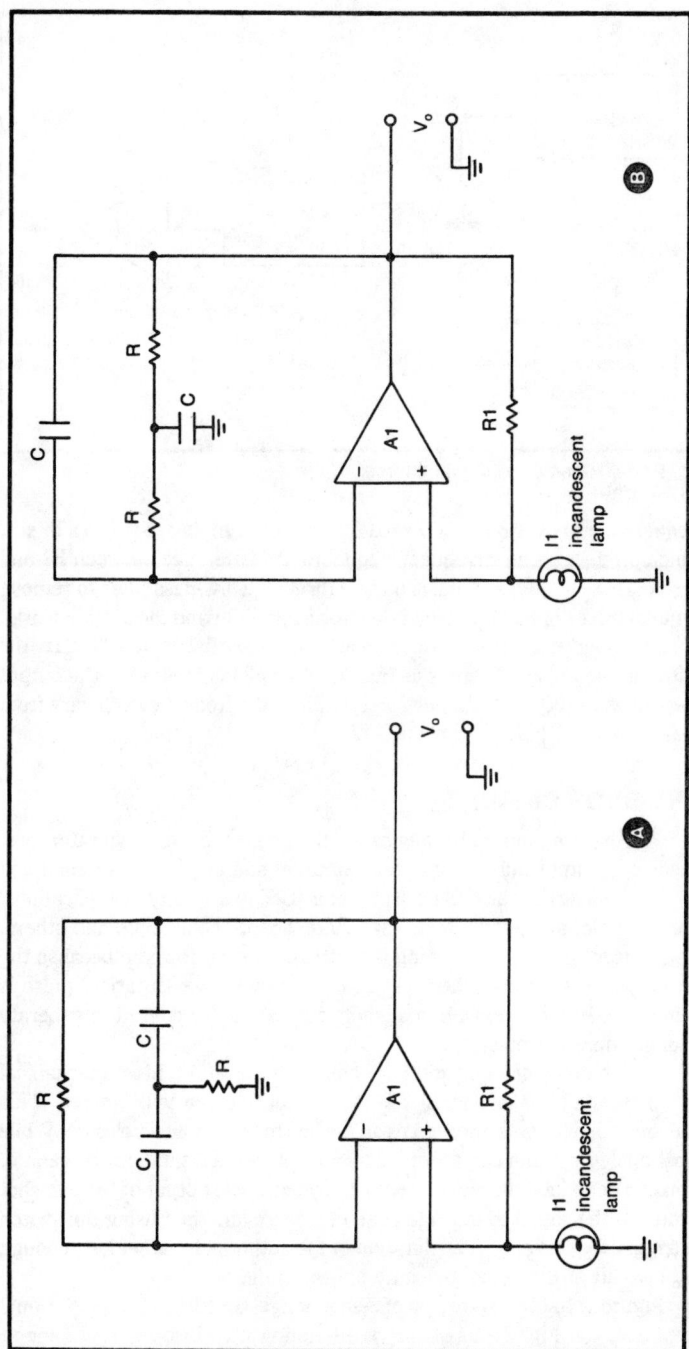

Fig. 24-18. Two versions of the bridge-tee oscillator.

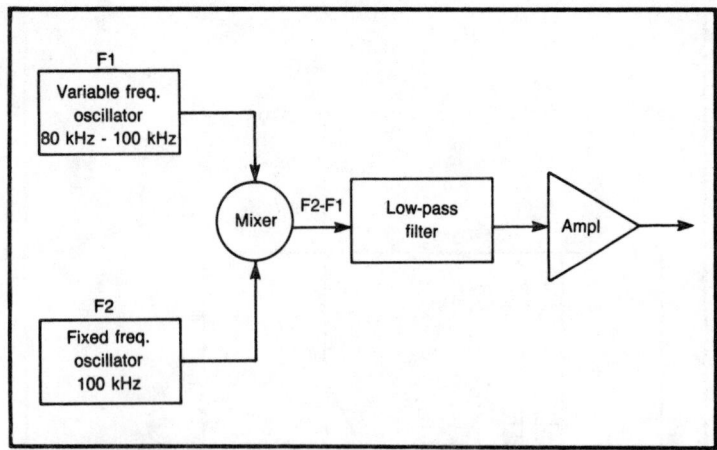

Fig. 24-19. Heterodyne signal source.

frequency (F1) over 80 to 100 kHz. These two outputs are fed to a mixer, which produces a new frequency equal to the difference between F1 and F2; i.e., F2 – F1. This signal is passed through a low-pass filter to remove residual traces of F1 and F2 and then to an amplifier and the outside world.

The operation of this circuit is simple. When F1 is at 100 kHz, the difference is zero; so there is no output. When F1 is at 80 kHz, the output frequency is (100 – 80) kHz, or 20 kHz. Thus, the frequency will vary from near zero to 20 kHz.

ONE-SHOT CIRCUITS

The *one-shot*, more formally called the monostable multivibrator, produces one output pulse of constant duration and amplitude for each and every input trigger pulse. Such a circuit is used in a variety of applications. For example, pulse-stretching, switch/keyboard-debouncing, and others. The concept of pulse-stretching is a little confusing to some because the actual pulse is not stretched, but rather a new (longer-duration) pulse is generated. In that application, a short-duration pulse is used to generate a longer-duration pulse.

Switch-debouncing is another common application. Most mechanical switches don't make and break cleanly, but rather will bounce. This phenomenon places several spikes on the line for each contact closure. While most analog circuits can absorb the extra pulses, a digital circuit cannot. You can debounce the switch contacts by using the output of the one-shot circuit as the signal to indicate contact closure and then using the switch to trigger the one-shot. The duration of the output pulse is set long enough to allow all of the contact-closure spikes to die out.

Figure 24-20(A) shows a one-shot stage based on the operational amplifier, and Fig. 24-20(B) shows the timing waveform. For a moment,

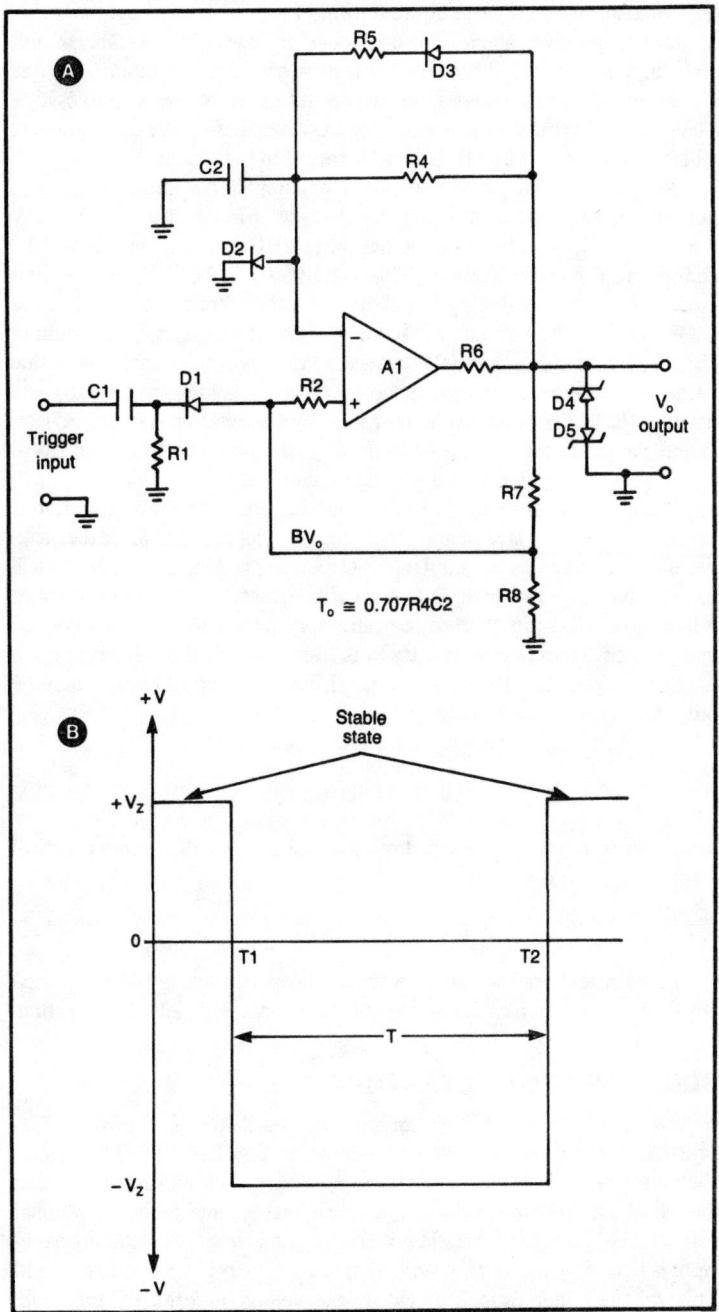

Fig. 24-20. Op-amp one-shot (A) circuit, (B) timing diagram.

let's concentrate on the timing waveform—keeping in mind the definition of the one-shot given above. The one-shot stage has only one stable output state, in this case $+V_z$ (hence the name monostable multivibrator). At time t1, the one-shot is triggered; so the output snaps to the unstable state $(-V_z)$. It will remain in this state for only a specified period of time, t, which is defined as $(t2-t1)$. Now let's return to the circuit, Fig. 24-20(A).

Normally, the output of the operational amplifier is HIGH (i.e., $+V_z$), but when a short pulse is applied to the trigger input, it will snap LOW (i.e., $-V_z$). Because the output has been HIGH, capacitor C2 will be charged to a positive voltage. This voltage would be $+V_z$, except that diode D2 clamps the voltage to about 0.7 volts. When the output snaps LOW, however, the output voltage is $-V_z$; so the capacitor will begin to charge toward that potential. But the voltage will not reach that value because the noninverting input of the operational amplifier is biased to some fraction (B) of the output voltage, as set by the resistor network R7/R8. When the capacitor voltage exceeds BV_o, the output of the operational amplifier will snap back to the HIGH (stable) state.

The period immediately after the output returns to the stable state is called the *refractory period*, and the circuit will not respond to further trigger inputs during this period. Unless something is done the capacitor will have to discharge under the influence of $-V_o$, R4, and C2, and that takes a long time. You can shorten the refractory period by placing diode D3 and resistor R5 in the circuit. If R5 is less than R4, the discharge cycle is much shorter than the charge cycle. Diode D3 conducts only when the output voltage is positive.

The duration of the output state is approximately

$$t = 0.707 \ R4 \ C2 \qquad (24.11A)$$

or, if I wish to rewrite the equation in a more reasonable, more practical form,

$$R4 = t/0.707 \ C2 \qquad (24.11B)$$

In the latter case I will select a standard capacitor value and the desired output duration, then calculate the resistance that will yield that duration.

SQUARE-WAVE OSCILLATORS

An operational amplifier square-wave oscillator is shown in Fig. 24-21(A), and its timing diagram is shown in Fig. 24-21(B). This circuit has no stable output states; so its output snaps back and forth between the HIGH and LOW conditions. The noninverting input of the operational amplifier is biased by a fraction of the output voltage, and the inverting input is biased by the voltage across capacitor C1 (from point A to ground). This voltage is determined by the output voltage and the RC time constant of R1C1.

454

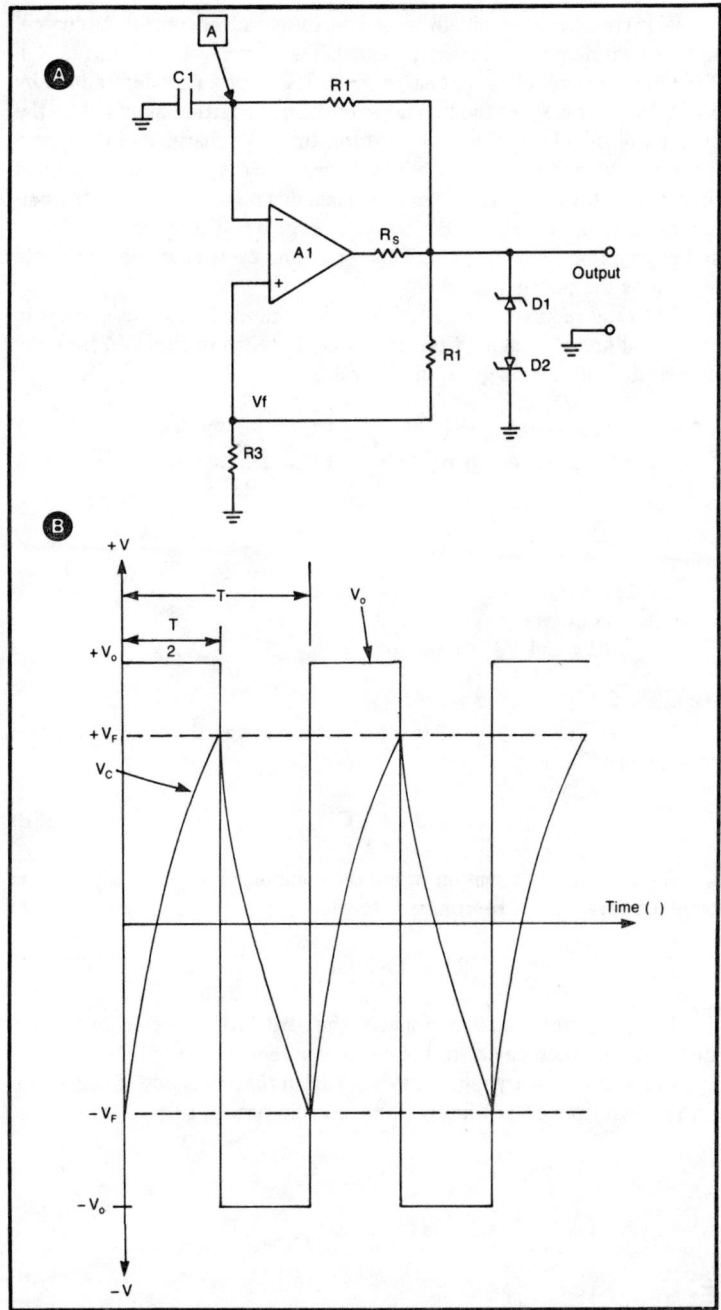

Fig. 24-21. Op-amp astable multivibrator (A) circuit, (B) timing diagram.

455

When this circuit is initially turned on, the capacitor contains no charge; so the inverting input is at zero potential. The output will be HIGH ($+V_o$). The capacitor will begin to charge toward $+V_o$ at a rate determined by R1C1. When it reaches the bias point of the noninverting input ($+V_f$), the output will snap LOW (i.e., $-V_o$). At this time, the charging of the capacitor reverses, and the voltage will discharge from $+V_f$ towards zero and then charge toward $-V_f$. When it reaches this negative voltage, the output again snaps HIGH, and the capacitor begins to discharge toward zero and charge toward $+V_f$ again. This oscillating cycle continues endlessly as long as the power is applied.

The time required for one cycle is determined by the resistances in the circuit and the value of the capacitor. In terms of Fig. 24-21(A), the period of oscillation is given in general by

$$t = 2 \text{ R1 C1 Ln} \left[1 + \frac{2 \text{ R2}}{\text{R3}} \right] \qquad (24.12\text{A})$$

where

   t is in seconds
   C1 is in farads
   R1, R2, and R3 are in ohms.

In the event that R2 = R3 (a common occurrence), the equation reduces to a simpler form

$$t = 3.2 \text{ R1 C1} \qquad (24.12\text{B})$$

Again, I have a situation where the equation is not as easily used as it might be; so I will rearrange it to solve for the resistance

$$\text{R1} = t/3.2 \text{ C1} \qquad (24.12\text{C})$$

Using this version I would specify the period (t) and a standard capacitor (C1), and then calculate the resistance required.

In the event that you want to work from the frequency of oscillation rather than the period, simply find the reciprocal of t

$$f = 1/t \qquad (24.13)$$

where t is in seconds and f is in hertz.

Example 1

What is the frequency of a 1-millisecond (0.001-second) oscillator?

$$f = 1/t$$
$$= 1/0.001$$
$$= 1000 \text{ Hz}$$

Similarly, you can calculate t from knowledge of f:

Example 2

What is the period of a 1500-Hz oscillator?

$$t = 1/f$$
$$= 1/1500$$
$$= 0.00067 \text{ seconds}$$

Now, let's work a practical design problem. Calculate the resistance needed to make a 1500-Hz oscillator (i.e., one with a period of 0.00067 seconds (per Example 2).

Example 3

Find the resistance needed to make a square-wave oscillator similar to Fig. 24-21(A) operate at 1500 Hz (0.00067 second) when capacitor C1 is 0.001 μF.

$$R1 = t/(3.2 \text{ C1})$$

$$= (0.00067)/(3.2 \times 0.000000001 \text{ } \mu\text{F})$$

$$= 209,375 \text{ ohms}$$

The *duty cycle* of the circuit of Fig. 24-21(A) is 50 percent; in other words, the output is HIGH for the same length of time as it is LOW. In some cases you might not want a 50-percent duty cycle but will want either the HIGH or the LOW portion to be shorter than the other part of the cycle. There are two strategies that will yield different HIGH and LOW durations (see Figs. 24-22(A) and 24-22(B)).

One alternative, shown in Fig. 24-22(A), is to use a different timing resistor for the HIGH and the LOW portions of the cycle. In this case, diodes are used for switching the resistors R_a and R_b. When the output is HIGH ($+V_z$), diode D4 is forward biased, and R_b affects the waveform timing. Similarly, when the output is LOW ($-V_z$), D4 is reverse biased, and D3 is forward biased; so R_a controls the timing of events.

The other method is shown in Fig. 24-22(B). Here I add a positive or negative biasing charge to the capacitor through resistor R_{off}. The polarity of the bias voltage across C1 is determined by the setting of potentiometer R2 and can vary from V − to zero to V +. You can monitor the output of

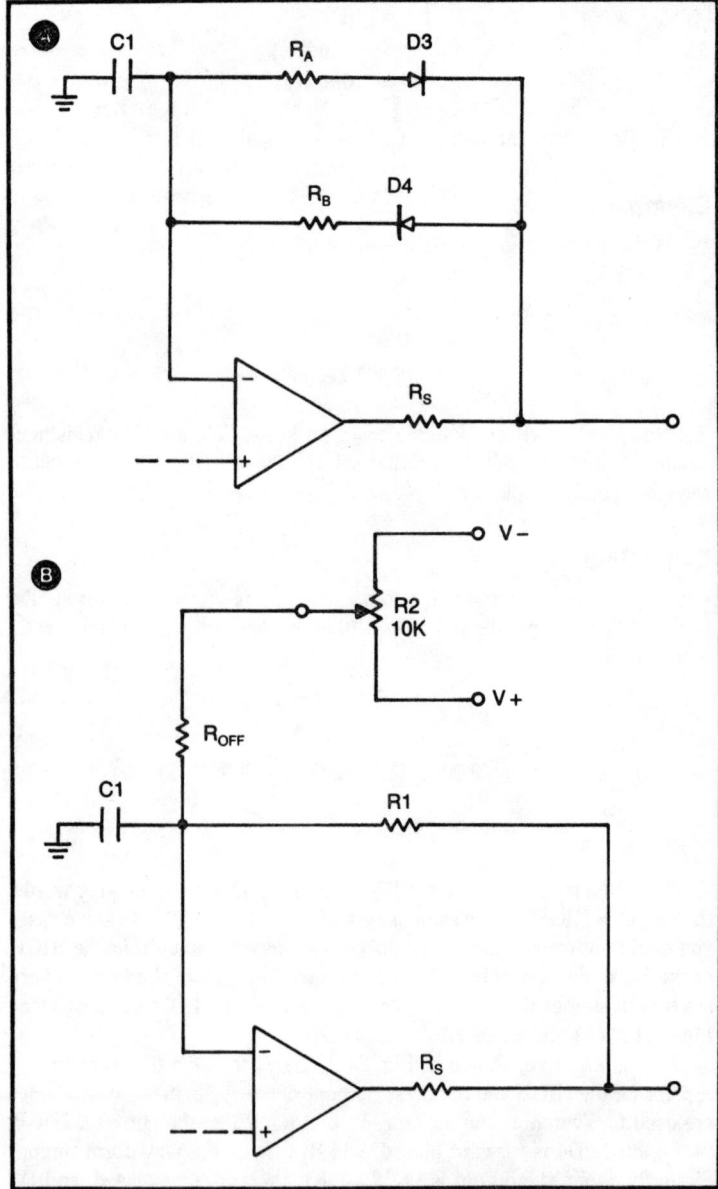

Fig. 24-22. Duty cycle variation (A) fixed ratio, (B) variable ratio.

the circuit on an oscilloscope while adjusting R2 for the desired duty factor. In most applications, the potentiometer will be a multi-turn trimpot mounted on a printed-circuit board unless front-panel adjustment is desired.

TRIANGLE-WAVE OSCILLATORS

A triangle-wave oscillator can be made by combining the square-wave oscillator with a simple Miller integrator (see Fig. 24-23(A)). There are other ways of making triangle and sawtooth waveforms, but they all seem to lack something compared with this method.

A simple operational-amplifier version of the Miller integrator is shown in Fig. 24-23(A). The slope of the output waveform depends upon the time constant R1C1. Figure 24-23(B) shows a typical waveform to expect when the time constant is long compared with the period of the applied square wave. When the square wave is LOW ($-V$), the integrator output will climb in a nearly linear manner in a positive-going direction. The linear slope is due to the fact that the input square wave has a constant amplitude during this time. Similarly, when the square wave goes to the opposite sense (phase), the output of the Miller integrator will be negative-going. This cycle continues as long as there is a symmetrical square wave applied to the input of the integrator.

An example of a triangle-wave generator based on this principle is shown in Fig. 24-24. Here I have a Miller integrator connected to a comparator circuit (A1). The frequency of oscillation will be

$$f = \frac{R2}{4\ R1\ R3\ C1} \tag{24.14}$$

CRYSTAL-CONTROLLED SINE-WAVE OSCILLATORS

For most common applications the most stable frequency-control method is provided by the piezoelectric crystal oscillator. Some crystalline elements, notably quartz, possess the property of piezoelectricity. This phenomenon is not found in all materials, only certain crystalline and ceramic materials. A material is said to be piezoelectric if it generates an electrical potential when deformed. When a piezoelectric element is mechanically deformed, therefore, a voltage appears across the faces of the crystal slab.

There is also an inverse version of this phenomenon. When you apply a voltage across the faces of a piezoelectric crystal slab, the crystal will mechanically deform. You can use this phenomenon to make the crystal oscillate at its resonant mechanical frequency.

If a pulse is applied to the faces of the crystal slab, the slab will vibrate back and forth in an oscillatory manner. Losses in the slab will cause the amplitude of these mechanical oscillations to die out exponentially. While the pulsed crystal slab is oscillating, however, it is producing an ac voltage with a frequency equal to the resonant frequency.

Figure 24-25(A) shows a circuit symbol for the piezoelectric-crystal resonator, and Fig. 24-25(B) shows the actual physical form the crystal will take. The cut slab of crystal material is sandwiched between two contact electrodes. In older crystals, these electrodes are attached with spring ten-

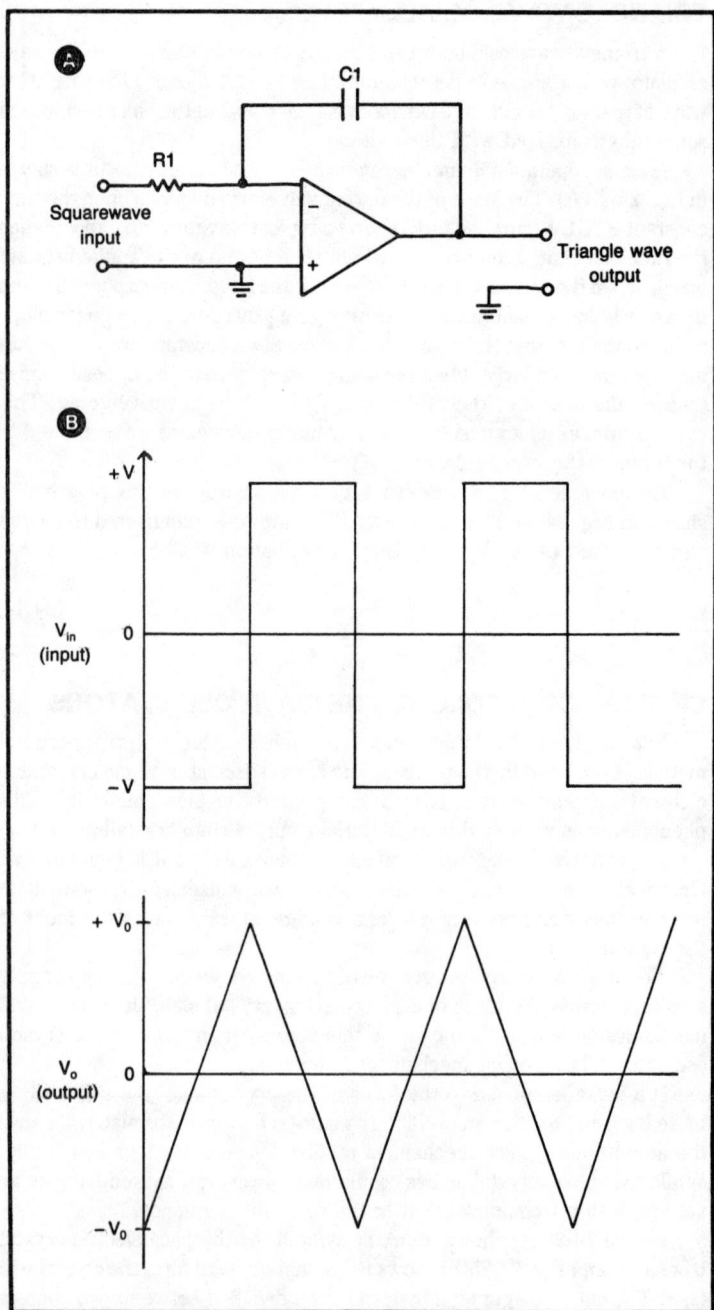

Fig. 24-23. (A) op-amp integrator, (B) waveforms.

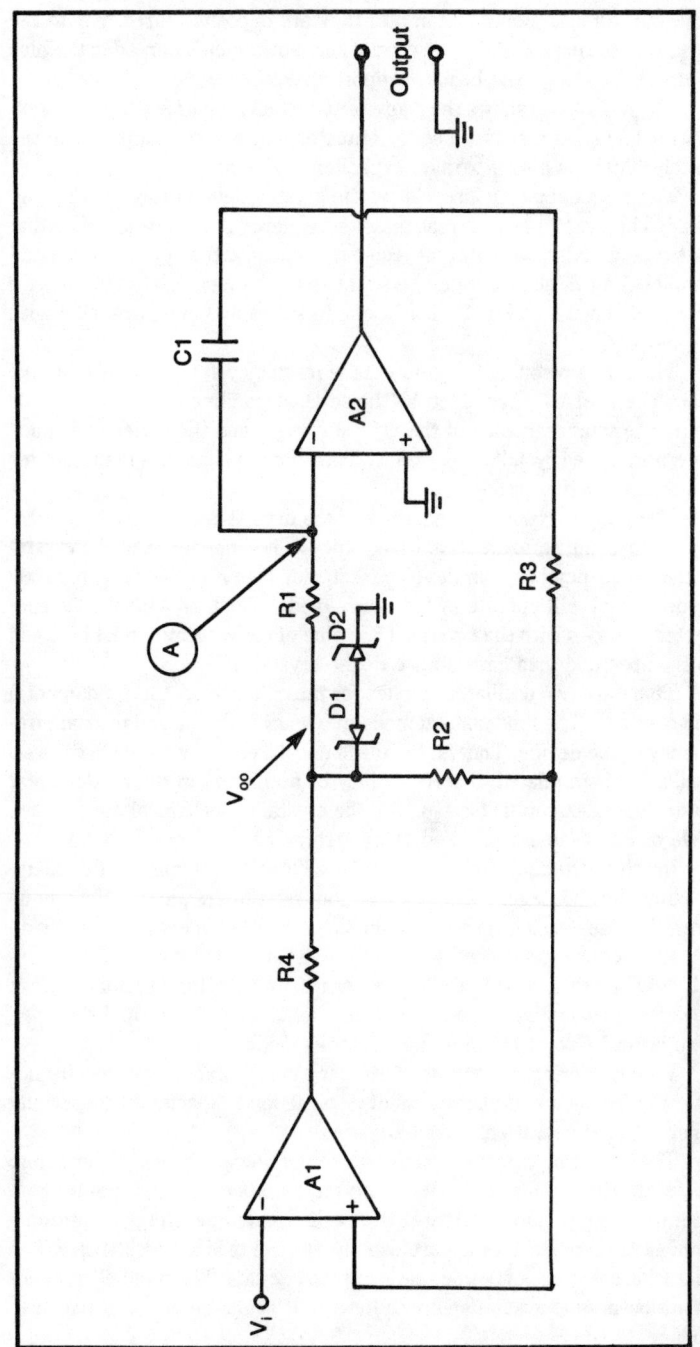

Fig. 24-24. Triangle waveform generator.

461

sion, but in more modern examples they are deposited directly onto the surface of the crystal slab. The crystal element is then mounted to the pins protruding through the header-support structure.

Figure 24-25(C) shows the equivalent circuit of a piezoelectric element. There is a series resistor, a series inductor, and a series capacitor in the circuit. There is also a parallel capacitor in the circuit.

The frequency-response plot of the crystal element is shown in Fig. 24-26. This graph plots the reactance-vs-frequency of the crystal. Note that there are actually two different resonant modes for the crystal: series and parallel. You might have guessed that from the fact that there are two capacitors shown in Fig. 24-25(C), one in series and one in parallel with the inductor.

The series-resonant frequency is the frequency at which the inductive reactance is exactly cancelled by the series capacitive reactance. At this point, the total reactance of the crystal is zero, and the series resistance determines the crystal's impedance. Thus the impedance is minimum for the series-resonant mode.

The parallel-resonant frequency of the crystal is usually 1 to 15 kHz higher than the series-resonant frequency. The impedance of the crystal at this frequency is maximum (see frequency f_p in Fig. 24-26). A parallel-mode crystal will operate in the series mode if a small capacitor is connected in series with the crystal. The value of the capacitor must be equal to the specified load capacitance of the crystal.

There are two oscillatory modes for the crystal element: fundamental and overtone. The fundamental-mode frequency is the natural resonant frequency of the device. That is, the mechanical frequency at which the slab oscillates if stimulated. The frequency of oscillation in the fundamental mode depends upon factors such as the crystal's mechanical dimensions, style of cut, temperature, and other factors.

In the overtone mode, the crystal oscillates at a frequency that is approximately an integer multiple of the fundamental frequency. Note, however, that the overtone is not a harmonic of the fundamental. If you divide the 5th overtone frequency by 5, the result is not necessarily the fundamental, but a number close to the fundamental. The overtone frequencies are always approximately an odd multiple (e.g., 3, 5, 7, 9, 11), and the overtone crystal always operates in the series mode.

When ordering crystals for frequency control, you must specify not only the frequency of operation but also the load capacitance (especially when parallel operation is contemplated).

The operating frequency of the fundamental mode is usually less than about 20 MHz; above 20 MHz the crystal slab becomes too thin for safe operation and may fracture under normal operating conditions. Fundamental-mode crystals are usually parallel-mode, except below 500 kHz where it is usual to order series-mode crystals. The parallel-resonant condition produces a 180-degree phase shift at the parallel-resonant frequency.

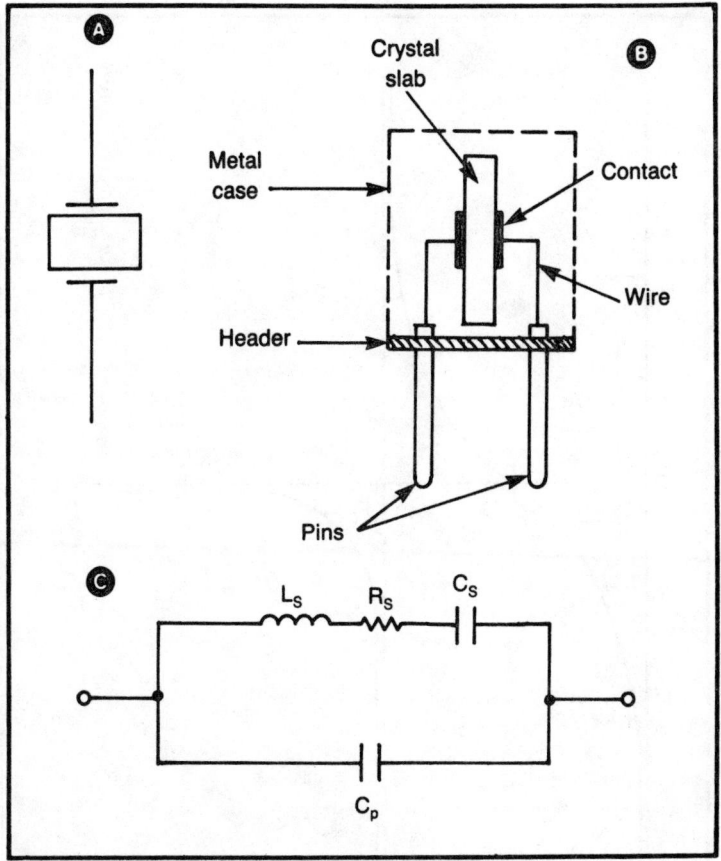

Fig. 24-25. Piezoelectric crystal (A) symbol, (B) construction, (C) equivalent circuit.

The power dissipation of the crystal must be limited in order to prevent fracture damage to the slab. It is the *equivalent series resistance* (ESR) of the crystal that determines the power dissipation at any given level of applied signal. In a practical crystal-oscillator circuit you control the dissipation by controlling the feedback-signal amplitude. Most fundamental-mode crystals will dissipate up to about 200 microwatts (μW), although it is usually considered good engineering practice to make the actual dissipation somewhat less than this.

Low-frequency crystals, i.e., those operating below 1000 kHz, usually have a dissipation rating of 100 μW. It is generally regarded as good practice to limit the dissipation to 50 μW in order to improve the frequency stability of the crystal oscillator. It is not good practice to obtain huge amounts of rf power from the crystal oscillator, despite the fact that 25 years ago amateur handbooks typically carried crystal-oscillator-transmitter

463

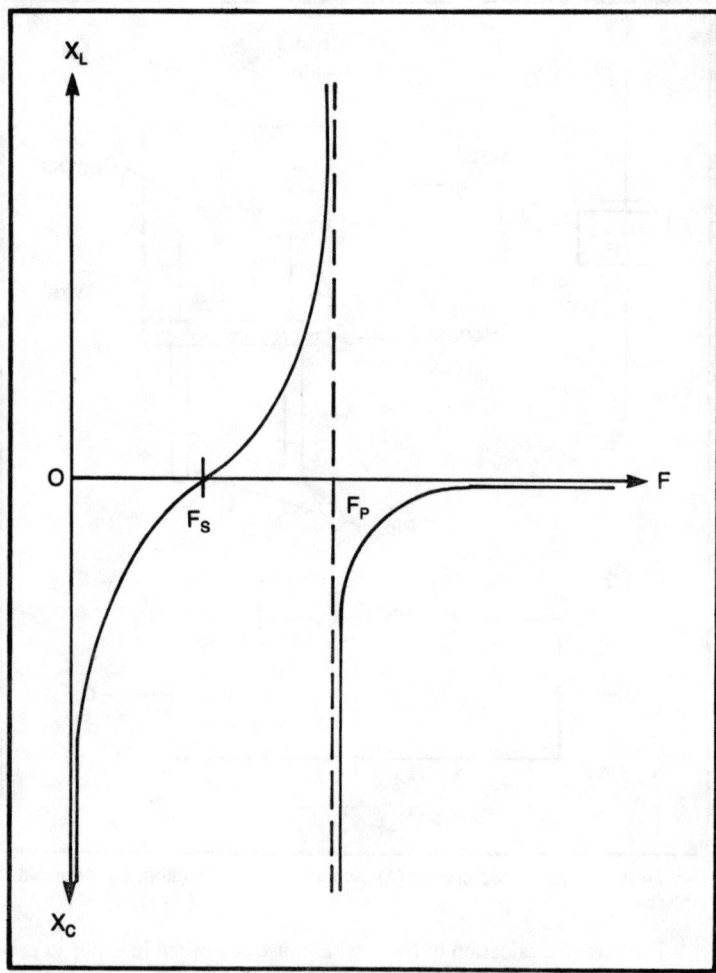

Fig. 24-26. Impedance-vs-frequency.

projects with up to 50 watts of power. It is far better to limit the dissipation of the oscillator circuit and then use an amplifier to build up the levels.

EXAMPLES OF CRYSTAL-OSCILLATOR CIRCUITS

The piezoelectric-crystal resonator is essentially equivalent to a complex LC circuit (with series and parallel sections), so can be used in many of the same circuits as ordinary LC resonant tank circuits. In many such cases, the sole difference between the circuits is that the crystal replaces the LC tank circuit.

Figure 24-27 shows a simple circuit for low-frequency crystals, specifically the 100-kHz *marker-frequency* crystal shown. The oscillator operates at a fundamental frequency of 100 kHz, but its harmonics go high enough to make the circuit useful for marker service in communications receivers and shop alignment.

The active element in this circuit is a bipolar npn transistor, such as the popular 2N2222 (the actual type is not as critical as one might assume). Almost any transistor with a beta in the 50-to-125 range and a gain-bandwidth product of 50 MHz or more will work. The reason for specifying the 2N2222 is that they are readily available in hobby and amateur-radio parts stores.

Note that the circuit configuration for the 100-kHz oscillator is similar to the familiar Colpitts oscillator seen earlier. Feedback level is a function of the capacitive voltage divider C1/C2. The ratio of these two capacitors is a trade-off between output amplitude, stability, and the dissipation of the crystal element. The collector load is provided by resistor R1, and the transistor is biased in the collector-base manner by resistor R2.

The crystal element Y1 shunts the feedback voltage divider and is in series with variable capacitor C3. This capacitor is used to set the actual operating frequency of the calibrator. Almost any crystal operating frequency can be *pulled* a little bit with a series or parallel capacitor.

The output signal is coupled to the load through a low-value capacitor, C4. This circuit should be lightly loaded in order to minimize changes in frequency due to load variations. When this is a critical concern, the solution is to follow the oscillator with a buffer-amplifier stage.

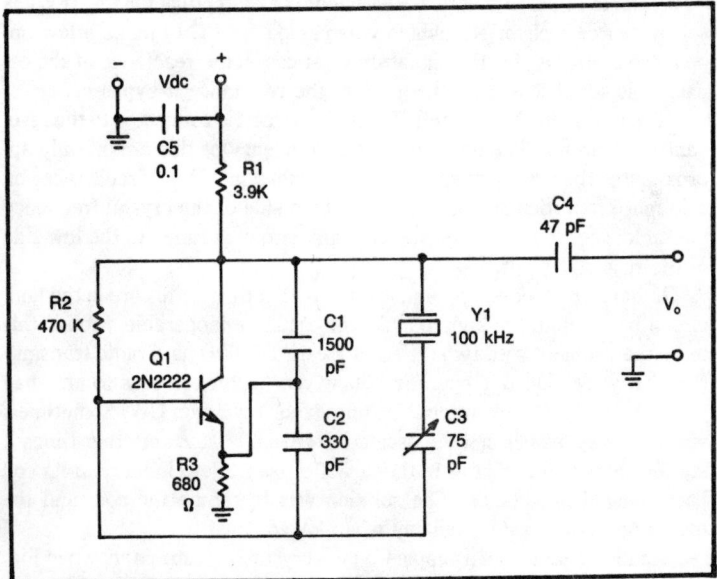

Fig. 24-27. 100 kHz oscillator.

There are several ways to calibrate the operating frequency of the oscillator. If you have a digital frequency counter, use it. Once upon a time this would have been a ridiculous statement in an article in this textbook because digital frequency counters were a multi-kilodollar investment. But today, there are a number of low-cost DFCs on the market that all but surpass the performance of instruments that were top-dollar only a decade or so ago.

The alternate way to calibrate the circuit is to use a radio station. If you lack a communications receiver, use an AM-broadcast-band radio. The radio should be tuned to a radio station that has a frequency divisible by 100. In my area, for example, I could use WTOP on 1500 kHz. If you have a communications receiver, use WWV in Fort Collins, Colorado or WWVH in Hawaii. These stations transmit accurate frequency checks on 5, 10, 15 and 20 MHz. Select the highest of these frequencies that produces a strong local signal. Because the error is multiplied along with harmonics, it is best to use the 15 or 20 MHz frequency, even if you have to time your work to when "skip" is in. Adjust C3 for zero beat with the radio station. If you have an *S-meter* on the receiver, zero beat is indicated when the meter-pointer wobble gets slower and slower, which means the two frequencies are closing on one another.

The classic Miller oscillator used a crystal resonator in the grid of a vacuum tube and a parallel, tuned, LC-tank circuit in the plate. In Fig. 24-28 you see the updated version using a junction field-effect transistor in place of the tube.

The crystal resonator is connected between gate and common in Fig. 24-28 and is in parallel with a 10-Megohm resistor. Bias for the JFET is set by source resistor R2, which is bypassed for RC to make a low impedance to ground for the signal. In most cases, the reactance of the capacitor is set at less than one-tenth of the resistance it bypasses.

The drain terminal of the JFET in this circuit is connected to the resonant tank circuit. The actual resonant frequency of this circuit only approximates the crystal frequency. The oscillation of this circuit takes on different properties depending upon which side of the crystal frequency the tank is set to. In most cases, the tank circuit is tuned to the low side of the resonant point.

If you own older equipment, it is likely that the ferrite core in the tank circuit has changed enough to make the oscillator inoperable. I found this to be the problem with two different Heath DX-60B ham-radio transmitters that were placed in operation many years after previous owners had given them up. The symptom was unreliable keying on CW. Sometimes, when the key was closed the oscillator refused to start; at other times it started fine. The oscillator in the DX-60B is a Miller circuit, and its coil had changed inductance. The solution was to retune the coil, and the oscillator began to start reliably when keyed.

Another example of a simple Colpitts crystal oscillator is shown in Fig. 24-29. This circuit is somewhat different from the other Colpitts configura-

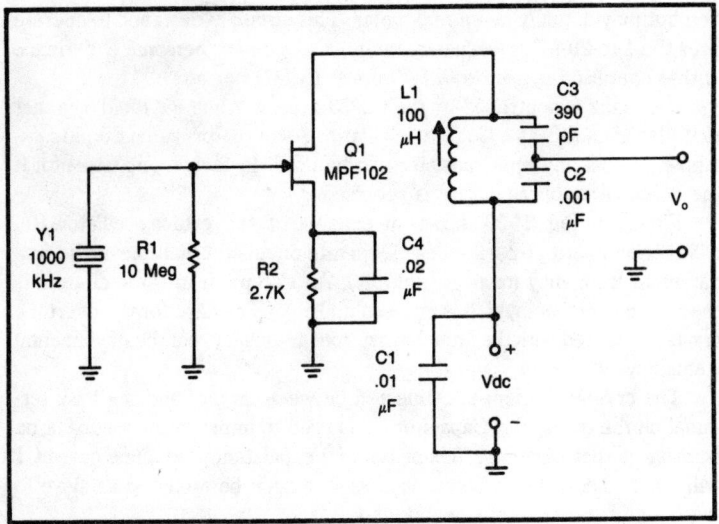

Fig. 24-28. Miller oscillator.

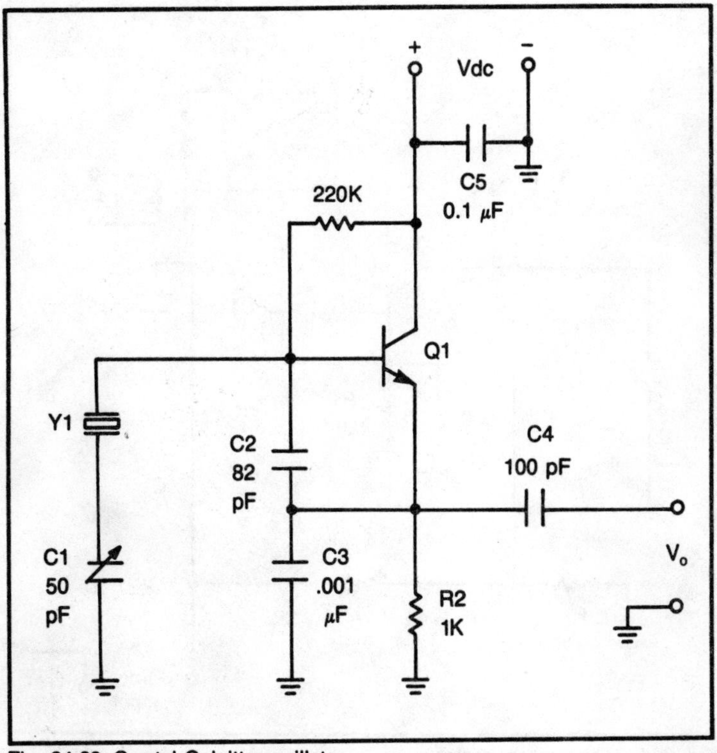

Fig. 24-29. Crystal Colpitts oscillator.

tion but may actually be more popular. This circuit is designed to operate over the 1-to-20-MHz-frequency range, although my personal experience is that operation is a little "iffy" above 15 MHz or so.

Feedback is controlled by the C2/C3 capacitor voltage divider, a fact that identifies this as a Colpitts oscillator. As in the previous Colpitts example, almost any npn transistor can be used. In fact, a pnp is useful if the dc power-supply polarity is reversed.

Finally, in Fig. 24-30 you see an example of an overtone oscillator. Recall that an overtone oscillator will operate on an odd multiple of the fundamental frequency (or near multiple). An exception in some circuits is the fourth overtone, which is an even multiple. The case for the overtone crystal is marked with the intended overtone frequency, not the fundamental frequency.

The crystal element is connected between ground and the base terminal on the transistor. Capacitor C1 is used to improve the feedback on transistors that have insufficient internal capacitance. In some cases C1 will not be needed, but where it is used it must be mounted as close to the body of the transistor as possible.

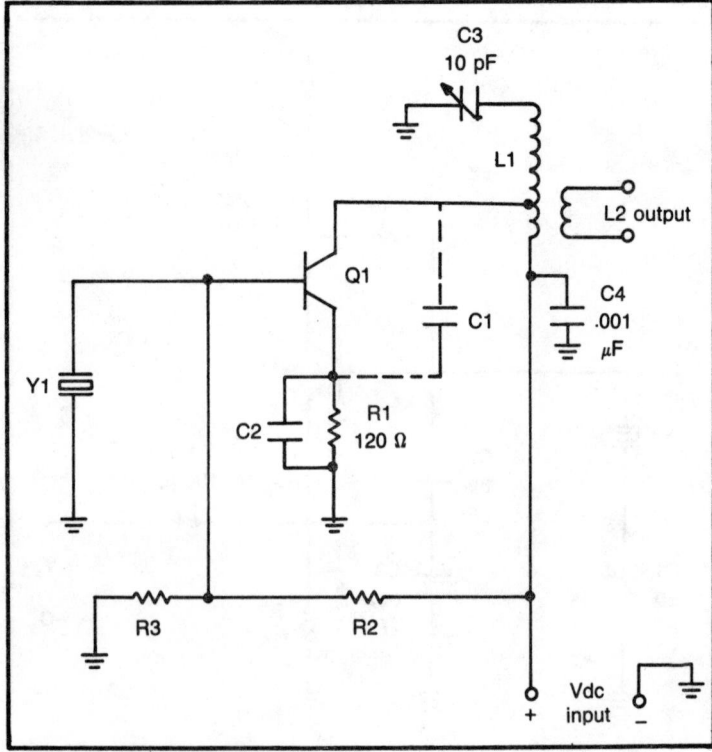

Fig. 24-30. Overtone oscillator.

The tank circuit (L1/C3) in the collector circuit of the transistor is tuned approximately to the desired overtone frequency. The emitter-resistor-bypass capacitor is designed to have a reactance of approximately 90 ohms at the frequency of operation; values for common frequencies are given below.

Frequency (MHz):	50	150	220	450
Capacitance (pF):	36	12	8	3

TTL CLOCKS

In this section I will discuss digital electronic "clocks." These circuits are not time-of-day clocks, but rather are used to create pulse trains that time or synchronize digital electronic circuits. In almost all cases the digital clock will produce either a square wave or trapezoidal wave. In this section I will deal with digital clock circuits based on transistor-transistor-logic (TTL) integrated circuits.

TTL BASICS

The TTL family was probably the first really successful family of integrated-circuit digital devices. Previous families (e.g., RTL and DTL) never really achieved the widespread use of the TTL devices. As in any standard logic family, the TTL device uses standardized, input and output circuits and standardized logic levels.

Digital electronics is binary in nature; that is, it permits only two possible states. These states represent the digits of the binary (base-2) number system, 1 and 0. These two states are sometimes called HIGH and LOW, a convention that I will follow in this article. Figure 24-31 shows these levels for TTL devices. The HIGH level (which represents 1) is attained when the input or output voltage is greater than $+2.4$ volts but less than $+5$ volts. The LOW (or 0) condition is represented by a grounded state, which means a potential between 0 and 0.8 volts.

Some literature will tell you that TTL devices will operate at supply and signal potentials greater than $+5$ volts dc, but I recommend against it. Such potentials are in what I call the "groan zone" and will drastically shorten the life of TTL devices.

Negative potentials are another area of concern because negative potentials will kill the device. Although it is possible for power-supply potentials to be applied backwards, the most likely case for oscillator builders is inappropriately-connected capacitors and other components that store energy. Negative potentials will destroy the TTL device, so must be avoided.

Figure 24-32 shows the standardized, input and output circuits of TTL devices. Any logic family works because inputs and outputs can be interconnected with only a conductor—no impedance-matching or other con-

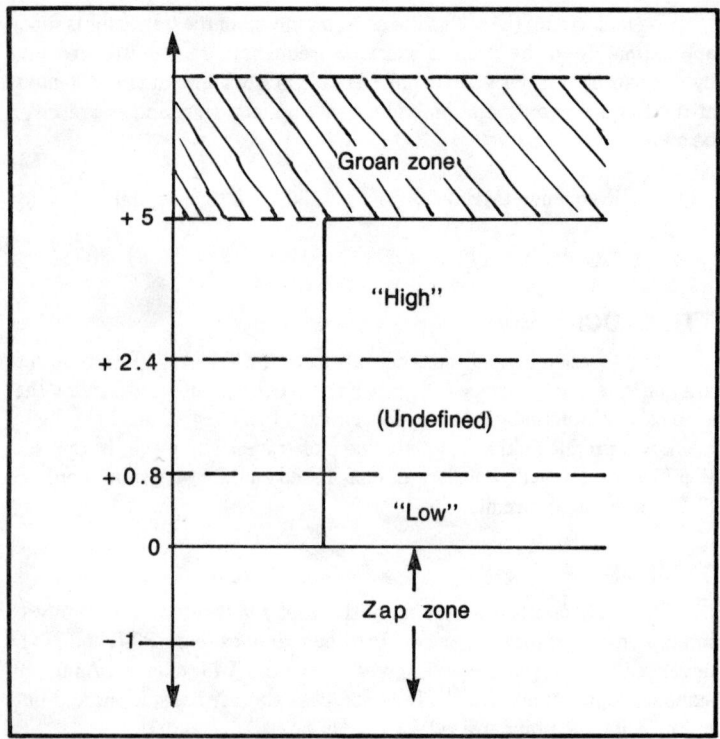

Fig. 24-31. TTL operating limits.

siderations are necessary. The TTL input acts as a 1.6-milliampere current source, while the output operates as a current sink. When you connect a TTL output to a TTL input, therefore, the circuits match perfectly.

Interfacing TTL devices becomes a matter of making sure that drive requirements are met. The requirements of a single standard input (1.6 mA @ +2.4 to +5.0 volts) is the interfacing unit. The input is said to have a "fan-in" of 1, which represents the requirements of a single TTL input. The output will have a fan-out number which defines the number of standard fan-in-1 TTL inputs it can drive. Thus, when we say that the standard TTL output has a fan-out of 10 we mean that it can successfully drive 10 standard TTL inputs.

Figure 24-33 shows a method for converting a standard, transistorized, oscillator circuit to TTL. The oscillator is a standard Colpitts oscillator built around an npn transistor (Q1). The feedback level is set by the usual capacitor voltage divider, C1 and C2. The oscillator frequency is set by Y1, a piezoelectric crystal. Some small degree of control is permitted by varying capacitor C3. Crystals oscillate at a frequency that is determined in part by the capacitance they operate into; so capacitor C3 will "pull" the operating of the crystal.

470

The output stage is an LM-311 comparator. This type of IC is basically a differential amplifier with too much gain. In any differential amplifier, the output voltage is a function of the difference between the two input voltages. Thus, when the voltages on pins 2 and 3 are equal, the difference is zero; so the output voltage also will be zero. But when these two voltages are different by only a few millivolts, the output voltage will be non-zero. Because the gain could be 10,000 to 100,000, the output will saturate any time the differential input voltage is non-zero. In Fig. 24-33 the pin-3 input is grounded so sees a zero potential. Thus, when the signal applied to pin 3 crosses zero or is negative, the output voltage will be zero. But when the input voltage is positive, the LM-311 output voltage is HIGH.

The LM-311 device has what is called an *open-collector* output stage. This means that it requires a pull-up resistor to V+. Although the LM-311 is able to operate with output potentials greater than +5 Vdc, I use +5 Vdc here to make the circuit compatible with TTL circuits. For higher potentials simply scale the resistor upwards in value proportional to the increase in voltage—up to about +15 Vdc.

Another way to do this same trick is to use a TTL chip called a *Schmitt trigger*. The circuit is shown in Fig. 24-34(A), and the operation of the Schmitt trigger is shown in Fig. 24-34(B). The operation of the Schmitt trigger follows this simple rule: the output will snap HIGH when a positive-going input signal crosses a certain threshold and will snap LOW when

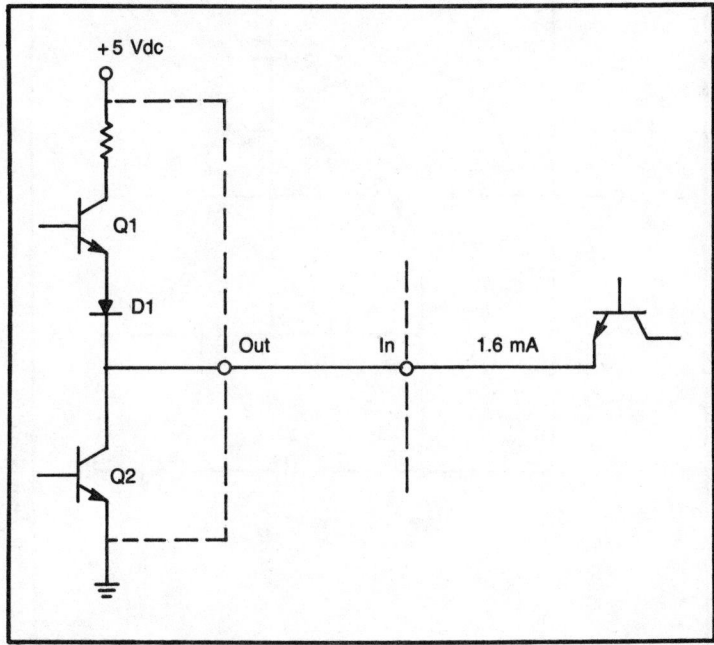

Fig. 24-32. TTL input/output circuitry.

471

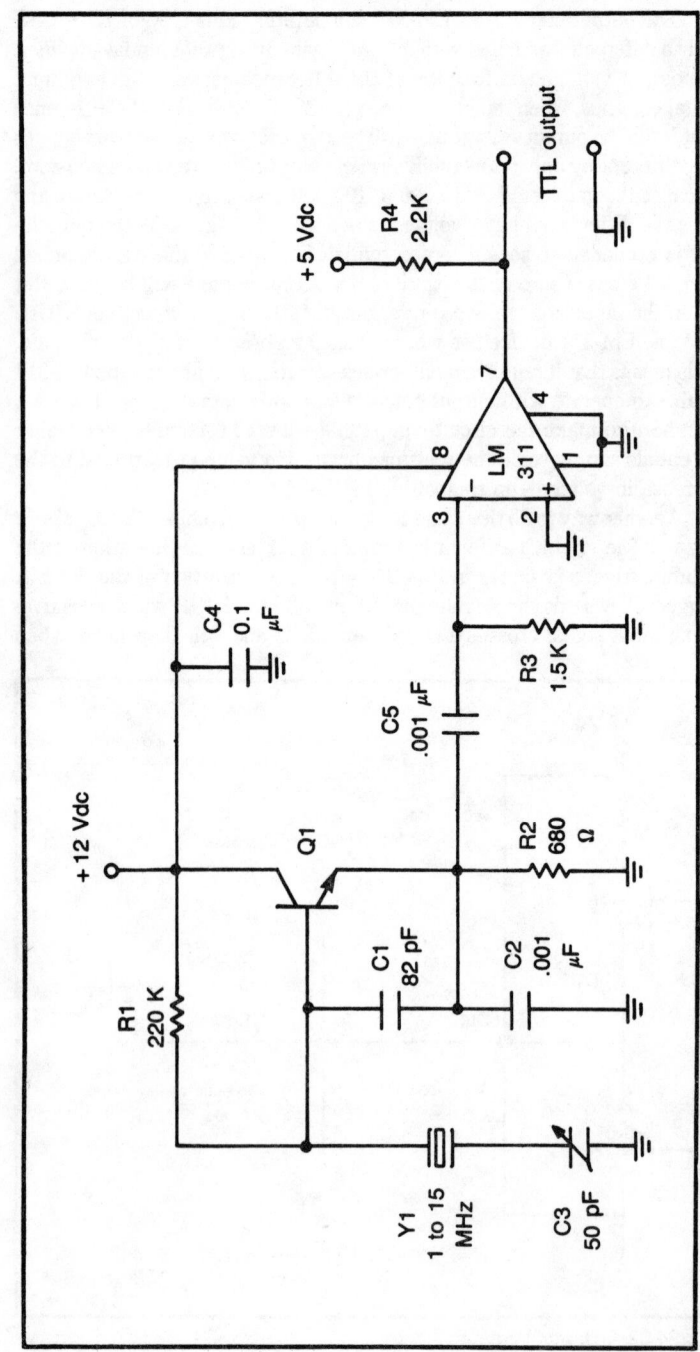

Fig. 24-33. TTL-compatible transistor oscillator.

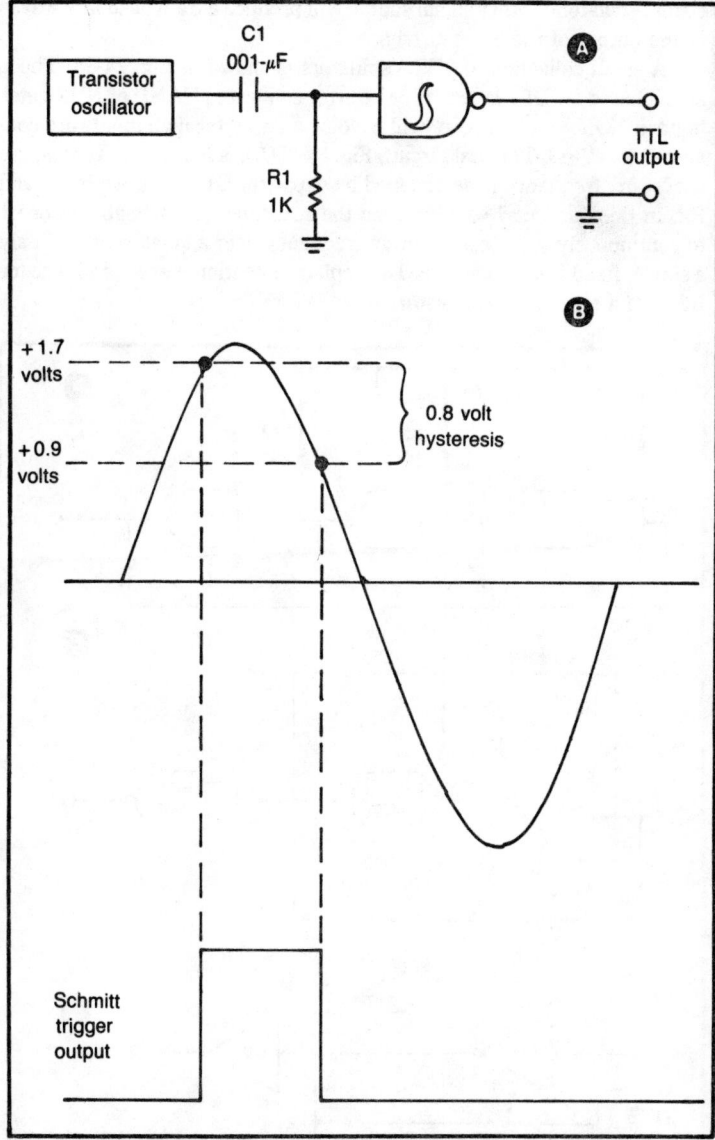

Fig. 24-34. Schmitt trigger TTL converter (A) circuit, (B) waveform.

the input signal crosses a lower threshold in a negative-going direction. Thus, when a signal from a transistor oscillator crosses +1.7 volts in the positive-going direction the Schmitt-trigger output snaps HIGH, and it will drop LOW again when the input drops to +0.9 volts.

In both cases, Figs. 24-33 and 24-34, the ac sine wave from a transistor oscillator or other signal source will produce a train of square waves at the output of the Schmitt trigger.

A small collection of TTL oscillators is shown in Fig. 24-35. These are all based on TTL inverters or inverter-connected NAND or NOR gates (note: a NOR or NAND gate will become an inverter if all inputs are connected together). The first circuit, Fig. 24-35(A), is RC-timed. That is, the oscillating frequency is determined by capacitor C1 and resistors R1 and R2. In this particular configuration the resistance is adjustable in order to continuously vary the operating frequency over a small range. If only a single, fixed frequency is needed, replace potentiometer R1 and resistor R2 with a single, fixed resistor.

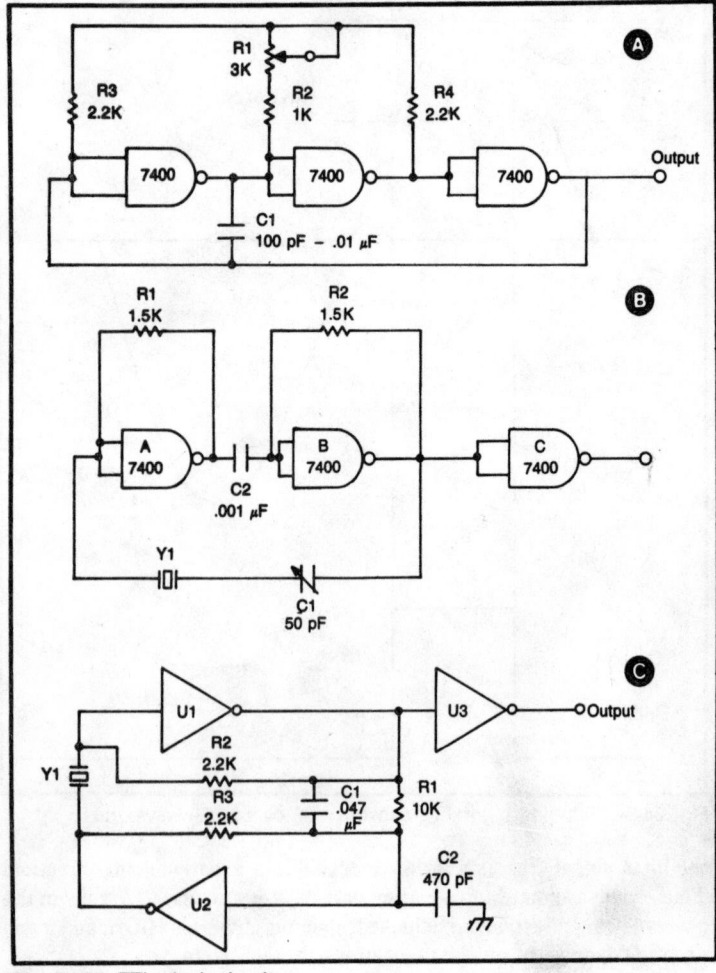

Fig. 24-35. TTL clock circuits.

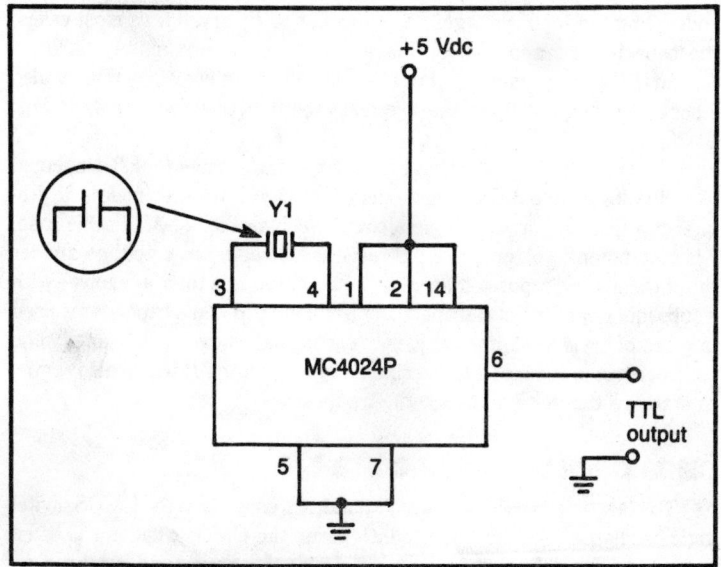

Fig. 24-36. Fixed MC-4024P circuit.

A disadvantage of any RC-timed oscillator is that the operating frequency is neither stable nor accurate. These problems are partially overcome using piezoelectric crystals, as in Figs. 24-35(B) and 24-35(C). The circuit in Fig. 24-35(B) is based on the 7400 TTL NAND gate. The 7400 IC contains four independent NAND gates, of which two are used for the oscillator and a third operates as a buffer stage; all three NAND gates are connected as inverters. The two gates (A & B) used in the oscillator circuit are each biased with a 1.5-kilohm resistor and are isolated from each other by coupling capacitor C2. The operating frequency is set by crystal Y1 and may be varied slightly with capacitor C1.

The version shown in Fig. 24-5(C) is similar and is based on TTL inverters. These stages are from a package of six called a *hex inverter* (e.g., 7404). Again, one stage is used as an output buffer, and the oscillating stages are self-biased.

SPECIAL TTL OSCILLATORS

There are several TTL oscillators available on the market, but some of them are a little hard for hobbyists and amateurs to obtain. In Fig. 24-36 you see a circuit based on the MC-4024P voltage-controlled oscillator. This Motorola circuit is not to be confused with the CMOS device called the 4024, which is a binary counter in another logic family. The MC-4024P device contains two TTL voltage-controlled-oscillator stages. In this circuit I show only one stage being used.

There are two ways to set the overall operating frequency (other than

475

controlling the input voltage!): by capacitor or by crystal. In most cases, the capacitor is used and will have a value of approximately $(300/f_{Hz})$ picofarads. For frequencies above 2000 kilohertz, however, you may also elect to use a crystal to set the frequency (both methods are shown in Fig. 24-36).

The circuit in Fig. 24-36 has the control-voltage input (pin 2) connected to +5 volts for fixed-frequency operation. If you want to make the frequency variable, you would use a circuit such as Fig. 24-37. In this case, I have a potentiometer and fixed resistor connected in a voltage divider to set the control input voltage at some potential less than +5 volts—with a subsequent pulling of the operating frequency. When a capacitor is used in place of crystal Y1, the frequency can be pulled over a 3:1 range. The control input is grounded for ac signals by capacitor C1, which also serves to stabilize the voltage (hence the frequency).

CMOS OSCILLATOR CIRCUITS

The last type of oscillator circuit that I will consider is the CMOS digital logic oscillator. Like its TTL counterparts, the CMOs oscillator is often used as a "clock" in digital electronic circuits for timing and synchronization. There are some significant differences between TTL and CMOS, however.

Perhaps the biggest difference, and one that drives the other differences, is that CMOS devices are made from metal-oxide field-effect transistors (MOSFETs) instead of npn/pnp bipolar transistors used in TTL devices. An implication of this fact is that CMOS devices draw considerably less current than TTL devices (microamperes instead of milliamperes), although generally at the cost of operating speed.

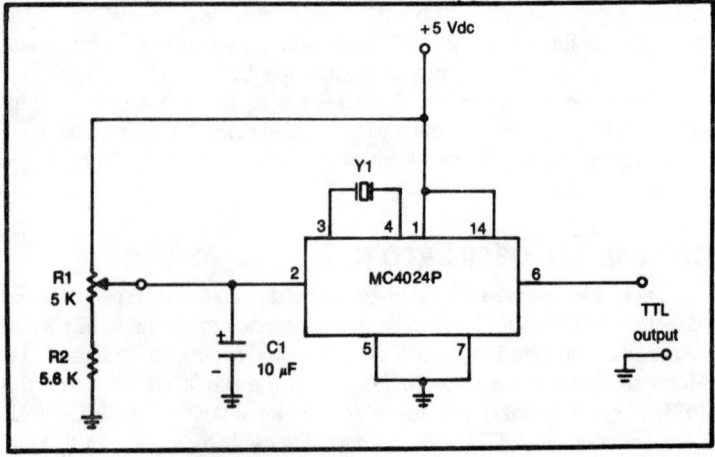

Fig. 24-37. Variable MC-4024P circuit.

Figure 24-38 shows the basic CMOS inverter circuit used to make most of our oscillator circuits. The basic inverter of Fig. 24-38(A) is representative of the older, A-series CMOS devices and consists of a pair of MOSFET transistors. Device Q1 is an N-channel MOSFET, and Q2 is a P-channel MOSFET. These devices operate in opposite "polarity" from one another: a HIGH applied to the common (parallel-connected) gate input will cause one of these transistors to saturate while the other turns off.

The newer, B-series CMOS device is shown in Fig. 24-38(B). There is a resistor in series with a gate and at least one diode shunting the gate. The reason for these components is electrostatic-discharge protection. Normal A-series devices have a gate-breakdown voltage of less than 100 volts. Because static built up on clothing and tools can reach several kilovolts (easily!), you can damage A-series CMOS devices by touching them. The B-series provides some protection in this matter.

Most CMOS devices have part numbers in the 4000-series. RCA devices are denoted as CD-4xxx, and Motorola uses MC-144xxx; other manufacturers use the 4xxx designations. The A-series devices will have either an "A" suffix on the part number or no suffix at all. For example, 4001 and 4001A both denote an A-series device. On the other hand, the 4001B is the same CMOS device in B-series and is preferred for most applications.

There are other differences between A-series and B-series CMOS devices. The B-series offers faster rise and fall times for square-wave pulses and will drive larger loads than most A-series counterparts. Although there may be some applications where the characteristics of the A-series provide some advantage, it is generally true that the B-series device is preferred.

The operation of the CMOS inverter can be seen in Fig. 24-38(C). The transistors in the CMOS device can be modeled as gated resistors. In Fig. 24-38(C) the channel resistance of Q1 is represented by R1, and the channel resistance of Q2 is represented by R2. The output is at the junction of the two resistances. CMOS devices operate from two power supplies: $V+$ and $V-$, both of which can be from roughly 4 to 18 volts. It is common to find $V+$ set to $+5$ volts and $V-$ set to 0 volts (i.e., ground); this method makes the device useable in circuits where TTL power supplies are already available. You could also select ± 5 volts, ± 12 volts, or some non-symmetrical potentials (e.g., -5 volts and $+12$ volts). Note that the operating parameters of most CMOS devices are measured at ± 10 volts and operation away from these potentials is somewhat different.

Let's take a look at what happens when a signal is applied to the input in Fig. 24-38(C). Assume that $V+$ is 5 volts (which also represents HIGH) and $V-$ is grounded (which represents LOW). When the input is LOW, the output is HIGH. This means that resistor R1 has a low value (less than 2 kilohms), and R2 is very high (1 megohm). In this case, the output terminal is effectively connected to $V+$; so the output is HIGH. Alternatively, when the input is HIGH, the resistances reverse. R1 becomes very high,

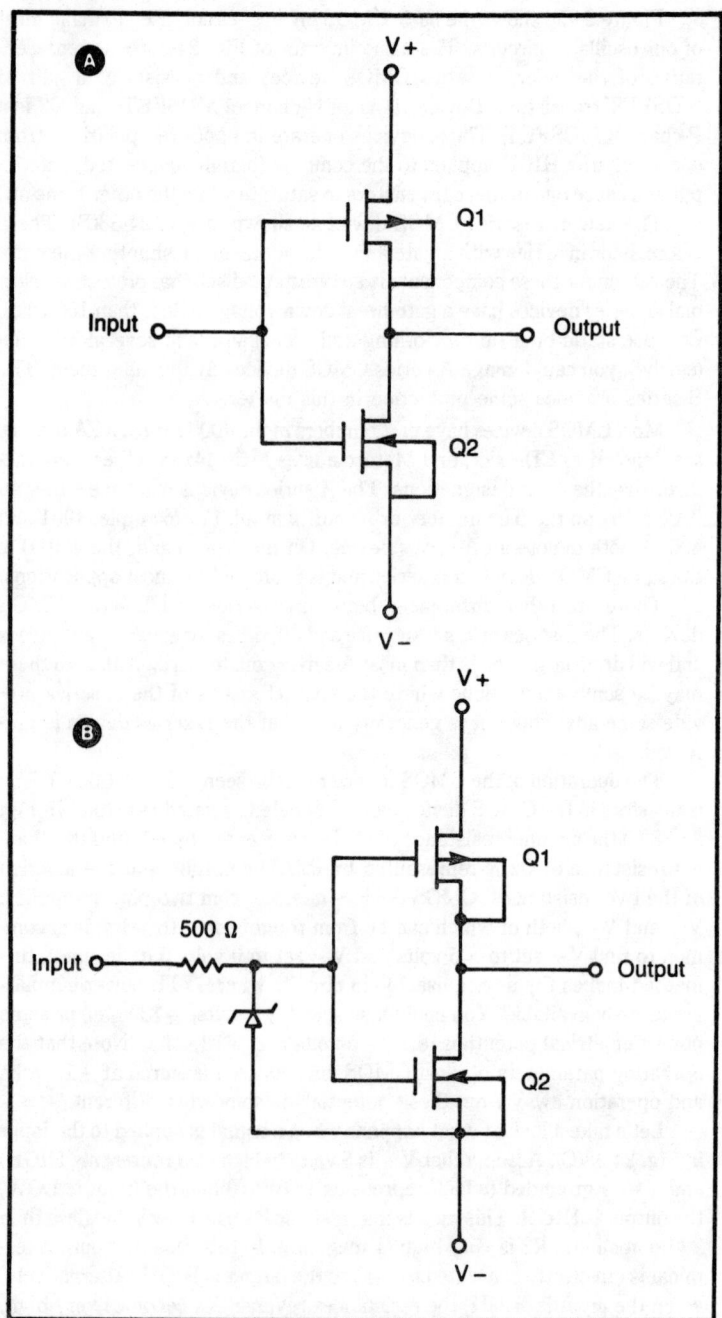

Fig. 24-38. CMOS inverters (A) A-series, (B) B-series.

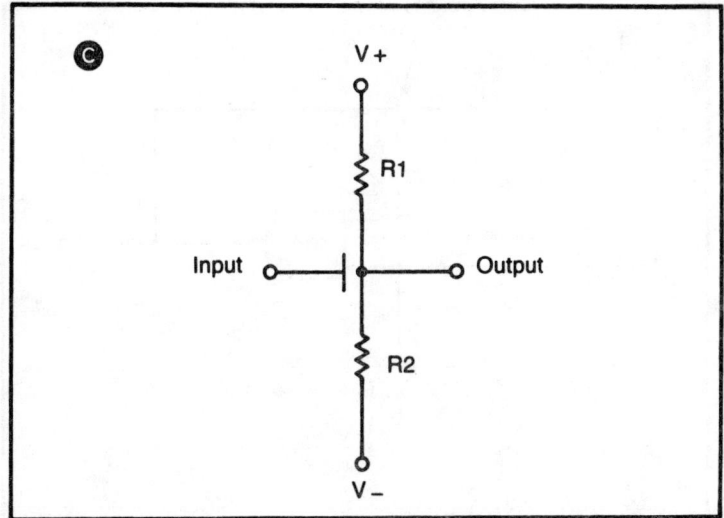

Fig. 24-38. (C) model.

and R2 becomes very low; so the output terminal is effectively connected to ground.

The reason for the very-low current requirement is that the path between V − and V + is always a very high resistance in series with a very low resistance. The only time that both resistances are moderately low is when the output is in transition between LOW and HIGH (or HIGH and LOW).

CMOS CLOCK CIRCUITS

Our first CMOS clock circuit is shown in Fig. 24-39; the circuit is in Fig. 24-39(A), and the timing diagram is shown in Fig. 24-39(B). The 4093 device is a Schmitt trigger, and four such triggers come in the 4093 package. A Schmitt trigger is a special device that operates only on specific input voltages. Assuming that the power supply is + 5 volts and ground, the inputs operate on positive-going signals that go above + 2.9 volts and on negative-going signals that drop below + 2.3 volts. The device must see a positive-going input prior to the negative-going-threshold crossing, or it won't work.

The device shown in Fig. 24-39(A) is a two-input NAND gate; so you can either tie the two inputs together or use one of the inputs as a control element. When the switch is open, the input is at V + so that the device will operate normally; when the switch is closed the input is grounded so that the device will sit there dormant with the output at HIGH. For my description of operation assume that the switch is open.

At turn-on, capacitor C1 is discharged (0 volts); so the input (point A) is at ground potential, or LOW. The output is HIGH at this instant (see

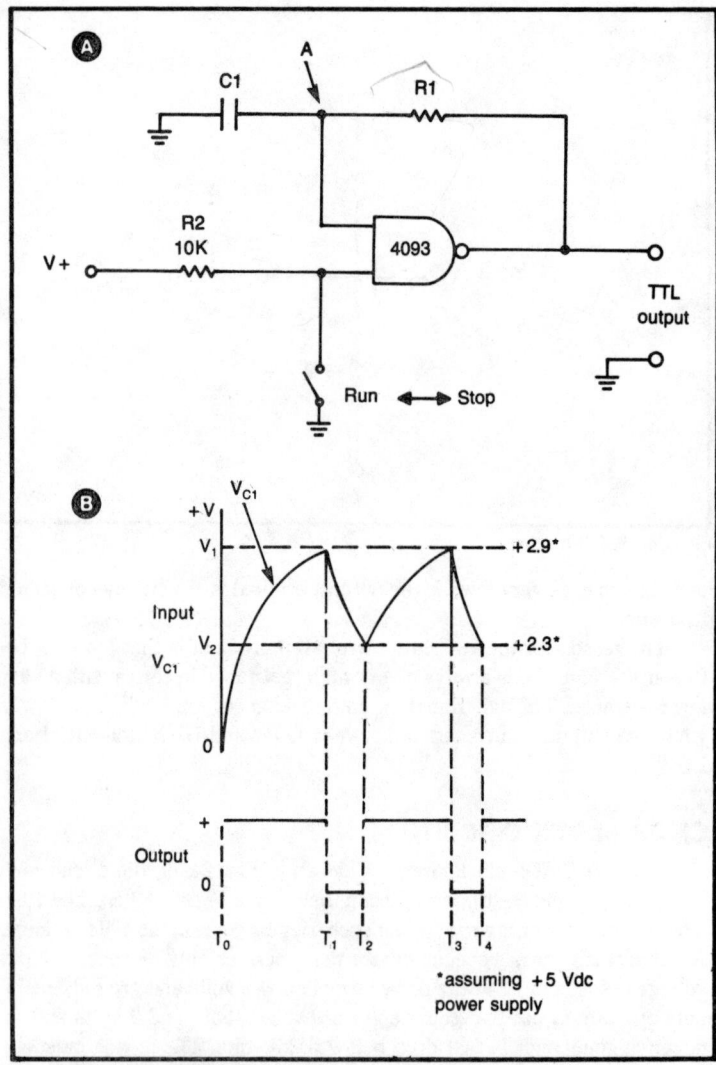

Fig. 24-39. Schmitt trigger astable (A) circuit, (B) waveform.

time T0 in Fig. 24-39(B)). The voltage across the capacitor begins rising as C1 charges from the output potential (through R1). Its voltage (V_{C1} in Fig. 24-39(B)) will eventually reach the positive-going threshold, V_1. At this time, the input sees a HIGH, so will cause the output to drop LOW. This change of condition removes the charge source from R1C1; so C1 will begin to discharge through R1 and the output terminal of the 4093. When the voltage decreases to V_2, the output again snaps HIGH, and the cycle repeats itself.

480

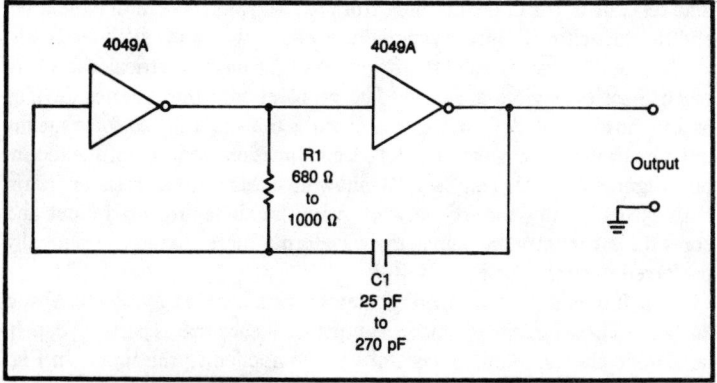

Fig. 24-40. Basic CMOS clock.

A second CMOS oscillator, this one a little more common, is shown in Fig. 24-40. This circuit can be made using a CMOS inverter or NAND and NOR gates connected as inverters (i.e., with both inputs tied together). The operating frequency of this circuit is set by the resistor-capacitor combination according to the approximate formula: $f = 1/(2.2 \times R1 \times C1)$.

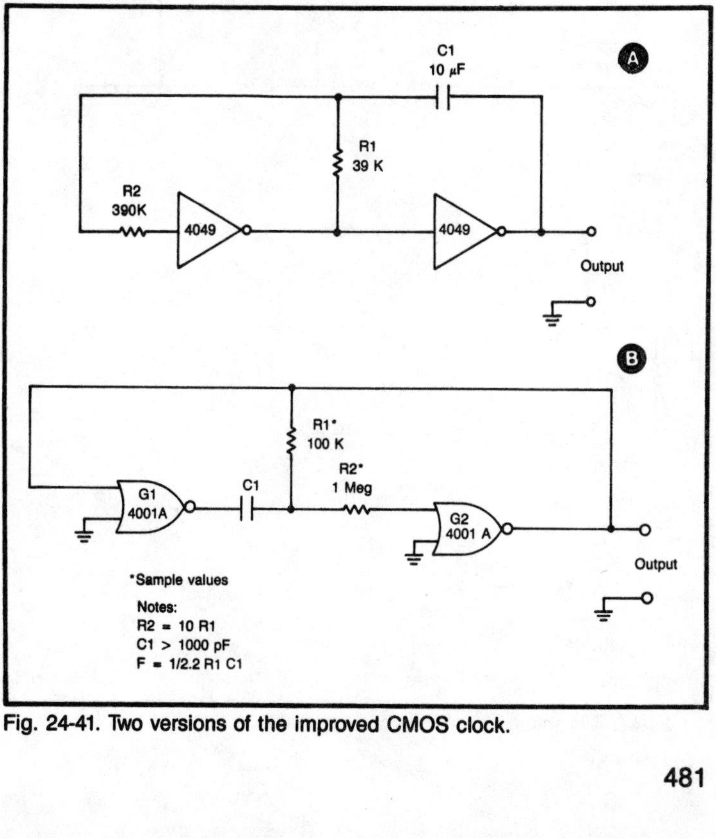

Fig. 24-41. Two versions of the improved CMOS clock.

481

The resistor (R1) can be anything from 3.3 kilohms to several megohms, and the capacitor (C1) from roughly the values shown up to about 10 μF.

The output waveform of this device is a little unsymmetrical, especially when B-series devices are used. The problem with the B-series devices is that the internal diode protection disturbs the situation; so some means must be provided to isolate the R1C1 combination from the protected input. Figures 24-41(A) and 24-41(B) show how this is done: resistor R2 (in both cases) isolates the RC network from the diode-protected input and cures the asymmetry problem in the waveform. These circuits are generally preferred over that of Fig. 24-40.

A pair of voltage-controlled oscillators are shown in Figs. 24-42(A) and 24-42(B). These circuits produce an output frequency that is partially dependent upon the value of the control voltage applied to the input. In Fig. 24-42(A) this voltage is applied in a manner to bias the input of the inverter (and the junction of the RC network) to some static value. The

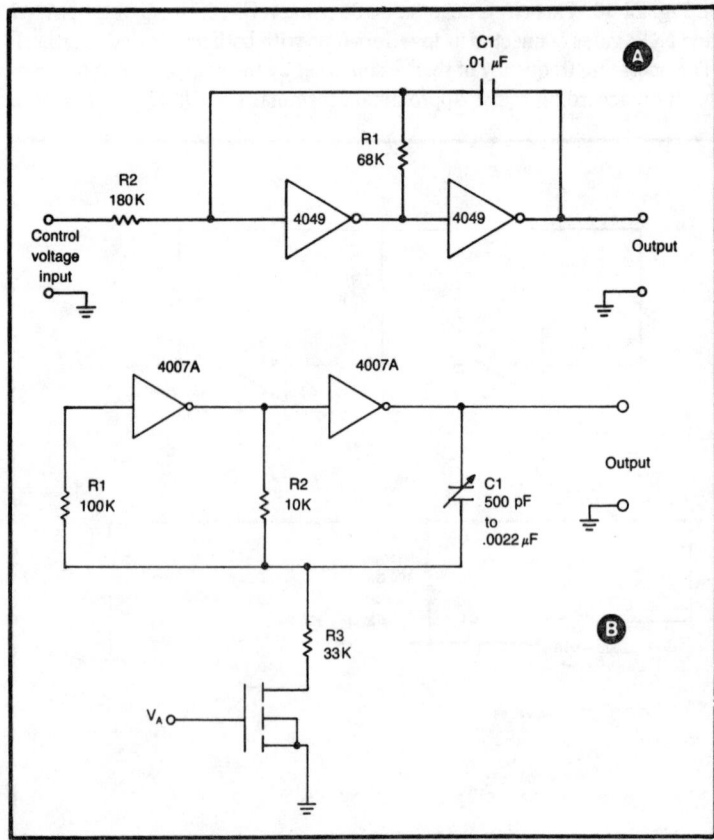

Fig. 24-42. Two voltage controlled oscillators (VCO).

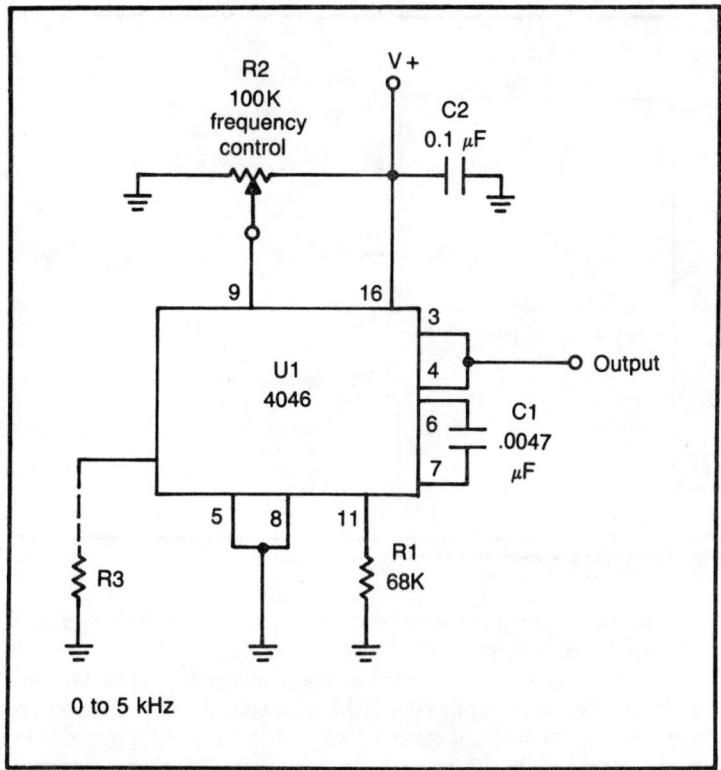

Fig. 24-43. 4046 used as VCO.

charge/discharge cycle that sets the operating frequency takes a longer or shorter duration depending upon the applied voltage. The V_{in}-equals-zero-set frequency is found according to the formula above.

The circuit in Fig. 24-42(B) works a little differently. In this case, a MOSFET transistor shunts the RC junction to ground, thereby affecting the charge/discharge timing. The control voltage is applied to the gate input of the MOSFET transistor.

The 4046 device shown in Fig. 24-43 is an IC, CMOS, voltage-controlled oscillator (VCO). The chip is listed in the data sheet as a phase-locked loop, but it can be operated as a VCO. The control-voltage input to pin 9 is provided by a potentiometer or other source. The operating frequency is determined by C1 and R1 and to some extent by the optional resistor R3. The value of R1 can be from 3 to 1,000 kilohms, and the capacitor may be from 100 pF to 10 μF. It is possible to make an oscillator that operates over a range of 1000:1 with the 4046 device.

The use of R3 varies the minimum and maximum operating frequencies. Ordinarily, with R3 open-circuited, the operating frequency will vary from dc to the maximum set by R1/C1 (to 1 MHz, or so). If R3 is used,

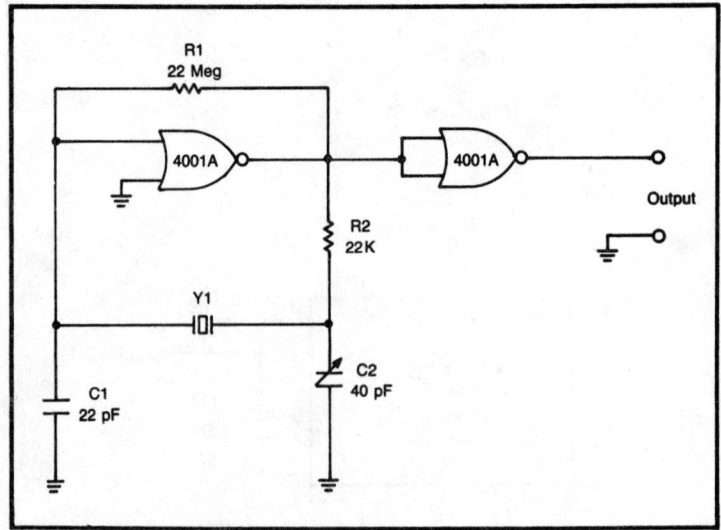

Fig. 24-44. Crystal oscillator.

however, the operating-frequency range changes; there will be a minimum and a maximum different from the above rule.

My final example is the crystal oscillator shown in Fig. 24-44. This circuit uses two sections of a 4001 NOR gate, both of which are inverter-connected (but by different means). One section of the 4001 operates as the oscillator, and the other as a buffer amplifier. The crystal oscillator is an exception to the digital types of circuits shown elsewhere in this section. The reason is that the 4001 is biased into the linear region by R1, so acts as an amplifier (gains of 10,000 are possible with buffered, B-series devices). The operating frequency is set by the crystal (Y1) operated in its parallel-resonant mode. The RC components are selected to provide a 180-degree phase shift at the crystal's parallel-resonant frequency. This circuit will provide good stability and accuracy of output frequency.

Appendix A

"Brute Force" Dc Power Supply Filter

T HE "BRUTE FORCE" FILTER CIRCUIT SHOWN IN FIG. A1 CONSISTS OF a single, large-value electrolytic capacitor in parallel with the load resistor (R_L) and the output of the rectifier. This program calculates the capacitance needed to achieve a specified ripple factor in the dc output, the ripple factor to expect from a given filter capacitor, or the input voltage required of a specified filter to achieve a given output voltage. Both full-wave and half-wave rectifier cases are considered.

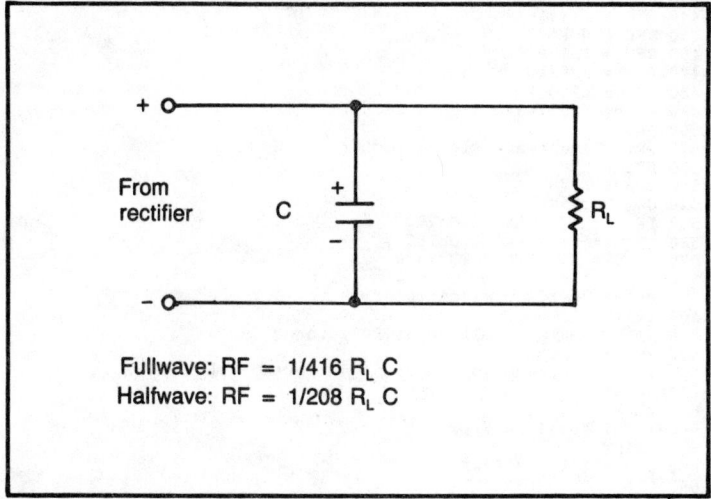

Fullwave: RF = 1/416 R_L C
Halfwave: RF = 1/208 R_L C

Fig. A-1. Brute force filter.

485

```
10   REM   The name of this program is BRUTEFIL.
20   REM   This program will calculate either the ripple factor
30   REM   of a known power supply filter, or, the capacitance
40   REM   required to achieve a specified ripple factor.
50   GOSUB 1240
60   PRINT "This program will compute either the capacitance"
70   PRINT "needed to achieve a given power supply ripple factor,"
80   PRINT "or, the ripple factor of an existing power supply."
90   PRINT "The type of power supply for which this program is"
100  PRINT "designed is the BRUTE FORCE type in which a single"
110  PRINT "large value filter capacitor is connected in parallel"
120  PRINT "with the load."
130  PRINT
140  PRINT
150  GOSUB 1280
160  GOSUB 1240
170  PRINT "Select type of calculation to be performed:"
180  PRINT
190  PRINT "1.  Ripple factor of a given power supply"
200  PRINT
210  PRINT "2.  Capacitance needed to achieve a specified"
220  PRINT "    ripple factor (r)"
230  PRINT
240  PRINT "3.  Input voltage to produce required output voltage"
250  PRINT
260  INPUT "Choice Please:",A
270  IF A > 3, THEN GOTO 170
280  ON A GOTO 290,580,860
290  GOSUB 1240
300  PRINT "Now, let's collect some information -- OK?"
310  PRINT
320  INPUT "Output Voltage at Full Load?",VO
330  PRINT
340  INPUT "Maximum load current (Amperes)?",I
350  PRINT
360  INPUT "Value of Filter Capacitor C1 in uF?",C1
370  PRINT
380  C = C1/(10^6)
390  RL = VO/I
400  RFH = 1/(208*RL*C)
410  RFH = RFH*100
420  RFH = INT(RFH)
430  RFH = RFH/100
440  RFF = 1/(416*RL*C)
450  RFF = RFF*100
460  RFF = INT(RFF)
470  RFF = RFF/100
480  GOSUB 1200
490  PRINT "Fullwave Ripple Factor:";RFF
500  PRINT
510  PRINT "Halfwave Ripple Factor:";RFH
520  PRINT
530  GOSUB 1280
540  GOSUB 1310
550  IF S = 1, THEN GOTO 290
560  IF S = 2, THEN GOTO 160
570  IF S = 3, THEN GOTO 1410
580  GOSUB 1240
590  PRINT "Let's collect information, OK?"
600  PRINT
610  INPUT "Output Voltage at Full Load?",VO
620  PRINT
630  INPUT "Maximum load current (Amperes)?",I
640  PRINT
650  RL = VO/I
660  INPUT "Desired Ripple Factor?",RF
670  PRINT
680  C1H = 1/(208*RL*RF)
690  C1F = 1/(416*RL*RF)
```

486

```
700 GOSUB 1200
710 C1H = C1H*10^6
720 C1H = INT(C1H)
730 C1F = C1F*10^6
740 C1F = INT(C1F)
750 PRINT "To achieve a ripple factor of ";RF
760 PRINT "use a capacitor as follows:"
770 PRINT
780 PRINT "Fullwave circuit:";C1F;" uF"
790 PRINT "Halfwave circuit:";C1H;" uF"
800 PRINT
810 GOSUB 1280
820 GOSUB 1310
830 IF S = 1, THEN GOTO 580
840 IF S = 2, THEN GOTO 160
850 IF S = 3, THEN GOTO 1410
860 GOSUB 1240
870 PRINT "Now let's collect some information"
880 PRINT
890 INPUT "Required Output Voltage Under Load? ",VO
900 PRINT
910 INPUT "Maximum load current (Amperes)? ",I
920 PRINT
930 INPUT "Filter capacitance being used (uF)? ",C
940 PRINT
950 C1 = C/10^6
960 VPH = VO + (I/(240*C1))
970 VPH = INT(VPH)
980 VPF = VO + (1/(120*C1))
990 VPF = INT(VPF)
1000 PRF = ((VPF-VO)*100)/VPF
1010 PRF = INT(PRF)
1020 PRH = ((VPH-VO)*100)/VPH
1030 PRH = INT(PRH)
1040 GOSUB 1200
1050 PRINT "Required Peak Pulsating DC Voltage:"
1060 PRINT
1070 PRINT "Halfwave case:";VPH
1080 PRINT "Fullwave case:";VPF
1090 PRINT
1100 PRINT "Voltage Regulation:"
1110 PRINT
1120 PRINT "Halfwave:";PRH;" %"
1130 PRINT "Fullwave:";PRF;" %"
1140 PRINT
1150 GOSUB 1280
1160 GOSUB 1310
1170 IF S = 1, THEN GOTO 860
1180 IF S = 2, THEN GOTO 160
1190 IF S = 3, THEN GOTO 1410
1200 FOR I = 1 TO 5
1210 PRINT
1220 NEXT I
1230 RETURN
1240 FOR I = 1 TO 20
1250 PRINT
1260 NEXT I
1270 RETURN
1280 PRINT "PRESS ANY KEY TO CONTINUE:"
1290 A$=INKEY$: IF A$="" THEN 1290
1300 RETURN
1310 GOSUB 1200
1320 PRINT "What's Your Pleasure?"
1330 PRINT
1340 PRINT "1.  Do Another of the same sort"
1350 PRINT "2.  Return to main menu to make another selection"
1360 PRINT "3.  Finished"
1370 PRINT
1380 INPUT "SELECTION?",S
```

```
1390 IF S > 3, THEN GOTO 1310
1400 RETURN
1410 GOSUB 1200
1420 PRINT "PROGRAM ENDED"
1430 END
```

Appendix B

RC "Pi-Network" Dc Power Supply Filter

F IGURE B-1 SHOWS A DUAL POWER-SUPPLY FILTER SECTION. OUTPUT voltage V_1 is higher than V_2 and is derived directly from the rectifier output as filtered by capacitor C1. This output uses the same sort of "brute force" filter as in the previous circuit. Output V_2 produces a lower voltage but with a substantially better ripple factor. This program calculates the capacitances and the value of R1 required to produce the voltages and ripple factors that you specify.

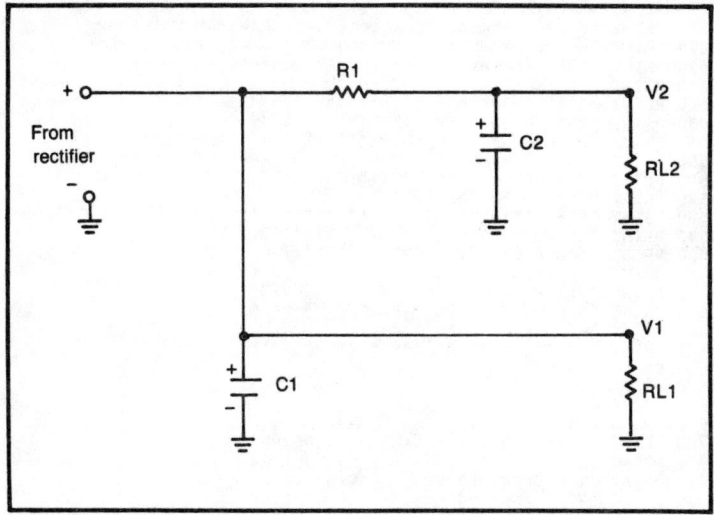

Fig. B-1. RC filter.

```
100  REM   The name of this program is RCFILTER.PS
110  REM   This program computes the values required for resistor
120  REM   and capacitor elements in an RC filter network in a DC
130  REM   power supply circuit.
140  GOSUB 830
150  PRINT "This program is used to select values for an"
160  PRINT "RC power supply filter circuit. YOU WILL NEED"
170  PRINT "to select the output voltages (V1 and V2) and
     currents"
180  PRINT "(I1 and I2), in addition to the desired ripple
     factors"
190  PRINT "for the two voltage outputs."
200  PRINT "In General, the ripple for the lower voltage output
     (V2)"
210  PRINT "is considerably lower than for the higher voltage
     output."
220  GOSUB 790
230  GOSUB 870
240  GOSUB 830
250  INPUT "Higher voltage output (V1)?",V1
260  PRINT
270  INPUT "Output current (in Amperes) for V1?",I1
280  PRINT
290  INPUT "Ripple factor required of V1?",RF1
300  PRINT
310  INPUT "Lower voltage output (V2)?",V2
320  PRINT
330  INPUT "Output current (in Amperes) for V2?",I2
340  PRINT
350  INPUT "Ripple factor required for V1?",RF2
360  GOSUB 830
370  RL1 = V1/I1
380  RL2 = V2/I2
390  C1 = 1/(416*RL1*RF1)
400  R1 = ((V2-V1)/I2) + (1/(120*C1))
410  R1 = -R1
420  C2 = (2*10^-6)/(C1*R1*RL2*RF2)
430  C1 = C1*10^6
440  C2 = C2*10^6
450  C1 = INT(C1)
460  C2 = INT(C2)
470  R1 = INT(R1)
480  PRINT "Capacitances given below are MINIMUM values"
490  PRINT "Select a Working Voltage DC (WVDC) rating that is"
500  PRINT "150-percent of the output voltage, or MORE"
510  PRINT
520  PRINT "*******************************"
530  PRINT "MAIN OUTPUT (V1):";V1;" Volts"
540  PRINT "MAIN OUTPUT CURRENT (I1):";I1;" Amperes"
550  PRINT
560  PRINT "Filter capacitor C1:";C1;" uF"
570  PRINT "Ripple Factor:";RF1
580  PRINT "*******************************"
590  PRINT "LOWER VOLTAGE OUTPUT (V2):";V2;" Volts"
600  PRINT "LOWER OUTPUT CURRENT (I2)";I2;" Amperes"
610  PRINT
620  PRINT "Filter Capacitor C2:";C2;" uF"
630  PRINT "Series Resistor (R1)";R1;" Ohms"
640  PRINT "Ripple Factor:";RF2
650  PRINT "*******************************
660  PRINT
670  PRINT
680  GOSUB 870
690  GOSUB 830
700  PRINT "Select one (1) from menu below:"
710  PRINT
720  PRINT "1.  Do another"
730  PRINT "2.  Finished"
740  PRINT
750  INPUT "SELECTION?",K
```

490

```
760 IF K > 2, THEN GOTO 710
770 ON K GOTO 100,900
780 END
790 FOR I = 1 TO 5
800 PRINT
810 NEXT I
820 RETURN
830 FOR I = 1 TO 30
840 PRINT
850 NEXT I
860 RETURN
870 PRINT "PRESS ANY KEY TO CONTINUE"
880 A$=INKEY$: IF A$="" THEN 880
890 RETURN
900 GOSUB 830
910 PRINT "PROGRAMED ENDED"
920 END
```

Appendix C

Design Of Simple
Regulated Dc Power Supplies

TWO SIMPLE VOLTAGE-REGULATOR CIRCUITS ARE SHOWN IN FIG.C-1; both are three-terminal, integrated-circuit regulators. The fixed-voltage version is shown in Fig. A3(A). In this circuit, the output voltage is fixed by the type of regulator inserted into the "REGULATOR" slot. There are several families of devices, and which one is selected by the program depends upon the maximum-output-current requirements. The adjustable-voltage version is shown in Fig. A3(B). This circuit is based on the LM-317 device. Potentiometer R1 determines the output voltage according to a formula.

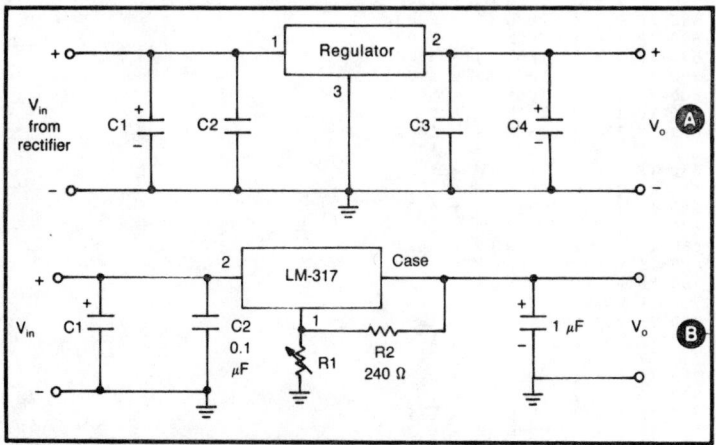

Fig. A-1. Voltage regulators (A) fixed, (B) variable.

```
100 REM    The name of this program is PSDESIGN
110 REM
120 REM    This program will select components for a standard power
130 REM    supply given your requirements.
140 DIM B$(15)
150 DIM C$(15)
160 DIM D$(15)
170 C$ = "K-"
180 B$ = "LM-340"
190 GOSUB 1480
200 PRINT "Select Value of REGULATED DC output voltage:"
210 PRINT
220 PRINT "1.    5-volts"
230 PRINT "2.    6-volts"
240 PRINT "3.    12-volts"
250 PRINT "4.    15-volts"
260 PRINT "5.    18-volts"
270 PRINT "6.    24-volts"
280 PRINT "7.    Adjustable (1.2 volts MIN, 35 volts MAX)"
290 PRINT
300 INPUT "SELECTION?",VO
310 IF VO > 7, THEN GOTO 200
320 GOSUB 1480
330 PRINT "Select MAXIMUM output current level:"
340 PRINT
350 PRINT "1.    100 mA"
360 PRINT "2.    500 mA"
370 PRINT "3.    750 mA"
380 PRINT "4.    1-Ampere"
390 PRINT "5.    1.5-Ampere"
400 PRINT "6.    3-Amperes"
410 PRINT "7.    5-Amperes"
420 PRINT
430 INPUT "SELECTION?",I
440 IF I > 7, THEN GOTO 330
450 GOSUB 1480
460 PRINT "PERMISSABLE REGULATOR TYPES:"
470 IF VO = 1, THEN GOSUB 980
480 IF VO = 2, THEN GOSUB 1060
490 IF VO = 3, THEN GOSUB 1160
500 IF VO = 4, THEN GOSUB 1230
510 IF VO = 5, THEN GOSUB 1300
520 IF VO = 6, THEN GOSUB 1350
530 IF I = 1, THEN C1 = 500
540 IF I = 2, THEN C1 = 1000
550 IF I = 3, THEN C1 = 1000
560 IF I = 4, THEN C1 = 2000
570 IF I = 5, THEN C1 = 3000
580 IF I = 6, THEN C1 = 5000
590 IF I = 7, THEN C1 = 10000
600 IF I = 1, THEN C4 = 10
610 IF I = 2, THEN C4 = 50
620 IF I = 3, THEN C4 = 100
630 IF I = 4, THEN C4 = 100
640 IF I = 5, THEN C4 = 150
650 IF I = 6, THEN C4 = 300
660 IF I = 7, THEN C4 = 500
670 IF I < 5, THEN C2 = .1
680 IF I > 4, THEN C2 = .47
690 IF VO = 1, THEN V = 5
700 IF VO = 2, THEN V = 6
710 IF VO = 3, THEN V = 12
720 IF VO = 4, THEN V = 15
730 IF VO = 5, THEN V = 18
740 IF VO = 6, THEN V = 24
750 IF I = 1, THEN IO = .1
760 IF I = 2, THEN IO = .5
770 IF I = 3, THEN IO = .75
780 IF I = 4, THEN IO = 1
790 IF I = 5, THEN IO = 1.5
```

```
800 IF I = 6, THEN IO = 3
810 IF I = 7, THEN IO = 5
820 IF VO = 7, THEN GOTO 1550
830 PRINT "C1:";C1;" uF"
840 PRINT "C2,C3";C2;" uF"
850 PRINT "C4:";C4;" uF"
860 PRINT
870 PRINT "Minimum input voltage to regulator:";V+2.5;" volts"
880 PRINT
890 PRINT "MINIMUM WVDC rating of C1:";(V+2.5)*1.5;" volts"
900 PRINT "MINIMUM WVDC rating of C4:";(V*1.5);" volts"
910 PRINT "C2 WVDC rating same as C1, C3 same as C4"
920 PRINT
930 PRINT "Power Supply Rating:";V;" Volts, @";IO;" Amperes"
940 PRINT
950 PRINT
960 GOSUB 1520
970 GOTO 1790
980 IF I = 1, THEN PRINT "LM-309H, LM-340LAH-05, LM-340T-05, or
    7805"
990 IF I = 2, THEN PRINT "LM-340T-05 or 7805"
1000 IF I = 3, THEN PRINT "LM-340T-05, LM-340K-05 or 7805"
1010 IF I = 4, THEN PRINT "LM-340K-05 or 7805 (K-package only)"
1020 IF I = 5, THEN PRINT "LM-340K-05, LAS-1505 or 7805
     (K-package only)"
1030 IF I = 6, THEN PRINT "LM-323K"
1040 IF I = 7, THEN PRINT "LAS-1905"
1050 RETURN
1060 PRINT
1070 IF I = 1, THEN PRINT "LM-340H-06 or LM-340T-06"
1080 IF I = 2, THEN PRINT "LM-340T-06 or LM-340K-06"
1090 PRINT
1100 IF I = 3, THEN PRINT "LM-340T-06 or LM-340K-06"
1110 IF I = 4, THEN PRINT "LM-340K-06"
1120 IF I > 4, THEN PRINT "CURRENT REQUIREMENT TOO HIGH FOR THIS
     SERIES"
1130 IF I > 4, THEN PRINT "OF VOLTAGE REGULATOR (select a lower
     current"
1140 IF I > 4, THEN PRINT "or use an adjustable regulator)"
1150 RETURN
1160 PRINT
1170 IF I = 1, THEN PRINT "LM-340H-12 or 7812"
1180 IF I = 2, THEN PRINT "LM-340T-12 or 7812"
1190 IF I = 3, THEN PRINT "LM-340T-12 or 7812"
1200 IF I = 4, THEN PRINT "LM-340K-12 or 7812 (K-package only)"
1210 IF I > 4, THEN PRINT "Use LM-317 or LM-338 Adjustable
     regulator"
1220 RETURN
1230 PRINT
1240 IF I = 1, THEN PRINT "LM-340H-15, LM-340T-15, LM-340K-15 or
     7815"
1250 IF I = 2, THEN PRINT "LM-340T-15, LM-340K-15 or 7815"
1260 IF I = 3, THEN PRINT "LM-340T-15, LM-340K-15 or 7815"
1270 IF I = 4, THEN PRINT "LM-340K-15 or 7815 (K-package only)"
1280 IF I > 4, THEN PRINT "Use LM-317 or LM-338 Adjustable
     regulator"
1290 RETURN
1300 PRINT
1310 IF I < 4, THEN PRINT "LM-340T-18, LM-340K-18 or 7818"
1320 IF I = 4, THEN PRINT "LM-340K-18 or 7818 (K-package only)"
1330 IF I > 4, THEN PRINT "Use LM-317 or LM-338 Adjustable
     regulator"
1340 RETURN
1350 PRINT
1360 IF I = 1, THEN PRINT "LM-340H-24, LM-340T-24, LM-340K-24 or
     7824"
1370 IF I = 2, THEN PRINT "LM-340T-24, LM-340K-24 or 7824"
1380 IF I = 3, THEN PRINT "LM-340T-24, LM-324K-24 or 7824"
1390 IF I = 4, THEN PRINT "LM-340K-24 or 7824 (K-package only)"
1400 IF I > 4, THEN PRINT "Use LM-317 or LM-338 Adjustable
```

```
            regulator"
1410 RETURN
1420 PRINT "ENDED"
1430 END
1440 FOR Q = 1 TO 5
1450 PRINT
1460 NEXT Q
1470 RETURN
1480 FOR Q = 1 TO 30
1490 PRINT
1500 NEXT Q
1510 RETURN
1520 PRINT "PRESS ANY KEY TO CONTINUE:"
1530 A$=INKEY$: IF A$="" THEN 1530
1540 RETURN
1550 PRINT
1560 IF I < 5, THEN PRINT "LM-317K"
1570 IF I > 4, THEN PRINT "LM-338K"
1580 PRINT
1590 IF I < 6, THEN R2 = 240
1600 IF I > 5, THEN R2 = 120
1610 INPUT "MAXIMUM output voltage (<= 35 VDC):",VOMAX
1620 IF VOMAX > 35, THEN PRINT "Voltage OUT OF RANGE!"
1630 PRINT
1640 IF VOMAX > 35, THEN GOTO 1610
1650 GOSUB 1440
1660 R1 = R2*((VOMAX/1.25)-1)
1670 VINMIN = VOMAX + 3
1680 PRINT
1690 PRINT "For adjustable power supply over the range"
1700 PRINT "1.2 VDC to";VOMAX;" VDC use the following:"
1710 PRINT
1720 PRINT "R1:";R1;" OHMS"
1730 PRINT "C1:";C1;" uF"
1740 PRINT
1750 PRINT "Minimum input voltage:";VINMIN;" Volts"
1760 PRINT
1770 GOSUB 1520
1780 GOTO 1790
1790 PRINT
1800 PRINT "What's Your Pleasure?"
1810 PRINT "1.  Do Another"
1820 PRINT "2.  Finished"
1830 PRINT
1840 INPUT "SELECTION?",D
1850 IF D > 2, THEN GOTO 1790
1860 ON D GOTO 190,1870
1870 GOSUB 1440
1880 PRINT "PROGRAM ENDED"
1890 END
```

Appendix D

Semiconductor Manufacturers

Advanced Micro Devices (AMD)
901 Thompson Place
Sunnyvale, CA 94086.

Analog Devices
Norwood, MA 02062.

Fairchild Semiconductor
464 Ellis Street
Mountain View, CA 94040.

GIC Microelectronics
Hicksville, NY 11802.

Intel Corporation
3065 Bowers Avenue
Santa Clara, CA 95051.

Intersil, Inc.
10900 North Tantau Avenue
Cupertino, CA 95014.

**Motorola Semiconductor
Products, Inc.**
P.O. Box 20912
Phoenix, AZ 85036.

National Semiconductor
2900 Semiconductor Drive
Santa Clara, CA 95051.

RCA Solid State Division
P.O. Box 3200
Somerville, NJ 08876.

Signetics, Inc.
811 East Arques Avenue
Sunnyvale, CA 94043.

Texas Instruments
P.O. Box 1443
Houston, TX 77001.

Zilog, Inc.
170 State Street
Los Altos, CA 94022.

Bibliography

Designing with Operational Amplifiers: Applications Alternatives; Graeme, J. G.; McGraw-Hill, New York, 1977.

Applications of Operational Amplifiers: Third Generation Techniques; Graeme, J. G.; McGraw-Hill, New York, 1973.

Operational Amplifiers: Design and Applications; Graeme, J. G.; Tobey, G. E. and Huelsman, L. P.; McGraw-Hill, New York, 1971.

Function Circuits: Design and Applications; Wong, Y. J. and Ott, W. E.; McGraw-Hill, New York, 1976.

Handbook of Operational Amplifier Circuit Design; Stout, D. F. and Kaufman, M. (*Ed.*); McGraw-Hill, New York, 1976.

Modern Operational Circuit Design; Smith, John I.; Wiley-Interscience, New York, 1971.

Digital Signal Processing; Stanley, W. D.; Reston Publishing Co., Reston, VA, 1975.

Operational Amplifiers: Theory and Servicing; Bannon, E.; Reston Publishing Co., Reston, VA, 1975.

Manual of Active Filter Design; Hilburn, J. L. and Johnson, D. E.; McGraw-Hill, New York, 1973.

Active Filter Cookbook; Lancaster, Donald; Howard W. Sams & Co., Indianapolis, IN, 1975.

How to Build and Use Electronic Devices Without Frustration, Panic, Mountains of Money, or an Engineering Degree; Hoenig, S. A. and Payne, F. L.; Little, Brown & Co., Boston, MA 1973.

Theory and Characteristics of Phototransistors, Bliss, John; Motorola applications note AN-440, 1973.

Applications of Phototransistors in Electro-Optic Systems, Bliss, John; Motorola applications note AN-508, 1971.

How To Use Photosensors and Light Sources, Bliss, John; Motorola applications note AN-561, 1972.

Isolation Techniques Using Optical Couplers, Christian, F.; Motorola applications note AN-571A, 1974.

Biophysical Measurements; Strong, P.; Tektronix, Inc. *Measurement Concept Series*, Beaverton, OR, 1976.

Practical Instrumentation Transducers; Oliver, Frank J.; Hayden Book Co., Rochelle Park, NJ, 1971.

Electronics for Scientists; Malmstadt, H. V. and Enke, W. A.; W. A. Benjamin & Co., New York, 1963.

Electronic Instrumentation & Measurement Techniques; Prentice-Hall, Englewood Cliffs, NJ, 1970.

Instrumental Methods of Chemical Analysis; Ewing, G. W.; McGraw-Hill, New York, 1975.

COMPUTERS

Microprocessor/Microprogramming Handbook; Ward, Brice; TAB BOOKS, Blue Ridge Summit, PA, 1975. Cat. #785.

Minicomputers for Engineers & Scientists; Korn, G. A.; McGraw-Hill, New York, 1973.

Computer Methods for Science & Engineering; Lafara, R. L.; Hayden Book Co., Rochelle Park, NJ, 1973.

The Bugbook-III; Rony, P., Larsen, D. and Titus, J.; E & L. Instruments, Blacksburg, VA, 1975.

Bugbook: Microcomputer Interfacing & Programming; Rony, P., Larsen, D. and Titus, J.; Howard W. Sams & Co., Indianapolis, IN, 1977.

An Introduction to Microcomputers: Volume-I Basic Concepts; Osborn, A.; Adam Osborn Associates, Berkeley, CA, 1976.

An Introduction to Microcomputers: Volume-II Some Real Products; Osborn, A.; Adam Osborn Associates, Berkeley, CA, 1976.

8080 Programming for Logic Design; Osborn, A.; Adam Osborn Associates, Berkeley, CA, 1976.

Programming Microprocessors; McMurran, M. W.; TAB BOOKS, Blue Ridge Summit, PA, 1977. Cat. #985.

Microprocessors: New Directions for Designers; Torrero, E. A. (editor); Hayden Book Co., Rochelle Park, NJ, 1975.

Modern Data Communications; Davenport, William P.; Hayden Book Co., Rochelle Park, NJ, 1971.

BYTE Magazine, Peterborough, NH (all issues)
KILOBAUD Magazine, Peterborough, NH (all issues)
EDN Magazine (see index)
ELECTRONIC DESIGN (see index)

Index

501

Other Bestsellers From TAB

Other Bestsellers From TAB

Other Bestsellers From TAB